THE WORLD PETROLEUM MARKET

By M. A. Adelman

THE WORLD PETROLEUM MARKET

Published for Resources for the Future, Inc.
by The Johns Hopkins University Press, Baltimore and London

Resources for the Future is a nonprofit corporation for research and education in the development, conservation, and use of natural resources and the improvement of the quality of the environment. It was established in 1952 with the cooperation of the Ford Foundation. Part of the work of Resources for the Future is carried out by its resident staff; part is supported by grants to universities and other nonprofit organizations. Unless otherwise stated, interpretations and conclusions in RFF publications are those of the authors; the organization takes responsibility for the selection of significant subjects for study, the competence of the researchers, and their freedom of inquiry.

This book is one of RFF's studies in energy and mineral resources, which are directed by Sam H. Schurr. The research was supported by grants to the Massachusetts Institute of Technology, where M. A. Adelman is professor of economics. Charts were drawn by Frank and Clare Ford. The index was prepared by L. Margaret Stanley.

RFF editors:
Mark Reinsberg, Vera W. Dodds, Nora E. Roots, Tadd Fisher.

Manufactured in the United States of America

The Johns Hopkins University Press, Baltimore, Maryland 21218
The Johns Hopkins University Press Ltd., London

Library of Congress Catalog Number 72-4029
ISBN 0-8018-1422-7 (cloth)
ISBN 0-8018-1561-4 (paper)

Originally published, 1972
Johns Hopkins Paperback, 1973
Second printing (cloth), 1973
Third printing (cloth), 1974
Second printing (paper), 1974

Library of Congress Cataloging in Publication data
will be found on the last printed page of this book.

Foreword

As the preeminent international commodity, petroleum can usefully be studied from several vantage points—economic, political, strategic. In this book, Professor Adelman examines the international oil industry from an economic vantage point, with what he calls a "narrow" focus on the real price of crude oil. Narrow as the focus may be, it is nonetheless of crucial importance to countries around the world, both those that sell oil in international markets and those that buy.

An outstanding aspect of the economics of the world petroleum market is the very wide margin (documented in Adelman's careful research) between the price at which oil has been selling and its cost of production in the major exporting regions of the world, particularly the Middle East and North Africa. For many years this wide difference between cost and price provided a strong incentive for sellers to offer oil at prices well below official postings. The result was a declining real price for oil on world markets.

A dramatic turnabout took place in 1971 when concerted actions by the major exporting countries succeeded in raising the posted price of oil and the per-barrel tax paid to the exporting countries by the international companies. This raised the costs (including tax) of oil—the actual costs borne by the international companies. The latter, in turn, increased their prices to consumers around the world. The real price of oil on world markets reversed its earlier downward trend, but the costs of production (excluding tax) remained as low as they had been before.

This change has been characterized in most of the current discussions of the subject as marking the advent of a "sellers' market," replacing the "buyers' market" which was said to have existed during the 1960s. The conventional wisdom in 1971—following the Teheran and Tripoli settlements in that year—tells us that the exporting countries, having discovered the strength of their bargaining position when acting in concert, will be able to exact ever higher prices despite the wide (and now even wider) margin between prices and the real costs of production.

If the prevalent view—that oil prices on world markets have started an irreversible upward movement—is judged to be valid, this could have a great effect on the energy policies that oil importing countries follow in the future, particularly on their efforts to develop domestic substitutes (e.g., nuclear power) for at least a portion of their imported oil.

This study provides an independent basis for understanding the past behavior of oil prices and for assessing future prospects: the results of the painstaking assembly and careful study of data on the real price of oil in the past and a skillful analysis of the underlying fundamentals of production and transportation costs and of industry structure that will mold price behavior in the future. Adelman's diligent pursuit of an understanding of the basic factors at work leads to an alternative view of the future, one with profound implications of its own for public policy both in the oil importing and exporting countries. The book appears at a propitious time.

SAM H. SCHURR
Director, Energy and
Mineral Resources Program
Resources for the Future, Inc.

March 1972

Acknowledgments

This study is the result of teaching the economics of industry to undergraduate and graduate students at M.I.T. An example is worth a thousand words, and an industry as large and varied as oil offered many examples. Graduate students wrote papers and theses and in time became colleagues; some are mentioned below or as authors. My thanks go first to my research assistants: Carl E. Beigie; Paul G. Bradley; Peter Eglington; Susan B. Foster; John M. Vernon; Gaby M. Gross, who stayed longest and endured most; Ching-chih Chen, then and now a devoted friend.

My eminent colleague Charles P. Kindleberger suggested the topic; he read large portions of the manuscript, as did Paul W. MacAvoy and Zenon S. Zannetos, with many substantial improvements. Endless typing and retyping were redeemed by the good humor and devotion of Inez J. Crandall and Beatrice A. Rogers.

The research was supported by grants from Resources for the Future and the Ford Foundation. At RFF, Sam H. Schurr gave good counsel. Orris C. Herfindahl, who had shown by example how to lay bare the relation between market organization and mineral depletion, read the whole manuscript with unfailing, sometimes dismaying, always helpful attention to principle and detail.

Thanks are due those who read parts of the manuscript or discussed the issues with me: in the United States, Harry J. Colish, Leslie Cookenboo, Jr., Nicholas G. Dumbros, Helmut J. Frank, Richard L. Gordon, Daniel C. Hamilton, Richard B. Heflebower, James W. Hanson, David V. Hudson, Stephen Hymer, Carl Larson, and Walter J. Levy–always kind and ruthless. The late Wallace F. Lovejoy and Stephen L. McDonald, Howard W. Page, Henry B. Steele, Thomas R. Stauffer, and George W. Stocking have all left their mark.

A year spent abroad was enriched by the hospitality of the Institut Français du Pétrole, particularly Jean Chappelle, Mme. Jacqueline Funck, Victor Henny, and Jean Masseron. Others in France include Lucien Gouni, Léon Kaplan, Jean Laigroz, Jean Lacoste, Maurice Mainguy, Yves Mainguy, Pierre Wack, and my dear friend Jacques Houssiaux, no longer living.

In Britain, John D. Ritchie read the whole manuscript. His contributions are everywhere in it; early and late he rid me of many a foolish notion. Paul H. Frankel has been teaching me oil economics, with varying success, for years, especially at the annual Northwestern University Petroleum Economics Conference. His partner Walter Newton was especially helpful on tanker rates, as were J. R. Curry and C. Grigg, E. E. Bowyer and P. Peters. Jack E. Hartshorn, Francisco R. Parra, and Edith T. Penrose, who have created a tradition of petroleum economics, often nailed a mistake.

Georges Brondel, Lucio Corradini, Jacques Hartmann, and Pierre Maillet made rewarding the visits to Brussels and Luxembourg. Søren Friis showed how good a window on the market could be found in Copenhagen, as did Mariano Gurfinkel in Caracas and Zuhayr Mikdashi in Beirut.

A visit to Japan in 1965 was a testing ground for many ideas. Chapters II and VII first took form as lectures there and were published in Japanese that year. I can only make inadequate mention of my closest collaborators on the visit there: Toshio Akatsu; Shizo Ebihara (thanks to whose cross-examination some Japanese translations are probably more clear than the English originals); Noboyuki Nakahara; Hiroo Tominaga; and most of all, Hidezo Inaba, who, rightly considering himself my friend, took upon himself the friend's prerogative of asking the most difficult, disturbing questions.

With all that good help, this should have been a better book. But some truths hold for both great and lesser men, and I hope the reader agrees with a certain Bishop Roncalli (later Pope John XXIII), when appointed to a difficult post: "Since they can't find a horse, they must ride on a donkey."

* * * * *

The cost of a long work project is borne more by the writer's family than by himself. Any who benefit by this book should thank Mil, Bobi, and Larry.

Newton, Mass.
July 15, 1971

M. A. Adelman

Contents

PART TWO
THE MARKET PRICE

PART THREE
SUMMARY AND CONCLUSIONS

TECHNICAL APPENDIXES TO THE CHAPTERS

LIST OF TABLES

APPENDIX TABLES

LIST OF FIGURES

THE WORLD PETROLEUM MARKET

Introduction and Plan of the Study

The lack of romance in my history will, I fear, detract from its charm. But if it is useful to those who want an accurate account of the past to help understand the future, . . . that is enough. —Thucydides

The long-run supply and price of crude oil are essential facts for investment decisions in energy production, conversion (mostly electricity), and use. This study concerns the "real price"—i.e., after allowance for general price-level changes. The "real price" of crude oil, like any price in any market, results from the interplay of demand, supply, and the degree of monopoly. The purpose here is to unravel the process and predict its future. It is a special, even a narrow, analysis. Those who seek a bird's-eye view of the industry as a whole are fortunate because there are already admirable books to be consulted.[1]

The official truth in the capitalist, communist, and third world is that crude oil is becoming ever more scarce, special measures are needed to assure its provision, and prices will rise. But the conclusions of this study are that crude oil prices will decline because supply will far exceed demand even at lower prices, and because—a separate issue—there will continue to be enough competition to make price gravitate toward cost, however slowly. The official truth has ruled for 25 years, and has not much resembled the facts. But its hour may have come; Chapters VII and VIII show the forces supporting prices to be strong. The greatest of these forces is inertia. Government policies provide most of the explanation of current oil prices and, particularly in government, each man assumes the official truth because everyone else seems sure of it, and so the prophecy may fulfill itself.

Of the three basic variables dealt with in succeeding chapters—supply, demand, and monopoly—demand gets the bare minimum of our attention. Our central problem is with supply. Petroleum is a mineral, always under public discussion and apparently subject to spasms of "glut" separated by fears of "scarcity." But raw materials are never free, hence always scarce. Scarcity is the strain imposed by consumption on available supply, registered in competitive markets by price. If a given market is not very competitive, the market price may be a poor or non-indicator of the cost of creating additional output. If

[1] J. E. Hartshorn, *Oil Companies and Governments* (London: Faber & Faber, rev. 1967; American title, *Politics and World Oil Economics*); Edith T. Penrose, *The Large International Firm in Developing Countries* (London: Allen & Unwin, 1968); Jacques Flandrin and Jean Chappelle, *Le Pétrole* (Paris: Edits. Technip, 1961); Jean Masseron, *L'Economie des Hydrocarbures* (Paris: Edits. Technip, 1969); Paul H. Frankel, *Essentials of Petroleum* (London: Chapman & Hall, 1946, 1969).

market price exceeds supply price, there is an incentive to invest to increase supply and a reward for some sellers and buyers to cut a new channel between them, to make sales at a lower price and wipe out the discrepancy. But defensive forces may contain the tension wholly or partly; price erosion may happen quickly, slowly, or never. To see ahead, even as through a glass darkly, we need to know how far apart prices and costs are today: we must measure each one.

This book is therefore an industry study and also a natural resource study, neither part possible without the other.

Competition, Monopoly, and Corporations

The petroleum industry is in two respects an extreme case of a typical phenomenon. First, oilmen live mostly in the future and decide on the basis of expected, not current, variables. Second, most of the world's oil is produced by the biggest corporations in the capitalist world. Ownership and management are completely divorced. The companies are staffed with large bodies of specialists. But their planning has no more replaced maximum profit (or, more precisely, maximum present value) than the jet aircraft has replaced motion. There seems to be no evidence that managers try to maximize growth, security, visceral satisfactions, utility functions in five variables, or whatever. One oil company president has said, his indifference lightly tinged with scorn, that there is nothing wrong with adding security and growth to profits as a corporate goal "... if you want three goals rather than one, but the latter two are largely implicit in [profits]."[2]

To analyze a market, we can make do with a simple theory of corporate action. Business decisions are made in the present, and commit for the future. Present discounted values are all we know and all we can calculate. Corporate managers seek to increase the owners' equity: the present value of future receipts minus future outlays. As new information comes in, hunches, guesses, and calculations are revised in the endless chase after the moving target. Depending on its expectations and its cost of capital, a firm may be better off to wait and give up profits today for more profits in the future, *or* to seize the happy moment. It is endless trial and error, particularly error. Two or more firms may react differently because they see the future differently; they cannot all be right, but they may be all wrong. Moreover, the old-timer with a high market share must worry more than the newcomer about feedback effects of higher output upon price.

With plenty of real problems, we will try to avoid false ones. Planning has not replaced The Market, because The Market does not exist: there are only many particular markets, including world oil. The degree of competition and monopoly in a market defines the choices open to the planners. Their problem is how to make the most of their opportunities and the least of their constraints.

A fashionable dogma today is that large corporations make their environment rather than merely exploit it; they administer their prices and manipulate their customers to encompass the profit and growth freely chosen. This is only a corporate executive's daydream, which, like other fantasies, can be turned against the dreamer. It cannot be applied to show why and how a given price exists, nor where it is going.

The ideas of competition and monopoly can be so used. A single seller (or a group acting as one) will let only so much output into his market as will give him the largest

[2] John E. Swearingen, "The Executive Decision," *Petroleum Management*, vol. 36 (1964), pp. 99–123.

possible profit, and will make due allowance for the lower price resulting from greater outputs. At the other extreme, a "large enough" number of sellers are forced to act in complete mutual independence. Competitors must ignore the feedback effect of increasing the supply, and will carry output to the point where industry incremental cost is equal simply to expected price. There will be greater output at a lower price than under monopoly.

"Large enough" does not mean infinite, and does not require small firms. To suppose that price cannot approach incremental cost without "atomistic competition" makes about as much sense as saying that we cannot use the radical *Pi* unless we carry it out to an infinite number of decimal places; rarely if ever do we need more than 10. The essential condition for competitive price behavior is that each firm is forced to seek its own salvation and cannot attempt to join in working out a solution that would be best for the group as a whole. The monopolist of course *is* the group as a whole. Whether a "large enough" number is 5, 10, 20, or more means: at that number, is there any statistical or other test that would show that price and quantity diverge significantly from the result reached under complete independence? Numbers are merely a proxy for independence.

In the oil market we find neither a single seller nor many, but the intermediate case of a few companies. Unfortunately, there is no general theory of small-group markets. "Oligopoly" is a meaningful term, but it is too loose to permit any deductions to be drawn from it. Moreover, between the "few" of theory and the actual ill-defined group occupying an imperfectly delimited portion of a vaguely bounded market, the gap is still too wide for general reasoning to cover. We must approach the problem in stages, and examine evidence about competition that clusters about the two poles of "structure"—number of sellers and the like—and behavior or performance. "These tests must be used to complement not to exclude one another."[3] My great teacher's advice is good because it is often hard to take: in trying to reconcile the two types of evidence we learn about the market under study.

"Competition" as a descriptive term or a conclusion may be a snapshot or a film. As a condition at any given moment, competition means a zero or negligible difference of price from incremental cost. But a price may decline toward (but not all the way to) incremental cost, and the evidence may not permit any explanation except independent rivalry. There may be paradox but no conflict if we need the hypothesis of monopoly to explain why prices are many times cost, and yet need the hypothesis of competitive change to explain why until recently prices have declined so. These are two aspects of the one market, which we cannot understand without a study of its history.

A Word of Caution

"The international oil industry is beset by myths, some of its own making, some imposed upon it."[4] The myth of oil as an industry of decreasing cost will be mentioned in the next section of this Introduction. Another is that vertical integration is somehow the only "right" way for the industry to be—it is no more inherently right than wrong.

[3] Edward S. Mason, "The Current Status of the Monopoly Problem in the United States," *Harvard Law Review*, vol. 62 (June 1949).

[4] Geoffrey Chandler, "Some Factors in Oil Company Decision Making," Lecture No. 15, *UN Interregional Seminar on Petroleum Administration*, Apr. 1968.

"Marginal" cost and revenue are important economic concepts. But in both English and French, "marginal" also means peripheral, transitory, secondary, *aléatoire*; hence the endless play on words, the dogged determination to disregard anything—a set of prices, costs, sales, or whatnot—if only it can be called "marginal." I have used "incremental" as being perhaps less easily misunderstood, though it is true that prices have sometimes been waved away because they were on "incremental sales." In fact, every single sale is incremental because the seller must decide whether or not the increment to receipts exceeds the increment to costs. Another myth, that short-term contracts, or spot sales, are inherently lower-priced than long-term contracts, survives even the commonly acknowledged fact that the short-term price known as the single-voyage or spot tanker rate can swing up above or down below the long-term rate.

The publicly known transactions are belittled as "marginal" and "unimportant," "brokers' gossip," "lunatic fringe" operations by "greenhorns" with "weak hands," and so forth. This is not mere invective. It is a fiction: that these markets are where odd lots are dumped at special, temporarily abnormal, low and uneconomic prices. But the *belief* that these markets are irrelevant to long-run demand and supply is important. The state of a man's mind is as much a fact as the state of his digestion. Indeed, a mental delusion can explain an upset stomach. The delusion, particularly among governments, that the true price of oil is something other and much higher than the price at which it is changing hands, goes far to explain why the price has not declined further.

Ambiguity is as hard to live with as myth. "Prevailing" or "normal market" prices or rates may mean what nobody has paid for years. When an oilman predicts that by the late 1970s supply and demand will be equated, his words mean that the price of oil will be a small fraction of what it is now. But surely he does not mean that. Probably he means that prices will be back where they were in 1957, wherein I think him mistaken. But nobody can really tell.

Throughout the study I have avoided any itemized polemic or parade of examples, trusting, rather, that careful explanation of what each concept is—and what it is not—serves the reader better. For several reasons the book is much longer than had been intended. Its primary audience consists of academic, government, and business specialists in energy economics, some of them economists by training, some engineers, some neither. Detail barely sufficient for one may be superfluous for another. The theses of the book are unorthodox, and the burden of proof is on the writer. Often much ground must be covered to settle even a narrow specific question. Hence, I have tried to make the work modular. Many readers knowing more than myself will at several points, I hope, substitute their own better knowledge to see what difference it makes in detailed figures or broad conclusions. Each calculation is a link in a chain of successive approximations: the production cost estimates, for example, are the fourth published, starting in 1963.

As to public policy, the view in these pages is one-sided. I pay it much attention because of its decisive effect on the market to be studied; but no policy advice will be found here.[5] The analysis may help on two—or more—sides of various bargaining tables. Knowledge is neutral stuff, which men use according to their lights.

It is assumed throughout that knowledge is better than ignorance, truth better than falsehood. But admittedly others must work under different rules, and often in public life one is forced to look to the results, not the truth, of what one says.

[5] The single exception: chap. VII, pp. 245–49, and the appendix to Part Three.

It was a man of great worldly experience, but no cynic—indeed, a saintly moralist—who said: "In counsel-giving unto the King, ever tell him what he should do, but never tell him what he is able to do. For if a lion knew his own strength, hard were it for any man to rule him." But Sir Thomas More was advising the statesman, not the scholar.

Summary of the Book

Chapter I sets forth the conditions under which a worldwide oil industry is at rest, with no internal stresses to make price and production patterns change. This will permit us to see why the industry has never been and is not now at rest, and what forces are changing it. We must first dispose of the mistaken but influential idea that oil and gas production is an activity with increasing returns or decreasing costs as more oil is produced. If this were true, the industry would be a "natural monopoly" and normal competitive rules would not hold. Allied to this mistake is the idea that once a field is discovered and "drilled up," oil pours out at little or no additional cost. In fact, the great problem is the steady-state development cost of the reservoir; the "drilled-up" field rarely if ever exists.

The cost of any mineral, including oil, registers the struggle between tight-fisted Nature and inquisitive man. The more is produced, the more must one draw upon higher-cost sources. But one can seek more and richer deposits, better methods of extraction, and more thrifty use. Exhaustion of known deposits tends to make the material ever more scarce and dear; greater knowledge tends to make it ever more plentiful and cheap. Some players enjoy the game, and it can fascinate the spectator. One cannot expect even uneventful progress, nor the same experience industry to industry. In general, over the long run mankind has done very well.[6] But this average comfort does not imply that any particular mineral resource at any given time is getting more plentiful. It depends on the unforeseeable growth of knowledge and on changes in the economy, which may radically displace supply or demand. In effect, we are studying a brief moment in a long history—looking backward and forward about 20 years each way.

Our task is to get the most out of what we know about the mineral resources and minimize the impact of guesswork about values that cannot be tested. We distinguish (1) oil-in-place, the fruit of discovery expenditures, from (2) reserves, that part of oil-in-place which development expenditures have converted into a ready shelf inventory. The categories used by geologists and reservoir engineers correspond essentially to this distinction. If this approach is sound, it permits us to calculate development costs for oil as a function of the amount to be produced. Unless we do this, we cannot analyze industry behavior in order to compare prices with costs.

But we have no supply-price relationships for oil and gas *new-field discovery*. It remains an unsolved problem in geology, physics, or economics. There is much particular knowledge, and new schemes of mathematical analysis look promising, but there is still no basis for even a loose industry-wide estimate of the relation between (1) the expenditure of a given amount of money now, and (2) the later addition to the stock of known

[6] Neal Potter and Francis Christy, *Trends in Natural Resource Commodities: Statistics of Prices, Output, Consumption, Foreign Trade, and Employment in the United States, 1870–1957* (Johns Hopkins Press for RFF, 1962). Vivian E. Spencer, "Raw Materials in the United States Economy, 1900–1966" (U.S. Bureau of the Census, Working Paper No. 30, 1969), table 12. In 1900–50, mineral prices increased more than all finished commodities (factor of 3.0 against 2.6), but in 1950–66, the increase was slightly less (factor of 1.23 against 1.30).

fields. Hunting for new fields is research–*recherche* is used for both in French–and the payoff is as uncertain here as elsewhere.

We seem to face a dilemma. To leave discovery cost out of the study would be *Hamlet* without the prince; to include it would put us to chasing his father's ghost: " 'Tis here . . . 'tis there . . . 'tis gone."

But there is an indirect solution. At the margin, development cost in known fields equals development cost plus finding cost in new fields. Higher finding cost drives up development cost. Zero new discoveries would eventually mean much higher development costs as old reservoirs were depleted. Discovery is necessary to stave this off. Hence the estimate of development costs rising as more and more of the oil-in-place is depleted gives us the Maximum Economic Finding Cost (MEFC), the penalty for doing nothing. If the cost of finding the new oil exceeds MEFC, the choice has been uneconomic and a loss has been inflicted on society. If the cost is less, there is a gain. MEFC gives us the worst that can happen and thereby sets an upper limit to costs and to competitive prices in the long run.

In Chapter II, actual numbers are set into the theory. Operating cost data are compiled from various sources. Development capital cost is supply price: the amount per unit produced that would just barely make the production investment worthwhile. The needed investment is equal to the present value of a diminishing stream of receipts (the production decline curve), discounted at a rate reflecting risk. Each receipt is partitioned into a variable quantity multiplied by a constant price, which is the development capital cost.

The most important datum is, of course, the investment needed per additional daily barrel of production in a developed field. In the late 1960s, a representative figure in the highest-cost Persian Gulf area whose production was needed to meet demand was under $100. (Elsewhere it may be 35 times as high.) Furthermore, real costs declined through the 1960s.

The real cost or incremental supply price of Persian Gulf crude oil (including a 20 percent return on all needed investment) is about 10 cents per barrel. If new discoveries were zero, growing pressure on the reserves in these fields would put costs up to about 20 cents ($1.50 per metric ton) by 1985. The difference (MEFC) is about 10 cents. That is the extreme upper limit, since there will be new discoveries and new technology. The aggressive policies of new competitors or old ones, or the misguided actions of companies or governments, etc., are only "the limbs and outward flourishes"; the central fact of world oil supply is the cost condition, which will persist for at least 15 years. Nobody knows, and nobody ought to claim to know, what will happen afterward. There will be deflection by forces not yet seen, some not even existing.

In Chapter III the structure of the world oil industry is sketched: the shares of the largest producers of crude oil, their vertical integration, and the size of the open market in crude. "Share" of market implies a definition or a setting of the boundaries: we want to exclude those producers and refiners whose output is not substitutable for that of the group we study. But substitution is basically a matter of price, so structure and behavior are mutually dependent. The boundaries drawn here are anything but precise; the best one can say for them is that others would not seem to make much difference. The concentration among the leading producers is an imprecise number because of the great Persian Gulf joint ventures, which facilitate uniform action and careful attention to the feedback effect. I would urge that vertical integration, in the particular cirumstances, has a similar effect. In any case, concentration has diminished since the early postwar period, and the open market in crude has increased.

Before we can confront this sketch of a changing structure with the price data, one other step is necessary. Most oil is produced far from where it is best refined, and transport accounts for a large part of the total cost and delivered price. Chapter IV therefore deals with oil tanker rates. The tanker trade is in one sense flesh of the flesh of world oil, yet it could hardly be more different. Ship owners and operators are many and relatively small, entry is easy and cheap; it is a purely competitive industry. Ships may be chartered for single voyages, for life, or anything in between. This parallels somewhat the price structure in crude oil and products, which may be sold at spot or under contract for longer or shorter periods. Our primary interest, consistent with the treatment of production cost, is the supply price of long-term service at any given time. This can be calculated and compared with delivered and f.o.b. prices. Analyzing tanker rates helps free one of the illusion that short-term transactions are somehow qualitatively different from long-term, and exist in separate worlds. In fact, they are imperfect substitutes for each other. Hence the price of a good or service at spot is a valid if often imprecise indicator of the price for long-run service.

The tanker industry is *purely* competitive, but very *imperfectly* competitive because massive innovation has been so little foreseen. Those more alert, skillful, and lucky have profited vastly. Since 1963, contrary to the general impression, there has been a chronic shortage of ships. Shipbuilding capacity has expanded under the forced draft of demand greater than supply, and is still expanding. Shipyards have raised wages and scratched for labor; ship prices have risen considerably. But long-term tanker rates have declined with the much more efficient new ships. In the face of the shortage, complaints of a tanker "surplus" and "depressed" rates are beside the point. In any case, a purely competitive tanker market is a piece of good luck, because it is itself a sensing mechanism to find and register the long-run supply price at any given moment. We need not do for tankers what must be done for crude oil, because the market has done it for us.

In hand now are the elements needed to analyze, in Chapter V, crude oil prices from World War II to Suez I. But the market process operates over time and cannot be explained even by a complex single formula. One cannot wind up a causal mechanism and then let it run to spin out the story once ordained. One must write history. Any impression from Chapter III of a gradual, continuing price retreat does not last long. The years 1948–49, when the sellers were fewest and closest knit, yet saw a drastic reduction in the price level and a revolution in the price structure. The United States government, as trustee for the customers–and home government to most of the companies–compelled them to do what would not otherwise have happened, at least not so much so soon. After 1950, when the pressure disappeared, we see an actual reversal, with prices rising by perhaps 20 percent through 1956. (The reader inclined to accept the writer's forecast of falling prices should reflect on this experience.)

To explain price stability or actual increase one must explain why there were no large-scale and increasing exports to the world's greatest market, the United States. Comparison of prices and costs disproves the hypothesis that incremental costs were the barrier to keep output from expanding. But curtailed U.S. imports are not explained by any variant of the hypothesis of monopoly. It follows that limited competition, or a substantial degree of market control, was *a nonbinding constraint*; some other tighter inhibition was at work. This is confirmed by an abundant documentary record of governmental action to restrict imports into the United States during the decade before the 1959 mandatory quotas. It is a classic case of opposing escalation: growing economic pressure to import versus growing governmental counter-pressure. The quota system was no new element; quite the contrary: it was the definitive stop to any change in the status quo. The idea that the 1959 quotas "threw the world industry into surplus" (which at one time this writer believed) is contrary to the facts. Here is where history joins with prediction: if U.S. import controls are relaxed in the years to come, world prices will weaken, not strengthen.

After Suez I, the road traced in Chapter VI was mainly downhill. We have two independent sets of price data: one for crude oil sales between unaffiliated buyers and sellers; the other for sales of refined products in Europe. Subtracting a margin for refining and for tanker transport yields a derived crude oil netback at the Persian Gulf. The two series are fairly close, each supporting the reliability of the other.

In Chapter VII a synthesis is attempted in the shape of a long-run forecast of the price of crude oil. The decline after 1956 is viewed as the very slow working of the competitive process, and the chief question is: Which forces impinging upon the market will act to reverse the decline, or, slow it down, or speed it up? The notion that rising demand will dry out the surplus has been a perennial and always disappointed hope, because it is based on a wrong theory. Higher demand only acts on price by raising incremental cost. But even on very generous assumptions, as is shown in Chapter II, incremental cost cannot be expected to rise above 20 cents from the current 10 cents. It is a floor so far below the current level as to give it no support.

Although even before the 1971 agreements the Persian Gulf price was about 10 times cost, pure competition there—say 100 independent producers—would wipe out only a small part of the cost-price gap. As the result of a curious historical sequence, the producing countries are getting nearly all of the monopoly revenue. They levy a tax, which in effect though not in form is in cents per barrel. The tax-plus-cost has served as price floor, even effective at a distance. No prudent management will commit for a price that is less.

Chapter VIII puts into perspective the events of early 1971, when the producing nations, with the approval or acquiescence of the consuming countries, drastically increased taxes while oil companies raised prices even more. The governments collect a pure monopoly revenue of about \$12 billion per year, programmed at \$40 billion or more in 1980. Were they to collect the profits and let the companies run the business they might long continue to do so. But they aim to absorb the rest of the profits and also to enter the oil-producing business. In both respects they are becoming owners, who must make price-output decisions. Unless they can operate a cartel to fix output and divide markets, they cannot escape acting like competitors. The temptation to shade taxes in order to let the concessionaire company shade prices, or to have the national company shade prices directly, will here and there be irresistible, because there is no assurance that other governments are not yielding. The price line will be breached and this will be unavoidable.

Thus we head for a series of difficult and dangerous confrontations. Governments are avowedly out to take everything over and above the minimum necessary return, and *also* firmly committed to actions that will tend to depress prices. Companies will need to consider, concession by concession, whether they can afford to go on at the terms laid down for them; governments will probe the limits of how much to demand of the companies. The appendix to Part III deals with an aspect—Eastern Hemisphere security—of government-company relations that is likely to engender or intensify a crisis. But this requires a look at the third participant—the consuming-country government.

Were there only the companies and host governments to consider, I would without hesitation forecast a continued and accelerated rate of decline in crude and product prices. But the action of consuming-country governments will over the next decade tend to slow down the price decline. Consuming-country governments, for a variety of reasons, want high oil prices. Protection of high-cost coal was long the most important reason; as coal is eased out, it is replaced, especially in Britain, by high-cost nuclear energy needing protection from oil. Various nations have direct producing interests in oil or gas and are moving to acquire more. Solicitude for an "independent" local refining industry leads to protection of local product prices and, indirectly, of crude prices. The expectation of stable or increasing oil prices causes the governments to make or subsidize investments in energy production. If oil prices decline, public companies or subsidized private companies lose money, a government loses face; this must not be. The issues are tangled and complex, and government policies form no consistent pattern. But the more they invest and make oil cheap, the more they will hold it dear.

Since both producing- and consuming-country governments want high prices, there may be a world commodity agreement, like the International Coffee Agreement, perhaps under the aegis of the United Nations. It might be effective in slowing price erosion. Oil resources will grow apace, needed or not.

Eventually, if some country's producing ventures in nuclear energy, oil, gas, etc., turn out so ill as to be past disguising, economic criteria may govern, and its government will move to the simple objective of minimizing the import bill and the cost of energy. If one large country or a few small ones did so during the 1970s, it would dovetail with the effort of one or more producing countries to gain more outlets even at lower prices. By the end of the decade, therefore, there may well be a rapid and disorderly price decline. But technology and demand may have changed so much by that time that a forecast is much less interesting. One thing is certain: there will have been a massive waste of resources invested in high-cost oil, coal, and nuclear power.

PART ONE

SUPPLY, PRODUCTION COSTS, AND THE MARKET STRUCTURE

CHAPTER I

Industry Behavior and Long-Run Supply

The object of this chapter is to expose a way of thinking: the economic theory of the world oil market. The principal theses may be baldly stated as follows, for later elaboration.

1. The crude oil industry, contrary to common belief, is inherently self-adjusting. Under competition, the level of output, and its division among various sources of supply, are set by the price acting upon the cost of bringing up more output. This incremental cost, for every individual unit and for the system as a whole, increases rather than decreases with greater output. It therefore serves as the governor of the producing mechanism in general and in any particular place. So long as incremental cost is less than anticipated price (or net revenue), there is a profit incentive to produce more; when the cost rises above the price, it chokes off further expansion. The spasms of the world oil market express both a strong adjustment mechanism and the very formidable barriers to the working of the mechanism.

2. The adjustment mechanism is always under strain, or always has work to do, because of changes in cost or demand. The prices and costs that count are always those expected in the future. Hence changes in expectations and changes in the rate of discounting must affect the relative costs of different supply areas and the industry as a whole.

Having explained why petroleum is similar to most other industries, let us consider where it is peculiar.

3. The discovery of new oil-in-place is unpredictable. The odds on finding a particular reservoir in a particular place are nearly impossible to calculate, and reservoirs vary greatly in quality. Moreover, it may take decades of additional work before the contents of a new group of reservoirs (a "field") are known. Hence, the very concept of discovery "cost" ex ante is too vague to permit calculation.

4. Development, the drilling and equipping of wells, carves out reserves from oil-in-place; "proved reserves," as reckoned strictly in the United States and loosely abroad, are the cumulative output expected to result from a development investment. Development is an investment activity, whose costs can be measured. The more intensive is development, the higher are costs and the greater the incentive to look for new oil-in-place.

5. Accordingly, one can formulate a very simple unified theory of production costs: the predictable increase in development cost is considered as the Maximum Economic Finding Cost (MEFC), which discovery can only improve.

Competitive and Monopoly Equilibrium

To understand movement and change, we need a concept of a system at rest, with no pressures to displace prices or outputs. By competitive equilibrium we mean a condition such that each seller (and buyer) is doing as well as possible for himself, ignoring the sellers as a whole. There is nothing he can do to change his prices or outputs to make him better off. Price equals incremental cost, including always the necessary return on investment, which in the short run may be zero. Higher or lower output would reduce profits.

By monopoly equilibrium we mean that sellers as a group are doing as well as they can. Actions taken for the good of the industry as a whole are the exercise of monopoly: the single seller can say, as apparently Louis XIV did not say, "The industry–that's me!" A monopoly price will be higher and output less than under competitive equilibrium. On both counts price will be above incremental cost. Thereby monopoly equilibrium is under a strain, and it may be undermined by one or more self-serving individuals who will try to appropriate the price-incremental cost gap.[1]

Competitive or monopoly equilibrium may be displaced or disrupted at any time by changed costs or demand conditions, coming from outside the industry or innovated inside. Monopoly equilibrium is under an additional strain; the price-cost gap is one more source of change, and it may be the most important.

Two metaphors may help. (1) The seas off Holland are above the level of farmlands on the other side of the dikes. Constant maintenance work is needed, failing which the land will be flooded. (2) An oil reservoir is a beautiful example of great force contained by trap rock for millions of years, which can continue for millions more–until breached by a well bore. Unless something is done, the first type of equilibrium will collapse; unless something is done, the second type will last forever.

If strong enough barriers can be set up against the force of competition, the noncompetitive market stays at rest. There is strain but no rupture, and current prices and supply patterns will continue. If the barriers fail but the break is very small relative to the forces suppressed, and the disequilibrium very great, the adjustment may take a long time until the force is spent. Theory can tell us no more–the question is: How great are the respective forces and the speed with which a given noncompetitive balance degenerates or runs down into competition?

As explained in the Introduction, competitive equilibrium as a condition is more briefly called "competition." The *process* of degeneration or entropy, from monopoly to competitive equilibrium, is usually also called "competition."

An Industry of "Inherent" Surplus? Surplus (or shortage) is a rather flexible term, whose meaning depends on the time horizon. "Surplus" may mean supply greater than demand at current prices, and a forecast of prices declining. Or it may mean that the lower price has already done its work in stopping investment and contracting capacity. If demand will henceforth increase, the forecast must be just contrary: price must rise. Sometimes the context makes it clear what "surplus" or "shortage" a writer means. But let us look at one particular example where the meaning is generally clear enough: the vision of the petroleum industry as having an incremental cost always and necessarily lower than the market price. Today cost is certainly much lower than price. Is this condition imposed by nature or is it the product of recent history? It makes a great deal of difference which explanation one uses.

[1] This holds equally well for oligopoly and monopolistic competition.

The common opinion, assumed everywhere, suffuses general discussion of the petroleum industry, yet is almost never explicitly stated. We only catch hints that oil production has "an inherent surplus" because at all times more is available than is wanted at the current price. Hence "it has been found necessary"–in the United States by law, elsewhere by industrial statesmanship–to restrict and control output. It is hard to come to grips with any theory so briefly mentioned or urbanely assumed. The nearest we come to a full statement is a United Nations publication.[2] Petroleum production is stated to be an industry of decreasing costs, or increasing returns. If so, the facts of nature compel a monopoly. There would otherwise be waste and extreme senseless price fluctuation, and anything except a monopoly must in time break down.

This vision of decreasing costs and natural monopoly in oil is held clear across the political economic spectrum. Perhaps it is just as well–or perhaps very bad–that we must reject the one opinion that seems common to all.

Decreasing Costs and Natural Monopoly. It is necessary to show explicitly why an industry of decreasing costs is a natural monopoly. First, let us dispose of a non-reason: that most oil-producing costs are overhead, and the fixed-to-variable ratio is very high. But so it is in agriculture, retail trade, and automobile repair. The fixed-to-variable ratio has no relation to economies of scale or decreasing cost. Furthermore, the fixed-to-variable ratio is often strikingly low in oil: in the Persian Gulf, sunk costs are in fractions of a cent per barrel; in a stripper well, operating costs are much higher than investment costs. There is no ratio of fixed-to-variable costs that can be called broadly representative of the industry, even in the loosest sense.

A decreasing-cost industry has unlimited economies of scale.[3] For example, within very wide limits the larger an electrical generating plant, the lower its unit costs, fixed and variable. Therefore additional capacity can always be had at a lower unit cost. If there were two or more independent concerns, each of them could always add to profits by building bigger, producing more, and selling it for anything even a little in excess of the cost of the increment. But since the cost of previous capacity is higher than that of the increment, each concern making a profit on the increment would be losing money on the

[2] UN-Economic Commission for Europe, *The Price of Oil in Western Europe* (E/ECE/205; Geneva, 1955). It is cited as "a lucid and convincing analysis" in M. G. de Chazeau and Alfred E. Kahn, *Integration and Competition in the Petroleum Industry* (Yale University Press, 1959), pp. 66–69, 375. See, however, the comments on an earlier version of this chapter: Yves Mainguy, *L'Economie de l'Energie* (Paris: Dunod, 1967), pp. 184–201; J. E. Hartshorn, *Oil Companies and Governments* (London: Faber and Faber, rev. 1967), pp. 63–65; E. T. Penrose, *The Large International Firm in Developing Countries* (London: Allen and Unwin, 1968), pp. 165–171. But see also the assertion that increasing costs are "irrelevant": Bruce C. Netschert, in American Institute of Mining, Metallurgical, and Petroleum Engineers (AIME), *Proceedings of the Council on Economics of AIME* (1966), p. 62, mis-citing Harold J. Barnett and Chandler Morse, *Scarcity and Growth* (Johns Hopkins University Press for Resources for the Future, 1963), particularly pp. 243–44. The "inherent surplus" theory is also explicitly assumed in the report by the Office of Science and Technology, *Energy R&D and National Progress* (Washington, 1965), p. 7.

[3] It is curious that Alfred Marshall's *Principles of Economics*, 1890 (1961 ed. London: Macmillan), to which we owe so much of the theory of industrial organization, has practically nothing on the connection of monopoly with increasing returns, except early and rather cursory remarks directed largely to the manager's character and personality (pp. 285–86; coming the closest to industry analysis at p. 397). One reason may be that Marshall often treats increasing returns as a trend over time rather than as a cost relation at any given time (pp. 318–22, 651). Even when he avoids this confusion (pp. 455–61 and Appendix H), the treatment is unclear. Concerned as he was not to confuse the cost function of a particular firm with the supply function of an industry, he yet never hit on the idea of making the latter an envelope curve containing the former.

rest of the output. Hence there must be a monopoly in any local area where electricity is generated and sold.

In most industries greater size brings lower cost—within limits. After a point, being bigger becomes a neutral factor, swamped by the usual run of management problems. In an industry of decreasing costs this brake on size is absent, and the management of every company is under constant compulsion to be bigger and produce more at all times, no matter how much is already being produced. But the more the industry produces, the more it pushes down prices, and the greater are the losses suffered by the group as a whole. This is "the well known economic theorem"[4] that in an industry of increasing returns, individual producers acting individually must operate at a loss. They must get together on prices and output, by an agreement not to produce past a certain point. Or else all try to live and let live and not produce and sell past the amount expected of them. But pressure is still there and must always be resisted. The most economic solution is simply one company. It is, of course, still confronted with the fact of nature—it could lower its costs by being larger—but the fact is no longer a market inducement or pressure. A monopolist can compare the cost of the additional output not with the current price nor with the expected lower price, but rather with the net increment or impact on revenues. Thus, if current costs are 100 and revenues 110, and if at doubled capacity costs would be 175 but revenues only 180 because prices would drop 10 percent, the incremental net revenue of 70 would be less than the incremental cost of 75, and there would be no inducement to build. If the company had been one of a number of independent ones, the management would have known that all the other companies would be trying to produce more, thus bringing down the price, and it would have no choice but to increase capacity and hence, by increasing output, help to bring about just the result it did not want.

The model of natural monopoly describes many persons' thinking about the oil-producing industry, except that they do not trouble to set it out. Indeed for many people this is a model of all or most industries: everywhere the bigger the better, or more efficient. But our business is only with petroleum—crude oil and natural gas.

Increasing and Decreasing Cost in Petroleum Production. Even experienced readers may need the warning that an industry of "increasing costs" does not necessarily mean costs rising over time (see Figure I-1).[5] A functional relation at any given time is not a trend over time. The petroleum industry has increasing costs, but not because most or all of the large low-cost fields have already been found. I venture no prediction, nor do I assume or imply that Persian Gulf costs will rise by 1975.

A look at two closely allied activities, pipelines and tankers, is a useful exercise in the theory of increasing and decreasing production costs. A single pipeline or a single tanker is a classic example of decreasing costs or economies of scale. For basic physical reasons the larger one builds, the lower the unit costs at optimum volume. A pipeline may be a decreasing-cost industry if and only if it is an industry entire of itself—if the market is so small as only to permit one or two to exist. Hence, in studying the natural gas industry in the United States it was necessary to analyze in some detail the special case of a pipeline

[4] UN-Economic Commission for Europe, *The Price of Oil in Western Europe*, pp. 16-18.

[5] Reproduced from my essay, "The World Oil Outlook," in M. Clawson (ed.), *Natural Resources and International Development* (Johns Hopkins University Press for Resources for the Future, 1964), p. 36. See chap. II for the demonstration that 1966 costs are lower than those of 1962.

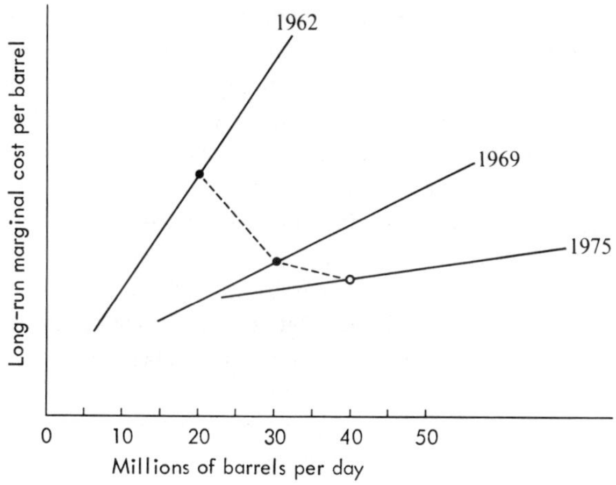

Figure I-1. Expected costs, 1962, 1969, and 1975.

as first having a buying monopoly in a producing area, then gradually losing it as more reserves were found and new pipelines were built into the region.[6] Decreasing costs in the single producing unit do not necessarily mean a whole industry or market of decreasing costs.

The world tanker fleet, which serves a single market (with the exception of U.S. coastal traffic), is one of the clearest examples of increasing cost at any given moment. Tankers form an array from lowest to highest cost. When, as during the winter, demand is especially high, more and more inefficient tankers are pulled out of lay-up or out of the grain trade; ships are run faster and consume more fuel per mile. The incremental cost and market price of the transport service is higher.

But given time enough to build more ships, the incremental cost tends to flatten out. We need to examine the supplying industry. A pressure of growing demand against laggard capacity may be transmitted to the shipbuilding yards, and ship prices may rise. Thus, greater output of tanker service may over a longer run still be at higher cost, though to a milder degree. Taking a still longer run: given time enough to adapt shipyards also, the tanker industry is one of approximately *constant* incremental and average cost. In time, as will be seen in Chapter IV, technical progress and cost reduction may dramatically lower the whole schedule.

The petroleum industry is like the tanker industry in that the variability among producing units makes the industry one of increasing cost. But unit costs also rise with increasing output within each producing unit. While tankers can be built to most efficient sizes, petroleum deposits cannot; we must take them as we find them.

Exploration is a process of spending money to find oil deposits, and the more barrels found per dollar the lower is the unit finding cost. At any given time there are a number of places to look, and the better places are preferred. A concessionaire deciding which area to turn back and which to keep is doing an exercise in increasing costs. So are the Russian oilmen who concentrate on shallow, permeable, porous reservoirs of low-viscosity

[6] M. A. Adelman, *The Supply and Price of Natural Gas*, Supplement to *Journal of Industrial Economics* (Oxford: Basil Blackwell, 1962), pp. 44–50. Paul W. MacAvoy, *Price Formation in Natural Gas Fields* (Yale University Press, 1962).

oil, who know about the more difficult (hence more costly) stratigraphic traps, but prefer to explore only structural indications as long as there are plenty of them, and who ascribe their exploration success to their concentration on the most favorable prospects.[7] The same has been true in the Middle East,[8] which is, indeed, mostly unexplored. The higher the output of oil aimed at, the more must the costlier prospects be explored. Conversely, the higher the price, the poorer the prospects it pays to evaluate and explore.

So far, taking the world as a whole, increasing knowledge of the geology of more of the earth's crust, and progress in technology–aided by chance–have outrun discoveries, thus bringing costs down. Where knowledge does not sufficiently grow, costs must go up. Hodges and Steele[9] suggested that in the United States new technology in exploration had not been important enough, after around 1940, to overbear the otherwise inevitable tendency to rising costs over time. Later data on oil and gas for the United States confirm their findings. Averages mean little when the dispersion is so great, but since about 1950 there has been a steep drop in the percentages, and even in the absolute numbers, of larger-size new oil reservoirs found.[10]

But at any given time, with exploration plans and commitments to be made according to what is then known, the more exploration, the more must it go into the less promising places. Over time, new and better places may be found (like Alaska in the United States), but if so the whole range of choices, from best to worst, is displaced–it is not inverted.

Development consists of finding the horizontal and vertical limits of the reservoirs in any given field, largely by drilling wells and equipping them for production. Most discoveries are noncommercial; few are developed in the United States. Only about 2 percent of all new-field wildcats discover a "significant" field, although about 10 percent produce some oil or gas.[11] The higher the price of oil or gas, the more it pays to develop the poorer discoveries.

A reservoir (or a whole field) is finite. As it is developed, the chances increase of getting to and past the edge. Hence, there is increasing risk or increasing cost as one gets closer to the edges and begins to get either thin and skimpy productive strata, or gets

[7] Robert Ebel, *The Petroleum Industry of the Soviet Union* (New York: American Petroleum Institute, 1962), pp. 41, 44, 53, 58. See also Adelman, *Supply and Price of Natural Gas*, p. 104n. For the greater desirability of structural traps, see A. I. Levorsen, "Exploration for Stratigraphic Traps," in *Economics of Petroleum Exploration, Development and Property Evaluation* (Prentice-Hall, 1961), pp. 17, 31, 36. On the rapid increase of drilling costs with depth, see Joint Association Survey (American Petroleum Institute, Independent Petroleum Association of America, and Mid-Continent Oil and Gas Association), *The U.S. Oil and Gas Producing Industry* (annual).

[8] Norval E. Baker, in *The Science of Petroleum*, VI (London: Oxford University Press, 1953), p. 83.

[9] John E. Hodges and Henry B. Steele, *An Investigation of the Problems of Cost Determination for the Discovery, Development and Production of Liquid Hydrocarbons and Natural Gas Reserves* (Rice Institute Pamphlet, vol. 46, Oct. 1959). Bruce C. Netschert, *The Future Supply of Oil and Gas* (Johns Hopkins University Press for Resources for the Future, 1958), noting what nobody would dispute, that existing statistics of "proved reserves" are far less than ultimate recoverable reserves from any given deposit, and also that there are a number of promising new techniques of exploration and development, concludes that the supply of oil and gas in the United States can in the future be greatly expanded at no increase in real cost. This conclusion is not absurd or implausible, but neither does it follow from the evidence adduced, which could at best prove that some cost-reducing factors will operate. On an important error in analyzing natural gas cost, see Adelman, *Supply and Price of Natural Gas*, p. 54.

[10] M. A. Adelman, "Trends in the Cost of Finding and Developing Crude Oil and Gas in the United States," in Gardner and Hanke (eds.), *Essays in Petroleum Economics* (Colorado School of Mines Press, 1967).

[11] L. H. Van Dyke, "North American Drilling Activity in 1967," *AAPG Bulletin* (American Association of Petroleum Geologists), vol. 52 (1968), pp. 904–8.

nothing.[12] In a large, simply structured, well-studied reservoir, the risk may be constant and low over a long interval.

A field may contain many reservoirs, each with several strata. The more productive are normally developed first. In one of the world's greatest fields, salt water is encroaching and being produced with the oil; the problem is temporarily and very cheaply solved by plugging back the lower perforations; but in time the production loss will be too great to accept and special separators at additional cost will be needed to handle wet crude.[13]

The productivity of a pool is less than proportional to the number of wells because past a certain point there is well "interference." The area over which oil migrates through permeable sands is very wide, so that one well could ultimately drain a very large reservoir.[14] If time—meaning the value of money—were no object, that would indeed be the best because it would be the cheapest way.[15] If the price is high enough, it pays to drill more wells into the reservoir and drain it faster, accepting the higher costs,[16] to the point where one may consider such further expenditures as fluid injection or development of another pool. Instead of just continuing to drill more wells, therefore, some Middle East operators combine it with fluid injection;[17] but elsewhere, as in the Khursaniya field, engineers advise against further developing for fear of excessive—which means expensive—pressure drop, so the company turns to developing Manifa.[18]

Well operating costs are considered over the remaining life of the reservoir, which at the beginning is the whole life. The greater the output, the higher the cost per barrel. Even neglecting well workovers, fracturing, acidizing, waterflooding, etc., the basic fact about oil and gas production is the production decline curve of the reservoir. The forces determining it are too complex for summary in a simple formula, as Muskat's treatise is at pains to show, but many empirical approximations are used, and there is a huge literature. The operator needs to know how much more production he can get out of a well or pool before reaching the "economic limit"—the point at which the unit cost of producing more is going to go above the price.[19] Many different formulas give results that in

[12] On the difficulty of distinguishing local from general structural dips, and mistaking the boundary, see Edgar N. Owen, "Petroleum Exploration—Gambling Game or Business Venture," in *Economics of Petroleum Exploration . . .* , p. 12. Again, in a given field, with past development costs at 92 cents per barrel, increments could be had at respectively \$1.48, \$1.71, \$2.01, \$2.40. With a price at \$1.50, the last three projects would not be worth doing; at higher prices, they would. (J. J. Arps and Marion S. Roberts, in *AAPG Bulletin*, vol. 42 [1958], pp. 2549–56.)

[13] A. F. Fox, "The Development of the South Kuwait Oilfields," *Institute of Petroleum Review* (London), vol. 15 (1961), pp. 373, 378.

[14] Morris Muskat, *Physical Principles of Oil Production* (McGraw-Hill 1949), pp. 591, 858–62, 899, and Rupert C. Craze, "Development Plan for Oil Reservoirs," in T. C. Frick and R. W. Taylor (eds.), *Petroleum Production Handbook* (McGraw-Hill, 1962), pp. 33-5–33-20.

[15] Muskat, *Physical Principles . . .* , pp. 897–99.

[16] Ibid., and C. C. Miller and A. B. Dyes, "Maximum Reservoir Worth—Proper Well Spacing," *Journal of Petroleum Technology*, vol. 216 (1959), p. 334.

[17] *Oil and Gas International*, Mar., 1961, pp. 29–33; Apr., 1961, pp. 22–25.

[18] Ibid., Sept., 1962, pp. 62–83.

[19] J. J. Arps, "Estimation of Primary Oil and Gas Reserves," in Frick and Taylor (eds.), *Petroleum Production Handbook* (McGraw-Hill, 1962), pp. 37-41–37-52. "The economic limit rate is the production rate which will just meet the operating expenses of a well (i.e.) . . . how much would actually be saved if the well were abandoned." Cf. J. L. Hopkinson, "Modern Method of Reservoir Assessment and Control," in *Planning for Productivity in the Oil Industry* (London: Institute of Petroleum, 1960), pp. 53–61, and references cited there. During 1961–62 the *Oil and Gas Journal* carried a series of over forty articles by E. T. Guerrero and F. M. Stewart, on the many kinds of production decline curves and practical ways of forecasting them at any given time. A particularly clear treatment is in William F. Stevens and George Thodos, "New Method for Estimating Primary Oil Reserves," *World Oil*, vol. 153 (Dec. 1961), p. 163.

practice are not far apart, and hence a simple exponential curve is generally used.[20] The usual example shows a fixed outlay per month and diminishing production per month; the operator forecasts the time to abandonment, which will come about when the value of the monthly production no longer exceeds the monthly cost.

Every development plan forecasts production over the life of the pool or well, estimating the production-decline curves. A convenient example of earlier-stage calculation is to be seen in a paper presented by an Aramco representative at the Second (1960) Arab Petroleum Congress.[21] This shows the cash flow and expenses anticipated "over a 15-year period at a declining production rate due to a normal decrease in reservoir pressure." Then the writer shows how a waterflooding project would put production on a much more steady basis and would increase total recovery; *but* waterflooding would cost additional money and the incremental cost per barrel would be some 19 percent above the non-waterflood design. The additional output is not worth this higher incremental cost because it ties up money in an unduly low-profit activity. If the waterflood had been worth its cost, the problem would have been to design it to a size where the additional cost was not worth incurring.

Of course, these calculations look more precise than they are. In practice, the operator must refigure and refigure as he proceeds, gaining more and more knowledge of the deposit as he drills more wells and produces more from them.[22] Porosity and permeability of an oil deposit are not like "a nice uniform sponge down below," holding the oil in place.[23] In so tremendously rich a deposit as the Burgan field in Kuwait, the increasing knowledge might seem trivial, and yet is discussed in a report on its development.[24]

An excellent example of how knowledge grows can be seen in a report on Saudi Arabia. The behavior of an oil reservoir can be simulated on a computer, with the parameters of the oil pool and its water drive supplied by trial and error: calculations from past behavior are compared for a time with actual behavior and then corrected. In the Abqaiq field certain persistent anomalies suggested an undiscovered hydrocarbon accumulation in the neighboring Ghawar field (or fields); and there were some less attractive possibilities. Four wells were therefore drilled on the Ghawar east flank. Because they were productive, the boundary had to be redrawn three kilometers farther east, thereby adding a billion barrels to published reserves. The anomaly was not yet altogether explained, and perhaps the flank delineation not necessarily complete.[25]

There is a stereotype of a reservoir (or even a field) "all drilled up" and needing no more investment to keep producing. But there is no such pool, except in its dwindling last years. Wells need to be periodically worked over either for cleaning or to improve the rate of flow by fracturing the zone around the well bore with explosives or treating it with acid. Even in the hugely productive Persian Gulf fields, a considerable fraction of the

[20] H. C. Slider, "A Simplified Method of Hyperbolic Decline Curve Analysis," *Journal of Petroleum Technology*, Mar. 1968, p. 235.

[21] Lawrence Ison, "Financial Analysis of Oil Projects."

[22] John P. Dowds, "Statistical Geometry of Petroleum Reservoirs in Exploration and Exploitation," *Journal of Petroleum Technology*, July 1969, p. 841.

[23] See the discussion following Hopkinson, "Modern Method of Reservoir . . . Productivity," pp. 61–67, for some fascinating examples.

[24] A. F. Fox, "The Development of the South Kuwait Oilfield," *Institute of Petroleum Review*, vol. 15 (1961), p. 373.

[25] W. R. Bartlett, "Saudi Arabian Oilfields Performance with an Electronic Model," paper presented at Third Arab Petroleum Congress, 1961, pp. 1–5, 15–19.

year's drilling is devoted to well workovers.[26] Were the fields not expanding so rapidly, the fraction would of course be larger. The additional investment needed for injection of fluids, usually water or natural gas, is almost too obvious to be mentioned.

To sum up: looking at any reservoir development-production programs ex ante, both total output and total cost are functions of time. If the expenses per unit of time are fixed (and a fortiori if they are increasing, as in fact they are), and if the production per unit of time decreases, then the production cost per additional unit increases the greater is the total production.

The industry as the sum of all reservoirs. We have examined the decreasing-cost hypothesis on its strongest and most familiar ground: the single pool, considered in isolation from all others. Even here it fails because it conflicts with the basic physical laws governing petroleum production. But even if it had been true that the individual field (like the individual tanker) operated under decreasing costs, it would still have been untrue of the industry as a whole, because production costs, even in the very short run, vary widely. The higher the price, the more higher-cost fields it pays to operate; at a lower price, only the very low-cost fields can pay.

It would be hard to find a better-documented example of increasing costs in the literature of economics. Why then is the error so firmly embedded in the thinking of the industry and of friendly, unfriendly, and neutral critics outside? I suggest that increasing returns are a scholarly gloss on revealed truth. When half a dozen companies made up the whole international industry, they *knew* they should be allowed, statesmanlike, to divide markets and set prices. Others *knew* oil was "too important to be left to oilmen"—which could mean all things to all men—and hence governments had to intervene.

Today there is some disillusioned understanding in industry, but less in government, that the good old days are gone for good, and that increasingly competition influences the market not because it should but because it does.

Whatever the reason, one must be grateful to those stating the idea in terms of conventional economic theory, which can then be compared with the facts to show how wrong it is—that is what theories are for. And those who know how agonizingly difficult or impossible it is to get any accepted idea clearly enough stated so that one can test it, will understand that our gratitude is not mixed with irony. The great merit of UN-ECE's *The Price of Oil* was that it was occasionally clearly wrong.

To sum up the preceding discussion: under any given conditions of knowledge, the more exploration is done, the higher—probably—will be the finding cost per unit of what is found. The more a given deposit is developed, the higher—almost certainly—will be the cost of additional development. Production cost always rises as a function of greater output.

Incremental Cost-Revenue Gap as the Engine of Supply and Price Change. A price above incremental cost is an incentive to produce and offer more, causing downward pressure on the price until and unless something is done to get rid of the excess supply. The average cost, whether of a given reservoir or area or the whole industry, is only an after-the-fact summary of what has been spent. It cannot be compared with price, and it has no implications for output. Reservoir A may have been very cheap to find and exploit, but if it is small, more output would raise incremental cost steeply, so a modest rise in price will give us nothing more from it. On the other hand, Reservoir B may be

[26]See chap. II, pp. 64-65.

more expensive but much bigger than A. Cost will increase only modestly even with a great deal of development, so a small price rise will give us a great deal more oil.[27]

Hence it is of little help to talk about "the cost" of oil or "the profits" of "the industry." At any time, the lowest-cost units are very profitable, others are mediocre, others just breaking even. Some are losing money if they take account of unrecoverable sunk costs, but breaking even or better on current account. The cost of the poorest facilities makes the price in a competitive industry and puts pressure on it in a noncompetitive market. For if the price is too low to keep the highest-cost facilities operating, these operations will shut down, supply will be less than demand, and price will rise. A demand surge, which sends up the price, signals that higher-cost facilities can be profitably operated; the profits of the best pools will be all the better. Conversely, if productive capacity exceeds demand, price will fall and squeeze out the highest-cost units no longer needed. That Persian Gulf costs are so low as to allow high profits may mean only that the best fields earn a big producers' rent. But if there is a great deal of available oil that can be cheaply developed, so that incremental costs will continue to be low over a long stretch and, consequently, that Persian Gulf oil could displace costlier oil and bring down the price, then there is a block to the competitive mechanism, which is now to be studied.

The Market-Sharing Mechanism of Relative Incremental Costs. The increasing-cost theory explains the coexistence of production centers with vastly different average costs. If development in each area is carried to the point where its incremental cost equates with that of every other area, account being taken of transport, then there is no further pressure to cut back in one place to make room for any other. Incremental cost then serves as a mechanism to divide markets in such a way that under competition it is not to anybody's interest to increase or cut back his offerings. If markets are divided any other way—considered "equitable" by governments or producers—then the incentive remains for some people to expand their output, potential supply exceeds demand, and the price is under pressure.

Because there would be coexistence under rigorous competition, it does not follow that the particular pattern of coexistence at any given time is thus explained. It may or may not be. But the fact that oil is produced in various parts of the world at widely differing average costs and profits needs no special explanation. Neither government regulation nor industrial statesmanship is required to explain why the United States and Venezuela continue as big producers; they will for many years. But if relative incremental cost were the governing factor, the United States would produce less, the Middle East more, and there is no telling about Venezuela. It is a question of more or less, not yes or no.

Transport cost is usually a large factor in comparative incremental cost. Higher-cost oil (or gas or coal) near to consuming centers competes with much lower-cost fuels from farther away. The earlier discussion of costs and prices was too simple in one crucial respect: strictly speaking, there is no world price of oil, but only a world price structure. Given competition, the structure cannot be stable unless three conditions are met. First, in any exporting area, f.o.b. prices must be uniform to all destinations; otherwise, there is

[27]This is not to deny the importance of the average expectation in searching for new deposits. It is of crucial importance for landlord-tenant relations. See below, p. 42. For a treatment of the problem of mineral deposits of varying quality, see O. C. Herfindahl, "Depletion and Economic Theory," in Mason Gaffney (ed.), *Extractive Resources and Taxation* (University of Wisconsin Press, 1967), pp. 78–79, 82–85.

an incentive to elbow your rivals aside (with their customers' help) in attempting to get to the more profitable markets, until all are equalized. Second, in any importing area price must be the same for any given quality of fuel regardless of point of origin; otherwise it is buyers (with some sellers' help) who have the incentive to concentrate on the cheaper offerings until they are all equalized. There is no cause to wonder, for example, why cheaper Venezuelan or other oil has always sold in the United States at the same price as dearer U.S. oil. Third, if the prices in all areas are in the neighborhood of the respective incremental costs, and differ approximately by the amount of transport cost, then the system is under no strain.

This last condition is the most difficult to set forth, but it is the unifying principle of the whole world market. There needs to be such a balance among incremental costs in the various production centers, plus transport costs to the various consuming centers, that there is no incentive to change the pattern of trade. Or, what comes to the same thing, to expand production anywhere would cost more than it was worth; to contract production would be to sacrifice profit.

This will be discussed at length in Chapter V; for the present Figure I-2 may serve. Assume that (1) it costs 80 cents per barrel to haul crude oil from Export Area to Import Area; (2) given time enough to build more ships if need be, there is no practical limit to the amount of transport service available at that price. (3) Production costs in Import Area are such that for every additional 10 cents added to the per-barrel price, one can have another million barrels daily average production. (4) In Export Area, every additional 10 cents in the supply price will give us not 1 million but 4 million barrels capacity. These conditions are shown in the two light lines labeled "domestic availability" and "foreign availability," respectively.

If demand is D_1, then production will be 3 million, all supplied by Import Area, despite its much higher production costs. Only when the price in Import Area goes above 80 cents is there any point in starting production in Export Area.

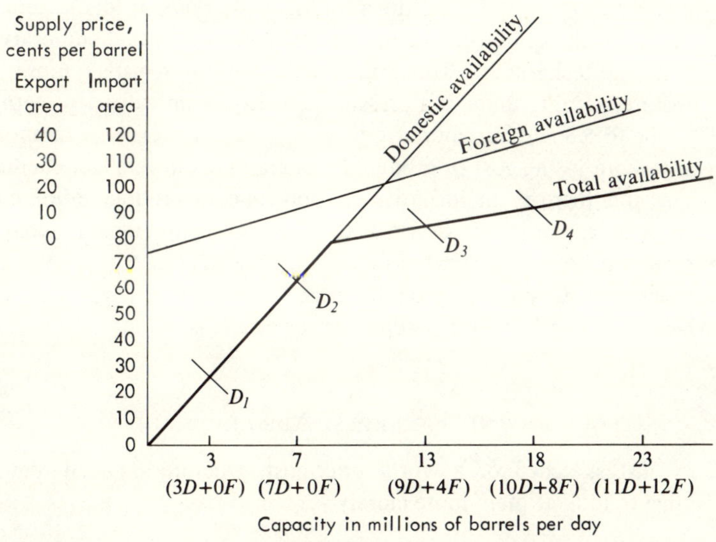

Figure I-2. Relation of supply to price level and structure.

If demand is D_3 or greater, we must add horizontally the domestic and foreign supply curves, to give us "total availability." At an Import Area price of 90 cents, 13 million barrels are supplied, of which 9 million are domestic and 4 million foreign. Foreign output is then "supplementing" domestic, and the rate of supplementation is governed by relative incremental cost, or relative supply price: four to one.

Changes in transport cost may have a substantial effect on total availability. For example, assume the transport cost to be not 80 but 60 cents. At a price of $1.00, total availability would be not 18 million but 26 million barrels, and production in Import Area must be reduced.

The world price structure and price level, the total of production and its share among various sources, are all interrelated. Relative costs may be transformed as output increases, and hence also world channels of supply. It was believed for decades that there was probably a lot of cheap oil to be found in the hills of the Middle East–the flat plains being long unjustly denigrated–if anyone wanted to spend the money to try to find it.[28] But we need no special theory to explain why there was little exploratory effort. Given the relatively low levels and therefore low cost of U.S. Gulf and Venezuelan output, and the very high cost of hauling Middle East oil as refined products in small tankers (for refineries were mostly crude-oriented and world trade largely in products), the delivered cost of Middle East oil in the large markets of the world was too high to compete.

Warning. A brief statement of the theory of a price structure cannot but give an impression that the process works neatly and precisely. It does not. Price parities among various sources of supply are made by sellers and buyers looking for a better deal, guessing ahead and often guessing wrong. Moreover, when the market mechanism works best it conceals itself. For example, importers do not necessarily bid against each other for a limited supply from nearer areas, gradually raising price to the point where product begins flowing in from farther areas. If the result is anticipated, prices rise immediately, the process is short-circuited and the final result appears quickly and seems "natural" to all concerned.

The theory is useful because it is not a realistic description of the world industry at any given time. By showing the conditions for stable markets, it lets us understand the destabilizers. Every such equilibrium as we see in Figure I-2 is the temporary result of a constantly changing technology. The price-supply pattern registers those changes, if allowed to operate. If not, there is a pressure toward the unrealized pattern because it would profit some to sell and buy more from one source at the expense of another.

A saving anywhere in the cost of production or transport lowers incremental delivered cost somewhere and there is an incentive for someone to produce more, changing the structure. Ten years ago anything over 50,000 tons was a supertanker; today most new tonnage ordered is in ships four times as large. A price structure in balance then would have become obsolete. Changes in production and transport costs are a set of incessant shocks to which the system is always adjusting, or else it is storing up tension that may later erupt.

Resources and Reserves: Adequacy and Cost

The theory just sketched of a world price-cost structure assumed that costs were known. It is time to look at them more closely.

[28] G. M. Lees, "The Oil Fields of the Middle East," in *The Science of Petroleum*, I (1938), p. 140, and VI (1953), p. 67.

The petroleum production process is essentially the building and depleting of inventories that are very large relative to current output. We need in the end to answer two questions: What does the discovery of new fields do to the stability of the cost-price system? What can we say about the adequacy of oil resources, which arouses so much worry, real and put-on? The questions must correspond to economic theory, and also to the available data, which are fragmentary and prepared for purposes other than those of this study.

Our thesis is that the underground stock must be considered at two stages. (1) Exploration outlays yield deposits of oil-in-place. (2) Development outlays are made to create producing capacity and reserves, the ready shelf inventory, which is only a fraction of oil-in-place. Reserves are not found; only oil-in-place is found; reserves are developed. This distinction must be made because there is a profound difference between the two activities. Exploration is a matter of reckoning small probabilities of future success. Because it is a probe of the unknown, past success or failure, low or high cost, it is of uncertain or often of no relevance in judging whether a new investment is worth making. In contrast, development is an investment process under only modest uncertainty, where future costs can be estimated from data on past operation, suitably adjusted for changes that are mostly predictable and regular.

But these two activities are linked because they are imperfect substitutes for each other. To overcome the constant tendency toward rising cost, as the industry carves more reserves out of the oil-in-place in known deposits, oilmen seek (1) better methods and (2) new deposits, and the proportion between them may change very much.

Setting discoveries of new oil-in-place to zero, rising development costs, as more reserves are taken out of a shrinking total of oil-in-place, are the worst that can happen. Therefore, we know the value of relief from scarcity, or the Maximum Economic Finding Cost. (See p. 39.)

Although our theory is simple, it can only be established by looking in detail at the process of enumeration: what kinds of wells are drilled for what purposes and how reserves are measured. In order to distinguish exploration from development, one needs to look carefully at the intermediate stages to see whether the distinctions that are drawn, or our attempted synthesis, are valid.

Concept and Measurement of "Reserves." Calculations of reserves began in the United States, and we have a continuous series going back many years with an explanation of how the numbers are derived.[29] In about two dozen areas, subcommittees of men familiar with local conditions meet once yearly to estimate the local crude oil reserves. Local estimates are forwarded to the American Petroleum Institute (API) Committee on Reserves and Productive Capacity.[30] There the subcommittee reports are combined into

[29]The following is based on Morris Muskat, "The Proved Crude Oil Reserves of the U.S.," *Journal of Petroleum Technology*, September 1963; Wallace F. Lovejoy and Paul T. Homan, *Methods of Estimating Reserves of Crude Oil, Natural Gas, and Natural Gas Liquids* (Resources for the Future, 1965); the American Petroleum Institute–American Gas Association annual reports, *Reserves of Crude Oil, Natural Gas Liquids, and Natural Gas in the United States and Canada*, particularly for 1966 and later years, and API Technical Report No. 1, *Definitions for Petroleum Statistics* (1969). (The writer has served for five years on the API Coordinating Committee on Statistics. He values highly his experience there, but neither the API nor his committee colleagues are necessarily in agreement with any statement made here.)

[30]In 1967 this committee had 16 members, of whom 13 were oil company representatives, the other three being a representative from the Indiana Geological Survey, the API Director of Statistics, and, as an observer-member, the head of the Office of Oil and Gas at the Department of the Interior. The twelve subcommittees contained 164 members (1967 API *Reserves* report, pp. 9–10).

estimates for states and also for subdivisions of some of the largest states. Thus even a company represented both on the subcommittee and the API committee cannot tell what any other company has. The committee gets the benefit of local expertise without sharing information on a company level, which might run afoul of the antitrust laws.

Oil-in-place, the source of proved reserves. As of the end of 1945, reservoirs operated in the United States contained 282 billion barrels of oil-in-place.[31] Of this original amount 41 billion barrels had been produced, and 241 billion were still in place, of which 20 billion or 8 percent were counted as "proved reserves." In other words, only 61 billion barrels, or 22 percent of the oil-in-place, had become reserves. At the end of 1968, cumulative production was 87 billion barrels and 31 billion were current proved reserves; hence 115 billion barrels had been taken for proved reserves from total discovered oil-in-place of 389 billion. Thus the fact that in 23 years an apparently small change in percentage of preexisting oil-in-place passed into reserves (22 to 30) is the sign of a drastic shift from new to old fields as a source of new inventory. We do not know how much of the remaining oil-in-place can be carved out into reserves, and at what cost.

Every API *Annual Report* says:

> The reserves listed as "proved" in this report, as in all previous API reports, are the estimated quantities of crude oil which geological and engineering data demonstrate with reasonable certainty to be recoverable from known reservoirs under existing economic and operating conditions.[32]

Through 1963 the phrase "beyond reasonable doubt" was used instead of "with reasonable certainty." But this change is in language, not substance, as is seen by the continued use of the phrase "as in all previous API reports," which was carried over unchanged.

Proved reserves are limited to those known to be recoverable "under existing economic and *operating conditions*" [emphasis added]. It is true that the proved area of an oil reservoir includes not only the drilled portion but also "the adjoining portion not yet drilled." But the area must be productive on the basis of "available geological and engineering data." Reserves coming from application of improved recovery techniques are included as proved only if the facilities have already been installed or at the very least there has been "successful testing by a pilot project."

The ratio of proved reserves to annual production has slowly declined for many years, from 16.3 years (in 1920), 15.1 (1930), 14.1 (1940), 13.5 (1950), 12.8 (1960), 8.9 (1970).[33] The decline has accelerated since 1959; the absolute amount of reserves has stayed within 2 percent of the nine-year average, while production is up 20 percent.

Economic concept of proved reserves. Let us suggest two definitions that will have a precise economic content and that will reconcile or be consistent with the facts stated up to now. Oil-in-place is a resource, obtained and paid for by exploration. Proved reserves are that small part of oil-in-place which has been developed for production by the drilling

[31] The estimate of 282 billion is in one sense an unavoidable exaggeration; it was made in 1967, and hence reflects the knowledge of that year, not of 1945. As will be seen later, the exaggeration is not great.

[32] E.g., the report for 1967, p. 15.

[33] Proved reserves of gas have declined from 39.6 to 15.9 years, but the economic significance is altogether different because of the distinction between nonassociated and all other gas. It cannot be presented here; see, however, Adelman, *Supply and Price of Natural Gas* (Alaska North Slope excluded).

and connecting of wells and associated facilities. It is the total of planned production from facilities already installed and paid for.[34]

It is possible to make a rough test of the definition. Reserves are defined as cumulative expected output,

$$R_0 = q_0 \int_{t=0}^{t=b} e^{-at}\, dt\,, \tag{1}$$

where q_0 indicates current output rate, b indicates time of abandonment when expected variable costs have risen to equal expected price; and a is decline rate. At time of abandonment, when the well reaches the "economic limit,"

$$q_b = q_0\, e^{-ab}, \text{ or } q_b/q_0 = e^{-ab}\,. \tag{2}$$

Substitute (2) into (1), obtaining:

$$R_0 = q_0 \left(\frac{1 - e^{-ab}}{a}\right) = q_0 \left(\frac{1 - q_b/q_0}{a}\right), \text{ or } a = \frac{q_0}{R_0}\,(1 - q_b/q_0)\,. \tag{3}$$

At the economic limit, the fraction q_b/q_0 is very small, since variable operating costs are low. In North America, a high price permits wells to operate at a very low percentage of original capacity; elsewhere the original capacity is much higher. Hence a differs only insignificantly from q_0/R_0.

But this relation does not always hold, for there are always some reservoirs which, over some period, show no decline rate. There may be a complete water drive. Or production may be restrained by lack of capacity in gathering or transport. Indeed, where a reservoir is isolated and dependent on a single outlet by land or sea, it does not make sense to build transport capacity to handle initial crude production, and then have it stand increasingly idle.

Therefore the true decline rate of a group of reservoirs, including some with stable (because constrained) production rates, should exceed the observed ratio of production to reserves.

As shown below, there has been one independent estimate of decline rates in U.S. crude oil and natural gas, for the year 1960.[35]

	Production (q_0)	*Proved Recoverable Reserves (R_0)*	q_0/R_0	*NPC Decline Rate*
Crude oil	2,471 BB	31,719 BB	.0770	.0868
Natural gas	13.02 TCF	261.2 TCF	.0499	.0550

The explanation in the NPC report is not altogether clear and may not be wholly consistent; hence the approximate agreement with my hypothesis must be viewed as tentative.

[34] One may suspect the definition is too narrow, but see note 51 in this chapter.

[35] Tabulated as follows: production and reserves, from API–AGA, *Reports on Proved Reserves*, annual; decline rates, from *Report of the National Petroleum Council Committee on Proved Petroleum and Natural Gas Reserves and Availability*, 1961, pp. 27, 36.

In 1969, a rough estimate of the decline rate was made by the Task Force on Oil Import Control: .125 compared with a 1969 p_0/R_0 ratio of .1075.[36]

Reserves are not measured as restrictively outside the United States. During 1970, Venezuelan production was 1,342 million barrels (MB); at a decline rate of .159, this would indicate proved reserves of 8.45 billion barrels (BB). Since "published proved reserves" were actually 14.0 billion,[37] this indicates an allowance of about 66 percent for oil expected to be produced by facilities not yet in place.

Proved reserves at any time are quite distinct from forecasts of what will be made into reserves later. Recent discoveries in northern Alaska were originally said to "contain" from 5 to 10 billion barrels. This was the amount expected to be developed in the course of years. Eight months later there was an implicit reserve forecast when it was announced that a pipeline built from the area would have a capacity of 2 million barrels daily (MBD), which indicates about 15 billion barrels forthcoming in 20 years. It would be a mistake to compare either forecast with proved reserves or with oil-in-place.

In order to understand the economics of investment in discovery and development, and avoid confusing the two, we need a brief description of the process.

How oil operations create new resources and reserves. The long-established Lahee classification of wells drilled during any given year is illustrated in Figure I–3. Development wells are "within or close to the limits of a producing or producible pool, as these limits are known at the time of drilling."[38] Lahee thereby implies what the API says explicitly: wells drilled after discovery of a pool "usually add to the proved area, thereby serving to increase estimates of proved reserves. The reserves credited to a reservoir because of enlargement of its proved area are classified as extension additions."[39]

The more development wells are drilled into a pool, the more is known about the character of the pool and the better become the estimates of what will probably be produced from it. The additions (or less frequently, subtractions) are known as "revisions." So are "increases in proved reserves associated with the installation of improved recovery techniques."[40] This confirms the unifying concept of proved reserves as the expected cumulative production from facilities already installed.

So much for the known reservoir under development. A company may now push some little way into the unknown, and try to find new pools "within the known (surface) limits of a pool," but shallower than the known pools, or deeper.[41] Or a well may be drilled as an outpost or extension in order to test the limits of the known pool. It may extend the area of the pool, or it may be a dry hole, or sometimes it may discover a new pool.

Thus outposts and shallower-pool and deeper-pool tests aim to extend the already known reservoir or find new pools close to it. These new wells will, if successful, prove up new reserves that will be also called "extensions," like the development wells' additions.

Thus an oil company faces a set of choices or trade-offs. It can increase its reserves in three ways: (1) in known reservoirs by development drilling, (2) around known reservoirs,

[36]Cabinet Task Force on Oil Import Control, *The Oil Import Question* (Washington. 1970), p. 222.

[37]For decline rate, see appendix II–D; reserves from *Oil and Gas Journal*, annual supplement "World Wide Oil."

[38]Frederic H. Lahee, *Statistics of Exploratory Drilling in the United States, 1945–1960* (Tulsa: AAPG, 1962), p. 132.

[39]API–AGA, *Reserves . . . 1968*, p. 17.

[40]Ibid., p. 18.

[41]Lahee, *Statistics of Exploratory Drilling . . .* , p. 134.

Objective of drilling				Classification when drilling is started	Classification after completion or abandonment			
					Successful		Unsuccessful	
				A	B		C	
Drilling for long extension of a partly developed pool				OUTPOST or EXTENSION TEST	EXTENSION WELL (Sometimes a new pool discovery well)		DRY OUTPOST OR DRY EXTENSION TEST	
Drilling for a new pool on a structure or in a geological environment already productive	NEW-POOL TESTS	Drilling within limits of proved area of pool:	For a new pool above deepest proven pool	SHALLOWER POOL (PAY) TEST	NEW-POOL DISCOVERY WELLS: (Sometimes extension wells)	SHALLOWER POOL DISCOVERY WELL	DRY NEW-POOL TESTS:	DRY SHALLOWER POOL TEST
			For a new pool below deepest proven pool	DEEPER POOL (PAY) TEST		DEEPER POOL DISCOVERY WELL		DRY DEEPER POOL TEST
		Drilling outside limits of proved area of pool		NEW-POOL (PAY) WILDCAT		NEW-POOL DISCOVERY WILDCAT (Sometimes an extension well)		DRY NEW-POOL WILDCAT
Drilling for a new field on a structure or in an environment never before productive				NEW-FIELD WILDCAT	NEW-FIELD DISCOVERY WILDCAT		DRY NEW-FIELD WILDCAT	
Drilling to exploit a hydrocarbon accumulation discovered by previous drilling				DEVELOPMENT WELL	DEVELOPMENT WELL		DRY DEVELOPMENT WELL	

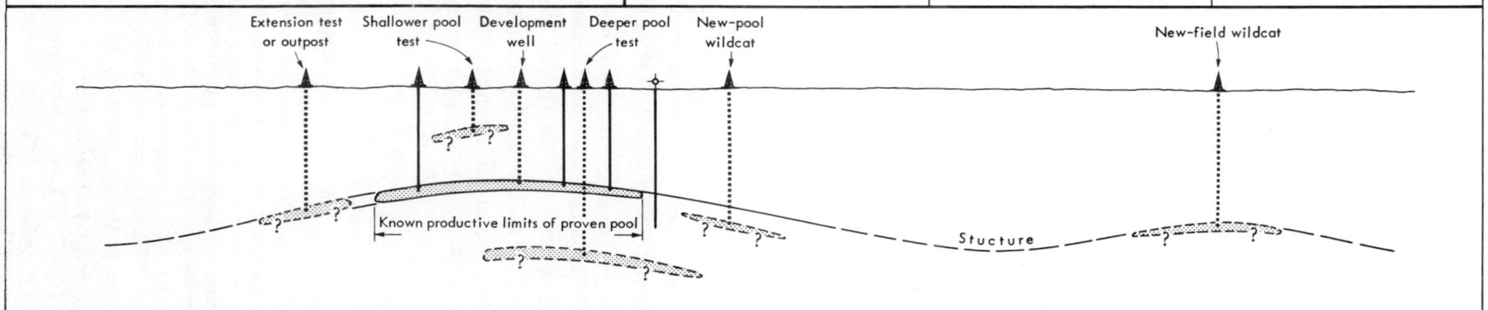

Figure I–3. Classification of wells. (Reproduced with permission from *AAPG Bulletin*, June 1970, p. 892.)

or (3) search for a new pool outside the limits of the old. Both (2) and (3) add to oil-in-place, (1) usually does not.

Finally there is the big jump into the unknown. Most risky and most profitable (if it succeeds) is a "new field wildcat." This brings us to the distinction between pools and fields. A pool can be defined precisely: "an underground accumulation of petroleum in a single separate natural reservoir . . . [is] a single natural pressure system so that production of petroleum from one part of the pool affects the reservoir pressure throughout its extent. . . . It is effectively separated [by rock, water, or other impermeable barriers] from any other pools that may be present in the same district or on the same geological structure."[42] But a pool may contain more than one zone, with different traits affecting drilling and operating costs.

A field may consist of a single pool, or of two or more pools on what is known as "the same geologic structure."[43] There is probably no rigorous way of defining a geological structure. Moody says simply: "I use the term 'field' to describe an aggregate of overlapping, contiguous, or superimposed pools."[44]

Pools and fields result from some set of geological disturbances which may extend over a wide area, or form what is often called a "trend" or "basin" which are to a field what a field is to a pool.

The oil company can thus keep on adding to its proved reserves for many a year without ever finding a new field or even a new pool, and with zero or modest additions to oil-in-place. But it must, all else being equal, run up against increasing development and producing cost.

The postwar shift from discovery to development in the United States. For a better understanding of the "reserves" concept, we review the American experience since World War II, although our study is not directly concerned with that country. Since the early fifties there has been a drastic decline in the productivity of exploration investment. Few large fields have been found, and much smaller amounts of oil-in-place for each unit of exploration expenditure. Yet improved technology[45] has, at least through 1963, offset the increase in developing cost that would be inevitable under stationary conditions.[46]

This trend is shown quite dramatically when we examine the figures on oil-in-place and proved reserves for the United States as a whole and for each individual state. It was not possible to do this before the recent improvements in petroleum statistics instituted by the API. When all we had available were the figures on current addition to proved reserves, they told practically nothing about discoveries because they only indicated what

[42] Ibid., p. 132.

[43] "Structure" is used here in a generic sense. It may also refer to a "structural" as opposed to a "stratigraphic" trap. The distinction is basic but not absolute. Structural traps result from a deformation of one or more underground strata, forming domes or dips or faults. These can be found by geophysical methods, although the presence or absence of oil can only be found by the drill. A stratigraphic trap results from a chance phenomenon; a permeable layer is bounded by nonpermeable layers, but with no systematic relationship of one to the other. In a structural trap, the ultimate block to further migration of the hydrocarbon fluid is at the highest point of the deformation ("structure"). In a stratigraphic trap, the permeable layer containing fluid is stopped or pinched out as the layer ends in contact with a nonpermeable layer. See Fred L. Oliver, "Reservoir Traps," in Frick and Taylor (eds.), *Petroleum Production Handbook* (1962), chap. 26.

[44] John D. Moody, address to the Southwestern section of the AAPG, Feb. 7, 1968 (mimeo.).

[45] National Petroleum Council, *Impact of New Technology on the U.S. Petroleum Industry, 1946–1965* (1967), part I, especially chaps. 2–4.

[46] M. A. Adelman, "Trends in the Cost of Finding and Developing Oil and Gas Reserves," in Gardner and Hanke (eds.), *Essays in Petroleum Economics.*

was being added to the current shelf inventory. New reserves developed during any given year are practically independent of new fields discovered that year.

But the new data make it clear that discovery deteriorated rapidly after World War II. (See Table I-1.) The 1968 finds in Alaska open a new oil province and a new finding history.

TABLE I-1. Crude Oil: Newly Found Reserves Related to Total New Reserves, United States, 1945-1965

(*billions of barrels*)

Years	Production	Net growth of reserves	Gross new reserves (col. 1 + col. 2)	Reserves (est. ultimate) from fields discovered during period	Col. 4/col. 3
	(1)	(2)	(3)	(4)	(5)
1946-50	9.35	+5.33	14.68	11.67	0.795
1951-55	11.48	+4.74	16.22	8.59	0.529
1956-60	12.46	+1.60	14.06	5.89	0.419
1961-65	13.01	−0.26	12.75	3.47	0.272

Source: API-AGA, *Reserves . . . as of December 31, 1969*, table III. Net growth is difference between, e.g., end-1945 and end-1950, etc. Years after 1965 excluded because of underestimation of reserves in newly found fields. This is not considered significant for even so late a year as 1967 (compare API-AGA, loc. cit., with *AAPG Bulletin*, June 1968, p. 910 and June 1969, p. 1170; the smaller the new field, the sooner its limits are established, and the less the degree of early-year understatement).

Unit discovery costs must have increased considerably, probably by several hundred percent, because total finding expenditures have been fairly steady over the last 10 years, while newly found oil-in-place has declined rapidly. The AAPG series show the decline in new field discoveries to be greatest in the largest size classes, which are lowest cost.

My estimates may be biased in making exploration look somewhat better than it really is. Plenty of new pools are still discovered whose reserves are added mostly to the oil-in-place of the existing field of which they are considered a part. As has been shown earlier, the distinction between new pools and new fields is a matter of expert judgment and not of any rigorous criterion–it is art, not science. A "geologically significant" new pool discovered in a known field may have its oil-in-place and proved reserves credited to that year and not to the earlier year in which the field was originally discovered.[47]

Improved technology made more intensive development increasingly good as compared with more exploration. But because exploration looked increasingly bad, more attention would have been given to intensifying development anyway. This is confirmed by the strenuous bidding for leases off Louisiana and to a lesser extent California, and now Alaska. The harder it becomes to find new fields, the greater the premium on those places where there is still hope of making good finds. The cheaper is development, the bigger the profit on any given new field and therefore the bigger the payments that oil companies are willing to make to obtain them. Thus the rising price of offshore leases, despite stable crude oil prices, suggests that in any given reservoir developing-producing costs have been going down. New flush fields are cheaper to develop than years ago. But in the United States generally, the lower costs approximately offset what would otherwise be a cost increase, and held the average approximately stable through 1963.

[47] API-AGA *Reserves . . . 1968*, p. 20.

An editorial in the *Oil and Gas Journal* of October 28, 1968 deprecates the bidding for offshore leases and the high prices offered: "Only explanation for the frenzy of bidding at recent sales is the pressing need for domestic reserves by nearly every company." But, as has just been seen, most new reserves come from development of oil-in-place, not from discovery of new fields. The high prices are paid because there is expectation of finding new fields so large and low-cost to develop that they will leave a sufficient surplus to pay for a risky activity.

We cannot be sure how much oil-in-place it will pay to draw into proved reserves. (Table I-2 summarizes a sample study of recovery efficiency.) The percentage recovered obviously varies a good deal, depending partly on nature, partly on investment. Water drive is usually best.[48] Moreover, as is noted in the study, the deeper the well the greater the average ultimate recovery, and the authors consider this as evidence of inducements to invest more. Since the sample is small and not necessarily representative, one ought not to generalize from it, but it checks loosely with nationwide figures. Of the original U.S. oil-in-place as reckoned at the end of 1968, 66 percent was in sandstone reservoirs, accounting for 73 percent of the ultimate recovery.[49]

TABLE I-2. Recovery Efficiency, United States, by Type of Drive and Rock

Predominant drive mechanism	Supplemented	Rock type	Recovery efficiency (percent OIP produced)		
			Low	Median	High
Water	...	S	28	51	87
Water	...	C	6	44	80
Gas cap	Yes	S, C	16	32	67
Solution gas:					
(a)	Yes	S	13	28	58
(b)	Yes	C	9	22	48
(c)	No	S	10	21	46
(d)	No	C	16	18	21

Source: J. J. Arps, Folkert Brons, A. F. Van Everdingen, R. W. Buchwald, A. E. Smith, *A Statistical Study of Recovery Efficiency* (*API Bulletin*, D14, 1967), tables 1-7.
OIP, oil-in-place; S, Sandstone; C, Carbonate.

Oil-in-place is estimated only for the United States and Canada, as shown in Table I-3. For the United States, proved reserves, the ready shelf inventory, were only a tenth of the remaining oil-in-place; in Canada, it was 22 percent. This is reasonable. Much of the U.S. oil-in-place is in small pools, where loss of the original reservoir energy, or formation damage, makes costs of additional recovery prohibitively high, and operators have gone on to other pools. One cannot say, however, how much of the 78 percent of oil-in-place remaining in Canada, or the 90 percent in the United States will be developed into proved reserves and produced; it depends on prices and costs. In one respect the two countries are very similar: "probable" reserves exceed "proved" reserves by 23 percent in the

[48] Oil is usually under pressure from an overlying cap of compressed gas, or from gas dissolved within it, or from water beneath it, or from a combination. Water is slightly compressible, and where its volume is very large relative to total reservoir fluids, pressure on the whole hydrodynamic system may be great. Water expansion may suffice to occupy the porous space opened by the migration of oil toward a well, with hardly any pressure loss for a long time. Much less frequently, the water is fed by streams recharged through rain, and since it lies below the oil, tends always to force it up by occupying the space voided by the oil movement.

Natural gas is very different; because it is highly compressible, expansion toward the well bore does the work, and normally the bulk of the gas-in-place is recovered. Water drive does help by occupying space and compressing the gas. A permeable uniform sand yields 75-85 percent of the gas-in-place. (Arps, in *Petroleum Production Handbook*, chap. 37, p. 41.)

[49] API-AGA, *Reserves . . ., 1968*, pp. 74-75.

TABLE I-3. Oil-in-place and Reserves: United States and Canada, 1968

(*crude oil in billion barrels*)
(*gas in trillion cubic feet*)

	Crude Oil		Natural Gas		
				Canada	
	U.S.	Canada	U.S.	Raw gas	Marketable
Original resource-in-place:					
probable	...	42.6	...	114.5	...
proved	389.1	41.6	...	104.2	...
Ultimate reserves:					
probable	124.5	14.3	...	83.2	65.1
proved	117.5	12.2	...	74.1	58.6
Already produced	86.8	3.8	...	15.0	10.9
Remaining resource-in-place:					
probable	...	38.8	...	68.2	54.2
proved	302.3	37.8	...	59.1	47.7
Current reserves:					
probable	37.7	10.5	...	68.2	54.2
proved	30.7	8.4	287.3	59.1	47.7

Source: API-AGA, *Reserves . . . , 1968*, pp. 11, 86; pp. 266–279 constitute "Report of the Central Reserves Committee of the Canadian Petroleum Association." The nomenclature is not uniform, and those wishing to use the data would do well to consult the original publication.

United States, by 25 percent in Canada. In the United States, the excess corresponds to improved recovery methods already installed but not yet evaluated, or to better methods not yet installed but similar to those proved by operating experience,[50] while in Canada the explanation is "a somewhat less conservative approach on such items as productive area and recovery efficiency."[51]

We have a rough estimate for oil-in-place in the whole non-Communist world: as of the end of 1962, about one trillion barrels, of which an expected half would be extracted from known pools with known methods.[52] This would indicate about 700 billion barrels oil-in-place outside the United States. But the 500 billion doubtless included some amount for the United States; it is not an estimate of an existing quantity but rather a forecast of future reserves taken from present oil-in-place. Total published proved reserves outside North America (see below) were about 245 billion barrels, of which the Persian Gulf had 195 billion. This suggests that in reservoirs outside the United States, proved reserves were about 35 percent of the oil-in-place—a higher figure than the Canadian, as the latter in turn was higher than the American. Again, the relation is as would be expected.[53]

Reserve estimates outside North America. For many years the *Oil and Gas Journal* and *World Oil* have made estimates of "published proved reserves"[54] from information

[50] Ibid., p. 16.

[51] Ibid., p. 266. The U.S. practice of not counting reserves even where certain facilities have already been installed is an offset to any downward bias one may suspect in my definition: cumulative production from everything installed.

[52] M. J. Rathbone, "Energy for the Future," speech to Independent Gas Association of America, Sept. 17, 1963.

[53] See chap. II, pp. 70–71, for a more detailed analysis.

[54] In its excellent annual compendium, *Statistical Review of the World Oil Industry*, British Petroleum defines "published proved reserves" as in effect identical with the API concept: "the volume of oil remaining in the ground which geological and engineering information indicates with reasonable certainty to be recoverable in the future from known reservoirs under existing economic and operating conditions."

furnished by companies operating in the field and checked against the opinion of neighbors and competitors. It is hard to say how closely comparable they are with those of the United States. For example, published papers make it clear that in Saudi Arabia very similar methods of estimation are followed for oil-in-place. As for proved reserves, the published Arabian-American Oil Company (Aramco) score is kept exactly like that of API: proved reserves at end of previous year plus gross additions during year less production during year equals net additions. But it does not follow that the ground rules are the same in Saudi Arabia or elsewhere. There are conflicting indications. The *Oil and Gas Journal* estimated the end-1954 Persian Gulf reserve at 94 billion barrels, while Wallace E. Pratt estimated the area's "proved reserves," defined as in the United States, at 230 billion.[55] More recently, Gulf Oil Corporation said of its half share of Kuwait: "Reserves as estimated in the following table could in time be recovered from its present wells plus the drilling of a limited number on what are considered to be proved locations. In some fields, however, additional wells will continue to be drilled in order to maintain or increase the rate of production or improve the recovery performance."[56]

This seems identical with API proved reserves. Kuwait reserves on this basis come to nearly 43 billion barrels,[57] or 68 percent of the 63 billion barrels of end-of-1962 proved reserves published in the *Oil and Gas Journal*. But more recently, the Saudi Arabian government commissioned a study by two well-known geologist-consultant firms, which estimated 1968 proved reserves in that country as 142 billion barrels rather than the 84 billion reckoned by Aramco.[58] This is 69 percent higher than "published proved reserves." The difference of opinion among reservoir engineers is not as great as it appears. Where oil-in-place is so large relative to current production, a very small difference in the interpretation of current production data must be multiplied by a huge factor to estimate ultimate recovery from existing facilities, i.e., proved reserves.

The Stability of the Finding-Development Process. Petroleum production begins with the finding of a new pool in a new field and continues with the finding of new pools in the field and the development of the pools as they are found. Wells drilled to provide new producing capacity add to reserves, though there is no strict proportion between the two. In the limiting case of pure "infill" drilling, additional capacity does not add to reserves.

We now ask whether this process is subject to any unusual instability, by which we mean the likelihood of wasteful over- or under-investment, such that either costs are higher than they need be, given available resources and knowledge, or else that too much or too little is produced, penalizing consumers and perhaps producers. The industry is said to be peculiarly subject to unforeseeable random shocks, which, had they been foreseen, would have caused producers to invest differently and more efficiently than in fact they did. "One cannot tell when new deposits will appear, and cause past investment, intelligently made on the basis of the best possible information, to be shouldered out of profitable outlets and run at a loss. . . . This randomness means that the industry's response to higher or lower price is inherently awkward and sticky. It operates like an automobile with a worn and rusted accelerator pedal."[59]

[55] See chap. II, pp. 70–71.

[56] Gulf Oil Corp., *Prospectus*, May 6, 1963, p. 7.

[57] Gulf reserves doubled to 37,250 million, then multiplied by the ratio 1:0.875 to allow for the royalty interest. Since additional "other interests" are said to be deducted, but not explained, it may be that I underestimate.

[58] *Petroleum Intelligence Weekly*, June 30, 1969.

[59] Adelman, *Supply and Price of Natural Gas*, p. 22.

This is the common opinion, which the writer previously shared. But it is largely erroneous because it confuses exploration with development. The strong random element in initial new-field discovery is buffered by new-pool discovery and by development. We must look carefully at each phase, and translate its technology into economics.

The random element in new-field discovery is indeed very great. There is only a small chance of finding a reservoir in any given unexplored area, and few of the pools found are worth finding. The results of any given discovery effort will vary radically: in one place a large money outlay is wasted; elsewhere a small expenditure brings in a big field. On some very general logical grounds the distribution of the sizes of fields (measured by oil-in-place) might be expected to follow the logarithmic-normal distribution.[60] Assuming this to be true of deposits in a large area, and owing their existence to a common geological force,[61] it is not so clear that the reasoning can be extended to a larger area containing a number of independent basins. Moreover, the reservoir-creating process includes oil and gas migration and accumulation, from the original source rocks to the final trap. "... Petroleum geologists [have] relatively little interest in looking for source-rocks [because] many large deposits of oil and/or gas [have] resulted from a lateral migration of as much as several dozen or even several hundred kilometers."[62] The Persian Gulf deposits usually are not assigned any source rocks at all, there being an embarrassment of possibilities.[63] At any rate, there seems no general or *a priori* reason why the migration and entrapment of oil and gas should follow the same statistical law as the formation.

But the log-normal curve or similar types are at least a good description of a radically unequal distribution. In the United States, to take the biggest sample, 200-odd fields out of 10,000 account for more than half of total reserves.[64] The effects of finding one more or less giant field are therefore great. The effect of chance is vividly shown in a single instance: "if Upper Cretaceous erosion [175 to 35 million years ago] had cut 100 feet deeper, the whole content of the Burgan field [about a fifth of the Persian Gulf] would have had free channels of escape to the surface." In fact, bigger ones may have got away.[65]

Complicating the distribution of deposits is the distribution of the ratios of recoverable reserves to oil-in-place. Here, too, small causes generate large results. The Agha Jari field of Iran, the Hassi Messaoud field of Algeria, and the Spraberry field of West Texas, all have a rather nonporous sand, and oil content per cubic foot is not high. But all three structures are so huge that they hold billions of barrels of oil-in-place. Hassi Messaoud is broken up into many pockets of permeable sands surrounded by almost impermeable rock. Unless one wants to wait until doomsday a fairly large number of wells must be

[60] Maurice Allais, *Evaluation des Perspectives Economiques de la Recherche Minière sur de Grandes Espaces* (Alger: Bureau de Recherches Minières de l'Algérie, 1957); J. J. Arps, "The Profitability of Exploratory Ventures," in *Economics of Petroleum ... Evaluation*, pp. 154–59; Gordon Kaufman, *Statistical Decision and Related Techniques in Oil and Gas Exploration* (Prentice-Hall, 1963); G. Mathuron, "Application des Methodes Statistiques à l'Evolution des Gisements," *Annales des Mines*, vol. 144 (1955), pp. xii, 52. Paul G. Bradley and R. S. Uhler, "A Stochastic Model for Determining the Economic Prospects of Petroleum Over Large Regions," *Journal, American Statistical Association*, vol. 65 (June 1970), pp. 623–30. Allais argues that the chemical reactions producing mineral deposits are subject to the law of proportionate effect and should generate log-normal distributions.

[61] Arps and Roberts, in *AAPG Bulletin*, vol. 42 (1958), pp. 2549–56.

[62] Jacques Flandrin and Jean Chappelle, *Le Pétrole* (Paris: Edits. Technip, 1961), pp. 17–18 (writer's translation). See Muskat, *Physical Principles ...*, and Craze, "Development Plan for Oil Reservoirs," pp. 33-7 to 33-14.

[63] Lees, "The Oil Fields of the Middle East," p. 61.

[64] *Oil and Gas Journal*, Forecast-Review (annual).

[65] Lees, "The Oil Fields of the Middle East," p. 71.

drilled. The Spraberry pockets are smaller and the barriers are greater; it seems today that only a tiny proportion of the oil-in-place (estimated by some at over 7 billion barrels) will ever be produced. Agha Jari is even worse than Spraberry but with a difference—the field is extensively fractured horizontally and vertically. Hence migration and pressure are so high that a successful well, which takes difficult and costly drilling, comes in as though somebody had accidentally struck a high-pressure pipeline. Thus Hassi Messaoud is a great oil field, Agha Jari is amazing, and Spraberry is a fiasco.[66]

The strong random element should therefore make us cautious about forecasting additions to oil-in-place or reserves in the years to come. But it is a mistake to suppose that an oil operator can suddenly find a huge resource that overnight adds greatly to the industry's inventories and induces a bargain-basement sale. All the operator knows in the first instance is that he has found what looks like a big new pool and that, given the surrounding structures, he has a good chance of finding some more big ones and more smaller ones—which takes time and more investment. The bigger the new pool or field, the longer it takes to find the horizontal and vertical limits.

It is an important economic fact that there is no rigorous definition of a field: it is a collection of pools "on or related to the same geologic structure . . . [But] all the pools on one major structure do not necessarily constitute *one* field. If the structure is long, several fields may be located on it."[67] It usually takes years to find them. In the Burgan field, the Magwa and Ahmadi extensions were found by exploration begun in 1948, 10 years after the original discovery well. It took over a decade before Ghawar in Saudi Arabia was recognized as one field. Khafji, found in 1960 offshore of the Saudi Arabia-Kuwait Neutral Zone, turned out to be an extension of Aramco's Safaniya (1951). The "Bolivar coastal fields" of the eastern shore of Lake Maracaibo are considered as essentially a single field, yet, as shown below, five distinct groups of pools were discovered at widely varying times, each showing very different characteristics.[68] After nearly 40 years of development, one was not yet delineated.[69] This is far from unique.

Pool	*Year of Discovery*	*Gravity ° API*	*Depth (000 ft.)*	*Daily average Production per Well, 1961*
Bachaquero . . .	1930	14–35	3.4–11.5	423
Cabimas	1917	20–26	2.2	n.a.
Langunillas . . .	1926	16–32	3.0–8.2	461
Pueblo Viejo . .	1940	14–19	2.0	n.a.
Tia Juana	1928	14–27	3.0	153

n.a. = not available.

In offshore Louisiana, "arbitrary definitions of field boundaries have already obscured the true size of some reservoirs. South Pass 24, for example . . . would extend con-

[66]On Hassi Messaoud, see J. J. Rousseau, in *World Oil*, vol. 152 (Feb. 1961), p. 23; C. de Lapparent, in *Oil and Gas International*, Jan. 1961, p. 43. On Agha Jari, *Oil and Gas International*, Apr. 1962, pp. 38–45. Spraberry is a sad, much told tale.

[67]Lahee, *Statistics of Exploratory Drilling* . . . , p. 132 (emphasis is in original). Even a pool is not a perfectly clear idea. Consider two or more accumulations linked by a common hydrodynamic system, floating as it were on the same underground water. If movement between the two is after a while determined to be improbable or slow, are they counted as one or two?

[68]*Oil and Gas Journal*, Dec. 26, 1955, Dec. 25, 1961. In the latter issue, Pueblo Viejo has been merged with Bachaquero.

[69]Ibid., Aug. 1, 1966, p. 81.

siderably . . . at several points except that other field names already claim the territory."[70] In Alaska, the first three discoveries may be only one.[71]

The result of all these unknowns is that the contents of a newly found pool and the chances of other such pools on the same structure can only be guessed at. As the pool is developed there is a probing on the same structure for new reservoirs. The gradual sequence of finding and then developing new pools takes years, and buffers the impact of new-field discoveries.

Let us subject this rule to the severest possible test—the impact of the Persian Gulf.[72]

The Persian Gulf as a Random Shock. If one arranges the largest fields in the United States, South America, and Africa into a distribution, the Persian Gulf fields are above and clear off the chart.[73] It seems impossible for so many huge oil-bearing structures to be grouped in one relatively small area, and yet there they are.[74]

But the impact of this huge unpredictable disturbance was extremely slow. The first big Iranian field was discovered in 1908. Thirty-six years later the DeGolyer report estimated Persian Gulf reserves, which are compared below with cumulative production from fields known before 1944.[75]

	1944		*1968*
Country	*Proved Reserves*	*Probable Reserves (in known but undeveloped fields)*	*Cumulative Production from Fields Known before 1944*
	(· · · · · ·	*billions of barrels*	· · · · · ·)
Iran	5–6	1	7.9
Iraq	4	1	4.2
Kuwait	4	. . .	9.9
Saudi Arabia . .	2	2.3	3.4
Qatar	0.5	0.5	1.0
	16	5	26.4

To find out how big those deposits really were took years of effort and large sums of money, as capacity built up and more was learned about the number, size, and behavior of the reservoirs. In the meantime, capacity and reserves were running down in fields elsewhere.

The production decline curve and the growth rate together moderate the impact of the evolving new field. In the United States, annual capacity loss was about 8 percent in 1960, assuming no restriction on output.[76] (It is surely greater today.) Demand has increased about 4 percent per year. In the Persian Gulf, growth is a trifle over 10 percent,

[70] Ibid., Nov. 18, 1968, p. 104.

[71] Ibid., Aug. 11, 1969.

[72] I avoid the term "Middle East" to exclude Syria, Turkey, and Israel. Egypt is included in Africa.

[73] Cf. George M. Knebel and Guillermo Rodriguez-Eraso, "The Habitat of Some Oil," *AAPG Bulletin*, vol. 40 (Apr. 1956).

[74] Vincent C. Illing, in "Factors Which Control the Assessment of World Resources of Oil and Natural Gas" (Sixth World Power Conference, Melbourne, Paper 185 I. 1/2, p. 21), surmises that the three crucial optima—source rocks, reservoir rocks, and cap rocks—are independent or even inversely correlated. The Middle East is the statistically unlikely spot where they coincide.

[75] *Oil and Gas Journal*, Mar. 23, 1944, p. 45, and Dec. 30, 1968, supplement.

[76] NPC, *Report of the National Petroleum Council Committee on Proved Petroleum and Natural Gas Reserves and Availability*, 1961, p. 27.

decline is (at least) 1 percent,[77] thus in both areas current producing capacity must always be increased by 12 percent to stay in place and prevent shortage.

Or we may consider the recent Alaskan discovery, announced in June 1968. If all goes as planned, by 1973 at the earliest there will be a production of 2 million barrels daily. Yet total American production in 1968 was 9 million barrels daily, and at 8 percent a five-year wastage would be 34 percent, while normal growth at 4 percent would add another 22 percent: a total gap of 5 million. Therefore, if there were a free market in oil, even this greatest of North American discoveries could be accommodated without any strain or need to cut back production.

Expectations and discounting. We have argued that even a random and irregular sequence of discoveries of new oil-in-place (as known ex post) need not have strong effects on proved reserves and current capacity (nor therefore on price). We now take a given amount of oil-in-place and of proved reserves, and ask what destabilizing effects may arise from changes in expectations. Assume first the expectation of lower prices. The operator of a given reservoir, who up to now considered his oil-in-place as about "right" can no longer do so. Previously the discounted present value (expected price less cost) of a barrel to be sold off in the future was equal to what could be got for it today; now it is less. It does not follow that he will produce more, increasing supply and depressing price.

At the expected lower price, some development plans will be shelved, as no longer profitable. In some lower-cost reservoirs development investment will be speeded up, but higher incremental costs will dampen the increase in investment. On balance, the expectation of lower future prices may raise or lower current output. Clearly there are factors working in both directions, hence promoting stability. (An actual price decline will provoke an immediate shutdown of those wells whose direct operating costs no longer are covered.)

Stability is reinforced in today's market by a condition that is not necessary: restricted competition. There is a general impression (which will be confirmed in Chapter II) that oil-in-place, and even published proved reserves, are in some parts of the world so large in relation to current output that they will not be depleted for many years. Hence the discounted present value of a very large fraction of the oil-in-place must be effectively zero. A change in price expectations will therefore make no difference. But this raises another question: Why are such large stocks carried? It suggests that the firms owning them are not governed by independent self-interest but by attention to group welfare; each refrains from selling off his excess inventory in the well-founded hope that all other large owners will do the same. (Or the host government will not permit it.) Thereby the number of new entrants is kept smaller, and the market is easier to control. This conjecture will be tested against the data of later chapters.

Let us now consider the effect of an increase in the rate of discount of future outlays and receipts. Like a drop in expected future prices, this would, all else being equal, tilt the balance, inducing more current and less future output. But the net effect would probably be quite the other way, because higher capital cost, which is the discount rate in another guise, is a disincentive to any new investment. To consider an extreme case: suppose that a group of oil companies fear they will be expelled from their concessions a year or two hence. The very high discount they put on future receipts would make them stop drilling and production would decline, tending to raise prices.

[77] As will be seen in chap. II, this is assumed as a minimum.

We have arrived at no elegant general conclusions on instability. But there seems to be no strong built-in tendency toward it, such that the petroleum industry is peculiar on that account. General answers may deceive; the reader may wish to reconsider the question after we have been able to measure costs in the next chapter.

A Unified Theory of Oil Operating-Developing-Finding Costs. At this point a simple theory is suggested, which reduces all three cost types to one: first for a single reservoir, then aggregated over the whole industry.

Consider a development plan. The operator decides how long a pool can produce until declining production drives the daily operating cost per barrel to equal the expected revenue. Once the time period is estimated, each unit of the expected operating cost is discounted back to the present. Far-off errors, even if substantial, are negligible because their present value is small. The sum of the present value of the future operating outlays is added to the development investment: the cost of drilling, equipping, and connecting the wells and doing all else needful to start producing. Supplementary development expenditures, workovers, waterfloods, etc., are added in, again with proper discounting. The result is a single figure of investment per daily barrel. Development has swallowed up operating cost, because now is the time to decide whether the total commitment is worth while.[78]

The more wells we drill, the more well interference and the lower the total output per well. The incremental investment cost in the new well must include not only the necessary investment in the well but also a feedback effect: the increased investment requirements per barrel of production forced on the existing wells by drilling the additional well. As development proceeds, an oil company faces the choice between (1) carving more and more reserves out of the reservoir's oil-in-place by drilling more wells at rising cost, (2) exploiting known higher-cost reservoirs, and (3) trying to find new reservoirs or whole new fields. At any given point of development the incremental development cost equals the MEFC. Spending less for finding than MEFC would leave us better off; anything more would not be worth spending.

Past expenditures on finding the deposit we now exploit are not relevant to our investment decision. Past costs may serve as a basis from which to estimate future outlays, perhaps by way of a probability analysis, but in no other way do they enter. Incremental development cost is the total production cost—including any necessary finding cost. The increase in incremental development cost is the maximum necessary allowance.

So much for the individual reservoir. The incremental cost for the system as a whole is that of the highest-cost reservoir at any particular level of output. The expected increase over time of incremental investment per new daily barrel is the standard against which a single operator or a large group or government needs to decide what is worth spending on research to find new fields, or competing sources of energy, or ways of economizing on energy use.

We may not be able to calculate the benefit from investment in any of these types of research, but we can reckon the penalty for doing nothing: the increase in incremental development costs. This is the only measure of scarcity. But future penalties or benefits must be reduced to present values.

[78]Chap. II supplies an explanation of how the investment figure is converted to a per-barrel cost; see especially table II-4.

"Exhaustible resources" in general. Petroleum resources and their measurement may be put into context: the economics of "exhaustible resources."[79] The classic problem is that of the optimum rate of production, or the best balance between present and future use of a stock on hand, either for a private profit-seeker, or for a public cost-benefit analysis. With an appropriate interest rate, we can convert every future price into a present-value equivalent, and can choose that profile of output which will realize the maximum present value of the stock.

But for any given mineral, including petroleum, the nature of the stock complicates the theory. At any given moment, we have a collection of imperfectly known deposits of oil-in-place. Some of these deposits have been partly converted into proved reserves, by investment in development. For those portions of the stock, major commitments have been made and the range of alternatives is simple: produce now or hold for future production. As seen earlier, current operating cost for a given pool must rise over time. The excess of price over operating cost must decline, or the price must rise; the question is how much. If the excess of price over current operating cost is expected to increase by no less than the current rate of interest, it is worth holding. Or, what comes to the same thing, the present value of the stock is the "user cost," which is sacrificed by producing it. If "user cost" is less than price, the stock should be sold; for the loss suffered in not producing at capacity, letting stand idle the machinery and in-the-ground material, exceeds the eventual gain.

But this is not an interesting or important problem, nor is user cost a helpful concept because it assumes the unknown we should be solving: the future price, which itself depends on future incremental cost. In any case, the size of the problem itself shrinks with declining output. Let us turn to a more important aspect: investment decisions.

At the limit, there are the completely developed pools where additional drilling will speed up production but not increase reserves. More importantly, additional proved reserves can be created, even without discoveries of new pools, again at rising cost. The problem is the best rate at which the stock oil-in-place should be developed into the stock proved reserves. The best size of the two inventories is when the present value of the next barrel to be impounded into reserves is barely less than that of the last barrel already impounded, which in turn is barely equal to its incremental cost. For such decisions, expected prices, costs, and discount rates are harder to foresee. One can no longer talk of the current "rate of interest," because the risk and uncertainty premium may be as large as or larger than the pure interest rate. Prices need not rise; we must reckon with forces pushing both ways.

The knowledge of rising cost in the known reservoirs induces the constant search for better extraction methods and new deposits to increase supply; more economic use, to lessen demand. Over time, unit cost may go up, down, or remain stationary. The outcome is uncertain not only in any given place but even in the large. That petroleum is "an exhaustible resource" is no help in answering the real questions.

[79] See my paper "Economics of Exploration for Petroleum and Other Minerals," *Geoexploration*, vol. 8 (1970), pp. 131–50. The works upon which this discussion relies are: Edward S. Mason, "The Political Economy of Resource Use," in Henry Jarrett (ed.), *Perspectives on Conservation* (Johns Hopkins Press for Resources for the Future, 1963); Richard L. Gordon, "A Reinterpretation of the Pure Theory of Exhaustion," *Journal of Political Economy* vol. 75 (1967), 274–86; Orris C. Herfindahl, "Depletion and Economic Theory," and Anthony C. Scott, "The Theory of the Mine Under Conditions of Certainty," in Mason Gaffney (ed.), *Extractive Resources and Taxation* (University of Wisconsin Press, 1967). Obviously, there is no necessary agreement among all these writers, nor of myself with them. But they have defined the issues as I see them, and given new life to some elegant but previously unusable theorems.

If the probability were known of finding a given size and quality of field or a method offering a given reduction in cost, the only risk would be in the individual project. Pooling of interests could reduce the impact on the individual firm to a very low figure, and for society as a whole it could be equated to zero. But the probability is in fact not known, and the resulting risk of wasted investment in exploration is much greater than for a project of developing reserves from known deposits with given technology. Our ignorance is both of basic geology and of future patterns of technology, consumption, and competing supply sources.

The uncertainty of oil exploration infects, as it were, the economics of investment in competing supply sources. For example, investment in nuclear power is reasonable given one set of expectations of oil discoveries and costs, and wasteful given another. But this works both ways: investment in oil finding is reasonable on one expectation of nuclear research results, wasteful on another.

We must push the uncertainty to its logical conclusion. No individual pool is ever completely drained, because it is more economic to shut it down and turn to another. But this may also be true for all pools taken together. Although we treat petroleum as a problem in exhaustible reservoirs, the resource oil-in-place, as known today, may be truly inexhaustible because the need may disappear before the resource. Nuclear power from fast breeder reactors[80] plus new supplies of uranium ore, may reduce consumption enough to stretch current oil-in-place out to the time when fusion and solar energy would open a limitless source. Even setting these attractive possibilities aside, the odds are good that much or most of the oil-in-place now known will never be used because cheaper oil will be found. For the United States, this is certain.

The expedient of MEFC replaces the massive uncertainty with a bias, and we reckon with the upper limit of cost under present known resources over some limited time period. But past the near term, MEFC has no relevance. Nor has any other known method of forecasting. For this reason, future benefits from investment in new energy sources must be discounted at high and increasing rates: development more than extraction, discovery much more than development. Past twenty or at most thirty years, supply-demand calculations become arithmetic, not economics. Arrays of figures stretching out beyond the year 2000 are at best an exercise in method; at worst, a vain presumption. As we get closer to any given date, we know more about likely costs and benefits, and we need to know more, because the present value of knowledge, or the present value of the penalties on ignorance, is greater.

So far, we have assumed competitive markets. Some degree of monopoly means that the price is higher (or lower, if the buyers have monopoly power). To reckon private or public cost-benefit, there must be an explicit calculation of the differential, and a forecast of the time path over which they will converge—if they ever do. Unfortunately, the monopoly problem is usually ignored in long-run forecasting.[81] As will be seen in Chapter VII, the neglect is costing Europe several billions of dollars annually. It is instructive to see economists, who barely admit that any market price can be a valid indicator of true social scarcity, accept without question a fictitious "price" set by a nationalized industry or private monopoly. Such a "price" is an artifact, the result of putting blinders on the decision makers, not letting them recognize any alternative supply sources.

[80] Paul W. MacAvoy, *Economic Strategy for Developing Nuclear Breeder Reactors* (M.I.T. Press, 1969).

[81] See Office of Science and Technology, *Energy R&D and National Progress* (Washington, 1965), chap. II, and OST, *Civilian Nuclear Power—A Report to the President*, 1962.

In competitive markets, current prices reflect current and future (expected) demands and costs, including interest rates. Prices embody all knowledge relevant to finding new deposits and developing new reserves. Prices are therefore a system of distant-early-warning signals. Competition is useful for the information it provides, regardless of any value judgments on what is best for whom.

Rents and Landlord-Tenant Relations. A price equal to incremental cost does not result in low or zero profits, as is too often assumed; it may mean very high profits for the more efficient and better located producing units. These amounts above bare production cost (including a minimum necessary return on investment) are known in economic jargon as *rents*, or surpluses, and they exist in all industries even under the most stringent competition. In general, a rent is an outlay that can be decreased without any supply response. It does not matter how much of the above-normal profits of the unusually low-cost oil go to the landowner, the producer, the man who put up the money at the right moment, or the government. In the very short run, everything paid above operating costs is a rent, but more must be paid to ensure continued production, and still more to ensure rising production: hence outlays on needed development or exploration are not rents but costs.

But the great uncertainties of finding cost lead to some predictable relations among these parties. In the beginning is a piece of completely unexplored or "rank wildcat" acreage. An oil company is not willing to pay much for the right to explore and produce. Since the few good discoveries must pay for the many failures, they are willing to offer at most an average value. If the landowner wants more, the company goes elsewhere. If the landowner accepts less and then learns others would have paid more, he is sorry he let the tract go so cheaply; contrariwise, if the nearest price was much lower than the price accepted, the landowner is glad and the producer sad.

Once the tract is drilled, uncertainty is greatly reduced, giving way to real knowledge. If there is no oil worth developing, the producer's money has been wasted. If there will be large development-producing profits, the landowner wishes he had held out for much more. With foreknowledge, the producer would willingly have paid much more.

This is the great divide of the petroleum industry: a rich discovery means a dissatisfied landlord. He knows that the tenant's profit is far greater than is necessary to keep him producing, and he wants some of the rent. If he gets some, he wants more. *L'appétit vient en mangeant.*

If oil explorers are deprived of the hope of making big profits occasionally, then the expected profit of good and bad discoveries taken together will fall and less capital and enterprise will be available, ultimately reducing supply. But what is true for the industry as a whole may safely be neglected in any one instance. A landlord could confiscate all the rent being gained on his acreage without any perceptible effect on the industry.

Whenever as a result of a bargain the alternatives open to the parties are transformed, the new situation puts a strain on the bargain itself. The classic example is the Pied Piper of Hamlin. In prospect (as with no oil found), the burghers were willing to pay a thousand guilders to get rid of the rats. But after the rats were gone (oil found) they wanted to renegotiate at 95 percent discount: "A thousand guilders? Come, take fifty!" With no higher authority around to enforce his contract, fifty guilders were all the Pied Piper could get. He retaliated so brutally that the burghers in turn would have paid much more than a thousand guilders to get their children back. Both parties were too hasty, too ignorant of the harm the other could inflict, and too passionately sure they were right to come to any agreement, and in the end both were much the worse off. This is the kind of showdown one does not like to see.

A Note to Chapter I

THE RULE OF CAPTURE AND THE BOGEY OF EAST TEXAS

Nothing, said Joseph Schumpeter, is so durable as a folk memory. And the influence on oil industry thinking of the slogan "Remember East Texas in '30; we'll have 10 cents a barrel if we don't watch out!" is extraordinary. Europeans who reproach Americans for thinking too much in terms of U.S. experience will in almost the same breath expound on East Texas. The UN memorandum treats it as a proof of its decreasing-cost theory;[82] others point out that oil has an inherent surplus, as witness East Texas in 1930, and hence "it has been found necessary" to take such steps as seemed necessary. Surely we should hesitate to generalize from a pattern so wildly unrepresentative: the Great Depression (output down 22 percent in 1929–32), a new field with reserves (of 2 billion barrels as they were then calculated) that were twice the annual U.S. output; and, most important, the law of capture, which holds nowhere but in the United States and which, having once made clear the inventory nature of oil production, we are now able to analyze.[83]

Imagine a group of retailers, each of whom carries a considerable stock. At any given moment, the inventory is already there and paid for, and operating costs are mostly fixed anyway, so that additional sales would be almost clear additional profit in the ultra-short run. But if one intends to stay in business he must reorder; the cost of the inventory thus prodigally sold off is the cost of restocking. Hence there is no tendency to throw goods into the market for next to nothing, and the trade can function and earn profits quite normally.

Suppose now that every retailer suddenly gains possession of the keys to the stockrooms of a dozen or so of his nearest competitors, and is told that he may legally and in good conscience help himself to as much of their wares as he and his employees can move away. Furthermore, he knows that every one of the other dozen retailers has a similar key to his own stockroom. Obviously, his cost of replenishing inventory is no longer the manufacturers' price; it is the much lower one of renting a truck and paying some overtime. To remain in business now, he must not be left holding the goods, he must sell off as fast as possible, at any price above zero, both his own and his competitors' stocks. Of course, the goods will be flung into the market or into the streets; nobody will want to reorder, so that manufacturing will slow down or cease. Finally, after the debauch is over, the goods are gone, and prices have shot far up, manufacturing must be started up again and some plans must be made to prevent the madness from recurring. But, had it been illegal or physically impossible for competitors to help themselves to each other's stocks—hardly an unusual state of affairs in capitalist society—the market could have been stable all along.

To complete the analogy, one must suppose that the retailers' trade paper is full of complaints about how strange and murderously competitive is the retailing business *in general*; and the editor, in collaboration with a professor at the local university, publishes a study showing with technical economic citations and terminology that retailing *in general* (and not merely because of the strange ground rules in this town) is an industry of decreasing costs which must break down into wild extremes of chaos, waste, and scarcity unless the merchants are allowed to get together to fix prices and regulate sales—or have the government do it for them.

[82] UN-ECE, *The Price of Oil in Western Europe* (E/ECE/205; Geneva, 1955).

[83] For a classic description and legal analysis, see Northcutt Ely, "The Conservation of Oil," *Harvard Law Review*, vol. 51 (1938), pp. 1209–44.

Given the rule of capture, therefore, and nothing more sweeping or pretentious in the way of economic assumptions, the discovery of every new field will mean a sudden and wasteful rush to overproduce. But this results from a legal system peculiar to the United States. Moreover, as was recognized by 1916, the waste is not only a burden on society but even on the producers themselves, and it is therefore in their interest to "unitize" operations in order to minimize the number of wells drilled and prevent any neighborly appropriation of oil.

The technical problems of reconciling unitization with private property in subsoil rights are not negligible. (Even in the Soviet Union there has been slant drilling from one side of a republic's boundary into a pool on the other side.) Unitization is best done soon after the pool is discovered, when there is still no adequate basis for determining the field's vertical and horizontal limits, variations in thickness and permeability, and water drives and the effect of differing rates of encroachment. This lack of knowledge means that one cannot determine precisely the relative values of surface rights. Estimates made on the basis of what is known are bound to be unfair to some owners without some kind of retrospective adjustment. But nobody has thought these problems insuperable,[84] and our federal system would have provided room for three or four separate trials with an eventual pooling of experience in a model statute. Had the United States not settled in the thirties for prorationing–which seemed then like a satisfactory halfway house between unrestrained drilling and compulsory unitization–the private interests of oil operators and of the states would have driven them to unitize, and the industry and the country would be much better off today. We would also have been spared much misunderstanding. For outside the United States the special situation of the rule of capture does not exist. Any theory based on that rule, however rich in corroborative detail, must be put in the same category with the Loch Ness Monster and the Abominable Snowman. People may have seen "it" all right, but they did not see what they thought they were seeing.

[84] Herman H. Kaveler, "Unitization," in American Petroleum Institute, *History of Petroleum Engineering* (New York, 1961), especially pp. 1170, 1173.

CHAPTER II

Crude Oil Production Costs

It is useful here to sum up the concepts of cost and supply dealt with in the previous chapter. The greater the cumulative output from a given pool, and the higher the current output, the higher the unit cost as pressure declines and a limited amount of liquid is drawn up with a larger investment. At any given moment, the industry is the aggregate of all pools everywhere. The array, from lowest to highest cost, of the many possible sources of petroleum defines the industry supply curve, at increasing costs. Over time, with greater knowledge of the earth's crust and of better finding and producing methods, the upward-sloping function shifts to the right. An increasing-cost industry does *not* mean that because of a finite earth there must eventually be scarcity, or that the big fields have all been found, or that real costs must increase over time—propositions of doubtful truth or relevance. But at any moment the greater the amount to be produced, the higher is the minimum attainable cost per unit. From period to period that cost may rise or decline.

The influential theories of high fixed-to-variable cost ratios, decreasing costs, inherent surplus, etc., are one and all wrong in themselves, or irrelevant, or simply never explained. If industry incremental cost were compelled by nature to be always or usually below average cost, then price would be under corresponding competitive pressure to fall below average cost also, and inflict losses on the industry; this "natural monopoly" does not hold for petroleum production.

In Chapter I, therefore, the relations between price and cost could be stated in general and qualitative terms, assuming production is at various points mostly distant from the main consuming areas. In this chapter, where we go from qualitative to quantitative analysis, costs are measured as later will be prices. And since the object remains price-cost comparison, the only kind of cost sought is the one strictly comparable with price.

Much of the data is grievously imperfect, and this must be shown in detail. However, no further apologies should be needed: better to be approximately right than precisely wrong. Few if any published cost estimates are explained; fewer are reproducible; nearly all are irrelevant. In order to render an accounting to government or partners, oil operators are obliged to define and reckon something called "costs" and "profits," and this may be published. Thus, in Libya in 1964 "the high development expenses carried on last year's books, matched against the still low production reported there, raised per barrel costs of the main independent to around 70 cents or more."[1] This is doubtless sound and necessary accounting. It is also so far-fetched as to be harmless. Lower, more plausible estimates are not. For example, when told that costs under a halfway price arrangement

[1] *Petroleum Intelligence Weekly*, Sept. 27, 1965.

are 38 cents, or about 45 cents for the whole country,[2] we can make nothing of such figures, however much we may respect the expertise of those who released them, because they are not explained. Quite a few such "cost" figures have been cited, which a third party cannot define or derive. As lawyers would say, they are void for vagueness. Let us therefore state what we seek.

Technology sums up what we know about nature at any given time. Given this knowledge, there is a set of best possible results that we can achieve. Or, what comes to the same thing, there are inescapable barriers to production, which we translate into money terms. Cost registers the *unavoidable* using-up of real resources in the production of new crude oil. Arrangements made among companies or imposed by government may involve higher expenses which are a burden to them and society. These non-cost expenses are a different problem, not touched in this chapter except as they affect risk. Government removes by taxes or royalties a part of the profit or net return above cost. To the business concern, one deduction stands on about the same footing as any other: wages, or taxes, or anything else. "Costs" are what they pay out, and profits are what they have left. But for anyone trying to understand the industry, and for the business concern trying to plan ahead, this is a dangerous mistake. If a government imposes taxes so high, or restrictive arrangements so tight, that the added expense, plus the real costs we measure, amount to more than the price obtainable in the market, the result is untenable. Something must give. Either some or all of the operations must cease, or the government or group must modify its arrangements. What men can privately or officially do is limited by costs imposed by nature, as mediated by technology, which is what we study. Our target is a body of knowledge on which anybody can draw, and everybody must draw, regardless of what he is trying to accomplish or his position at any bargaining table.

We shall observe the usual three-way division of production costs.

1. Operating costs, assuming all equipment in place.
2. Development costs—the installation of new productive capacity in pools already known, whether newly found or already partly developed. With the exception of exploration, *all production investment is included*, below and above ground, including the costs of such facilities as marine terminals and jetties.
3. The costs of discovery of new fields and of new pools in old fields.

After each class of cost has been covered, it will be necessary to look back and reconsider the traditional three-way scheme.

Operating Costs

United States. In 1962 total producing expenditures, including overhead but excluding taxes, were $1.4 billion (lines 11 + 15 in Appendix Table II-A-1). Since production for the year was 2.6 billion barrels, average producing expense was 51 cents a barrel. However, this is not a cost because it contains more than unavoidable expense. It is based on the output of all wells in the United States. The great bulk of oil wells, however, are superfluous—a political curiosity, not an economic asset. They result from a peculiar kind of regulation in the United States, and the prices it supports.[3] Furthermore,

[2] Ibid., Dec. 7, 1964. Libyan oil sold by Grace to Esso at $1.30 is said to be a halfway price between cost and posted price. Since the latter is $2.23, "cost" must be 38 cents. A less complicated explanation is that $1.30 was simply the price.

[3] M. A. Adelman, "Efficiency of Resource Use in Crude Petroleum," *Southern Economic Journal*, Vol. 31 (1964), pp. 101, 120.

they are so far outside any statistical regularities that if only for the sake of getting an estimate of the non-marginal wells, here and abroad, the marginal ones should be eliminated.

Table II-A-1 explains the calculation of expenditures on non-marginal wells. These should not be related to actual production, which is mostly restricted far below capacity, but rather to what wells could produce if oil operators were free to employ them properly.

"Productive capacity . . . is defined as the average rate of production from existing wells that could be maintained for a period of from six to twelve months without further development and with no significant loss in ultimate recovery."[4] What we are trying to approximate is a condition which in 1944 had been attained and should have been kept: "Severe production curtailment . . . is now behind us. . . . [P]roration for reasons other than prevention of underground waste is gradually disappearing. More and more wells are, or will be, producing at capacity or at their optimum rates, as determined by sound engineering practice."[5]

In the early 1960s, the average U.S. non-stripper well had a capacity of 47 barrels daily, and my best guess is that it cost $9.28 daily to operate, or 19.8 cents per barrel. (See Appendix Table II-A-3.)

Outside the United States. To estimate operating costs outside the United States, we conjecture that costs per well vary as the square root of output, since the basic operation is that of putting the liquid through a cylinder. Cost of the cylinder is proportional to the radius of the cylinder, but cubic contents to its square. The relationship is tested against U.S. data (see Appendix Table II-A-5) and seems to hold fairly well inside a range of about 25 : 1.

Extending such a rule to Persian Gulf production means applying it to wells from 100 to 500 times as productive. But so long as nothing better is available, we can make do with the rule because the absolute cost per barrel is so small at high well capacity that even large percentage errors will not matter much.

No adjustment has been made for higher costs of doing business outside the United States or other home country. Undoubtedly these are substantial, although they should by no means be confused with the higher costs of exploration, where it can be a matter of several thousand percent. The big supply and logistics problems arise in probing into the unknown; one must expect to face one costly interruption after another, while one hunts and waits for the one missing part or person or expedient. But in a developed project operated as a going concern there is no question of frequent unexpected stoppage. The cost of operating far from home is largely transportation cost, the need to maintain a somewhat larger inventory, and the employment of labor that is paid much less than in the United States but that may be very much less productive.

But there is a more serious gap. Our U.S. estimates are costs of operating wells. What we seek abroad, however, is the cost of moving oil not only to the wellhead above ground, but thence into a stock tank, gas-oil separator, gathering system, pipeline, and

[4] Independent Petroleum Association of America, *Report of IPAA Productive Capacity Committee*, Apr. 1965. See also National Petroleum Council, *Proved Reserves and Productive Capacity of Crude Oil, Natural Gas, and Natural Gas Liquids in the United States*, 1965, p. 3.

[5] J. J. Arps, "Analysis of Decline Curves," American Institute of Mining, Metallurgical, and Petroleum Engineers (AIME), *Transactions*, 1945, pp. 5, 160, 228. Reprinted in Arps (ed.), Petroleum Transaction Reprint Series No. 3, *Oil and Gas Property Evaluation and Reserve Estimates*, Society of Petroleum Engineers (SPE), n.d. (apparently 1960).

finally a marine loading dock. Table II-1 presents approximate operating cost data for three countries, from which is derived a computed total in cents per barrel; an actual total for each is also given.

TABLE II-1. Operating Costs: Kuwait, Iran, Libya

Area	Average daily production per well (*TBD*)	Well operating costs, by formula	Gathering system	Pipeline, 0.7 cents per 100 miles	Marine terminal	Computed total	Actual total
		(.	. . . .	. . . *cents per barrel* . . .	. . . .	. . . .	. . . .)
Kuwait.	4.5	2.0	0.7	0.2	1.5	4.4	4.8
Iran Consortium. . .	16.0	0.8	0.7	0.7	1.5	3.7	5.3
Libya	3.0	2.2	0.7	2.5	1.5	6.9	8.4

Source: Appendix II-B.
TBD = Thousand barrels daily.

Based on the figures shown in Table II-1 and the detailed data in Appendix II-B, we shall use the following approximations. Kuwait, Saudi Arabia, and Iraq are each reckoned at 4 cents per barrel. Because these three countries are put on a Persian Gulf basis for comparison,[6] the costs of the Kirkuk-Mediterranean pipeline can be disregarded for the moment. The Iran Consortium is reckoned at 5 cents and Libya at 8.5 cents. (The latter cost includes the operation of a 350-mile pipeline, which for the time being is much longer than in any particular field.)

One final aspect of operating costs must be considered in detail now, and it serves as the bridge to development cost. The estimates of operating costs are as of one particular year. They are relevant to price and production decisions for that year and a short time later. But the well lasts longer, and the operator must take into account the effect of production today on costs tomorrow in order to figure the real cost of producing today.

In practice, operating costs are part of a development decision; an oil operator forecasts his costs over the life of the project. But the more productive and lower-cost period comes earlier and should therefore be given a greater weight, depending on the value of money to the operator. In the electric power field this is called "levelized" cost, and the calculation is sometimes called "present-worthing."[7] It is analogous to a level-term insurance premium. Until we solve the problem of reducing expected future outlays to present values, operating cost estimates are biased downward. We now turn to this problem.

Development Cost in Theory

Many estimates have been made of petroleum development costs in the United States. Essentially, they match a given year's expenditures with its gross additions to reserves. One should not belabor the many difficulties that affect the precision of the estimates. The most important is that reservoirs vary so much that a barrel of reserves in one reservoir cannot properly be compared with or added to a barrel in another. But there is a basic logical defect. Although development is the investment of a sum of money today for production over a long period in the future, no allowance is made for the return

[6] I avoid the term "Middle East" to exclude Syria, Turkey, and Israel. Egypt is included in Africa.

[7] For an excellent example, see Paul Dragounis, J. Cademartori, and S. Milioti, "Estimating Nuclear Fuel-Cycle Costs," *Nucleonics*, Jan. 1966, pp. 40 ff.

on the capital involved. These deficiencies were well stated by Hodges and Steele[8] and by Lovejoy and Homan.[9]

Since World War II the literature on the managerial economics of oil production has been transformed by two new developments: first, capital budgeting and the discounted cash flow method, i.e., the application of long-established economic theory to management practice;[10] and second, the art and science of reservoir engineering. In the early 1940s several techniques for predicting reservoir performance were worked out, and the independence of ultimate recovery from well spacing was established.[11] In 1949 basic ideas were synthesized in Muskat's *Physical Principles of Oil Production.*[12] Thereafter, many new methods were found for prediction of reservoir performance and the profitability assessment of a development project. As Pirson says, "Reservoir engineering may be defined as the art of forecasting the future performance of a geologic oil and/or gas reservoir from which production is obtained according to probable and preassumed conditions."[13] Once prediction became a regular method it was possible to apply formal methods of discounting, and there is now a considerable literature on the subject. In 1960 J. J. Arps, who had helped lead the way with his 1944 article on the analysis of decline curves,[14] edited a volume of reprints of articles issued by the Society of Petroleum Engineers, and this is brought up to date in the more recent works by him and by W. S. Eggleston.[15]

Today a project is planned or a property is evaluated broadly as follows. A forecast is made of the production profile and of the declining rate of output over time; a price is assumed and multiplied by the declining production profile, to yield an income stream; and each installment of this stream is discounted by such a rate as just to equate the sum of the present value of the forthcoming barrels to the expenditures necessary to install the producing capacity. In other words, we take the rate of discount as the unknown and solve for it. The next step is to compare this rate with the firm's cost of capital, which is defined as either (1) the price the firm needs to pay as a borrower and equity partner, *or* (2) the opportunity cost: the return the firm could expect from the best available alternative use of the money.

We proceed as follows: instead of taking price as known, and solving for rate of return, we use the company's cost of capital, treat price as the unknown, and solve for it.[16]

[8] John E. Hodges and Henry B. Steele, *An Investigation of the Problems of Cost Determination for the Discovery, Development, and Production of Liquid Hydrocarbons and Natural Gas Reserves* (Rice Institute Pamphlet, vol. 46, Oct. 1959).

[9] Wallace F. Lovejoy and Paul T. Homan, with Charles O. Galvin, *Cost Analysis in the Petroleum Industry* (Southern Methodist University, 1964).

[10] Ezra Solomon (ed.), *The Management of Corporate Capital* (The Free Press of Glencoe, 1959).

[11] R. C. Craze and Stuart E. Buckley, "A Factual Analysis of the Effect of Well Spacing on Oil Recovery," in *API Drilling and Production Practice*, 1945, p. 144.

[12] Morris Muskat, *Physical Principles of Oil Production* (McGraw-Hill, 1949).

[13] Sylvan J. Pirson, *Oil Reservoir Engineering* (McGraw-Hill, 1958), p. 2.

[14] Arps, "Analysis of Decline Curves."

[15] Arps, "Valuation of Oil and Gas Reserves," in Frick and Taylor (eds.), *Petroleum Production Handbook* (McGraw-Hill, 1962), vol. II, chap. 38, p. 23; Arps, "The Profitability of Exploratory Ventures," in *Economics of Petroleum Exploration, Development, and Property Evaluation* (Prentice-Hall for Southwestern Legal Foundation, 1961); and Eggleston, "Methods and Procedures for Estimating Fair Market Value of Petroleum Properties," *Journal of Petroleum Technology*, May 1964, pp. 481-86.

[16] For a parallel treatment, see Paul G. Bradley, *The Economics of Crude Petroleum Production* (Amsterdam: North Holland, 1967).

Suppose capital costs were zero. Then a price that just covered current expenses—including, of course, the services of management—would make it just barely worthwhile to produce. Similarly, a price that makes it barely worthwhile to invest capital in new producing facilities, in preference to using it in the next best employment, is the cost of capital.

In the early years of high-level production the capital cost per unit seems to be very low; in the later years, much higher. But it is illogical to look separately at early-year barrels and late-year barrels, since they are really joint products. One cannot have capacity for the early years without also having capacity for the later ones. Hence the project must be evaluated as a whole. Since the investment is all being made today, the production must all be reduced to present equivalents.

Note that nothing is said about future prices or revenues. They must be estimated independently of cost for a management decision. Minimum cost is not maximum profit or present worth. Development is optimal when carried to the point where the incremental cost is pushed up to equal the price (or incremental revenue, as the case may be).

It was once realistic, at least in the United States, to project current prices a long time into the future. That day has surely passed. Management should today be making explicit price forecasts.

The basic variables. To every conceivable development plan, there corresponds an investment I, which is necessary to establish the initial output, q_0, which will decline at a percent per year. The discount rate, r, is the operator's direct or opportunity cost of capital. That is, there exists a stream of revenues, R_t per year, whose present value is exactly equal to I. In symbols:

$$I = \int_0^T R_t \, dt = \int_0^T (p_t q_t) e^{-rt} \, dt \, . \qquad (1)$$

Let p_t = a constant, p_c . (2)

The production profile is: $q_t = q_0 e^{-at}$. (3)

$$\text{It follows that: } I = p_c q_0 \int_0^T e^{-(r+a)t} \, dt \qquad (4)$$

$$\text{and: } p_c = I/q_0 \int_0^T e^{-(r+a)t} \, dt \, . \qquad (5)$$

It is convenient to define the expression (I/q_0) as the investment per initial *daily* barrel. The integral is the present-barrel-equivalent (PBE) factor. Multiplying it by 365 gives the number of present-barrel-equivalents (PBEs) corresponding to the development plan, just as I is the corresponding investment. Dividing (I/q_0) by the total PBEs gives the capital cost in cents per barrel.

Table II–2 gives PBE factors corresponding to various combined discount-plus-decline rates and various assumed project lives. Henceforth 25 years will be assumed, but at realistic discount rates this hardly makes any difference.

TABLE II–2. Present-Barrel-Equivalent Factors

Discount factor $(a + r)$	PBE factors when life of project is: 20 yrs.	25 yrs.	∞
10%	8.93	9.52	10.40
11	8.39	8.88	9.53
12	7.91	8.30	8.79
13	7.47	7.80	8.16
14	7.08	7.34	7.62
15	6.72	6.94	7.15
16	6.39	6.57	6.73
17	6.09	6.24	6.37
18	5.82	5.94	6.04
19	5.57	5.67	5.75
20	5.34	5.43	5.48
21	5.13	5.20	5.25
22	4.93	4.99	5.03
23	4.75	4.80	4.83
24	4.58	4.63	4.65
25	4.43	4.46	4.48
26	4.28	4.31	4.33
27	4.15	4.17	4.18
28	4.02	4.04	4.05
29	3.90	3.92	3.93
30	3.79	3.81	3.81
32	3.59	3.60	3.60
34	3.41	3.41	3.41
36	3.24	3.25	3.25
38	3.10	3.10	3.10
40	2.97	2.97	2.97

Source: Jerome Bracken and Charles J. Christenson, *Tables for Use in Analyzing Business Decisions* (Irwin, 1965), table 1.

Note: Discounting is usually stated in terms of the "effective" or discrete rates. The "effective" compound discount rate is $(1 + r)(1 + a)$, e.g., $(1.20)(1.01) = 1.212$. But since oil production and sale is a daily activity, it is best described by continuous compounding. The present value of one dollar per year in a continuous stream for t years discounted at rate $(r + a)$ is:

$$PV = \int_0^T (1+r+a)^{-t}\, dt = \frac{1 - (1+r+a+ar)}{\log_e (1+r+a+ar)},$$

e.g.;

$$\int_0^{20} (1.212)^{-20}\, dt = 5.13.$$

Corresponding to the "effective rate" 0.212 the "nominal" (continuous) rate is 0.19062, by the formula: nominal rate $= 1 - \log_e(1 +$ "effective" rate). Then the present value is:

$$PV = \int_0^T e^{-.191t}\, dt = 5.13 .$$

In Table II–2, therefore, the discount factor given is "effective" or discrete, but the three columns are generated by continuous corresponding at that rate. The discussion in the text is in terms of the "nominal" rate because it is easier to explain.

The cost of today's investment is its future income, available for reinvestment at the rate designated here. (The farther off a mistake is in time, of course, the less it counts.) The unit cost of capital is equal to the price that would just barely provide that income.

Present-worth operating costs. We can now "levelize" our operating costs (above, p. 46) to take account of the higher costs imposed by decline rates. If operating costs for a well or reservoir are a constant K per year, and initial output is q_0, the initial operating cost per barrel is K/q_0.

The present value of the future expenditures is the equivalent of an investment made today, denoted I':

$$I' = K \int_0^T e^{-rt}\, dt = K \frac{1 - e^{-rT}}{r}, \tag{6}$$

where r is the relevant discount rate and t is time in years.

It is consistent to state the unit operating cost as a unit investment cost, by substituting equation (6) into equation (4):

$$c' = K \frac{1 - e^{-rT}}{r} / q_0 \frac{1 - e^{-(a+r)T}}{a + r} = \frac{K}{q_0} \cdot \frac{a + r}{r} \cdot \frac{1 - e^{-rT}}{1 - e^{-(a+r)T}}. \tag{7}$$

Let t increase without limit, and (3) becomes simply $(K/q_0)\frac{a + r}{r}$. This is the maximum value. Convergence to the limit is rapid, and we do not normally consider any other value.

Cost of capital. Of our four basic parameters for each area–investment, capacity, decline rate, and cost of capital–the first three must be discussed separately for each region, since the data are largely independent. But cost of capital must be discussed for all regions together, largely because risk is so large an element in the cost of capital, which, in turn, is so much larger than the decline rate that there is hardly any point in speaking of it except comparatively.

The locking-up of resources in durable goods, whose use is realized only in small doses over a long period, has always been recognized in economic theory as a cost. This is being recognized with painful slowness in the socialist states, as ideology proves weaker than truth.[17] Even in a world of certainty a rate of interest would be needed to enable us to evaluate the current contribution of a stock of goods or services to the current flow of income. The nearest approach in actual economic life to a riskless interest rate or pure cost of money is approximated by the lowest-yield government bonds. In mid-1967, U.S. government long-term bonds bore around 4.7 percent; over the preceding 10 years they had varied between 3 and 5 percent, but by mid-1970 they were over 7 percent.[18] Elsewhere rates have been generally higher, partly because capital is more scarce and more productive at the margin, partly because some capital markets are relatively small and inefficient. It is hard to disentangle these two elements from yet a third basic commercial

[17]Chukhanov, Z. F., "Economic Effectiveness of Thermal and Hydroelectric Power Stations," *Teploenergetika*, 1961, No. 12; Gregory Grossman (ed.), *Value and Plan* (University of California Press, 1960).

[18]Federal Reserve System, Board of Governors: *Federal Reserve Bulletin* and *Annual Reports*.

risk, to be discussed shortly. Hence, in the United States industrial bond yields on the very best issues (Moody's Aaa) were around 3.25 percent a decade ago, but over 5.50 percent in mid-1967 and over 9 percent in mid-1970. Similar companies paid over 6 percent in 1967 in Western Europe; the average paid by a sample of large well-established firms in the mid-1960s was over 7 percent in Germany and France, nearly 9 percent in Italy, but only around 5 percent in Switzerland. The lender received from half a percent to 2.7 percent less, the remainder being partly cost, partly disguised risk allowance.[19] In 1967-68 Euro-dollar bonds yielded nearly 7 percent, while in the following two years the yield was closer to 9 percent. We are not concerned with short-term money rates because our interest is in capital expenditures.

Over and above the basic interest rate or the bare cost of money, and of much greater importance, is the element of uninsurable risk in determining the cost of capital.

Risk allowance–in general. Risk and uncertainty continue to fascinate economists because they elude complete understanding, but for our purposes the elemental notions are enough. In contemplating any proposed investment a prediction is made of the return. Only by improbable chance will the actual return be just as expected–it will be higher or lower. The chances of doing better or worse than predicted can be conceived as a probability distribution. Let us consider the very simple case of a completely symmetrical distribution, where the chances of doing x percent points better than predicted are exactly equal to the chances of doing x percent points worse. Anyone lending money for this project knows that he will get no more than the contractual payments if it succeeds, while he may get nothing if failure is complete. Obviously he will not lend without some compensation. But consider those who will bear the profit or loss. The fact of risk and uncertainty, which are expressed in such a probability distribution, makes the investment less attractive. Hence it will not be undertaken unless the expected return is greater, by what we call a risk premium, than the return that would suffice to draw in the needed funds for an equal riskless return. And the greater the dispersion of probabilities about the expected return, the greater the premium that needs to be paid over and above the return.

Risk allowance–United States. In 1953 when basic interest rates were around 3.25 percent[20] Terry and Hill of the Chase Manhattan Bank used 8 percent for oil development, remarking that there was "nothing sacred about [it] . . . but it appears to the authors to result in a reasonable estimate of the fair market value" for a property to be sold.[21] They defined the discount factor for an equity participant, not a lender: "that rate of return or yield on the purchase price which is sufficient to induce a producer to risk his funds in the particular project evaluated rather than in safer investments offering lower yield." After the early 1950s, the basic interest rate, as was seen earlier, tended to drift upward, perhaps by as much as 1.25 percent points by the early 1960s, when a former president of Gulf stated that in purchasing oil and gas reserves, his company would not be interested in a 6 percent rate of return on a discounted cash-flow basis, would regard 7.9 percent as borderline, and would be sufficiently interested in 9 percent

[19] Jean-Paul Delacour, "Remarques sur Certains Aspects du Financement des Investissements Industriels dans les Pays de la CEE," Chronique No. 942, *SEDEIS* (Paris), Feb. 1, 1966, p. 39.

[20] Federal Reserve System, Board of Governors: *Federal Reserve Bulletin* and *Annual Reports*.

[21] Lyon F. Terry and Kenneth E. Hill, "Valuation of Producing Properties for Loan Purposes," *Journal of Petroleum Technology*, July 1953, reprinted in Arps (ed.), Petroleum Transaction Reprint Series No. 3, Society of Petroleum Engineers.

to make a thorough investigation.[22] For exploration and development combined, he thought an individual investor would require at least 25 percent while a large company, which could do considerable averaging, would require perhaps 10 to 12 percent (after taxes). This would point to something below 10 percent for development alone for a company or consortium large enough to average.

Offshore Louisiana exploration and development combined have been calculated as returning only 7 percent, but this is contradicted by what the same source calls "the fierce price competition for leases."[23] Hence I would consider 7 percent as too low a return for U.S. development. Another discussion of offshore operations states: "We have chosen 9 percent as representing a reasonable rate of return we like to expect on future invested capital for the exploration and production business."[24]

Eggleston suggests that successful companies will buy producing properties on the basis of 9 to 10 percent, but that a yield of less than 9 percent is acceptable if the property also holds a possibility of unconfirmed but at least not unlikely additional reserves. This fits our need precisely: a rate of return on a property which has been judged ripe for development but which in the nature of the case can be only incompletely known. I propose to follow Eggleston's recommendation[25] for "a framework within which the appraisal engineer may operate [with] . . . the cost of money at one end of the scale and the rate of return on invested capital as experienced by the more successful companies at the other end of the scale." The limits set in this fashion are between 5 and 13 percent; we will use 9 percent as the mid-point. Our time-reference is the middle to late 1960s, before the extreme stringency of 1968-70.

Commercial, geological, technical risk–United States and elsewhere. We might call "basic commercial risk" the premium that needs to be paid by even the most solvent and profitable firms. Costs may turn out to be higher or receipts lower than expected, because of faulty planning or unforeseen changes in demand and supply. Risk of this kind exists also in a socialist economy. Dams may be worth building today but not in 5 or 25 years, depending on progress in thermal and nuclear power generation, and in power transmission. An expected service life may be halved by some superior instrument. No matter what kind of economic system we live under, there is no escape from the pain of choice and of forecasting. Risk premia crystallize the fact of human error compounding unforeseeable changes.

Looking beyond basic commercial risk to those types of risk associated with oil development operations, we may distinguish three: (1) geological risk, as compounded or amplified by (2) engineering risk; and finally (3) political risk.

Geological risk arises from development as pushing into the unknown. A deposit is established by drilling a well, or several wells. From a small sample, the operator must predict future performance of the structural or stratigraphic trap. The studies cited earlier analyze those basic geological variables that determine the chance of the deposit lensing out a few feet away or going on for miles; of the daily rate of flow being one barrel or 20,000 barrels, etc. A drainage sale of a lease adjoining tracts with crude production

[22] Sidney A. Swensrud, testifying before the Federal Power Commission in September, 1961. Quoted in Eggleston, "Methods and Procedures . . . Properties."

[23] *World Oil*, Apr. 1967, p. 26.

[24] J. R. Wilson (vice president, Shell Oil Co.), "Economics of Offshore Louisiana," talk at Louisiana-Arkansas division, Mid-Continent Oil and Gas Association, Sept. 12, 1967.

[25] Eggleston, "Methods and Procedures . . . Properties."

would be a much lower risk venture in the mid-continent area than in a highly faulted though rich area like offshore Louisiana.

Intertwined with geological risk is engineering or technical risk. At one extreme would be human error, such as mistakes in interpretation or reading of logs, failure to keep up with current literature; at the other extreme is the inspiration one man has and another has not. Some excellent examples of technical risk are cited by Essley, who notes: "Reservoir engineering is more of an art than an exact science [because] most observed reservoir facts, phenomena, or symptoms, are subject to more than one logical interpretation. . . . It is analogous to . . . having more unknowns than equations and obtaining multiple solutions." (Hence he defines a reservoir engineer as one who "takes a limited number of facts, adds numerous assumptions, and arrives at an unlimited number of conclusions.")[26] Obviously, engineering risk is only a multiplier or amplifier of geological risk; if the data are so clear as to admit of few errors in interpretation, even the unskilled will make few.

Finally, there is political risk. Government may change the rules unfavorably—and sometimes favorably when they recognize the limits to what they can do without injuring themselves. But the high or low chance of unfavorable government action certainly will make capital less or more freely available for investment in a given type of project, and this is a necessary element of supply price.

Geological (and technical) risks are very much less outside the United States. Reserve-to-production ratios are half again as high in Venezuela as in the United States, and many times as high "in the Middle East [where] the big simple structures (definable geologically or geophysically), large concessions, and few operators allow a comparatively large area to be considered well-delineated after relatively few wells have been drilled."[27] Many very large fields can be expanded simply by drilling and equipping more wells in obvious places. There is much less chance of unpleasant surprises in the shape of higher-cost oil. Indeed, as knowledge accumulates, the risks tend constantly to be revised downward as the size of the deposit is revised upward. This same increase in knowledge, of course, operates in the United States, as the National Petroleum Council studies amply prove.[28] But where fields are relatively small, as in the United States, the oil operator learns more and more about less and less. Where fields are very large, he simply learns more. Technical risk is also less. Because it pays to spend large sums of money to test a given hypothesis about a big field, mistakes can be revealed and corrected.

If we could set aside political risk, the acceptable development rate of return for an oil company with access to the U.S. capital market would be considerably lower on operations outside the United States than inside. But in view of political risk, in the calculations to follow I have assigned a necessary rate of return of 20 percent for Africa, Venezuela, and Asia. Perhaps like Hamlet's wicked uncle, I have been "frighted with false fire" in guessing, without documentary support, that oil companies have had some such figure in mind.

Income tax. There is no allowance for income taxes, royalties, or the like as part of cost. A government can only adapt to the cost set by nature and the price obtainable in

[26] P. L. Essley, Jr., "What is Reservoir Engineering?" *Journal of Petroleum Technology*, Jan. 1965, pp. 20–21.

[27] D. C. Ion and W. Jamieson, "Reserves in Relation to Demand," in Peter Hepple (ed.), *Petroleum Supply and Demand* (London: Institute of Petroleum, 1966).

[28] NPC, *Report of the National Petroleum Council Committee on Proved Petroleum and Natural Gas Reserves and Availability* (Washington, 1961).

the market, however monopolized or competitive that may be. But government action can increase the degree of monopoly (as will be seen in Chapter VII) or raise development cost by increasing risk, and hence raising the cost of capital. The 1966 events in Libya marked the passage into the era of unilateral action by host governments. As *The Economist* put it, the companies in Libya "were rolled in carpets and all their bones broken"; they gave in "when it became obvious that it did not matter whether they were going to or not."[29] The law of January 5, 1966 gave the Libyan government the power to order the companies to change the concession agreement on taxes. No one expected this to be the last such action, just as no one was greatly surprised when it happened—in effect it had been discounted.

Suppose that the increasing risk raises the cost of capital from 20 to 30 percent. That is, it would now take a 25 percent yield on the investment to make it barely as good as it used to be at 20 percent. Referring to Table II-2: the PBE factor drops from 5.43 to 4.46, and costs per barrel rise by 5.43/4.46, or 22 percent. This is a true increase in cost.

Let us now consider the costs of the national company of a producing nation. Since they do not face any political risk, their cost of capital is governed by commercial and technical risk added on to what it costs them to borrow, or what they can expect to get by lending. Here the governments vary greatly. There is an indication that the government of Iraq believes a rate of about 5 percent is a proper cost of capital for them.[30] This is surely much too low. For a government like Kuwait, the current long-term rate for borrowers of the highest standing is applicable; in Europe in the mid-1970s it has been around 9 percent. Let us use a round 10 percent to allow for commercial risk, in which case, all else remaining equal, the PBE factor rises to 9.52 from 5.43, lowering cost by over 40 percent. The greater the political risk, the greater is the competitive advantage of a national company.

Measuring Development Capital Cost

Development expenditures: United States. Development expenditures in the United States are calculated from the estimates of the Joint Association Survey for the years 1959-68. I consider it best not to use their earlier estimates because of the likelihood of substantial under-reporting during those years.

I have computed the number of development dry holes, and have calculated their expense in order to add it to development costs. Appendix Table II-C-1 shows how this was done for each year, and permits the reader to identify and appraise any important discrepancies. The theory is that development dry holes express engineering risk. This is, of course, debatable. There is much merit in the argument that a dry hole conveys information and so belongs with exploration. However, I would argue that all wells, productive or dry, also yield some additional information, which for drilling as a whole is as much a product as hydrocarbons. If an oil operator proposes to develop a known pool, he must face the odds that some development wells will be dry. Therefore, to consider only the costs of productive wells would be to underestimate substantially the number of wells and the expenses needed to develop the deposit.

U.S. capacity and cost estimates. The calculation of new installed capacity in Table II-3 takes place in two steps. The first is simply subtraction of capacity at the beginning

[29] *The Economist*, Jan. 14, 1966, p. 232.

[30] *Platt's Oilgram News Service*, Nov. 30, 1965, p. 3; but compare *Middle East Economic Survey*, Dec. 24, 1965.

TABLE II-3. United States: Oil and Nonassociated Gas Producing Capacity Growth, 1951-1963

(crude oil and natural gas liquids in barrels)
(nonassociated gas in millions of cubic feet)
(expenditures in millions of dollars)

Years inclusive	Net increment in producing capacity	Sum of annual reserves	Cumulative production (daily rate)	Decline rate (col. 3/col. 2)	Capacity lost and replaced (col. 3 × col. 4)	Gross new capacity (col. 1 + col. 5)	Development investment expenditures
	(1)	(2)	(3)	(4)	(5)	(6)	(7)
			1. CRUDE OIL				
	TBD	*BB*	*TBD*	*Percent*	*TBD*	*TBD*	*$000,000*
1951-52 ...	738	55.4	12,250	8.9	1,090	1,828	n.a.
1953-56 ...	1,402	118.9	26,100	8.8	2,297	3,699	n.a.
1957-59 ...	718	92.5	20,300	8.8	1,786	2,504	n.a.
1960-63 ...	1,005	125.8	27,780	8.9	2,472	3,477	7,901
			2. NONASSOCIATED GAS AND LIQUIDS				
	MMCFD	*TCF*	*MMCFD*	*Percent*	*MMCFD*	*MMCFD*	
1960-63 Gas	24,139	775.9	110,000	5.9	6,489	30,628	3,293 (gas and liquids combined)
	TBD		*TBD*		*TBD*	*TBD*	
1960-63 Liquids ...	759	...	2,027	5.9	120	879	

Sources and Notes: Capacity–NPC, *Report of the National Petroleum Council on Proved Petroleum and Natural Gas Reserves and Availability*, 1961, tables 5,6; and id., 1965, tables 4,5 and p. 18. Oil capacity loss 8.4% of output–id., 1961, p. 27. This is 1.10 times the production/reserves ratio, which factor is applied to the earlier years. Gas capacity loss reckoned by NPC (id., 1961, p. 35) applies to 1959 production, which is reckoned at 24.16 bcf daily. The latter figure is from Adelman, *The Supply and Price of Natural Gas* (Oxford: Basil Blackwell, 1962), table 4-B. Nonassociated gas liquids are assumed to have the same decline rate as nonassociated gas. Combined liquid equivalent is 5,020 TBD, assuming 6,100 mcf of gas to equal one barrel of hydrocarbon liquid.

of any given year from the capacity at the beginning of the previous year, or previous date. But this increment is a net figure, representing the gross new capacity built during the year *plus* the loss of capacity during the year. In order to calculate this loss, we need a decline rate. The National Petroleum Council estimated that in the first year of capacity operation, the rate for the nation as a whole would be 8.44 percent.[31] This figure is 1.10 times the simple production-reserves ratio for oil, 1.14 times that for nonassociated gas, and we apply these factors to previous years' ratios. Decline rate multiplied by actual production yields the capacity lost and replaced.

Having now an estimate of $2,272 investment per initial daily barrel from Table II-3 (column 7 divided by column 6), it remains to compute the PBE factor. Since the 1960-63 decline rate is approximately 9 percent and we are assuming a 9 percent rate of return needed for strictly development work, the "combined discount rate" is 18 percent. At a 25-year life, the PBE factor is 5.94, PBE total is 2,168, and capital cost per barrel is $1.048.

[31] NPC, *Report of the National Petroleum Council Committee on Proved Petroleum and Natural Gas Reserves and Availability* (Washington, 1961), p. 28 and Table IV.

For 1960–63 inclusive, development investment for nonassociated natural gas and liquids was only about $656 per additional daily barrel of capacity. However, since the average capacity factor of nonassociated gas production is roughly 50 percent because of seasonal variations and the normal substantial delay between development and pipeline building, the effective cost of gas-and-liquids should perhaps be doubled to the neighborhood of $1,300. But even this is far below the cost of oil development, and shows again what a mistake it is to speak of "the U.S. petroleum industry" as though it were a homogeneous block.

It may be that the assumed decline rate is too low. A recent publication[32] suggests an average decline rate of 12.5 percent per year. Estimates by other authorities considered there are said to range from 8 to 14 percent. However, in a study of Louisiana[33] a decline rate of 200,000/3,290,000 or 6.1 percent is suggested (assuming 100 percent capacity production). This is consistent with an 8 percent national average.

The higher decline rate works both to raise and on balance lower the per-barrel cost. The new capacity q_0 is estimated using the following definition:

$$q_0 = q_t - q_{t-1} + aq_{t-1}\,,\ a \leqslant 1$$

where q_t is output in year t, and a is the decline rate. Obviously, the higher is a, the greater is q_0. At the limit $a = 1$, the total production would be newly installed, i.e., $q_0 = q_t$. As we saw earlier, the capital cost per barrel

$$c = I/q_0 \int_0^T e^{-(a+r)t}\,dt\,.$$

Obviously, the higher is a the greater is q_0, hence the lower the investment per initial barrel per day; but the greater the combined discount factor $(a + r)$ the lower the PBE factor, and hence the higher the cost. Thus the higher decline rate works in both directions, but always lowers cost in the relevant range of values.

Development costs outside the United States: in general. For the three areas to which we now turn, Venezuela, North Africa, and the Persian Gulf, we have altogether different basic data, which enforce different methods of estimation.

Appendixes II-D and II-E contain details of calculation; here the guiding principles are explained. First, we exclude lease acquisition expenditures. The compilation in Appendix Table II-E-1 may be too low, but, given the sparse and uneven data, we are unable to check it. The United Nations publication *Economic Conditions in the Middle East* lumps together both recurring and nonrecurring payments. The initial bonus and other payments represent not cost but the host government's first installment on the anticipated profits of finding and developing. These payments are generally small.[34]

[32] Cabinet Task Force on Oil Import Control, *The Oil Import Question* (Washington, Feb. 1970), appendix D, pp. 222–23.

[33] *Oil and Gas Journal*, Apr. 13, 1970, p. 39.

[34] O. C. Clifford, Jr., Joseph S. Pluta, and Joseph A. Mehan, "Overseas Economic and Exploration Challenge and Change," preprint (subject to revision) of a paper presented at the 35th meeting of the Society of Exploration Geophysicists, 1965. Figure 7 gives "offshore acreage costs, foreign free world" for 1961–64, also 1958 for Iran.

Geological and geophysical work is obviously exploration, not development. Dry hole expenditure is more complex. For the United States, the cost of dry development holes was calculated, but there are so few such holes in the countries of the Persian Gulf and North Africa that all dry holes there were reckoned as exploration.

For Venezuela, expenditures by the Ministry of Mines and Hydrocarbons[35] check very well with those estimates by the Chase Manhattan Bank.[36] For the Persian Gulf and North Africa, the chief problem is to allocate the regional expenditures among countries in proportion to indexes of development activity. The procedure consists of the segregation and then the addition of three costs: drilling, for all wells; completion, for all wells other than dry holes; and equipping leases, for oil, gas, and service wells. Our basic assumption is that the three types of cost of drilling and equipping wells outside the United States are governed by the same physical factors as in the States, hence are systematically related in the same way. Of course, depth is not the only cost-determining factor, but it is the only one covered in the Joint Association Survey (JAS).[37]

Appendix Table II-E-5 illustrates the method used to obtain the country-by-country estimates of 1947-67 Persian Gulf development capital expenditures shown in Table II-E-6. Much of the year-to-year fluctuation is meaningless, and the reader will usually be better off to average several years together, as is done in Appendix Table II-E-8. But some changes are significant, and certainly a trend is better appreciated by not suppressing some variations at the start.

An alternative method of allocating capital outlays (Appendix II-F) is simpler and more accurate, but it requires data that often are not available. The total amount of rig-months is known with fair accuracy for large areas and for most important countries and concessions. Rig-time is affected not only by depth but also by hardness of sediments, lost circulation, and other drilling difficulties that absorb time and money. Furthermore, exploration and development rig-time can often be separated, making it unnecessary to estimate by allocation.

We are able to check against actual expenditures in four cases. The rig-time method yields much lower figures in a new area (Libya) probably because time spent on successful oil wildcats is excluded from development time, as it should be. In Iraq, the well sample is so small as to incur a huge sampling error, and the rig-time method rests in effect on a larger sample. In Kuwait, the difference is not surprising because there were dramatic reductions in drilling time in 1963[38] reflected in rig-time, but not in wells.

Appendix Tables II-F-5 to II-F-9 present a summary comparison of various estimated expenditure amounts where it is possible to check against actual expenditures. In Iran and Libya there is considerable underestimation, as there should be, since our method is designed to exclude gas or gas-liquids capital expenditures, which have been considerable in both countries. Elsewhere, the error of estimate is more likely to be upward.

[35]Venezuelan Ministry of Mines and Hydrocarbons, *Petróleo y Otros Datos Estadísticos* and the annual reports.

[36]For details of the comparison, see M. A. Adelman, "Oil Production Costs in Four Areas," in *Proceedings of the Council on Economics,* AIME, 1966.

[37]American Petroleum Institute, et al., *Joint Association Survey*, Part I, *Industry Drilling Costs*; Part II, *Estimated Expenditures and Receipts of the United States Oil and Gas Producing Industry* (issues for 1960-63).

[38]Kuwait Oil Co., *Annual Report of Operations*, 1964, pp. 6, 16, 61.

Having calculated the inputs–development expenditures–we now turn to the outputs–new producing capacity. One method is to compute average production per active oil well in the country or area, and multiply by the number of new oil wells completed that year. This assumes that the average new well is as productive as the old. There are two opposing errors. The more wells drilled, the lower is average output, other things being equal. The true incremental cost of the new well must include any lower production which it forces on the old wells. But there is an opposing error: the new wells are drilled with greater knowledge of the reservoir, and should be better located and more productive. There may be very wide variation among wells in a given reservoir. In heterogeneous reservoirs the distribution appears to be lognormal, with the better half of the wells in a sample contributing 78 percent of output.[39] Elsewhere variation is much less, but it must be substantial. Moreover, old wells sand up, develop leaky casing, start going to water, and age in other ways. They are usually worked over, but it might be more efficient to put some out of production in favor of new wells which would be drilled in the most favorable possible locations.

Complicating the picture are the large-scale fluid injection projects–notably in Venezuela, Algeria, Iraq, and Saudi Arabia–which have increased both expenditure and capacity per well.

There are alternatives to the well method of computation. In the case of the United States, we have estimates of producing capacity and of the national average decline rate. By adding the net new capacity to the estimated annual loss during any period, we arrive at gross new capacity installed (see Table II-3 above).

For Venezuela, the Ministry of Mines and Hydrocarbons kindly supplied me with unpublished data on the average capacity of all new oil wells completed each year during 1946-68. Multiplied by the number of completions, this gives a hard figure of gross new capacity installed that is more reliable than the figure for any other country. But in addition we are able to compute a decline rate by using the following equation:

$$w_t y_t = q_t - q_{t-1} + a q_{t-1}\ , \text{ or } (w y_t - q_t + q_{t-1}) / q_{t-1} = a$$

where w_t is the number of completions and y_t the average capacity: the gross capacity increment $w_t y_t$ equals the net increment $(q_t - q_{t-1})$ plus the capacity lost and made good. This loss, as a percent of production, is the decline rate; over 1947-68 it is 15.9 percent. This may seem high, but it is close to the earlier informal estimates of oilmen working in Venezuela, who had suggested 15 percent.

Elsewhere, there is no direct information on capacity or on decline rates. Consequently, we shall use actual production as an approximation to capacity. But net changes in productive capacity are the result of the gross increase and the decline. Neglect of the decline is a source of error, the importance of which varies greatly from country to country: the greater the growth rate of production, the less the error. In general, we can describe the annual output as equal to capacity and proportional to e^{nt}, where n is the annual net growth rate in percent, and t is time in years. If a is the annual decline rate, then the capacity lost in any given interval of time i is:

$$a \int_{t-i}^{t} e^{nt}\, dt / (e^{nt} - e^{n(t-i)}) = a/n\,.$$

[39] J. E. Warren, "The Performance of Heterogeneous Reservoirs," SPE 964, at the 39th (1964) Annual Fall Meeting of the Society of Petroleum Engineers, Oct. 11-14, 1964.

Thus, with (say) a 1 percent decline rate and a 10 percent net growth rate, one-tenth of the year's growth is replacement of lost capacity, and gross new capacity installed is 1.1 times the net growth. As the net growth rate becomes smaller the error involved in neglecting the decline rate becomes greater. In the extreme case of zero net growth, the fraction becomes impossible.

For countries other than the United States and Venezuela, the following crude expedient has been used for our calculations: the inverse of the ratio of proved reserves to production, taken to the nearest whole number. For example, in the Persian Gulf, reserves have been about 75 times production; hence we assume that one-seventy-fifth is lost each year, or 1.3 percent, rounded to 1 percent.

Almost certainly this decline rate is too low, but, as has been explained above, the error works both ways: it causes an overstatement of the investment required per additional daily barrel of capacity, and also an overstatement of the number of PBEs resulting from that investment. For the relevant range of estimates, as we saw in the case of the United States, assuming a higher decline rate would lower the cost per unit. Hence the net result of the error is to yield development cost estimates somewhat higher than the true ones. The error is small and swamped by some other factors to be shortly considered; but there is a real gap in our knowledge–the zero-investment decline rate.

Minor sources of error and bias. 1. At any moment in time there is a substantial "overhang" of work in progress, plus facilities that are finished but not quite ready to be put into operation and are awaiting supplementary equipment. A developed field may be waiting for a pipeline, or both for a loading jetty, or all of them for a refinery. Since our expenditure figures for the year or longer are always complete but capacity is counted only when operations begin, the investment per daily barrel and the cost per unit barrel is overstated. This bias is the least for Venezuela, substantial in the Persian Gulf, and largest in areas being intensively explored and in the early stages of development. The first roads, buildings, etc., must be provided, at a cost small in relation to the massive expenditures to come, but high in relation to zero or very small production. Thus, in Libya the British Petroleum-Hunt concession spent $168 million mainly in 1965-66,[40] but only began producing in 1967; Occidental spent $65 million mostly in 1967,[41] but only began producing in 1968.

2. Another upward bias arises from treating all productive completions as development wells. But, of course, many of them are exploratory wells, especially in growing areas, and their cost is many times higher than that of a development well.

3. The rig-time method of calculating investment tends to reduce this error by leaving out exploratory drilling altogether. But since exploration carries with it a larger complement of expense per rig-month than does development, the allocation according to rig-time normally credits too much to development. If all countries in an area had the same exploration-development mix, this would not matter. These declining investment requirements must be expected as a field becomes better known. Attention was drawn to this fact as reducing risk (see p 55); it also tends to reduce cost directly, although the effect soon flattens out.

4. A bias unimportant in the Persian Gulf, but not in North Africa, occurs in the rig-time method of calculation, which counts gas-well drilling as development. This limits our use of the rig-time method in the case of Algeria, and may make it progressively worse in the case of Libya. Moreover, early in the life of the Zelten field in Libya, a

[40] *Platt's Oilgram News Service*, Feb. 13, 1967.

[41] *Wall Street Journal*, Apr. 5, 1968, p. 8.

250-mile, 36-inch pipeline was built to supply seawater for pressure maintenance. This was a mistake; the Zelten natural water drive turned out to be much better than had been expected.[42] Fortunately, the field produces very large amounts of associated gas, and so, on the signing of a large contract with Ente Nazionale Idrocarburi (ENI) late in 1965, Esso converted the line to natural gas.[43] It also spent large sums on a liquefaction plant at Brega.

5. During 1964-68 (data for earlier years are not available), development plus workover and service wells accounted for a little over 94 percent of all Venezuela rig-time, and exploratory wells for under 6 percent.[44] Therefore, in assigning production investment wholly to development, the latter is overstated by about 6 percent, or even less, because some of the exploratory wells are outposts, and few are new-field wildcats.

6. Production may fluctuate for seasonal or political or other reasons, and the change from year to year may overstate or understate the change in capacity. Where the net growth is small, the error becomes troublesome. In two instances, during 1966-68, growth from peak month to peak month was used as a proxy for capacity increase rather than growth from period to period.

Trends in development cost. Table II-4 makes a comparison of development investment requirements for 1959-60, the first two years for which there is sufficient rig-time data, and 1968-69, the most recent. (Single years are more easily distorted by fluctuations in numerator or denominator.) The reduction is all the more striking because the general price level increased by about 12 percent.[45]

Venezuela shows a similar strong trend since 1954,[46] but no such comparison can be made for Libya. As Table II-5 shows, the early years' averages are unreliable because the amounts are so small and development costs are understated because of a large element of "overhang." The 1968 leap in output also benefited by a large investment pressed into service as soon as possible. There may have been a rising cost trend, but this is uncertain because of large natural gas processing, liquefaction, and loading facilities outlays.

Without doubt, most of the decline at the Persian Gulf was the fruit of improved technology. The decade was an impressive one in the United States, where few new fields were found yet development costs per well or per barrel of capacity were steady or declining.[47] But it is difficult to separate out the effects of learning the particular field from those of constant improvement in drilling and completion techniques, or in field processing. In the Bibi Hakimeh field of Iran, average drilling time per well fell from 44 to 36 days from the first complete year to the second.[48] In Saudi Arabia, in 1966, all the field drilling records were broken.[49] In Kuwait, completion time per well was "reduced

[42] *World Oil*, Aug. 15, 1964, p. 134.

[43] *Oil and Gas Journal*, Newsletter, Dec. 13, 1965, p. 2.

[44] Rig-years totalled: workovers and service wells, 213; development well drilling, 70; exploratory drilling, 17. See *AAPG Bulletin*, respective years, issue "Developments in Foreign Fields."

[45] The implicit price deflator for total GNP and for nonresidential structures and equipment increased respectively by 22.2 and 12.1 percent from 1959-60 to 1968-69. See *Annual Report of the Council of Economic Advisers* (Washington, 1970), table C-3.

[46] See appendix table II-D-2.

[47] M. A. Adelman, "Trends in the Cost of Finding and Developing Oil and Gas Reserves in the United States, 1946-1966," in S. Gardner and S. Hanke (eds.), *Essays in Petroleum Economics* (Colorado School of Mines, 1967).

[48] Iranian Oil Operating Companies, *Annual Report, 1966*, p. 9.

[49] *AAPG Bulletin*, vol. 50 (1966), p. 1634.

TABLE II-4. Oil Field Development Investment per Daily Barrel, Total Persian Gulf and Largest Producing Countries, 1959-1960, 1968-1969

	1959-60	1968-69
Middle East, Total		
1. Rig-months	1,260	1,242
2. Production capital expenditures (millions)	$ 525	$ 630
3. Additional capacity (TBD)	1,060	2,249
4. Investment per daily barrel (line 2/line 3)	$ 495	$ 280
Four Largest Producers (Iran Consortium, Iraq, Kuwait, Saudi Arabia)		
5. Development rig-months	365	308
6. Development capital expenditures (line 2 × [line 5/line 1])	$ 152	$ 156
7. Additional capacity (TBD)	986	1,829
8. Investment per daily barrel (line 6/line 7)	$ 154	$ 85

Sources: Lines 1 and 5, appendix table II-E-2; line 2, Chase Manhattan Bank; lines 3 and 7, *Oil and Gas Journal*–output 1961 less 1959, output 1970 less 1968.

Notes: Drastic slump in Iraq both in development well drilling and in capacity may distort tables downward. No allowance for decline made good makes lines 4 and 8 too high, but does not distort trend.

The total production cost of Kuwait oil (presumably Kuwait Oil Company only) is given by the Ministry of Finance and Oil as 12 cents per barrel in 1950-51; 11 cents in 1952-53; 10 cents during 1954-59; 8 cents in 1960; 7 cents in 1961-64; and 6 cents in 1965-68 (Central Bank of Kuwait, *First Annual Report for the Year ending 31st March 1970* [Kuwait, 1971], table 8, p. 19). There is no explanation of concept or method, but an annual decline of a little over 4 percent does not conflict with our estimates.

by half on introducing the new techniques" in 1963, and further improvement took place in 1964. Moreover, as shown below,[50] the working staff decreased while output rose.

Concession	*Working staff* 1962	*Working staff* 1970	*Output (TBD)* 1962	*Output (TBD)* 1970	*Increase in output/man (percent)*
KOC	5,757	3,720	1,832	2,734	130
Aramco	13,573	10,353	1,521	3,549	205
IOOC	12,966	7,277	1,300	3,520	171

Cost as function of development intensiveness. Looking backward, development cost per barrel *was* some definite figure. Looking forward, it is no constant but a variable. To find the best development plan for a given reservoir, the operator needs to calculate, and compare with expected price, the cost per barrel as it varies with the intensiveness of drilling.

The decline rate of a well or pool is a function of the number of wells. Nature gives us a schedule of output per well per day for any given number of wells we drill. The decline rate will increase if many more wells are drilled; but it can be decreased by investment in pressure maintenance, well workovers, fracturing, acidizing, etc. All have been instituted in large Persian Gulf fields.

For all reservoirs under all types of drives, the effect of additional wells is diminishing returns and increasing costs. If time were no object, with zero rate of return or cost of capital, one well would drain a whole reservoir at lowest cost.[51] All the operator can do is

[50] Tabulated from annual reports of KOC, Aramco, and IOCC.

[51] B. Mostofi, "Petroleum Production and Reservoir Behavior," UN, *Interregional Seminar on Techniques of Petroleum Development, 1963* (UN, 1964, Sales 64, II, B.2), p. 148.

TABLE II-5. Libya: Field Development Investment, 1961-1970

Year	Gross increase in output (*TBD*)	Development rig-time as percent of African rig-time	Total Africa production capital expenditures (*million* $)	Libyan development expenditures (col. 2 × col. 3) (*million* $)	Investment per initial daily barrel (col. 4/ preceding yr. col. 1)
	(1)	(2)	(3)	(4)	(5)
1961....	18	3.5	400	12	70
1962....	171	6.0	335	20	66
1963....	302	8.8	275	24	59
1964....	404	11.5	325	37	96
1965....	384	15.4	340	52	165
1966....	314	13.4	355	47	165
1967....	284	18.5	355	65	72
1968....	908	19.8	575	118	205
1969....	577	24.1	575	139	213
1970....	653	n.a.	n.a.	n.a.	n.a.

Sources: Col. 1: 1961-69–from *Oil and Gas Journal* annual "World Wide Oil," adding 3% allowance for capacity lost and replaced. 1970 (April)–from *Petroleum Intelligence Weekly*. Col. 2: app. table II-F-2. Col. 3: Chase Manhattan Bank.

n.a. = not available.

to speed up the rate at which he gets this oil out of the ground, and this acceleration costs more money.

Additional wells will give us a higher initial production rate and nearly always a higher decline rate. Let us, however, take the unusual case of a complete water drive and no pressure drop. Even here, the greater the reservoir output, the more average output per well will fall. Depending on whether there is a side or bottom water drive, edge wells or all wells begin producing more water and less oil. But let us disregard this, and assume that no matter how many wells we drill, daily output is constant, with zero decline rate until like the deacon's one-hoss shay, our project stops for good. We also ignore well workovers. Then with every new well drilled, the investment I is increased by exactly the same proportion as the initial daily output q_0. Consider a pool with 1.825 billion barrels, which we assume can be exploited *ad libitum* without any decrease in output per well or any decline rate (see Table II-6).

From the beginning, and then at an increasing rate, average and incremental costs rise. How far to push development depends on the price (or incremental revenue) expected.

TABLE II-6. Cost as Function of Development–A Reservoir With No Well Interference or Decline Rate (discount rate 20 percent)

Number of wells	Investment expenditure (*million* $)	Initial daily output (1,000 barrels)	Time to depletion (*years*)	Present barrel equivalent (*millions*)	Average cost (*¢/PBE*)	Incremental cost (*¢/PBE*)
(1)	(2)	(3)	(4)	(5)	(6)	(7)
100	20	100	50	200	10.00	10.00
200	40	200	25	396	10.10	10.20
300	60	300	16.67	570	10.53	11.49
400	80	400	12.50	718	11.14	13.51
500	100	500	10	840	11.90	16.39

Col. (6) = (2) / (5).

Col. (7) is increment of col. (2) divided by increment in col. (5), e.g., second line is

$$\frac{40 - 20}{396 - 200} = 10.20.$$

Thus, even under the most extreme of favorable assumptions there is no escape from rising costs. But in fact, even under a complete water drive, not only does output per well tend to decrease as more oil is produced, but there may be well interference. Wells must be inspected, worked over, and sometimes abandoned. Of 636 Kuwait wells drilled through 1967 for oil production (excluding injection and observation wells), 128 were abandoned, suspended, or otherwise not connected. In the Aramco concession, of 494 wells completed as oil producers, 10 were abandoned, 59 standing, and 19 shut in. In the Iran Consortium, 18 rig-months out of a development total of 60 were for workovers. In Kuwait, in 1968, workover rig-months were 21, total development rig-months were 57.[52] Workover outlays were probably between a fourth and a third of development investment outlays (on the assumption that a workover rig-month is cheaper than a rig-month spent drilling a new development well). It is a popular illusion that once a field is "all drilled up" it requires only current operating costs. Even in the largest and most prolific fields, diminishing output per well can only be avoided by investing more—which means diminishing returns per dollar of investment. The more oil to be produced, the more must we reckon with such costs. For less efficient driving mechanisms than a complete water drive, all this is true *a fortiori*.

But in the early stages, cost per producing well falls, often drastically. This is in part a spreading of overhead: access roads, supply dumps, living quarters, etc. But the time to drill and complete the twentieth well may be one-tenth of the time needed for the first. Therefore, the inevitable tendency to decreasing returns for each additional well must over an initial interval be overborne by a sharply diminishing cost per well. This is quite aside from spreading overheads or fixed costs, and it takes no account of exploration costs. But in almost every reservoir, the cost per PBE will decline and touch some minimum point before rising with the more intensive rate of development and higher output rates. And so long as incremental cost of output is less than what can be gained by producing it, more should be produced.

So much for the single reservoir. Oil production in any field or area comes from the whole group of reservoirs, and marginal cost for the field or area is the result of the *intensive* margin within each reservoir and the *extensive* margin found by arranging all reservoirs in order of efficiency. Computers have made it possible for oil operators to make refined calculations of how to get any given increment of supply at lowest possible cost.

Hokail[53] gives an ideal life history of the East Texas field as a scenario replayed under optimum management. Initially, 50 wells would be needed to delimit the field. Assuming that desired production was 400 TBD, and "an average well production of 5 TBD initially, eighty producing wells would be required . . . When wells on the west flank started to produce water they would be plugged back for dry oil production, and later converted to water-advance observation wells or water-injection wells. New wells would be drilled along the eastern edge of the field as advancing water moved up the structure . . . [and]

[52] *Sources*: Kuwait Oil Company, *Annual Report, 1967* (more recent data not available). Other numbers calculated from *AAPG Bulletin*, vol. 52 (1968), p. 1556; and vol. 53 (1969), pp. 1768, 1772, 1776. Workovers are occasioned by water leaving the formation and reducing pressure, or oil leaving the formation before entering the well bore. They always involve redrilling of some of the well, and usually the application of dilute acid or explosives, to improve permeability by removing scale, sand, or other obstructions; or by opening or enlarging the pores in the rock.

[53] A. M. Hokail (Aramco), "Modern Concept of Oil Field Development," paper presented at Sixth Arab Petroleum Congress, Mar. 6–13, 1967.

well density would be increased." Initially, as production built up, investment per well would be $210,000; later, injection would require another $30,000. Still later the need to treat wet oil and dispose of water would require another $60,000. Ultimately 500 wells would be drilled, at a total cost of $150 million, or $300,000 per well. At the limit some 250-350 wells would each be producing 500 barrels of oil daily with 1,500 to 2,500 barrels of water. This calculation assumes a high enough price of oil to pay operating costs. If the price were lower than contemplated, this last most expensive increment of oil would never be produced. A decline rate of 4.87 percent annually is implied, but this is obviously by choice–it could have been greater or less.

Incremental Development Cost in the Long Run

The foregoing estimates are all historical, even if the history is recent. They might be called long-run costs in the short run, since they measure for recent years the price just necessary to recoup a given incremental investment in long-lived facilities. We turn now to the long-run incremental costs in the long run to answer a more interesting question. What is the best estimate of operating-developing costs under conditions of greater output? This is the practical business problem, just as it is the more interesting one for the economic theorist–the costs and benefits of change, or (what comes to the same thing) the incremental cost of output.

In the whole vocabulary of economics there is probably no more misunderstood and misused word than incremental (or "marginal") cost. For nearly all who are not economists, and unfortunately even for many in the profession, it sets off a reaction like that in a salivating Pavlov dog. Incremental cost is said to be the cost of producing that infamous "incremental barrel"; it is low, near zero, and hence "when product is sold at incremental cost it is sold at a loss, undermining price." Such views show misunderstanding of a basic economic concept. There is no such thing as incremental cost standing by itself. The term itself should warn us that we cannot measure or discuss it until we say where the increment begins. A little additional output from existing equipment, *if* it is operating at a low percent of capacity, will have a very low incremental cost. If the facilities are already being used at a very high percent of capacity, the incremental cost will be very high. More output supplied from new facilities presents an altogether different question, and the answer can go either way.

Along with the stereotype of incremental cost being necessarily low is that of its being a *part* of cost; thus, one often hears that prices must cover "not only incremental costs but also fixed costs," if a business is to prosper. But incremental costs are not part of cost at all. They refer to the difference between two amounts of total cost corresponding to two amounts of output, or two methods. Where we need to know how to raise capacity from current to a higher figure, then the incremental cost includes all necessary investment plus the other factors needed to operate the facilities, plus–and this is most easily overlooked–the feedback effect of the incremental operation on the preexisting operation.

An example may help. Suppose that we can build a one-story building, with a certain area of useful space, for a given sum. If instead we build a two-story building, the incremental cost is not simply the labor, materials, and so forth, needed for the second floor. There must also be substantial redesigning of the first floor–heavier foundations, less useful space because of the need for such facilities as stairways and heating and plumbing systems. Hence the incremental unit cost is the difference between (a) the total cost of building two stories and (b) the total cost of building one, divided by (c) the *net* increment of floor space. It may be higher or lower than the unit cost of one floor.

In oil production, as noted earlier, incremental cost is related to the extensive margin among reservoirs and the intensive margin in each reservoir. Given a highly accurate cost system and cost-output formula, we could calculate within tolerable error the least-cost source of the next increment from anywhere in the world.[54] Some companies are doing this today. But our data do not permit it, except in the roughest way. For the most part, we can only compare one country with another as though each were a single reservoir, with constant costs per additional unit. But unless we keep these comparisons within narrow limits, they will surely mislead.

An example may help. In late 1965 two executives of Standard Oil (New Jersey) said that Jersey's Libya production was its "most profitable" of anywhere in the world.[55] This confirms the relative costs estimated here. But they indicated that since Esso's Libya fields were "approaching their maximum production" the company expected its growth in output to come "largely from the Middle East" in the next several years.[56] In other words, average cost (including transport) was lower in Libya, but incremental cost was higher. Given 1966 conditions and knowledge, more oil could only be developed in Libya at a higher cost than during 1962-65, and higher than in Persian Gulf countries, because Jersey's Libyan reserves were small relative to Jersey's Persian Gulf reserves. For other companies, such as British Petroleum or Mobil, the intersection between the two lines of incremental cost would come at a different point. Others, like Marathon and Continental, might have no Persian Gulf production of any consequence.

United States incremental costs. The situation here is unique. Since the regulatory system has built up excess capacity (which is much diminished, however), output could be expanded within those limits very cheaply. Hence an increase in output, say through decreased imports, would be extremely profitable.

But we might take a longer time perspective and ask, first, how much could be saved if the U.S. producing industry decreased by some given percentage, assuming either effective competition or intelligent planning, so that the most expensive output gets eliminated first; second, how costly would it be to expand output along the least-cost path?

The forecasts of Schroeder[57] and the econometric studies of Davis and of Fisher[58] are all consistent with the notion that higher output is more costly. But they do not have the data for cost calculation. Fisher is concerned with the response of wildcat drilling to price, and only indirectly, if at all, with the response of supply to price over the long run.

[54] Cf. A. S. Lee, and J. S. Aronofsky, "A Linear Programming Model for Scheduling Crude Oil Production," *Petroleum Transactions, AIME*, vol. 213 (1958), pp. 389-92.

[55] *Platt's Oilgram News Service*, Nov. 8, 1965, p. 4.

[56] *Petroleum Intelligence Weekly*, Nov. 8, 1965.

[57] W. S. Schroeder, "Fuel Consumption and Availability for 1975 and 2000," in *Background Material for the Review of the International Atomic Policies and Programs of the United States*, Report to the Joint Committee on Atomic Energy, Congress of the United States, 86 Cong., 2 sess., vol. 4, commonly cited as the "1960 McKinney Report," pp. 1456 ff.

[58] Warren B. Davis, "A Study of the Future of Productive Capacity and Probable Reserves of the United States," *Oil and Gas Journal*, Feb. 24, 1958, p. 114; and Franklin M. Fisher, *Supply and Costs in the U.S. Petroleum Industry, Two Econometric Studies: I. The Supply Curves of Wildcat Drilling and of New Petroleum Discoveries in the United States. II. Measuring the Effects of Depth and Technological Change on Drilling Costs* (Washington: Resources for the Future, 1964).

In my opinion, without information on the distribution of fields and wells by capacity and cost, no long-run incremental cost function can be reckoned.[59]

Venezuela incremental costs in the mid-1960s. For Venezuela as for the United States, one may ask at what costs output could be increased in the better deposits at the same time that it was being decreased in the poorer ones. Published proved reserves have been about 17 billion barrels, which are comfortably in excess of the necessary inventory to support current output levels. The president of Shell Venezuela has estimated that current reserves could be doubled by the use of new pressure maintenance methods, though he does not discuss the costs.[60] There seems also to be agreement that many of the existing deposits could be more intensively drilled and produced with but little increase in the decline rate or, therefore, in cost. Data are lacking to say anything more precise. It is an important statistical gap for Venezuela.

However, it is possible to make some calculations permitting a check both on our average Venezuelan figures and on the costs in the better areas. In 1964 oil well development expenditures (see Appendix Table II-D-2, col. 15) were $160 million for 560 oil wells, or $286,000 per well.

Let us now consider the sections of the great Bolivar coastal field designated as the Bachaquero and Lagunillas subfield. The average depth of well (weighted by production) was only 3,180, as compared with a national average of 5,375. By the factors in Appendix Table II-E-2 a well should cost on the average only 48.6 percent of the national average. Since an average Venezuelan well cost $286,000, a well in this field would be expected to cost $136,000.

A drilling program announced by Creole[61] for 1966 permits us to check this. The company intended to spend 84.5 million bolivares, or $19.2 million, in drilling 134 wells offshore. The cost per well comes to $143,000, which is close to expected. Assuming a 16 percent decline rate for the new wells, the combined discount rate is 36 percent, PBE factor is 3.25, and there are 1,185 PBEs for every initial daily barrel. What we do not know is the average capacity of those new Maracaibo wells. The average for wells in the Bachaquero and Lagunillas fields are in the 300-400 range, but new ones must be much more productive. In 1966 the national average was 710 barrels daily (see Appendix Table II-D-2) per newly completed oil well. Lake Maracaibo wells are more productive than the national average: new wells produce between 1,000 and 5,000 barrels daily.[62] If we assume that the distribution is highly skewed, with most new wells nearer the lower limit, it is conservative to reckon the investment per initial daily barrel as $143,000 per 1,000, $143 per daily barrel, or 12.1 cents per PBE. By the square root formula, operating costs are 4.2 cents; multiplying by 1.8, which is the value of $(a + r)/r$, yields a levelized operating cost of 7.6 cents. Adding gathering costs and terminalling at about 3 cents, the total for the new wells is about 12.1 + 7.6 + 3.0 = 22.7 cents per barrel operating-development costs.

As shown in Table II-7, average Creole cost is not much over half of average Venezuela cost (see Appendix Table II-D-2).[63]

[59] See, however, the Cabinet Task Force on Oil Import Control, *The Oil Import Question* (Washington, 1970), part II, table C, p. 39.

[60] *Petroleum Press Service*, Oct. 1965, p. 395.

[61] Creole Petroleum Corp., *Carta Semanal de Noticias*, Dec. 31, 1965.

[62] M. E. Lynch and C. B. Lepak in *World Oil*, Aug. 1, 1966, p. 81.

[63] Yet Creole may have encountered sharply rising marginal costs in 1970, since large-scale drilling was expected to maintain output (*Petroleum Intelligence Weekly*, Aug. 10, 1970). The expected 451 BD per new well is far below the national average.

TABLE II-7. Creole Development Capital Costs, 1965-1969

(*millions of dollars*)

	1965	1966	1967	1968	1969	Total
1. Lease and well investment	15.6	17.3	19.7	33.6	42.4	128.6
2. Industrial, camp, and drilling facilities and equipment	12.8	11.0	18.0	14.3	11.4	67.5
3. Construction in progress and miscellaneous	10.1	...	...	...	...	10.1
4. Subtotal	38.5	28.3	37.7	47.9	53.8	206.2
5. Development rig-years	3.1	3.1	4.2	5.3	5.3	17.4
Total rig-years	4.7	3.9	4.6	5.6	5.8	24.6
6. Field development expenditures (line 4 × line 5)	25.4	22.5	34.4	45.3	49.1	176.7
7. Natural gas liquids production facilities	0.4	0.0	0.3	2.9	[a]	3.6
8. Pipeline and terminal storage facilities	1.0	2.4	3.2	4.1	14.3	25.0
Total development expenditures	26.8	24.9	37.9	52.3	63.4	205.3
Production in TBD	1,350	1,287	1,431	1,555	1,574	1,580[b]

Development capital cost per barrel, assuming 16% decline rate, 20% discount:

Capacity: new plus lost and replaced: 246 + (7,197 × 0.16) = 1,396.

Investment per daily barrel: \$205.3 million/1,390 TBD = \$147.5.

PBE factor (assuming 15 years) 3.22, PBEs 1,185.

Cost per daily barrel $= \frac{\$147.5}{1{,}185} =$ 12.5 cents.

If decline rate 8%, cost would be 17.1 cents per daily barrel.

Source: Creole Petroleum Corp., Annual Reports.

[a] Not shown separately; assumed zero.

[b] First half 1970.

However rough the calculations, one conclusion is plain: national averages understate the real competitive strength of Venezuelan crude oil. For, as seen in Appendix Table II-D-3, about three-fourths of the nation's output is in fields equally or more prolific than Bachaquero and Lagunillas. Precise comparisons are impossible without access to detailed source material, but the bulk of Venezuelan output would appear to be cheaper to produce and deliver to the United States than is oil from the Big Four of the Persian Gulf, and some considerable fraction can compete even at Rotterdam. If there were unrestricted competition and comparative cost governed, only Venezuelan crude would now be consumed on the U.S. East Coast. But comparative cost does not seem to be at work inside Venezuela either, since high-cost crude continues to be produced when lower-cost crude is available. Companies with the higher-cost crude prefer using their own to buying it from others, because even less economic crude yet costs less to produce than to buy from others.

Persian Gulf incremental costs in the long run. The crucial problem for incremental analysis is: at what cost could the large deposits of the Persian Gulf expand output if they were called upon to do so by the growth of the world energy market and also their displacement of higher-cost sources of fuel?

I have assumed[64] that Persian Gulf output, 13.65 MBD in 1970, grows at 11 percent per year in 1971-85 inclusive, faster than previously. Cumulative output over the 15-year period is 170 billion barrels.

[64] For details of the forecast by T. W. Nelson of Mobil Oil, and my use of it, see M. A. Adelman, *Oil Production Costs in Four Areas*, reprinted from *Proceedings of the Council of Economics*, presented at Annual Meeting of the American Institute of Mining, Metallurgical, and Petroleum Engineers, Feb. 28-Mar. 2, 1966.

Reserves in known fields. We must now ask from what approximate reserve total this cumulative production must be subtracted. I assume zero new discoveries anywhere, in order to take one thing at a time, and deal separately with operating, developing, and finding costs. At this point, our concern is with development costs in known fields and the expansion of reserves in those fields.

The National Petroleum Council[65] showed that the estimates made in 1945 of crude oil reserves in fields discovered during 1920–44 had to be increased by 60 percent by 1963. If we applied the same factors to the Persian Gulf fields, we would reckon with 345 billion barrels at the end of 1969, plus 44 billion already produced, or 389, multiplied by 1.6, then subtracting 44 again: fields operated in 1970 developing by 1985 cumulative reserves of 578 billion barrels. To see whether this is too hopeful or too conservative an estimating procedure, let us consider how reserves are created out of oil-in-place.[66]

The more that drilling emphasizes new field wildcats and discovers new fields, the more potential is accumulated for increasing reserves by drilling development wells and new-pool wildcats. This potential is consumed as reserves are expanded under the pressure of the need to exploit deposits more intensively. The estimates made for the United States in 1945 followed on four years of war, when discovery was at a minimum and production at the maximum efficient rate. All else being equal, we would expect a higher potential expansion of crude reserves in deposits at the end of a long period of great overabundance, *and* active search for new fields, as in the Persian Gulf. Hence to expect an increase of the order of 60 percent in current proved Persian Gulf reserves seems not unreasonable.[67]

As a check, let us recall that Nelson and Burk[68] estimated 1965 proved reserves of the U.S. continental shelves at about 3.5 billion barrels of liquids and 22.8 trillion cubic feet of gas. They expected that an additional 3 billion barrels of liquids and 27 trillion cubic feet of gas "probably await drilling in known reservoirs," aside from larger amounts which "remain to be discovered." Their coefficient of reserve expansion therefore is 86 percent of current proved oil reserves (and 118 percent of proved gas reserves), as compared with our 67 percent. But Nelson and Burk confine themselves to known reservoirs, whereas here we have also included new reservoirs in old fields, and will soon reckon with increased costs on that account. Our projection, then, is more conservative to the extent that Persian Gulf and U.S. Gulf conditions are comparable. The continental shelves in this country have been developed efficiently, in large tracts, like fields abroad. Hence the Nelson-Burk estimates do not assume any effect of improved regulatory practices.

Let us try another check. A decade ago, Wallace E. Pratt[69] estimated Middle East end-of-1954 proved reserves at 230 billion barrels, out of 572 billion barrels oil-in-place.

[65]NPC, Report of the National Petroleum Council Committee . . . , 1961.

[66]The reader may wish to refer to chap. I, pp. 24ff.

[67]At the end of 1965, following this procedure, the writer projected net reserve growth from 215 BB to 400 less 135, or 265 BB in 1980, thereby increasing at the rate of 1.4 percent per year. In fact, they increased by 10 percent annually over the next five years, to 344 BB. The increase in Libya, equated by assumption to zero, was in fact another 9 percent of 1965 Persian Gulf reserves. It would be interesting to keep annual score.

[68]T. W. Nelson and C. A. Burk, "Petroleum Resources of the Continental Margins of the United States," in Society of Petroleum Engineers, *1967 Symposium on Petroleum Economics and Evaluation*, Dallas, pp. 116–33.

[69]W. E. Pratt, "Peaceful Uses of Atomic Energy," *Background Material on the Impact of the Peaceful Uses of Atomic Energy*, Report to the Joint Committee on Atomic Energy, Congress of the United States, 84th Cong., 2d sess., vol. II, commonly cited as the "1956 McKinney Report," pp. 89–97.

He defined proved reserves as "reserves already proved by drilling which is recoverable by methods currently in use and under present economic conditions. Proved reserves represent working stocks or inventory only." He based his estimate on 40 percent recovery of oil-in-place, which he thought would in time be surpassed. Pratt's estimate drew some skepticism at the time, which later experience has silenced. Considering Pratt's opinion that this estimate was conservative, and that the assumed recovery percent of 40 is well below the 44 to 87 percent attainable in water-drive fields,[70] it does not seem unlikely that these reserves would be increased by 60 percent in the 31-year period 1954-85; a much less promising group of fields in the United States were increased in that proportion in only 18 years. This would give an estimate of 368 billion barrels, less 23 billion produced in 1954-65, or 345 billion barrels made available by 1985 in those fields that Pratt was appraising in 1954.

We must now consider the fields added since. As of the end of 1970, the *Oil and Gas Journal*[71] shows 20 Persian Gulf fields discovered through 1950 (allowing two years for the first estimates of their reserves), 23 more through 1960, and 31 added since then. In 1970 these fields discovered since 1950 accounted for at least 5.0 million barrels per day, or 38 percent of Persian Gulf output. (Kuwait does not publish field production and we set to zero the post-1950 fields that account for half the wells.) If they accounted for a corresponding portion of reserves, this would amount to 130 billion barrels at the end of 1970. If they grow by a factor of 1.5 in 15 years, this would add 195 billion barrels to the estimated 345 billion in Pratt's universe, for a total 1985 amount of 540 billion. The estimate of reserves is based on a period of chronic oversupply. If 54/74 of the fields have 37.5 percent of the reserves, the average field discovered after 1952 was only half of the average discovered through 1952. This is by no means impossible, since the dispersion of field sizes is so great.

Eleven years after Pratt, Persian Gulf oil-in-place was estimated by T. A. Hendricks and Daniel C. Ion[72] at 928 billion barrels, an increase of 356 billion. We cannot tell how much of the difference is ascribed respectively to new fields, new pools, and changed estimates. In 1963 M. J. Rathbone of Esso estimated over one trillion barrels in place in the non-Communist world, of which over 500 billion barrels would be recovered by current methods while a much larger amount would be recovered with expected new methods in deposits known today.[73] He did not, it seems, cause any eyebrows to be raised. Of the then over 500 billion, surely 400 billion were in the Persian Gulf.[74]

Rough as they are, these estimates point to reserves in 1970 Persian Gulf deposits that, through development and new-pool drilling, can be turned into proved reserves of about 550 billion barrels, given sufficient time and incentive to develop between 1970 and 1985. I doubt if anyone considers this estimate to be excessive. Subtracting 170 billion would leave us 380 billion. With 1985 Persian Gulf output projected at 65 million barrels

[70] See above, p. 32.

[71] *Oil and Gas Journal*, "World Wide Oil," Dec. 8, 1970.

[72] Daniel C. Ion incorporates the estimates of Hendricks as of the end of 1965 in a paper for the Seventh World Petroleum Congress, "The Significance of World Petroleum Reserves–1967."

[73] M. J. Rathbone, "Energy for the Future," speech to Independent Natural Gas Association of America, Sept. 17, 1963, p. 9.

[74] Cf. Vincent C. Illing, "Factors Which Control the Assessment of World Resources of Oil and Natural Gas" (Sixth World Power Conference, Melbourne, paper 185, I. 1/2), p. 19: "No doubt this process [assessment of reserves] is continuing in many of these Middle East fields at the present time." He is confident of plenty "for at least a half century ahead, in spite of increasing consumption. Further than that we cannot see, for the future will involve fields not yet discovered. . . ."

daily, or nearly 24 billion barrels per year, 1985 reserves would be nearly 16 times production. Sir Maurice Bridgman of British Petroleum and John P. Berkin of Royal Dutch-Shell[75] regard 15 : 1 as adequate, and in the United States it is around 9 : 0 today.

Now the problem becomes one of estimating development and operating costs at 1985 production volume, after assumed 1970–85 depletion, under the technology of the mid-1960s. First, we assume the decline rate as 6 percent, the reciprocal of the reserves : production ratio. This has two effects.

(1) The lower PBE factor (see Table II–2) raises capital costs in the ratio 5.20/4.31, a factor of 1.206.

(2) Let q_t be the amount produced in any given year t. Then on our assumptions $q_{1971} = q_{1970}/(1.01), q_{1986} = q_{1985}/(1.06)$, etc. We will assume continuous change in the decline rate a by a constant factor, such that $f^{15} = 1.06/1.01$, whence $f = 1.0032$. The decline rate speeds up by that amount each year, so that $q_2 = q_1/1.01(1.0032)$, $q_3 = q_2/1.01(1.0032)^2 = q_1/1.01(1.0032)^3$, etc. More generally, $q_t = q_1/(1+a)f^{\Sigma(1+\ldots t)}$. In our case, $q_{15} = q_1/1.01(1.0032)^{120} = q_1/1.47$. That is, by 1985 average well productivity is lower, and unit cost is higher, by a factor of 1.47.

These two corrections, by a combined factor of 1.78 (=1.47 × 1.21), are required under our assumptions. We add two which are not. New pools in old fields require some modest semi-exploratory work. We adjust, however, to include even more than total exploration. First, we assume that what the Chase Manhattan Bank calls "exploration," i.e. geological-geophysical activity, and lease rentals, are both necessary, though the latter is not even a cost, and little of the former applies. During 1966–68, Middle East "exploration" expenditures were $150 million, and "production capital expenditures" were $845 million. Hence the appropriate ratio is (845 + 150)/845 = 1.18.

Production capital expenditures consist of both exploratory and development drilling. Having separated out the exploratory segment either by the well method or the rig-time method, we now add it back again for each of the Big Four. The Iranian adjustment is more complex and is set out in Appendix II–I. But for all four concessions the adjustment is the ratio of total rig-time to development rig-time. This is summarized in the tabulation below.

Capital Cost Adjustment Factor in Respect of:

Concession	*Higher decline rate (lower PBE factor)*	*Lower 1985 well productivity*	*CMB "exploration"*	*Exploratory drilling*	*Total*	*1985 capital cost (¢/barrel)*
Iran Consortium	1.21	1.47	1.18	[a]	[a]	10.6
Iraq	1.21	1.47	1.18	1.00	2.10	5.2
Kuwait	1.21	1.47	1.18	1.11	2.33	14.0
Saudi Arabia	1.21	1.47	1.18	1.31	2.75	11.3

[a]See Appendix II–I.

The adjustments for operating cost increases are simpler. By the square root formula, operating costs must be multiplied by a factor of 1.213 (= $\sqrt{1.47}$). Furthermore, the

[75]Bridgman, "Worldwide Production–Its Prospects and Problems," speech to American Petroleum Institute, Nov. 1963, mimeo.; and Berkin, *The Advancement of Science*, vol. 22, Sept. 1965, p. 12.

levelizing present-worth factor $(a+r)/r = 26/20 = 1.30$. Hence all operating cost estimates must be multiplied by 1.577 (=1.213 X 1.30). Again a separate adjustment must be made for Iran because of the large pipeline factor. For Iraq, Kuwait, and Saudi Arabia, the effect is to raise operating costs from 4.5 to 7.1 cents per barrel. For Iran, the 1985 estimate is 6.1 cents (see Appendix II-I).

As a crude approximation to development-operating incremental cost, we may take cost in the most expensive country, Kuwait, to be 21.1 cents as of 1985. None will be deceived by the seeming precision of these numbers, nor of the adjustment processes. The results are a gross though not wild exaggeration; all adjustments were logical, but some were far overstated, and the dice were loaded in one direction at every throw.

As a prediction, 20 cents cost is very high. First, our assumed output increase would generally be considered improbable,[76] and costs are an increasing function of output. Second, new deposits will be discovered in the Persian Gulf and elsewhere which will add greatly to the reserves available, and decrease pressure on known reservoirs. Third, greater efficiency in developing and operating was set to zero. Also set to zero was the increasing competition from African crudes, natural gas, and nuclear power, which will occupy more of the energy market than comparative costs would justify. The estimate of 20 cents is in theory an estimate of what it would have cost in 1966-70 to produce from the big Persian Gulf fields once they were depleted to the point where the reserve : production ratio was down to about 16. Given that assumption, the method used here exaggerates the consequences. Hence the result forms a convenient maximum for any prediction of what long-run developing-operating cost is likely to be.

Maximum Economic Finding Cost and Total Long-Run Supply Price

We come now to the final part of our task: How much must be allowed for finding cost in order to have a total supply price which, when paid, will be just adequate to bring forth a given level of output?

We cannot estimate finding costs in the same way as operating and development costs. Future costs of developing existing deposits can be estimated because of continuity: costs tomorrow grow out of costs today, changing only as the depletion process continues. But this continuity assumption is precisely the one we cannot make about finding costs. The discovery of a new field is a unique event, and there is no bridge from the reporting of past costs to the estimate of future ones. If in retrospect Persian Gulf reserves have cost practically zero cents per barrel to find, we have no right to expect this in the future. "Past finding costs are often unknowable, usually unknown, and always uninteresting."[77]

Discovery in one place is really a joint product with nondiscovery in another; the cost of some failures is part of the cost of the success. Cost estimates are only relevant as predictions, as guides to decisions. In order to make a rational estimate of the cost of new-field reserves, we must have some knowledge of the total and size distribution of new fields. But so far we have no knowledge of what will be found.[78] Finding-cost estimating

[76] At an increase of 13.4 percent per year, 1970 output would have been 15.4 MBD, but actually was only 13.8 MBD.

[77] M. A. Adelman, "The World Oil Outlook," in Marion Clawson (ed.), *Natural Resources and International Development* (Baltimore: Johns Hopkins University Press for Resources for the Future, 1964), p. 67.

[78] John M. Ryan, "Limitations of Statistical Methods for Producing Petroleum and Natural Gas Reserve and Availability," Society of Petroleum Engineers 1256 (Oct. 1965), preprint.

is a reckoning of gambling odds, and we cannot start it until we have some notion of the chance of finding a pool of a given size. The work of Arps[79] and Kaufman[80] is confined to the United States, and to provinces where very large samples of past exploratory results gave them some basis for prediction of future results. Bradley[81] has applied essentially the same technique to the Persian Gulf on the basis of 1955 fields as given by Knebel and Rodriguez.[82] His application of the method is well worth the reader's attention, but the sample is so small and the variability so great that I refrain from any actual estimates drawn from it. Considering the importance of the world oil trade to companies in it, the industry and consumers they serve, and the many governments, it is discouraging that we have almost nothing on the size distribution of oil and gas fields the world around.

Nevertheless, it is possible to estimate a maximum value, such that the true supply price of new discoveries will fall below it. Let me pose a simple-minded question: Why do oil operators spend money on exploration? If they did not, they would in time have no more reserves to develop, and would be out of business. But one might as well say that people eat because if they did not they would eventually die. That would take many days. But after only a few hours without eating, a person becomes uncomfortably hungry. If the fast continues he becomes distressed. More and more effort is needed for even the simplest task, and with increasing physical weakness, after a while he can do no work. The economic cost or penalty of noneating is prohibitive long before starvation is fatal.

The analogy proves nothing, but may clarify the thesis: to consider exploration merely as a means of staying in the oil business is not wrong, but neither is it useful. Exploration is needed *to prevent an otherwise inevitable rise in developing-operating costs.* The oil operator seeks new reserves to keep his costs down. This is why the incorrect notion of crude oil production as an industry of decreasing costs is harmful, for it blocks recognition of the importance and role of finding costs.

If production rises at the high rate assumed above (pp. 69–70) and development-operating costs rise by 10 cents, then the supply price of discovering equally good new-field oil will also be 10 cents per barrel. This is considerably more than it has cost to discover Persian Gulf oil up to now. Up to that limit, it would be rational conduct to keep looking for new deposits even if they were expected to be no better, i.e. no cheaper to develop and operate, than present deposits. For it would stave off (a) the falling reserve : production ratio, which increases costs, and save (b) the expense of proving up new reserves in old fields. These are analogous to the feedback effect in my earlier example: where building the second floor subtracted from the useful space on the first floor.

[79] J. J. Arps, "The Profitability of Exploratory Ventures," in *Economics of Petroleum Exploration, Development, and Property Evaluation* (Prentice-Hall for Southwestern Legal Foundation, 1961).

[80] Gordon Kaufman, *Statistical Decision and Related Techniques in Oil and Gas Exploration* (Prentice-Hall, 1963), and "Statistical Analysis of the Size Distribution of Oil and Gas Fields," in Society of Petroleum Engineers, *1965 Symposium on Petroleum Economics and Evaluation* (Dallas, n.d.), pp. 109–24.

[81] Paul G. Bradley, *The Economics of Crude Petroleum Production* (Amsterdam: North Holland, 1967); and (with R. S. Uhler), "A Stochastic Model for Determining the Economic Prospects of Petroleum Exploration over Large Regions," *Journal of the American Statistical Association*, vol. 65 (June 1970), pp. 623–30.

[82] George M. Knebel and Guillermo Rodriguez-Eraso, "The Habitat of Some Oil," *AAPG Bulletin*, vol. 40 (Apr. 1956).

Therefore even companies with abundant Persian Gulf reserves do well to hunt for more reserves, there or elsewhere, if they think there are good enough odds of finding oil at a cost no greater than the higher developing-operating costs which they would otherwise incur. Even that most crude-rich of operators, British Petroleum, has saved money by exploring in Abu Dhabi, Libya, and Nigeria, and relieving pressure on Kuwait.

A company may go oil hunting for the sake of diversification or because it is nearly costless, after tax. Indeed, diversification may be considered as a means of reducing political risk and lowering the minimum acceptable rate of return. Contrariwise, the high risk allowance may be considered a proxy for diversification, which does not come free. In any case, given increasing costs, we need assume no more than ordinary good business sense to explain why oil-rich companies hunt for new oil.

Suppose an oil operator contemplates an exploration program on some wildcat acreage, to cost him one million dollars the first year, two million the next, and so on up to five million dollars in the fifth and terminal year, after which there will either be a deposit to develop, or the end of a fruitless search. As a proxy for the odds against finding anything, we assign a very high rate of return, which would diminish rapidly as accumulating knowledge points to rapidly improving odds, as shown below.

	Exploration Period					*Development Period*	
Year	1	2	3	4	5	6	7
Rate of return (percent)	70	60	50	40	30	20	20
Expenditures (in millions of dollars, development equivalent)	$1(1.7)^5$	$2(1.6)^4$	$3(1.5)^3$	$4(1.4)^2$	$5(1.3)^1$		
Total	$51.8 million						

At the end of the fifth year, the current value of the expenditures is $51.8 million equating to that sum spent at the start of development. The relevant discount rate would henceforth be the same 20 percent as for other development projects. If the deposit yielded, say, an initial 300,000 barrels daily, declining at 5 percent over 25 years, then by our formula exploration cost would be $(52.1 \times 10^6)/(365 \times 4.46 \times 300 \times 10^3) = 10.7$ cents per barrel. If the alternative to this discovery were to use a source where incremental operating-developing cost would increase by over 10.7 cents, the investment was at least barely worth making.

Although one cannot predict finding costs, it can be stated that Persian Gulf development-operating cost today fixes the supply price of oil, *including necessary finding cost*, or Maximum Economic Finding Cost (MEFC) for the whole world.

Retrospect: A Unified Cost Theory and Cost Measure

This chapter has been organized along the conventional three-way division into operating, development, and finding cost. In trying to put numbers into the abstract scheme, we have been forced to reexamine each. The result has been to combine three types of cost into one: development cost has swallowed operating cost on the one side and finding cost on the other.

Operating costs have no significance for comparison with prices except under a special condition never yet fulfilled: when output is expected to decrease at least as much as the zero-investment decline rate. The stereotype of a field "all drilled up" and henceforth needing no more than operating cost has been a harmful illusion. (See above, pp. 64-65.) If we cannot assume zero-growth investment, operating costs have no independent significance for price, and only make sense as part of a decision on development investment. To the investment made today must be added the present value of all the future expenditures to which we commit ourselves today. Price must at least cover this combined amount.

As for finding costs, the effort that it pays to put into the finding effort is nothing but a reflection of the expected rise in development-operating costs in known fields. Finding costs are only a special case of acquiring new knowledge about possible development. Producible reserves are created by a mixture of new reservoirs in known fields, and more knowledge of new fields. These three types of research and investment are broadly substitutable for each other. Every management must choose a best mix of the three, and there will be many disagreements that only time can resolve.

Summary and Conclusions

Table II-8 sums up a long, complex reckoning, and should therefore be read cautiously with our many qualifications in mind. But I think better statistics better used would confirm the conclusions set down on page 77.

TABLE II-8. Development-Operating Costs and Long-Run Supply Price

(dollars)

Area and years	Development investment per daily barrel	Cost per barrel: Development	Operating (including pipelines)	Total	Operating plus development (mid-1980s)	Freight advantage over Persian Gulf (1980)	Long-run supply price (1970-85)
	(1)	(2)	(3)	(4)	(5)	(6)	(7)
United States							
1960-63	2,280	1.048	.168[a]	1.22	...	...	...
Venezuela							
1966-68	417	.351	.101	.462	...	.42	.64
Africa							
Libya 1966-68	129	.074	.085	.159	...	.34	.54
Algeria 1966-68	293	.180	.100	.280	...	.37	.57
Nigeria 1965-66	165	.094	.070[a]	.164	...	.26	.46
Persian Gulf							
Iran Consortium 1963-69...	90	.047	.050	.097	.14	...	.20
Iraq 1966-68..	47	.025	.045[a]	.070	.12	...	.20
Kuwait 1966-68	114	.060	.045	.105	.20	...	.20
Saudi Arabia 1966-68...	78	.041	.045	.086	.18	...	.20

Sources and Notes (by column):

(2)—United States, table II-3; Venezuela, appendix table II-D-2; Nigeria, appendix table II-F-10; Iran, appendix II-I; all others, appendix II-F tables. Algeria includes exploration expenditures.

(3)—Col. (1) divided by PBE factors at the following discount-plus-decline rates: United States, 17%; Venezuela, 36%; Libya, Nigeria, 23%; Algeria, 25%; Iran, see appendix II-I; other Persian Gulf, 21%.

(4)—United States, appendix table II-A-3; Venezuela, 6.5 cents well plus 2.1 cents gathering plus 1.5 cents terminalling; Algeria and Nigeria, square root formula; all others, table II-1 and explanatory paragraph.

Table II-8. (Continued)

(5)–Col. (2) plus (3).
(6)–See text.
(7)–Delivery to U.S. East Coast assumed for Venezuela; to Rotterdam for all other suppliers; from Persian Gulf, around Cape. World Scale 40 assumed.
(8)–Col. (5) plus (6).

[a]Excludes pipelines.

For at least 15 years we can count on, and must learn to live with, an abundance of oil that can be brought forth from fields now operated in the Persian Gulf at something between 10 and 20 cents per barrel at 1968 prices and in some other provinces at costs even lower when account is taken of transport.

Price is many times cost, and no matter who gets what percent of the profit, it will pay (as it does now) not to buy from the abundant available supply, but instead to spend large amounts on exploration and development in the reasonable hope of recouping the investment and making a profit after even a few years of operation at current or even much lower prices. If oil exploration continues at a brisk pace, new fields will come in and available reserves will exceed the minimum figure used here. The supply price will rise less than is calculated here; it may well decrease in the future as it has in the past.

CHAPTER III

Market Structure: Concentration and Integration

In this chapter we examine the anatomy of the world oil market: the number and size distribution of suppliers and the degree of vertical integration. At the outset it should be noted that the statistical tables do contain some minor inconsistencies. In part, these reflect the condition of the data and the formidable task of supplying the few pieces that are still missing from the global picture—a task disproportionate, I think, to its contribution to the overall picture.

But this difficulty is symptomatic of one more subtle and serious. The ultimate object, now as always, is to explain and predict price behavior. The fewer the sellers in any given market, the better the chance of cooperative action. We therefore need to count in all firms and all capacity which bears on total supply and price, and to exclude all output which does not. The reasons for choosing some or all countries and companies, and excluding others, are not self-evident. Indeed, the market boundaries set forth in this chapter should be considered as provisional. The reader may take them largely on faith for the time being, but return later to see what change he would prefer to make.

The whole problem of market boundaries, which are at best uncertain and debatable, is one to which no altogether satisfactory solution can be found. All we can hope to do is to arrive at an expedient relevant to the problem at hand.

CONCENTRATION

Market Boundaries. Tables III-1 and III-2 exclude the Communist blocs, and also North America. The Mexican industry has been a government monopoly since 1938, and there is very little importing or exporting. At no time has the United States been open to imports from the rest of the non-Communist world.[1] It would exclude too much to disregard the American (and Canadian) market completely, as though no crude or product went there at all; there has, in fact, been a substantial regular flow since shortly after World War II. But the total amount has always been under a severe constraint. Hence, to include U.S. production in the total market would be a far worse distortion (and would lower the degree of concentration considerably) than to exclude it as we do from the two tables, and so understate the number of suppliers. Inclusion would imply that American output is in competition, say, with Iranian or Arabian output in the same way that they are with Venezuelan or Iraqi output.

[1] This is a controversial statement that can certainly not bear its own proof, and will be explored in chap. V. There has been no restriction on U.S. and Canadian exports, but they would only be available at prohibitive cost.

For the same reason we exclude from Table III-2 about one-tenth of non-Communist production outside North America. If it were included, the difference would be imperceptible. The areas excluded are: South America and the Caribbean outside of Venezuela; Western Europe; and Africa outside the three countries shown in the table. In the South American-Caribbean area, there is the considerable total of 870 thousand barrels daily (TBD). About one-fourth of this amount is accounted for by five of the large companies. But nearly all of it, especially in Argentina, Brazil, and Chile, is for local consumption, and mostly by government monopolies. There is very little export and no competition between locally produced and imported oil; the division of sales between foreign and domestic is in no way determined by relative prices and incremental costs. As in the United States, it is national policy to provide a large amount of the oil from domestic sources. One cannot therefore speak of competition between Argentine oil and imported oil. The same is true of Western Europe where, if it were not for government policy, little if any oil would be produced in 1970. The situation is quite different as to North Sea oil and natural gas, which will certainly be competitive when its production becomes greater.

As for the excluded African nations, their oil also was used mostly for local consumption, although the combined output of Egypt, Tunisia, and Gabon had reached significant proportions by 1969. The Middle East outside the Persian Gulf—Turkey, Syria, and Israel—also produced mainly for local consumption.

Hence it can be argued that the exclusion of one-tenth of the non-Communist production does not appreciably distort the market shares indicated in Tables III-1 and III-2. Perhaps most subject to criticism is the inclusion of Algeria, two-thirds of whose output goes to the protected market in France. For firms operating in Algeria, sales outside France were a poor second choice given world prices in the 1960s. Hence it might be more accurate to exclude about two-thirds of Algerian output, and this would tend somewhat to increase the degree of concentration. But to balance this, there is the inclusion of Nigeria, whose 1966 production—the great bulk of it by BP-Shell—had no close relation with the fields recently discovered, and whose development was interrupted by the Nigerian civil war of 1967-69.

Again, while it would be indefensible to exclude either African production on the one side or Indonesian production on the other, competition between these two areas is limited to a significant extent. African production does not compete in South Asia and East Asia, while Indonesian production does not go to Europe nor indeed does much of it "haul back" to India and its neighbors. The Persian Gulf is the great bridge which unites the east and west extremities to make them one market. Shortage or surplus in North Africa is transmitted quickly to the Persian Gulf because of direct competition, and thence to Indonesia. But this intervening point of supply not only transmits the impulse but may dampen it. To the extent that output can be increased at the Persian Gulf in response to a shortfall in African production at a lower production-plus-transport cost, Indonesia is unaffected. The impulse may not be damped if Persian Gulf output is for any reason already flat out. This consideration has been of no great moment, but the principle is important whenever considering a market composed of rival areas separated by transport cost.

Tables III-1 and III-2 compare 1950 with 1957, 1966, and the first half of 1969 because those years are most representative of a slowly changing market structure. By 1950, the postwar pattern had become established. By 1957 the first Suez crisis had been liquidated and the interruption of output in Iran had ceased; data for the intervening years 1951-56 would have been somewhat distorted. Thereafter the picture is one of a

TABLE III-1. Production of Principal Crude Oil Companies and Producing Countries in the World Oil Market, Selected Years, 1950-1969

(*thousand barrels daily*)

Country	Esso	Mobil	SoCal	Texaco	BP	Gulf	Shell	CFP	Others	Total
					1950					
Iran	...	...	...	...	665	...	...	...	...	665
Iraq	16	16	...	...	32	...	32	32	8	136
Qatar	4	4	...	...	8	...	8	8	2	34
Kuwait	...	...	...	...	172	172	...	...	...	345
Saudi Arabia	174	58	174	174	...	...	...	...	...	579
Subtotal	194	78	174	174	877	172	40	40	10	1,759
Indonesia etc.	3	3	16	16	...	...	41	...	...	79
Venezuela	818	48	15	...	...	231	378	...	10	1,500
Total	1,015	129	205	190	877	403	459	40	20	3,338
					1957					
Iran	50	50	50	50	289	50	101	43	36	720
Iraq	54	54	...	...	108	...	108	108	22	452
Qatar	16	16	...	...	33	...	33	33	7	138
Kuwait	...	...	...	...	570	570	...	...	...	1,140
Saudi Arabia	297	99	297	297	...	...	...	...	...	990
Others	...	...	...	...	...	...	...	...	230	230
Subtotal	417	219	347	347	1,000	620	241	184	295	3,670
Indonesia etc.	16	16	87	87	...	...	223	...	...	429
Venezuela	1,150	115	92	45	...	405	751	...	285	2,843
Total	1,583	350	526	479	1,000	1,025	1,215	184	580	6,942
					1966					
Iran	138	138	138	138	795	138	278	119	230	2,110
Iraq	165	165	...	...	330	...	330	330	70	1,390
Qatar	22	22	...	...	44	...	134	44	14	280
Abu Dhabi	30	30	...	...	127	...	60	93	15	355
Kuwait	...	...	...	...	1,138	1,137	...	...	...	2,275
Saudi Arabia	719	240	719	719	...	...	...	...	...	2,395
Other Persian Gulf	...	...	...	...	...	...	...	...	445	445
Subtotal Persian Gulf	1,074	595	857	857	2,434	1,275	802	586	774	9,250
Libya	489	86	41	41	5	...			838	1,500
Algeria	...	...	...	...	...	...	85	132	498	715
Nigeria	...	...	...	...	173	51	173	...	23	420
Indonesia, etc.	28	28	152	152	...	...	96	...	94	550
Venezuela	1,240	145	60	104	...	384	937	...	500	3,370
Total	2,831	854	1,110	1,154	2,612	1,710	2,093	718	2,727	15,805
Total non-Communist production outside North America[a]										17,550
					1969 first half					
Iran	208	208	208	208	1,192	208	417	178	394	3,223
Iraq	181	181	...	...	361	...	361	361	76	1,520
Qatar	23	23	...	...	45	...	203	45	9	348
Abu Dhabi	49	49	...	...	236	...	99	167	21	620
Kuwait	...	...	...	...	1,254	1,254	...	...	...	2,508
Saudi Arabia	870	289	870	870	...	...	...	...	...	2,898
Omar	...	...	...	...	...	...	256	...	45	301
Other Persian Gulf	...	...	...	...	...	...	...	...	453	453
Subtotal Persian Gulf	1,331	750	1,078	1,073	3,087	1,462	1,335	751	999	11,871

TABLE III-1. (Continued)

(*thousand barrels daily*)

Country	Esso	Mobil	SoCal	Texaco	BP	Gulf	Shell	CFP	Others	Total
Libya	683	126	181	181	155	. . .	119	. . .	1,561	3,006
Algeria	. . .	. . .	. . .	. . .	. . .	. . .	85	249	602	936
Nigeria	. . .	. . .	. . .	. . .	168	200	168	. . .	. . .	536
Other Africa . .	. . .	. . .	. . .	. . .	. . .	36	63	. . .	406	505
Indonesia, etc. .	25	25	285	285	. . .	. . .	138	. . .	90	848
Venezuela	1,472	112	56	171	. . .	383	916	. . .	424	3,534
Total	3,511	1,013	1,600	1,692	3,433	2,081	2,824	1,000	4,082	21,236
Total non-Communist production outside North America[a]										24,400

Sources: Company annual reports, *AAPG Bulletin, World Oil, Oil and Gas Journal*, and *BP Statistical Review*. There are numerous small discrepancies among all these sources, and it is impossible to reconcile them fully; hence the failure of detail to add to totals is not wholly due to rounding.

Notes: Table excludes data on Communist countries and North America (see text).

Iran includes output outside of concession area; Saudi Arabia includes Bahrein.

Mobil estimate for Libya is half of concession with Gelsenberg (GBAG); Shell credited with one-sixth of Oasis.

Indonesia includes Brunei, Sarawak.

[a]Includes small self-sufficient countries.

more or less steady progression interrupted in 1967 by the June war. The years 1966 and 1969 bracket that disturbed year.

In the tables, the eight large companies are arranged as much as possible according to their important joint ventures. Standard Oil Company (New Jersey) and affiliates (Esso) and Mobil Oil Corporation are placed side by side because of their participation in Iraq, Saudi Arabia, and the Iranian Oil Consortium, as well as in Indonesia. Similarly, Standard Oil Company of California (SoCal) accompanies Texaco, Inc., because they operate jointly in some parts of the world; Caltex is being dissolved as to marketing activities but not production. British Petroleum (BP) and Gulf Oil Corporation are joint operators of Kuwait. For each country, we can see how its output is divided by the amounts to which

TABLE III-2. Concentration Ratios and Number of Firms-Equivalent, Selected Years

		1950	1957	1966	First half 1969
		A. Production Shares			
Largest four:	Esso	30.4	22.8	18.0	16.6
	BP	26.3	14.4	17.0	16.1
	Shell	13.8	17.5	12.9	13.3
	Gulf	12.1	14.8	10.8	9.8
	Subtotal	82.6	69.5	58.7	55.8
Lesser four:	Texaco, SoCal, Mobil, CFP	17.4	22.2	24.7	25.0
All other:		0.0	8.3	16.5	19.2
	Total	100.0	100.0	100.0	100.0
		B.			
Herfindahl measure of concentration		.2039	.1319	.1116	.1046
Number of firms-equivalent		4.9	7.6	9.0	9.6

Source: Table III-1. The Herfindahl measure is the sum of the squared market percentages of the firms in the market. "Firms-equivalent" is the reciprocal of H-measure; see M. A. Adelman, "On the H-measure as a Firm-Equivalent," *Review of Economics and Statistics*, Feb. 1969.

Note: Table excludes data on Communist countries and North America (see text).

each company is entitled if it participates in a joint venture, or—as a second choice—simply what it has produced during that year. The four great Persian Gulf producers, Iran, Iraq, Kuwait, and Saudi Arabia, are divided as shown in Table III-1. Iraq Petroleum Company (IPC) also operates in the nearby sheikdom of Qatar and onshore Abu Dhabi. Offshore Abu Dhabi is conceded to BP and Compagnie Française des Pétroles (CFP) in a two : one proportion. Outside of the Persian Gulf countries there are some other joint ventures, but in no case does a single joint venture operate for a whole country. Even in the Persian Gulf this pattern is slowly disappearing, but very slowly indeed, as is shown by the column headed "Others." By looking down any column, one can see the importance of the holdings of any of the large eight companies in any particular producing area.

The figures here are not strictly comparable with those anywhere else because the production rates in the joint ventures represent not what each company is actually taking out of the field, but rather what it is entitled to take out. Some partners in a joint venture may overlift and others may underlift, i.e., take more or less than their proportionate share. We will need to analyze these arrangements in detail, even though the amounts of overlifting and underlifting are generally unknown. The most important difference is in Kuwait, where Gulf Oil takes not 50 but about 55 percent on the average. But entitlement is a better measure of command over resources than is lifting. A more serious theoretical criticism might be that what looks like long-term purchase contracts with Gulf are really equity shares, but the evidence will not support that proposition.

In Table III-2 the data of Table III-1 have been summarized to show the shares of the four largest companies, of the second largest four, and of all others. The measure of concentration shown at the bottom of the table means that in 1950 the eight large companies were the equivalent of not quite five equal-sized companies. One would expect them to act more like a market of five sellers than of eight. By 1957, companies other than the big eight had appeared, and the large eight were somewhat more evenly matched; hence there would now be the equivalent of 7.6 equal-sized sellers. By 1969, the tendency had continued and the number of firms-equivalent was now 9.6.

The degree of concentration has diminished very much over 16 years, but it is still high. Before trying to draw any implications, however, we must look first to the association among these largest companies in their joint ventures, chiefly at the Persian Gulf. As Table III-1 shows, their principal concessions in the Persian Gulf accounted in 1966 and 1950 for the same 53 percent of world market output; in June 1969 it was down to 51 percent. And considering the involvement of the eight large companies in the other areas, it is obvious that market control, ability to coordinate investment and production plans, and therefore prices, cannot have decreased as much as the simple concentration figures would suggest.

The Persian Gulf Joint Ventures: The Problem Stated. Most of the sparse information on the Persian Gulf joint ventures that has come into the public domain is to be found in the Federal Trade Commission report.[2] That document contains much valuable information and therefore is useful, despite an overemphasis on the possibility of conspiracy to bring about illegal price fixing. If, to take an example, it were significant that SoCal and Texaco each had a director on the board of a New York bank,[3] this would surely imply that we

[2]FTC, *The International Petroleum Cartel*, Staff Report to the Federal Trade Commission submitted to the Subcommittee on Monopoly of the Select Committee on Small Business, U.S. Senate, 82nd Cong., 2d sess. (Committee Print No. 6, 1952).

[3]Ibid., p. 32, chart 19.

were fairly safe from conspiracy until and unless they got together and found nobody else around.

The real problem is more complex. In order to see what effect the joint ventures have had on output and prices, I propose to set up, as it were, a zero milestone by suggesting a scheme whereby a joint venture would have no effects on output, and companies would behave as if they were entirely independent and drew from the common store of a joint venture with no regard for anything but their own individual profit. We will then see how far the actual practice of the joint ventures diverges from this "null hypothesis." Thereby we will have some way of appraising their importance, even if only in general and qualitative terms.

Since the heart of the problem is the control of output, we need to ask what a joint venture does about bringing the production plans of its members into line. Can each partner order out as much as he likes without being penalized for it, and regardless of the others' opinion that he may be spoiling the market?

Transfer price: the "zero effect" model. A little reflection shows that control or noncontrol of output cannot be separated from the transfer price from the joint venture to each partner. Suppose that each partner could nominate in due course unlimited amounts of oil at anything less than the market price, not to mention bare cost. This would be a bonus or subsidy to order out huge amounts. For on every barrel of oil a given offtaker would realize without additional investment the difference between a transfer price and the market price. Transfer at less than market price (or incremental net revenue) in a joint venture requires that there must be some kind of arbitrary output decision by the group rather than by individuals. Output must be limited by joint action, overt or tacit. The various partners are not free to produce and sell as much as they please.

But transfer at market price is no "zero effect solution" either. In practice, the market price has been and still is a difficult number to find. The estimates to be presented later will only hold within margins for error which may be acceptable for a market analysis but would engender endless haggling and litigation among private parties. But even if market price were known, it would not represent the amount of additional net revenue which each company could make on additional sales, for they might tend to beat prices down if they competed. Finally, transfer at market price would allow each offtaker on every barrel of oil produced, only his pro rata share of a profit from his production. Let us take IPC as an example. In ordering out crude, BP, which has 24 percent of the concession, would be working for the other offtakers 76 percent of the time; Esso or Mobil, 88 percent. Obviously this would not offer as great an incentive to order out as each offtaker would have if he were operating his own concession and received 100 percent of the profits. Suppose that Esso were able to take away a contract from BP. If BP accordingly cut back its IPC nominations to that extent, total IPC output would be no greater and Esso's producing profit would be no greater. Hence there would be no incentive for Esso to compete with BP in selling crude or products.

Let us try the other extreme: transfer at bare cost of production (which need not be strictly defined). In that case there would be no profits to the joint venture nor to the partners as stockholders. The joint venture would become a mere conduit. Each offtaker in reselling or refining would realize a profit proportional to his takings as a buyer, not to his investment as a stockholder. The joint venture would be a cooperative, each member receiving a coop dividend on every barrel. But then partners taking more than their purchase percentage would be making money on the investment contributed by others, who of course would not stand for it.

But transfer at bare production cost can be supplemented to arrive at the same result as the purely competitive one. Each offtaker could be allowed to have as much oil as he wished, provided that he paid the operating costs of the additional output, and *also* put up any additional capital required, in proportion to its nominations. This would not be a loan repayable in money, nor would it increase his equity. Instead, over the estimated life of the facilities there would be additional offtake for the company which wanted it, with no effects on the takings or the profits of the other partners.

Physically and financially, the new facilities would be comingled with all others. If it turned out that the over lifter had been overoptimistic and really could not use any oil in excess of his equity share, he would have saved money for his partners by providing some of their share of facilities which all used. Conversely, the increased output might be even more profitable than expected. These are customary business risks.

This would leave only the property right in the unproduced oil itself or, strictly speaking, the valuable ownership share of the right to produce the oil. If one company took more than its ownership percentage, it would be depriving its partners of the present value of the oil which the partners could otherwise produce many years hence. In theory, there could be royalties as compensation. But, if the reader will recall, the reserve-to-production ratios in Chapter II (pp. 70ff.) showed as neglible the present value of a barrel that will not be available in 20 or 50 years because one has been removed today.

Hence a transfer price at the cost of operating new facilities plus the investment in those facilities would give a fully competitive result. Each seller would have an incentive to keep expanding output so long as the incremental cost of the oil was less than what he might realize from selling it. He would have an incentive to compete aggressively even to the extent of taking business away from other partners or offtakers, for the additional profit on additional sales of oil would go to the successful aggressor rather than be shared among all the partners. Such an arrangement would not, of course, guarantee that the partners would go all out in competing (even supposing it desirable that they should). But a joint venture need not in theory put any obstacle in the way of competitive behavior, and the situation can be the same as if each of the partners owned his own concession. But if the joint venture is more efficient than the separate partners could be on their own—and this can hardly be doubted—then more capacity would result from a given total expenditure than if there were separate concessions.

So much for the theory. Let us see wherein the concessions depart from it, and hence change the result.

Joint venture provisions. If we look at the Middle East joint ventures from the point of view just explained, they appear to form a pattern. The joint venture in the *Kuwait Oil Company* (KOC) with only two equal partners, Gulf and BP, fits our "zero-impact solution" quite well. In the other joint ventures the actual pattern of liftings is a well-kept secret; the KOC partners have been publishing it for many years. The transfer price is cost of production including interest on the capital employed.[4] Either partner could order out unlimited amounts of oil, requiring new wells, gathering systems, etc. so long as he put up the money. There have been no restrictions on output.

At an early stage, there were substantial marketing restrictions. Apparently Gulf bound itself not to upset its partner BP's trade position directly or indirectly, and by its

[4]FTC, *The International Petroleum Cartel*, p. 133, n. 23.

own statement Gulf considered these provisions "restrictive and objectionable." Apparently they were eliminated in 1951.[5]

Next to KOC, the *Arabian American Oil Company* (Aramco)[6] has the smallest number of partners and the least restricted output. According to the articles of incorporation, Aramco is to be run for its own benefit as a separate entity, i.e., for maximum profit. This would imply transfer at market price with owners receiving their return as dividends rather than especially low prices. But this principle was violated in the early years by the specially low transfer price of $1.43 to the offtakers. (See Appendix III–A.) If this provision was later eliminated, and Aramco now sells or transfers to its constituent offtakers at market price, it would be interesting to learn how that market price is determined. My conjecture would be that the problem is insoluble, and that the "separate entity" principle has therefore not been followed. For if it were, each of the offtakers would be free to buy as much from Aramco as it wished, since that would increase Aramco profit. The Aramco board would have no problem in reconciling the intentions of their constituent companies—they need only add up the orders.

It is quite clear even from what little is in the public domain that this has not been the case. Nominations in excess of prorated equity share (overlifting) and less than the share (underlifting) are only permitted within limits. Presumably there are even limits on crude ordered out by the parties over and above the sum of the proportionate nominations plus the deficit of the underlifters. According to Hartshorn,[7] there is no penalty for underlifting but a partner can persistently overlift subject to a penalty payment. "The rules of this arrangement have been modified over the years; details of the present 'incentive dividend plan' are a nightmare to anybody except a devoted accountant. Broadly, however, each partner gets part of its return from Aramco as a proportionate share of half of what the Aramco profit would be if all oil were sold at posted price." Beyond the "part" we know nothing. Moreover, this arrangement has, of course, made less and less sense over the years as actual prices deteriorated not only in the crude market but more importantly in the final products market. Hence changes have been forced on the Aramco arrangements, but as usual they are not made public. However, it is known that in late 1965 the penalty on overlifting was reduced, and Esso liftings under so-called special arrangements increased sharply for 1966. It is clear that Aramco has moved in the direction of greater freedom for its partners. Moreover, apparently there are no limitations on disposal of the oil.

Five partners are engaged in the operations of the *Iraq Petroleum Company* (IPC), or six if one counts the Gulbenkian interest.[8] The governing provisions, insofar as known, seem considerably more restrictive than Aramco. Each partner must file his requirements five years in advance, and these are firm with no changes permitted. He must also file for

[5] Ibid., p. 134, n. 25.

[6] Sole operator in Saudi Arabia (excluding Neutral Zone) at present, though it has relinquished very large areas to the government, which has in turn let new concessions, so far without production. Aramco includes Texaco, SoCal, and Esso with 30 percent each, and Mobil, with 10 percent.

[7] J. E. Hartshorn, *Oil Companies and Governments* (London: Faber and Faber, rev. 1967, American title, *Politics and World Oil Economics*), p. 181.

[8] BP, Shell, and CFP each have 23.75 percent; Esso and Mobil, each 11.875; the Gulbenkian estate, 5 percent. The company was stripped of all acreage outside of fields being operated in 1962. Some of these areas have been awarded as concessions. The North Rumaila field, containing an estimated 10 billion barrels, will be developed by Iraq National Oil Company (INOC), with Soviet technical aid. See chap. VII.

the following five years, but these are tentative and he is not committed to them. This, of course, is a help to collective action in that each party is now aware of the plans of all other parties and must take them into account. Each party knows that if he orders heavily the others may retaliate by doing likewise, increasing supply and lowering price.

The transfer price to each party is cost of production, undefined as usual, plus British income tax, plus one shilling per ton or something under two cents per barrel. Obviously this price has always been far below current market value. It is therefore a bonus or subsidy to order out huge amounts. Unless there were some kind of limit any of the partners could profit hugely by taking sales away from its partners. As a partial damper to the incentive, one partner is permitted to sell another part of its entitlement at a "half-way price," between the transfer price and the world market price–however that is defined. Here, as always, the person seeking to learn what actually happens is bedeviled by a multitude of terms which may have a precise meaning to insiders but can mean anything or nothing to the outsider. "World market price" could mean posted price, posted price less a "reasonable" discount, or–more likely in recent years–some approximation to a real third-party price of which there could be several varieties.

What is interesting is that the parties need this kind of special permission even to trade their output shares. Such a restriction, of course, lessens the inducement to take out unlimited amounts but cannot remove it altogether. Hence we know that there must be an output decision by the group as a whole (vested in the board of directors or chief executive officer) rather than the mere sum of individual actions.

As for disposal of oil there are apparently no restrictions in the IPC articles. The FTC report[9] does speak of "limitations" but this is neither supported by evidence nor even explained.

We know rather less of the *Iranian Oil Consortium*[10] output agreement and transfer pricing, the details of which appear to be secret even from the Iranian government. But it follows from the large number of participants, of whom some have very small market shares, that the output rules must be extremely rigid and restrictive. The large companies' output would in any case tend to be restrained by the knowledge that if they competed hard enough to put even mild pressure on prices, the incremental net revenue would be small or negative. Thus, if a company's sales are 100 units at a price of $1, an additional 10 units sold at its rivals' expense at 90 cents would mean a revenue of ($0.9 × 110) only $99, and so would be a losing game even if costs and taxes were zero.

But for the relatively small participants in the consortium, this is not relevant. Their potential sales are many times their current sales. Even if the price were driven down considerably it would still be far above cost. A small participant would profit immensely from ordering out more than his 30 thousand barrels daily (TBD) or so of Iranian output. If he could sell not 30 but 300 TBD, the price impact would probably be negligible; if at (say) 500 TBD the price effect was substantial, of the order of 10 percent, the incremental net revenue would still be positive and large–far in excess of costs-plus-taxes. The smaller suppliers, and perhaps even some of the larger ones who wanted more crude to

[9] FTC, *The International Petroleum Cartel*, p. 107.

[10] BP 40 percent; Shell 14; Gulf, Mobil, Jersey, Texaco, SoCal 7 each; CFP 6; Iricon Group (seven smaller American concerns) 5 percent. In practice, the Consortium produces and refines, delivering each partner its share of crude and refined products. Production is in a limited area in South Iran. Elsewhere the National Iranian Oil Company (NIOC) is a 50-percent partner with various smaller companies.

supply current sales levels would therefore desire ever-increasing output, while the larger sellers would not.

No inside information is needed to know the inner tension of the consortium. Without a rigid ceiling on output, it would increase very greatly to the damage of other concessions, and above all to price stability. On balance, Iran might well gain; the Persian Gulf industry would lose heavily.

What little is known of the offtake arrangements of the Iranian consortium makes it appear the most restrictive of the Persian Gulf joint ventures. "Each shareholder proposes a total program for the output of the consortium for the following year; these are ranked in order, and the program chosen is the 'lowest total nominated that will cover the estimates put in by shareholders representing a given percentage majority of the shares in the consortium'." The total output thus programmed is divided by ownership shares. "Whether or not a very few companies have had the power to determine the total output of the consortium thus depends on what this critical percentage is, for a low nomination by a few groups can be defeated only if those that want more oil own a percentage of shares in the consortium which equals or exceeds this critical cutoff point, which probably is between 60 percent and 79 percent of total shareholding."[11] Dr. Penrose' explanation for the 60–79 percent spread is ingenious and convincing. It does almost certainly permit the companies with very large resources elsewhere (BP, Esso, Texaco, and SoCal) to block the crude-short companies from ordering out much larger amounts than they might like.

Starting in 1966, the government of Iran exerted increasing pressure to expand output, possibly through relaxed permission to overlift following the Aramco 1965 change. But for reasons just shown the market reaction would be much greater in Iran than in Saudi Arabia. To some extent, additional output could be had without letting partners overlift and sell at shaded prices. A special arrangement provided for sales to the National Iranian Oil Co. (NIOC) at a special low price; the Iranian government waived royalty, since it gained through NIOC profits. This incremental oil was not to be sold but used only for barter arrangement with Eastern European countries. Thus Iran got the greater output, and therefore additional revenue though less per barrel, with minimum risk of market disruption and lower prices.

But plainly the Iranian government did not consider this as more than a small concession, and has since pressed for modification of the offtake agreement among the consortium members, as well as for a pledge to increase the output rather substantially over the next few years. This controversy is permanent and of course is a most difficult one for all concerned. If Iran obtains a guarantee of greater output, all Persian Gulf governments will press for the same. But they have not the same bargaining power. With the withdrawal of British forces from the Persian Gulf in 1971, Iran becomes the dominant military power there. This will not make them more amenable.

Appraisal: The Competitive Effects of Joint Ventures. Let us sum up our conclusions thus far about the joint ventures. Each of them considered alone could be operated without any restrictions on the competitive process: each offtaker free to produce and sell as much as he wished, paying the operating and investment cost of what he wanted. In practice, only the two partners in Kuwait seem to be on this basis. In theory, borne

[11] Edith T. Penrose, *The Large International Firm in Developing Countries* (London: Allen and Unwin, 1968).

out by experience, the more partners in a joint venture, the more strict are the production controls.

From this angle we must examine the significance of exchange of information and coordination of production plans among the eight large firms. For example, each Aramco offtaker becomes aware of the nominations of the others, and knows in advance what they intend to sell. Since Aramco nominations are one-year firm and three-years tentative, each offtaker has not only a precise idea of his partners' short-run plans but at least a good idea of their long-run plans for permanent expansion. Compounding this effect is the overlapping membership in joint ventures. If the production plans of, say, Gulf are known to its partner BP, they presumably must also be known to BP's partners in IPC. Therefore the others can at least get a good idea of the net impact of the Gulf plans. Two of BP's partners in IPC are Esso and Mobil; hence the plans of Gulf and of BP must presumably be known also by SoCal and Texaco. All these firms are in the Iranian Consortium which includes practically everybody else in the Middle East. As a result, each of the Persian Gulf producing companies takes account of the action of all others. It would be an exaggeration to say that the companies confer on their production plans. At least among the five American companies, antitrust prudence would not permit that. But each can be assured that nothing is contemplated to threaten an excess of supply and a threat to the price. Each can hold back on output in the almost certain knowledge that all others are doing the same. This may not explain the price level, but it does help (to anticipate Chapters V and VI) to explain why the decline has been so slow.

If we try to explain the pattern of output and its expansion, we can get a much better idea of what is really happening than by trying to choose the best name for it. Thus, Hartshorn[12] says: "No cartel agreements survived to clamp uniformity onto the business behavior of the major groups after the war. None was needed during that first fabulous decade."

Granted that there was no cartel it is difficult to pin down the idea that "none was needed." Does it mean that output was expanded as fast as was technically possible, and hence it made no difference whether competition was limited or not? Evidence of restriction on output, and therefore on the price, can be cited as early as 1952.

Month and Year	*Saudi Arabia*	*Kuwait*	*Iraq*	*Total*
	(· · · · · · · · · · · · · · · · · ·	Output in TBD	· · · · · · · · · · · · · · · · · ·	·)
December 1951 . .	854	673	192	1,719
June 1952	899	764	420	2,083
December 1952 . .	733	710	514	1,957

It seems clear that as Iraq output expanded, following completion of the new pipeline to the Eastern Mediterranean, Esso and Mobil made room for it by cutting back Aramco production, and to a lesser extent BP did the same with Kuwait production. If no IPC partner had had any share in either Kuwait or Aramco, there would have been no cutbacks to make room for expanding IPC production. With higher output the price might have been forced down, or exports increased, especially to the United States.

It would therefore appear that the overlapping of joint ventures permitted the limitation of Persian Gulf output. To be sure, the argument is incomplete. Had there been independence on the part of all of the partners, no incentive to limit output in any joint

[12] *Oil Companies and Governments*, p. 178.

venture, and higher exports to the United States, the Congress would probably have acted to limit all imports drastically, as they did years later. But the power of Congress was not yet required because the operators themselves limited the supply. As this one example proves—and it will be confirmed in Chapter V—the restraint on competition made a difference even during the first postwar decade. The companies limited the increase in available output to the increase in Eastern Hemisphere demand, at prevailing prices. For this, truly enough, no cartel was needed. There are no instances of the breakdown or suspension of any arrangement, and output shares stayed rigidly fixed within each concession. With no opportunity to increase sales—because no more crude could be obtained—there could be no rivalry in price.

Vertical Integration

Up to this point we have treated the Persian Gulf oil companies as sellers of the crude oil they produced. It is time to examine the pattern of vertical integration in the world oil market. Everyone knows that most of what the companies produce they also refine and market. But for a significant volume of oil there is a prior stage, that of long-term contracts among some of the eight large companies.

Table III-3 gives an approximate picture of the disposition of crude oil in the world market in 1950, 1957, 1966, and 1968. It is subject to error up and down. Some crude sources are excluded—e.g., lower South America and some parts of Africa—as not being in the crude market, because local production is largely reserved to local use. Yet some is sold abroad, and this tends to understate the volume of market crude. Moreover, in some of these countries the international companies operate refineries whose throughput is counted into line 6, and cannot be separated out. This results in the absurdity of negative 1950 open market sales. Moreover, we set to zero sales among the 31 (as of 1968) large companies. On the other side: there are some integrated companies other than U.S., British, French, and ENI. To say, for example, that 20 percent of world crude production is sold as crude obviously does not mean that the 20 percent is available for sale during any given month, day, or even year. Some of it is obviously being sold under long-term contract, even excluding those among the Persian Gulf producers. But except where the sales contracts run indefinitely with a life of capital facilities, as in Japan (see line 7), the two parties are free to look elsewhere for better terms when the contract expires. Thus if sale contracts were typically for three years, one-third of the contract oil would be in the market during any given year, and as a practical matter both parties would be looking to improving their lot for the expiring second and final thirds. Moreover, the world oil market expanded by 11.1 percent per year during 1950-66. With an average contract term of three years, 44 percent of nonintegrated crude would be coming into the market fresh every year (33.3 plus 11.1).

Table III-4 summarizes the surplus position of the eight largest crude oil companies. For each company, refinery throughput is subtracted from crude production; the remainder, if positive, means crude sold; if negative it means crude bought. This tends to understate the amount of market crude, though I think not substantially in this case. The reason is that only occasionally and by chance will a given company have its production and refinery locations in perfect balance. Then they are better off selling some of their production in one place and buying crude in another. A company which produced 100 barrels in one place, sold it off, and refined 100 barrels in another, would appear as a zero seller in our table, and 200 barrels actually bought and sold would be lost.

TABLE III-3. Nonintegrated Crude Oil in the World Market, Selected Years, 1950-1968

(*thousand barrels daily*)

	1950	1957	1966	1968
1. Crude produced for world market	3,338	6,942	15,805	18,971
2. East bloc exports	10	36	510	530
3. Total crude available (lines 1 and 2)	3,348	6,978	16,315	19,501
4. U.S. imports (excluding Canada)	487	911	876	831
5. Canadian imports (excluding U.S.)	215	274	395	475
6. Crude refined outside U.S. and Canada by eight largest companies and 30-odd other companies	2,415	4,356	11,119	13,361
7. Japanese imports tied to supply contracts	...	200	666	775
8. Total arm's-length crude (line 3 less lines 4+5+6+7)	231	1,238	3,259	4,039
9. Line 8 as percent of line 3	7.2	17.6	20.0	20.7
10. Long-term supply contracts at Persian Gulf (Gulf to Shell; BP to Esso and Mobil)	300	500	1,286	1,612
11. Line 10 as percent of line 3	9.0	10.6	12.1	12.4

Sources and Notes (by line):

1. Table III-2.

2. 1950, *International Petroleum Trade*; 1957, *Petroleum Press Service* (PPS), Mar. 1958, p. 100; 1966, *PPS*, Dec. 1967, p. 452; 1968, *PPS*, May 1969, p. 164.

4. U.S. Bureau of Mines, *Mineral Industry Surveys, Petroleum Statement, Annual*, 1950, 1957, 1966, 1968.

5. Dominion Bureau of Statistics, *Trade of Canada, Imports by Commodities*, Dec. 1956, Dec. 1966. *BP Statistical Review of World Oil Industry*, 1968. Imports from United States assumed negligible in 1968; see *PPS*, Dec. 1968, p. 457.

6. For 1950, company annual reports and *World Oil*, July 15, 1951; for 1957, 1966, and 1968, Chase Manhattan Bank, *Financial Analysis of a Group of Petroleum Companies*, respective years. Also annual reports of Shell, BP. Added: ENI crude production, ELF-ERAP refining, Gelsenberg and Libya production.

7. For 1957, the following approximations, in TBD: Koa (Caltex) 40, Nippon Sekiyu (Caltex) 55, Maruzen 20, Showa (Shell) 35, Toa Nenryo (Esso, Mobil) 50.

For 1966, according to *The Petroleum Industry in Japan, 1967* (Tokyo: Japanese National Committee of World Petroleum Congress), p. 24, "oil companies with foreign affiliation and subsidiaries of international oil companies accounted for . . . 62.8% of the total refining capacity during the fiscal year April 1, 1966-March 31, 1967." A list of foreign affiliations is given on the same page, and a list of refineries, giving ownership and capacity is on p. 22. The refineries thus identified, plus Maruzen, which was still affiliated in 1966 [for reference, 2-4, 5-7, 10-12, 14, 17, 18, 20, 21, 23, 24, 27, 31, 33, 36] add to 1,412 TBD or 63.8%, indicating an error of 23 TBD which cannot be isolated. It is assumed that the foreign participants include their ownership shares of the refineries as their share of output in their published world refinery output. They are assumed to have crude supply rights for the rest, and this is the entry for line (7), calculated as follows. To each of the numbered refineries we apply the coefficient $(1-s)$, where s = ownership share. The total is 30.1% which, applied to actual Japanese output for the calendar year 1966, is 666 TBD.

For 1968, *The Petroleum Industry in Japan, 1968* was the latest available, but the error is small because ownership percentages are as of July 1, 1968 and refining capacities of Mar. 31, 1968. There is unavoidable inaccuracy in this method, since we cannot tell to what extent the foreign companies reckon their Japanese refinery participation differently from our assumption. It is impossible to test the method by reference to company annual reports, which are not consistent and usually incomplete.

10. Company reports less estimate in 1966 for overlifting.

So far, this loss should not be substantial. First, the producing areas are few and so are the chances of these discrepancies. Second, because prices are so much higher than costs, a company will normally try to produce and refine its own crude if it is physically possible, even if there are strong social diseconomies, such as cross-hauling. But for the future, if prices and tax-plus-costs continue to come closer together, this inhibitor will get weaker, and there will be more selling and buying. Moreover, because of the doubling effect noted above, the error of omission can become quite large. In future calculations, therefore, it may not be safe to assume that the net surpluses or deficits calculated here are the gross transfers we seek.

In Table III-4, "availability" means production plus receipts or minus deliveries under the long-term sales contracts involving BP and Gulf as sellers and Esso, Mobil, and Shell as buyers. These will be examined in more detail below. The difference between aggregate production and aggregate availability of the eight companies should be zero. The 177-TBD discrepancy in 1966 is a statistical mirage arising largely from noncomparable figures, which are set forth in detail in Appendix III-B. But the estimated surplus is practically the same on either basis, as it should be.

Table III-4 is not strictly comparable with Table III-3 because it includes production in countries which we do not count in the world market. Therefore, part of the surplus in Table III-4 is apparent, not real. But the trend of the surplus is real and significant. Although their total production doubled during this time, the amount not retained in their individual integrated channels increased by less. But one finds completely diverse behavior patterns among the eight. CFP went from almost no open-market crude sales to 328 TBD; BP and SoCal doubled; Texaco increased much less; while Esso and Mobil increased only very modestly. Shell, which had been the largest net seller in 1957, sold almost precisely the same amount nine years later. Moreover, it had gone from a surplus to a deficit in its own production, and this at a time when the comparative advantage in the long-term sales contracts had actually decreased. The Shell stagnation was not planned. Their unsuccessful attempts to expand output are common knowledge. The simplest and best explanation of Table III-4 is that each company tried to expand as far as it could in all possible markets, but had varying degrees of fortune in getting either new crude supplies or the consent of their joint venture partners to expand—Esso obviously chafed under an Aramco constraint in 1965—*and* in obtaining sales contracts at prices they were willing to accept. The larger a firm's market share, the greater its degree of statesmanship in refusing otherwise attractive new business because of its threat to price levels.

TABLE III-4. Summary of Largest Companies' Crude Oil Surplus Position, 1957 and 1966

(*thousand barrels daily*)

	Esso	Mobil	SoCal	Texaco	BP	Gulf	Shell	CFP	Total
Production									
1957.........	1,782	392	529	617	977	774	1,497	183	6,751
1966.........	3,352	953	1,166	1,440	2,423	1,633	2,380	829	14,176
Total availability									
1957.........	1,887	437	529	617	827	362	1,853	193	6,705
1966.........	3,608	1,016	1,166	1,440	2,130	886	3,241	866	14,353
Surplus (production basis)									
1957.........	393	98	252	216	335	628	290	1	2,213
1966.........	349	109	562	382	838	1,182	–213	291	3,500
Surplus (total availability basis)									
1957.........	498	143	252	216	185	216	646	11	2,167
1966.........	604	172	562	382	545	435	648	328	3,676
Increase									
1957–66......	106	29	310	166	360	219	2	317	1,509

Source: Appendix table III-B-1.

Note: Production *plus* receipts on long-term sales contracts (or *minus* deliveries on long-term sales contracts) equals availability. Subtracting refinery throughput leaves surplus available for sales to third parties.

The Long-Term Sales Contracts. Let us first note briefly the main provisions of long-term sales contracts (details of their amounts are estimated in Appendix Table III-B-1), and then consider their significance in a highly concentrated industry. To what extent are they arm's-length sales, to what extent a form of vertical integration that precludes a market interface? The answer changes over time because of lower concentration and lower prices.

The Gulf-Shell contract, providing for deliveries through 2026,[13] involved one-fourth of Kuwait oil reserves as they were known in 1947, which of course became a decreasing proportion of total reserves. In 1957 about 52 percent of Gulf Kuwait output was still going to Shell; by 1966 this was down to 46 percent.

In the early years at least, the sales contract penalized Gulf for reducing prices if they tended to reduce Shell prices. Furthermore, if Gulf were a successful competitor against Shell either through lower prices or any other method, it lost part of its Shell business. The very fact that the contract needed to tie Gulf profits to Shell profits, and therefore provide an additional sanction to keep Gulf from competing with Shell, shows that the assumption of automatic tacit collusion was not necessarily true even in the late 1940s. Whether the marketing limitations were afterwards removed from the contract is not known. The two later Gulf prospectuses do not mention them.

The contracts between BP (then called Anglo-Iranian) and Esso and Mobil took about one-fourth of BP's 1949 output. BP's loss of Iran soon thereafter undoubtedly forced a revision downward. By 1957 the contract percentage was slightly over 15 percent; by 1966 it was below 12 percent. The price provisions of these sales differ significantly from the Gulf-Shell sales. Esso paid BP cost plus a fixed money profit; Mobil paid cost plus one-third of the margin to the "market price," which was not explained. Unlike the Gulf-Shell deal, price was unaffected by anything Esso or Mobil did with the oil, and there was no incentive or penalty for the buyers to avoid the markets of the seller. The restrictive feature of these contracts lies in the commitments by the buyers not to sell in certain areas; Esso could not market the oil east of Suez, and Mobil was to ship mainly to the United States. Again, whatever changes have since been made are not known.

In general, a long-term sales contract standing by itself has no particular significance, and is of no interest except to the management and stockholders of the particular companies. Particularly in the oil industry, there is never a nicely matched equivalence of available crude on the one side and available refining-marketing outlet on the other. Given the great distances separating production and consumption, not to mention the variety of crudes desirable in any finished product market, one must expect to see a considerable amount of trading going on. Some companies sell excess crude, some buy it. The striking fact about the large eight producers in 1950 and even 1966 is that they were all surplus in crude—even including Shell down through the middle 1960s.

The sales contracts prove that there is no inherent technical or cost necessity for vertical integration. (There may be others.) No matter what the pattern of ownership, the logistic problems still remain: how best to arrange to get the right amount and type of oil to the right place at the right time. The task is a huge and complex one, and the international oil companies are rightly proud of how well they perform it, sometimes under very trying circumstances. But the cooperative effort would be just as necessary and could be just as efficiently done if no oil producer owned a refinery, and no refiner produced oil. In any case, all hands must cooperate to fulfill instructions or programs no

[13] Gulf *Prospectus*, May 1963.

matter whether first decided by intercompany *negotiation* or intracompany *administration* (and negotiation). Whether cost is less within an integrated channel or by buying and selling depends on the particulars of the operation, which is to say, on operating efficiency and chance.

Once free of the error that integration is somehow inherently necessary in the oil business—or that it is somehow inherently evil—we can understand the significance of the contracts in a concentrated industry. In a market of many sellers the decision to integrate or to buy (or sell) is simply a cost decision. But in a market of few sellers one must consider the effect of any given arrangement on the market price and the incremental net revenue. BP, for example, must decide whether to sell the additional oil to Esso and Mobil or to find its own markets for it. A decision not to sell would mean that Esso and Mobil would seek to develop their own sources; the volume of offerings would be larger, competition stronger, and prices lower. Hence BP has reason to prefer a sales contract to refining and marketing its own crude. For the same reasons, Esso and Mobil would prefer to buy BP crude. But this mutual interest is not quite a sufficient condition—there remains the matter of price. To help stabilize the market, perhaps Esso and Mobil would be willing to pay a little more for BP's crude than it would have cost them to produce it themselves. But surely they would not be prepared to pay much more, for the cost of being industrial statesmen and warding off competition would fall wholly on them, while the benefit would be shared by all. Hence the price charged by BP to them had to be not too far above their own incremental cost-plus-taxes. But the matter is even more complex because Esso, Mobil, and Shell, unlike BP and Gulf in Kuwait, were by no means free to produce as much as they wanted from their own shares of other joint-venture concessions. For, as has been shown above, there were and are rather stringent limits to what any one partner can take. The sales contracts have been to some extent a supplement to, or safety valve for, the joint venture agreements.

Some reflection on price confirms that market structure and market behavior are so interdependent that one cannot describe either one without taking account of the other. In 1950, or 1956, a buyer outside the eight large companies had to pay a price which was so much higher than that set under the long-term contracts that there could be no question of any competition between the two transactions.

By 1957, perhaps one or two large buyers had been awarded prices not too far from those named in the Persian Gulf sales contracts, but we have no hard evidence on this. By 1966 the situation was quite changed. Although production costs had fallen, government take had greatly increased, while crude and product prices had considerably decreased. As will be seen in Chapter VI (pages 188-90), prices on the long-term contracts are only about 10 to 15 cents per barrel below the lowest prices on third-party crude sales, and only about 10 cents below net realizations from sales of products through integrated channels. Therefore sales among the large eight companies have become close alternatives to, or competitors of, sales of crude or refined products to those outside the group of eight. A sale among the eight largest at a "half-way price" or near it is no longer so different in degree as to differ in kind from a sale of crude to a third party. It would overstate the nonintegrated crude market to include these sales. It would also be an error to exclude them. The truth lies somewhere between.

Marine Transport. Integration by oil companies into tanker transport is of no great importance to competition and prices. The tanker segment is very large, of course—the real costs of transport are several times the real cost of production. But, as will be seen in the next chapter, the tanker industry is one of pure competition. Any amount of marine transport can be procured with no initial investment and at no higher cost than that incurred by anyone else—setting aside special skill. There is no restriction on the offers of ships for hire; most of them are owned and operated by relatively small firms. Most important, anyone can buy a ship. Even in 1971 a 250,000 tonner sold for about $35 million, and it was a negligible proportion of the total supply. A firm need not have many ships to be efficient. A shipper can easily and cheaply integrate into tanker transport, or arrange with one of many firms doing a transport-connected service, to expand their service, with either or both sides sharing the investment.

Hence, despite its huge size and heavy aggregate capital investments, the tanker trade has no role in checking or moderating competition either in crude or in products. Again we see how a phrase like "integration" may contain not only a double meaning but a contradiction. Because firms are effectively free to integrate into a particular part of the trade, there is no significance to the fact that many concerns are already heavily integrated into it. The decision to do so or not is purely a cost decision. Minimizing of transport cost requires that most of the fleet be available long enough in advance of any given voyage to let the voyage be carefully planned as part of a program of offtake and delivery. Whether the ship is to be bought or leased on long-term charter is a cost choice not a market choice. An oil company's marine department may do a better or poorer job of operating the additional tankers needed. There is also a choice between two methods of financing the operation—on-balance sheet versus off-balance sheet, stated versus contingent liabilities. In the postwar period, the oil companies' share of tanker ownership has decreased considerably, but this has none of the significance that we must attach to the largest oil companies' decreasing share of total oil production. Rather, it arises from the fact that independent ship owners are on the whole somewhat more enterprising and better equipped to grasp new possibilities than the oil companies.

Refining-Marketing. This brings us to the refining stage of the industry, whose market structure is admittedly difficult to analyze. Table III-5 shows worldwide refining, and Table III-6 gives shares of refining capacity in various countries, but for several reasons these are not good indicators of market share. First, there is a substantial amount of processing by some companies for the account of others. Second, there are a number of joint refining or joint marketing operations, or both together. In Britain, for example, Shell and BP refine separately but market jointly under the trade name of "Shell-Mex and BP." In Japan, the producing companies have a large number of joint ventures with Japanese firms, which give the international producing firms the crude supply rights. Hence the refining capacity of the internationals is an incomplete indicator of their real share of the market.

Moreover, much of the refining not done by the eight large companies is in the public sector. Some of these refineries buy crude in the world market; others are supplied exclusively by domestic crude. The public sector is very large in the less-developed countries, and also in France, where rigid marketing quotas minimize competition, and margins and profits are the highest in Europe. Finally, the ease or difficulty of shipping across national borders varies greatly in time and space. Here the oil market has seen contradictory developments. On the one side, refineries have been nearly all built as close as possible to market, and international shipments of refined products have stagnated while world trade in crude has multiplied nearly tenfold since World War II. This tends to make the consuming countries more self-sufficient and each one to constitute more of a self-contained market. On the other hand, the integration of Western Europe has been considerable, however slowly it has proceeded. There is much more intershipment among

TABLE III-5. Summary of Largest Oil Companies' Crude Runs, 1957-1966

(*thousand barrels daily*)

	1957	1958	1959	1960	1961	1962	1963	1964	1965	1966
I. Crude Runs										
Texaco	401	446	586	650	703	737	729	817	950	1,058
SoCal	277	283	295	340	373	405	433	482	543	604
BP	642	730	780	940	1,009	1,134	1,230	1,369	1,505	1,585
CFP	182	221	237	278	316	335	389	438	479	538
Esso	1,389	1,446	1,625	1,759	1,912	2,181	2,341	2,526	2,799	3,003
Mobil	294	337	391	433	487	525	582	676	757	844
Shell	1,207	1,289	1,486	1,624	1,753	1,898	1,983	2,245	2,441	2,593
Gulf	146	163	177	217	231	305	331	379	427	451
Total, Eight Companies	4,538	4,915	5,577	6,241	6,784	7,520	8,018	8,932	9,901	10,676
Total, World-wide Crude Runs[a]	6,729	7,475	8,379	9,536	10,521	11,537	12,659	14,263	15,916	17,329
II. Largest Oil Companies' Crude Runs as Percentage of World Total										
Texaco	6.0	6.0	7.0	6.8	6.7	6.4	5.8	5.7	6.0	6.1
SoCal	4.1	3.8	3.5	3.6	3.5	3.5	3.4	3.4	3.4	3.5
BP	9.5	9.8	9.3	9.9	9.6	9.8	9.7	9.6	9.5	9.1
CFP	2.7	3.0	2.8	2.9	3.0	2.9	3.1	3.1	3.0	3.1
Esso	20.6	19.3	19.4	18.4	18.2	18.9	18.5	17.7	17.6	17.3
Mobil	4.4	4.5	4.7	4.5	4.6	4.6	4.6	4.7	4.8	4.9
Shell	17.9	17.2	17.7	17.0	16.7	16.5	15.7	15.7	15.3	15.0
Gulf	2.2	2.2	2.1	2.3	2.2	2.6	2.6	2.7	2.7	2.6
Total, Eight Companies	67.4	65.7	66.6	65.4	64.5	65.2	63.3	62.6	62.2	61.6

Source: Appendix table III-B-1. Where refineries are jointly owned, they are allocated on the basis of annual reports.

Note: Detail may not add to total because of rounding.

[a]Excluding North America and Communist countries.

the countries of the European Common Market, and even Great Britain has not remained unaffected—to the embarrassment of its refiners. It would be difficult, therefore, to make any kind of general statement about market shares and the degree of market control in the particular areas where oil is refined and consumed.

But even in the wider markets of the late 1960s, the integration of refining with production is a barrier to competition. The exchange of information, noted earlier as an inevitable result of overlapping ventures in the Persian Gulf, was greatly reinforced when the producers met also in many markets as refiners and marketers. They had a close check on their rivals' plans for expansion, and could see whether their hopes of cooperative conduct were justified or not. Since a company could not increase Persian Gulf output without building refineries and distribution terminals in other countries, the implied pledge not to expand output faster than the group interest would approve was reinforced by the lack of any place to put additional crude.

Furthermore, although the boundaries of the refining markets were and are extremely hard to define, each is always narrower than those of the worldwide producing market. Hence, market shares tend to run much higher, and sellers fewer. Tacit or open collusion—possibly sanctioned or required by law—was often not even necessary, for the impact of any increased sales in a narrow market would have depressed prices severely. Even in the large British market, Shell and BP were one seller (45 percent of gasoline) with only Esso (27 percent), Texaco (11 percent), and a fringe to consider. For a long

TABLE III-6. Refining Capacity, Largest Companies' Proportion by Areas, 1957, 1966

	1957			1966										
	Total	8 largest companies		Total	8 largest companies		Esso	Mobil	Shell	SoCal	Texaco	Gulf	BP	CFP
	(TBD)	*(TBD)*	(%)	*(TBD)*	*(TBD)*	(%)	(.	. . . . .	. . . . .	. . . . .	*percent*	. . . . .	. . . . .	. . . . .)
Persian Gulf	1,145	1,064	93	1,662	1,363	82	7	4	4	13	13	10	30	2
Other Middle East	116	23	20	293	117	40	1	16	8	4	4	4	2	2
North Africa	77	14	18	303	47	15	5	1	3	0	0	0	1	4
Other Africa	21	21	100	402	274	68	3	12	20	5	6	0	20	2
Australia, N.Z., Malaysia	257	242	94	758	632	83	6	14	26	7	7	0	22	0
Japan[a]	430	101	23	2,211	482	22	5	3	5	4	4	0	0	0
Japan[b]	...	...	...	2,211	1,125	51	...	...	...	...	...	...	...	...
Other Asia	401	330	82	986	345	35	14	5	7	3	3	2	1	0
European Economic Community	2,083	1,433	69	6,551	3,858	59	16	4	16	2	3	1	9	7
Other Europe	820	529	65	2,976	2,030	68	22	3	20	2	2	1	18	0
Venezuela/N.W.I.	1,406	1,346	96	1,989	1,945	98	46	4	36	2	3	5	0	0
Other Latin America	1,238	363	29	2,294	773	34	9	0	7	0	15	2	0	0
Total World	7,994	5,465	68	20,426	11,867	58	16	4	15	3	5	2	9	3

Sources: Appendix Tables III-C-1 and III-C-2, and *The Petroleum Industry in Japan, 1967*.

[a] Including ownership shares in joint ventures.
[b] Including crude supply rights in excess of ownership shares.

time there was no incentive to expand output to take more of the British market, but matters changed greatly in the 1960s.[14] This will be discussed in Chapter VI.

The market would have looked and behaved very differently had refining been done by large firms independent of the producing interest. Such firms would have been able to look at the whole range of possible supplies, offer to buy large and profitable blocks of incremental output under short- or long-term contracts, and use the lower price granted by any seller as evidence that others could do the same—indeed, had better do so for fear of losing business.

This conclusion does not depend on the empty formula that large firms have some kind of "bargaining power" or "countervailing power" and can therefore push prices down simply because they have power. Nor does it depend on the unlikely supposition that a group monopoly of refiner-buyers might beat down the price. An independent refining industry would have provided a world market in crude oil, with producers selling as producers in a wide market rather than as refiner-marketers in many narrow markets. Competition would have been much more intense in such a market. Why did it never come into existence?

Entry to Refining. The lack of clear market boundaries makes it more difficult to judge the ease and cost of entry into refining. As of the mid-1960s, the minimum efficient capacity of a new refinery was probably upwards of 100 TBD. The average for existing refineries in Europe and Japan was smaller, of course. But, as Appendix VI-C shows, the average size for refineries installing new capacity in 1966 (including new refineries) was 105 TBD. The threshold would be even higher in Japan. We use 120 TBD to allow for later growth. The same appendix estimates the investment for such a refinery at around $500 per daily barrel, or $60 million at 1966 prices. This is a large sum, of course, but can be taken in two or more increments and, given the hope of an adequate profit, there is no reason to fear an inadequate response.

An important possible barrier to entry is the disincentive of lowering prices by providing more competition. At the extreme, if there is only one seller in the market, and he is at minimum efficient scale, the entry of a rival must double capacity and therefore drive down the price severely.

Table III-7 shows the impact of a 120-TBD refinery in Europe and Asia. If the Common Market were really one, a new refinery would have added only two percent to capacity during 1963-68. But entry would really have been much easier, because the swift growth of consumption meant that more capacity was needed to keep the price from going up. The entrant could therefore dismiss any need to worry about price effects from his own entry; he would have to worry lest all the existing and prospective refiners together were planning to add too much capacity for the good of the industry's profits. Of course, the air was always stirred by warnings that exactly this was happening.

In each of the Common Market countries (counting Benelux as a unit, as seems realistic), the growth of the market required between one and three large economic refineries each year. Even in the relatively small Australasian market (where counting Australia and New Zealand together is probably *not* realistic), one was needed every third year. Hence it would seem that relative entry barriers were low or not significant, to the point of having no effect on competitive behavior.

Let us now look for the presence of any necessary factors not available to all comers.

[14] *Petroleum Press Service*, vol. 35 (June 1968), p. 209.

TABLE III–7. Impact of a 120-TBD Refinery in Eastern Hemisphere Markets, 1963–1968

Area	Average 1963–68			A new refinery (120 *TBD*) as percent of:	
	Refining capacity (*TBD*)	Annual growth			
		Percent	Amount (*TBD*)	Avg. capacity	5-yr. growth
	(1)	(2)	(3)	(4)	(5)
Britain	1,470	9.5	140	8	17
Belgium	390	18.0	70	31	34
Netherlands	656	10.0	66	18	36
France	1,420	13.5	185	8	13
Germany	1,670	13.0	216	7	11
Italy	1,780	17.0	302	7	8
Common Market	5,916	13.0	769	2	4
Other Western Europe	1,130	19.5	212	11	11
Total Western Europe	8,600	14.0	1,200	1	2
Japan	1,580	12.0	190	8	13
Australasia	438	9.5	42	37	57

Source: BP, *Statistical Review of the World Oil Industry*. Average refining capacity is the geometric mean of the five-year period.

1. No refiner builds his own plant; there is a separate and well-known group of construction firms whose name is legion around the world. Many important processes are patented, but they are all duplicated by competing processes, and royalties are low. The established firm has the advantage of experience with operating a plant, and also of what the market is likely to take. This is a start-up cost which a wholly new firm must surmount, here as everywhere else.

2. The relation of refining with marketing is obviously important, but little documented. Because marketing is often an intensely local activity, markets may be very small and easily monopolized for the same reason that refining cannot be: additional competition would knock down prices. Certainly, this must be true in numerous gasoline service-station markets. One must beware of exaggerating its importance. Gasoline can be sold in large amounts to fleet owners, without the need of any marketing investment or know-how. "Gasoil," or "middle distillates," covers a wide range of products: aviation jet fuel, or railroad diesel fuel, with each airline or railroad as, in effect, a fleet owner; space heating fuel (mostly residential and commercial), where a considerable network of dealers is needed, but much less than with motor gasoline. Finally, there is residual or heavy fuel oil, much of which is sold directly to large industrial users, and which requires little distribution apparatus.

In Europe, motor gasoline and heavy fuel oil are each respectively about one-fourth of the value of the barrel; middle distillates the other half (see Chapter VI). There are unaffiliated chains of distributors, including gasoline service stations, to whom a refiner can sell. In Britain and on the Continent, these groups have tended to gain increasing shares of total sales volume.

This very brief summary of marketing can only suggest the huge variety of local conditions, which need not favor the integrated producer-refiner and do not exclude the independent.

3. So far, we have found no entry barrier in capital requirements, market impact, or distribution. Matters are different with respect to crude oil. The integrated producer-refiner companies are uncomfortable neighbors, but not because they are mostly so large or so diversified, offsetting losses in one place with profits elsewhere. This is unsupported legend. The smaller integrated companies are tougher competitors because they have smaller market shares and hence need worry less about impact on either crude or refined

product prices. But to the integrated producer-refiner a failure to cover even refining costs (including return on investment) is simply a subtraction from the net value of crude oil produced. If, for example, the sales value of a barrel of products is around $2.50, and freight and refining cost each 50 cents, then the realized f.o.b. value of the crude is $1.50. But if producing cost plus taxes are clearly less than $1.50, the barrel is still worth producing, shipping, refining, and selling. The integrated producer-refiner will indeed be more cautious about cutting refined product prices in any one place, because if the reduction spreads, he will be more heavily penalized for it than will an independent local refiner. The bitter resentment felt in Europe against the international companies for their supposed belligerence in starting "price wars" could not be more misplaced; the large companies are forced in their own best interests to be timid, and this has cost them market share. Not size and not even integration per se is the key to the riddle, but, rather, integration into *crude produced with wide profit margins.* A producer-refiner with large crude oil profits has greater staying power when product prices decline; the large can follow to the letter Polonius' advice: "Beware of entrance to a quarrel; but, being in, bear't that the opposed may beware of thee."

In fact, the persistence of the independent refining sector is in part to be explained by the reluctance of the integrated companies to cut prices; particularly in Britain, their persistence in high prices and old-established distributing channels has been damaging to them.

But the independent refiner or marketer who wins a place in the market through his processing and marketing skill thereby becomes a valuable asset to a crude producer, who can count on a large increment to producing profits if he can assure the outlet of the refiner by buying him out. It becomes worth his while to pay a large premium above the mere reproduction cost of the assets. Hence survival and prosperity for the individual refiner or marketer may represent decline or disappearance for the class. Furthermore, the independent refiner is well advised to find a source of low-cost crude if he can; by purchase if possible, but ownership is even better because it is permanent. Thus independent refining as an activity is always in jeopardy: unsuccessful refiners fail, while successful ones may be bought up or may integrate. Indeed its survival depends in part on public-sector companies who are not under market constraints, and in part on the opportunities that exist in a large industry composed of a mosaic of individual competitive opportunities.

The structure of oil refining in Europe and Asia leads to a curious result. The one inhibition on entry—ownership of low-cost, high-profit crude—is so powerful as to keep most refining in the integrated circuits. But the nature of this inhibition is the known willingness of the integrated companies to take, if they must, even less than the barely competitive margin to cover refining costs. As we shall see in Chapter VI, the structure of the various product prices is that of a competitive market. Furthermore, since by subtracting competitive refining-transport costs from product prices one can be fairly successful in predicting crude oil prices, it would appear that refining margins are indeed no more than competitive. If the reader accepts the arguments of both these chapters, a refining structure that is noncompetitive in one essential respect has generated a margin (price of its services) not detectably different from the competitive one. However, the price data of Chapter VI are generated at the crossroads of the market, where competition is most intense. Particularly in Great Britain and France, prices have been higher. Hence the margins have been higher than our calculations show. One cannot say how much higher, but it is clear from developments in Great Britain, and perhaps in France, that those margins have tended to narrow where they were formerly most fat.

A Look Backward and Forward. As is true in many industries, the current structure is to be explained by history plus inertia. Before World War II, producers of crude were few outside the United States because entry was risky. The few practiced an extremely restrained competition, and during 1928–39 joined in a cartel which kept price well above cost, thus generating a constant desire on everyone's part for more volume which could only be had by building refining-marketing networks. Furthermore refining-markets were so much smaller and closed off that, by the test of Table III-7, barriers to entry were very high. Consequently, refining was a good operation to stay out of, except for those who had, or hoped to have, cheap crude, or for those who had entered refining early. Buyers (refiners) did not care to deal with few sellers, any more than sellers (producers) cared to deal with few buyers. Whether oil was ever a "natural oligopoly" in 1938, when world production was 4.9 million barrels daily (MBD), it was not one in 1969 at 33.4 MBD. But oligopoly in each market maintained it in the adjacent market.

Thus, a concentrated integrated market structure perpetuated itself. The history of the industry since World War II is that of a gradual loosening: less concentration in crude production; less also in refining, due largely to market enlargement and unification of the Western European economy; and less vertical integration. The basic explanation, if I may be excused for trying to reduce everything to one cause, is a high growth rate.

As to the future: if one is correct in thinking that the chief barrier to entry is in the wide profit margin of the integrated companies on their own-produced crude, that barrier is being gradually reduced. Over the years there will be more scope for independence in refining, less incentive for producers to go into it. If the rate of growth of consumption slows down, as it must eventually, that will exert an effect the other way. But this will in turn be offset by the growth of needed replacement capacity. In sum, it is hard to see any change in the current condition where the most efficient size of plant, and even of firm, plays a very small part in the markets as they are and will be.

Conclusion on Market Structure

The structure of the industry was, and is, a barrier to competition. Crude oil producers were few, and entry was extremely slow. No producer would sell off any part of his reserves, although their present value was nil, so that any price would have been clear gain to him. He preferred to keep an inventory which might be 50 to 150 times sales, in the well-founded hope that every other producer would do the same, thereby keeping large volumes of oil out of new competitors' hands. Entry was therefore difficult and unpromising, and although some succeeded, none of them did so on a scale to rival the older established producers. The huge resources of the original Persian Gulf concessions were effectively locked up, and only by the late 1960s was there a comparable rival, in Libya. In Venezuela, concessions were granted only once in the entire postwar period, in 1956. Although some good discoveries were made, no great volume of oil was available to compete with the Persian Gulf producers, whose overlapping joint ventures allowed the coordination of output and hampered individualistic sales and production policies. Vertical integration with refining-marketing precluded any class of large and well-informed buyers who would have helped induce competitive conduct. It also permitted mutual surveillance, for refining provided a check on any given company's producer plans.

But in examining this system, we have become aware of its incompleteness. That there will be substantial barriers to competition does not imply *how far* price will stay above cost. We saw that the size of the unintegrated crude market could not be measured without understanding price relationships. In the following chapters, we turn from anatomy to physiology, and see what happened to prices.

PART TWO

THE MARKET PRICE

CHAPTER IV

Tanker Rates

In this chapter the cost of moving oil by tanker is calculated in order to derive delivered (c.i.f.) prices from f.o.b. prices, or vice versa. If one does not know the tanker rate, what looks like a price observation may be only noise. Many c.i.f. sales are reported at a "discount" from we know not what, hence must be discarded as neither true nor false, simply without content.[1]

Much reliable tanker rate information is publicly available. Observations have almost invariably a specific precise time reference: single-voyage (spot) charters; consecutive voyages (usually from three months to two years); and time charters, which may last 20 years.

Tanker rates form a structure composed of interrelated long- and short-term transactions. Hence this chapter may also help to explain markets where a product or service is bought for long-term supply by some and short-term supply by others. Once the reader has gone through the analysis of tanker rates, he may have a better idea of what can be done with the less satisfactory data on crude and product prices.

Our primary interest is, of course, in long-term tanker rates. Together with estimated production and refining costs, they afford a coherent set of long-term supply prices of crude and product which can be compared with market prices to help see future trends. But let us first look at how rates are quoted.

The rate for a single voyage or a set of consecutive voyages is quoted as a payment per long ton of cargo shipped from loading to discharge port. The rate is therefore immediately comparable with a crude oil price at each end of the trip, to be added or subtracted. We can speak of "the spot rate" with little inaccuracy as between one size ship or another. "The consecutive-voyage rate" is often though not always a pardonable over-simplification. But a term charter, unlike a spot or consecutive charter, is not the sale of a specific transport service, but the hire of a ship.[2] The owner must keep the ship fit to travel wherever the charterer wants it to go. He pays the crew and for routine

[1] For example, it has been reported that Kuwait crude was being discounted by 35-45 percent in 1964. [Pierre Clair, *L'Indépendence Pétrolière de la France*, vol. 1, ed. Cujas, Paris, 1968, p. 97, citing *Financial Times* (London), Mar. 4, 1964.] This would imply an f.o.b. price of $0.88-$1.04 per barrel, which is far-fetched. It obviously refers to a discount from delivered prices, but since no tanker rate is named in the report, we cannot tell what the 35-45 percent is discounted from. Much or most of two Indian reports (p. 114, n. 14) falls in this category.

[2] Occasionally one finds a "contract of affreightment," whereby an owner contracts to transport a given amount of oil over a given period at a pure price which can be stated on Intascale or Worldscale (see below).

maintenance, periodic overhaul, and insurance;[3] the charterer pays the particular expenses of sending the ship to one or another place—fuel and canal and port dues.[4] These expenses must be added to the ship hire and then the total outlay must be compared with the load carried in order to obtain the cost of carrying a ton of oil over a specific route. This amounts to computing a spot or voyage-ton equivalent. References to spot-rate equivalents are common in the trade press. Some ship brokers publish tables or charts for quick approximate conversion of the hire of any given size and type of ship to its spot equivalent. The trade press will often publish estimated spot-equivalents of recent term charters. They show the level of the market with enough accuracy for most purposes. For recent years we need to be a little more exact.

Scale, Intascale, and Worldscale are the successive schedules of tanker rates issued by International Tanker Nominal Freight Scale Association, Ltd., London.[5] Scale was in effect from the end of the war to May 1962, then succeeded by Intascale (IS), then by Worldscale (WS) in September 1969. Rates are calculated by using a hypothetical vessel with a given rate of fuel consumption and other defined characteristics, and a given ship hire, and calculating a total cost per ton for a very large number of port-to-port (and some multiple port) voyages. These rates are in no sense average or typical or suggested. Indeed, no single rate means anything. But the whole set of rates is a schedule permitting immediate and fairly precise comparison of any two or more spot charters. If a ship was fixed yesterday, Aruba to London at $2.06 per long ton of cargo, and another today from the Persian Gulf to Japan at $2.64, a ship broker need report only that they were fixed at Intascale less 55 (Worldscale 56) and Intascale less 60 (Worldscale 48) respectively. The discussion will be wholly in terms of these schedules, to whose publishers everyone interested in oil economics is much indebted.

We wish to analyze the structure of tanker rates according to ship sizes and charter time periods. But market prices may diverge radically from supply prices when a market contains important non-competitive elements. Such a divergence sets up a force tending to push prices up or down in the future. In order, therefore, to predict future oil prices, it would be necessary to predict not only future tanker supply and cost conditions, but also the future of the tanker market structure. One is spared this difficult task if it can be established that the tanker market contains no important elements of monopoly, and that one can therefore treat market prices as supply prices.

Pure Competition in Tankers

Oil companies own a fourth (or possibly as much as one-third) of non-U.S. private carrying capacity. The eight large companies own about 20 percent (see Table IV–1). Roughly 10 to 15 percent of world tanker tonnage is in the spot market at any given time. If we assume that the 65 percent not owned by oil companies is under consecutive-

[3] Insurance is a difficult item to classify and the statement in the text is not wholly correct: no brief summary can be. Insurance was formerly paid by the owners (and included in the ship hire), on the theory that it was independent of the particular voyages. But in recent years the growth in ship sizes, and some spectacular accidents, have both raised insurance rates and tailored them to the lanes travelled by the ship. In consequence, some recent charters do not provide for insurance, and some oil companies, confronted by high rates to cover large but unknown risks, have become self-insurers.

[4] A "bareboat" charter obligates the charterer to pay all operating costs.

[5] These schedules are the successors to and improvements upon a schedule set during World War II by the U.K. Ministry of Transport, and known as MOT. A corresponding schedule drawn up by the U.S. Maritime Commission was continued with some changes by a group of New York brokers as American Tanker Rate Schedule, or ATRS, but was never as widely used. Worldscale replaced both Intascale and ATRS. (See appendix IV-B.)

TABLE IV-1. Ownership Distribution of Tankships Registered Outside United States, December 31, 1951-1968

(*in T-2 equivalents*)

Company or class	1951	1955	1960	1966	1968
Esso				304	378
Shell				234	503
BP				162	231
Mobil				106	104
Gulf				79	82
Texaco		(not available)		67	111
SoCal				66	94
CFP				42	56
Subtotal				1,040	1,559
Other oil companies				370	381
Total oil companies	351	565	904	1,430	1,940
Independent owners	579	1,223	2,394	4,200	5,151
Total private	930	1,788	3,298	5,629	7,091
Government	68	113	203	459	598
Total	998	1,901	3,501	6,088	7,639

Sources: For 1951, Petroleum Administration for Defense, *Transportation of Oil* (1951), date of tabulation Apr. 1; otherwise, Sun Oil Co., *Analysis of World Tank Ship Fleet*, respective years. (Sun Oil Co. was also the ultimate source of the PAD data.)

Notes: Because the U.S.-registered fleet is high cost and noncompetitive, it plays no part in the world market.

A T-2 tankship of 16,765 tons and 14.5 knots is used as the unit of carrying capacity.

There is an important discrepancy between the Sun and other tabulations of ownership. As of the end of 1968, John I. Jacobs & Co., *Tanker Fleet Review*, and BP annual *Statistical Review* gave oil company ownership at about 33%.

In a personal letter E. E. Bowyer of Shell International has kindly set out and explained the disagreement:

		As reckoned by:	
Year 1966: percent held by:	*Sun*	*Jacobs* (and *BP*)	*Shell*
Oil companies	25.2	35.9	29.9
Independents	66.0	61.3	61.4
Governments	8.7	2.8	8.7

Sun and Shell regard a completely nationalized company, including a Soviet company, as government; Jacobs looks upon it as an oil company. Sun regards ships held under Demise Charter as owned by the nominal owner; Jacobs and Shell classify them as owned by the oil company chartering. Although I would regard the Shell method as nearest to the mark, there is obviously no "right" or "wrong" theory. In any case, the Sun tables are the only continuous series for the period, and the only reproducible ones–i.e., they show individual company ownerships, from which anyone may compute totals.

voyage or long-time charter for an average period of five years, then about a fifth of this group, or 14 percent of the fleet, reentered the market in any given year. Thus in any given year about 25 percent of the world privately owned tanker fleet would be available for charter. With a 10 percent growth rate, about 35 percent would be available in any one year.

Chartering is done through brokers centered chiefly in London, Oslo, and New York. Each individual ship available for spot charter is, in effect, like a separate firm, and the worldwide market allows no protected enclaves outside the United States, which we need to ignore.[6] In any given month, several dozen ships are offered for oil-company use all

[6]See note to table IV-1. See also Zenon S. Zannetos, *The Theory of Oil Tankship Rates* (M.I.T. Press, 1966).

over the world by several hundred owners, none with over 5 percent of total tonnage. Tacit collusion would be impossible, and no attempt at open collusion has been made since World War II, with the exception of the mild and abortive Intertanko scheme.[7] The single-voyage "spot" charter market therefore seems purely competitive.

The time-charter market is linked to the spot market at one end, and at the other to the cost of creating new capacity. Here entry is open and cheap. In 1950 a "large" ship was 25,000 tons, and it cost a little over $6 million; in the early 1960s a "large" ship was twice as large but cost only about $5 million. By the late 1960s a 200,000-tonner was considered a "very large crude carrier" (VLCC) and it cost about $17 million. The explosive ship market of 1970 carried it to about $25 million (see Table IV-6 below). Moreover, there are no strong economies of scale in ship operations. Many owners have only one ship—an oddity until one realizes that many firms perform services incidental to tanker shipping, and take advantage of some particular opportunity to buy one or a very few ships. But to say that many competent firms cluster on the boundaries of the industry, and that the minimum capital requirements are low, is to say that entry is easy and market control impossible.

With many ships available in the short run, and easy entry for the long run, what possibility is left for control in the meantime? Little if any in theory, and none can be observed in practice. Tankship owners, oil companies, and independents cannot control the long-term supply even in concert, for anyone contemplating a production or refining investment and needing the transport service has time to charter a ship or buy a new one. As an independent corroboration we have Zannetos' demonstration that long-term charter rates are almost completely explained by the cost of buying and operating. The data leave much to be desired, but no systematic appreciable discrepancy can be found between what has been observed and what would be expected under pure competition.[8]

Since the tanker market is purely competitive, one would expect that a given oil seller or buyer can have all the tanker service he wants at the current price. This is not strictly true of the single-voyage rate. The number of ships and of bids available within a few days is so small, and the penalties—of an idle ship on the one hand, or an idle refinery or field on the other—are so great that neither owner nor charterer can afford to wait. Only by improbable chance would the number of ships available on any given day be precisely equal to the number of ships demanded that day. Hence the rate fluctuates considerably from day to day and week to week. As in other such markets (e.g., fresh fruits and vegetables), a given buyer or seller may find himself unwillingly a monopolist or monopsonist, discomfited by his little brief authority. An owner (or a charterer re-letting) who has the only two or three ships available at the moment has a possibly unpleasant choice between (a) offering both or all of them at sharply lower rates, or (b) offering, say, one in order not to break the rate, and holding the others in idleness. Similarly, a would-be charterer who needs several ships and is the only one in the market that day must either (a) try to fill all his needs and possibly drive up the price, or (b) must wait and risk incurring higher costs in production or refining.

The short-term market is therefore fascinating to watch and rewarding to the skilled, but we shall now see that for our purposes its constant restless dance is not important.

[7] *Petroleum Press Service*, Apr. 1964, p. 149.

[8] Zenon S. Zannetos, "Time Charter Rates," in Sun Oil Co., *Analysis of World Tank Ship Fleet*, 1966.

The Term Structure of Rates

Our interest is in the competitive relation of f.o.b. and delivered prices. The delivered price must equal the f.o.b. price plus the cost of current available transport. For if the delivered price were higher, sellers would not be willing to buy oil delivered. It would pay them or others to hire tonnage and pick up cargos f.o.b. Conversely, if the delivered price were lower than f.o.b. price plus current transport cost, tanker operators would need to accept a lower rate for this voyage than for others, or the producers would need to accept a lower return. Either or both of them would turn away to the normally profitable destinations until the price came back to equal them. Hence a persistent divergence either at loading or unloading ports between delivered prices and the sum of f.o.b. plus current transport rates is a sign of a restraint on buyers or sellers.

Which rate is "the current transport rate" depends on the transaction of which it is a part.

The single-voyage or "spot" rates are the least important for our purposes, but *not* because they cover only a small part of the total tonnage in use. With tankers as with crude oil or products, or any other samples from a large population, the absolute size of the sample is what counts; the proportion sampled does not.

But the single-voyage rate is a measure only of the delivered cost of that one cargo. There are few data on single-cargo transactions, but even if there were, it is often difficult to connect up a given rate with a given shipment. For example, if a shipment arrives on December 1 at a given port, the relevant rate is not that observed on December 1, nor even on the date of loading, which may be three weeks or more earlier, but rather the rate which the parties agreed or counted on when they made the agreement to buy and sell the cargo.

More interesting are contracts which name the delivered price of a given quantity to be delivered over some specified period, e.g., October–March inclusive. It will probably not do to reckon the netback f.o.b. loading port by subtracting a spot rate, nor even the average spot rate over those six months as known after the fact. For unless we assume that the spot rate was forecast with tolerable accuracy, which is possible, we cannot tell what calculations or commitments, and at what price, were made by buyer or seller to acquire tanker service. For such a transaction the consecutive voyage rate for October–March as set in the previous quarter (June–September) is our best measure. In effect, it is the current market prediction of average spot rates over the six-month life of the charter. The shipper has provided in advance for the tonnage needed, at that price. If he happens to be using his own tankers, he is giving up the consecutive voyage rate or the average of expected spot rates, which they could otherwise earn. It measures his opportunity cost. Most consecutive voyage charters appear to be signed with some particular delivery sequence or program in mind.

However, our principal interest is in a period long enough for an investment decision in production and refining. We seek the amount per barrel which an integrated oil company nets back to the lifting port, or which an independent refiner can pay and stay in business. We need therefore the best arrangements available to oil companies making large commitments for years ahead, perhaps enough to amortize the producing and refining investment. This would ordinarily be a time charter unless, as sometimes happens, the company expects lower costs with a succession of short-term contracts.

Time-Charter Data, 1948–1952.[9] For this period, there are two sources. The first is an "award" issued in April and October during 1948–53 by the London Tanker Brokers Panel (LTBP): the time-charter rate at which a 12,000-ton vessel could be hired for two years, translated into voyage-ton (Scale) equivalents. This award was generally known as "the London award," a misleading title and thus avoided here. Since World War II the LTBP has made hundreds of "awards," which are their estimates of rates that would be contracted for some particular service if there were enough buyers and sellers bargaining over it at any particular time. Designation of any one as "the" London award is unfortunate; to avoid ambiguity we will call it the Two-Twelve award. It covered a two-year period, nearly long enough in those days to build a new ship, thereby expressing most if not the full range of alternatives open to a shipper.

The Two-Twelve is correct in theory—it names the current price of a regular flow. But it is biased upward in its use of too small a ship at too high a cost. During 1948–52, the war-built T–2, of 16,675 tons, was approaching a marginal condition where it would have been the highest-cost price-determining facility. The force of this qualification is not constant, however, but grows through time. As of October 1952, the world tanker fleet contained 1,727 T–2 equivalents, of which 15 percent were ships under 12,000 tons and another 13 percent between 12,000 and 16,000 tons.[10] But the marginal rate-determining ship is the highest-cost one which is actually being bargained over for current use. As early as 1953, the T–2 may already have been getting past marginal status; there was a report that one or two T–2s had been laid up because of the collapse of tanker rates to as low as USMC less 40 percent for certain long runs.[11]

But if the Two-Twelve figure is too high, the next approximation is somewhat too low. This is the five-year time-charter rate of ships newly chartered at that time. Its apparent precision is misleading; most entries are interpolations. These rates represent the best course of action for an integrated oil company, or for a buyer or seller committed to a five-year contract. However, we have no assurance that sales contracts actually were this long. Hence, to charter a ship this far ahead might be a reasonable risk for a large integrated producer-refiner of crude oil, but not for an arm's-length buyer, for the former might find himself with an expensive ship which had nothing to do. Under those circumstances he would turn to rechartering, in which case the Two-Twelve award or, in the extreme case, the spot rate might be the relevant one.

In general, the true tanker rate would lie somewhere between the Two-Twelve award and the five-year time-charter rate if we assumed that most of the oil traded at arm's-length was sold on contracts which usually exceeded a year but fell well short of five.

Tanker Rates Summarized, 1960–1967. For the later period, the relevant tanker rates are first summarized to give a general view. We will then examine the calculation of the time charter rate which determines the maritime transport cost comparable with oil prices at a Persian Gulf loading port and a Northwest Europe unloading port. Table IV–2 shows that it makes a difference which rate is used as a reference.

In 1960, spot and consecutive rates were some 20 percent below time charters; over the next two years they converged. In 1963, spot rates rose strongly but consecutives declined mildly. Over the next three years all series declined, time charters least. Over the

[9] At this point only the uses and limitations of these data are explained. For the numbers and application, see chap. V.

[10] Sun Oil Co., *Analysis of World Tank Ship Fleet*, 1953.

[11] *New York Times*, Mar. 6, 1953.

TABLE IV-2. The Term Structure of Tanker Rates: Principal Types, 1960–1970

(Intascale less percent shown, Worldscale as percent shown[a])

Year	Single-voyage (spot) rate: Mullion *IS*	Mullion *WS*	Platou *IS*	Platou *WS*	Consecutive voyage rate (50,000 *dwt*) *IS*	*WS*	Average T/C rate Persian Gulf–Rotterdam, by incremental ship (through Suez Canal) *IS*	*WS*
1960	–53	n.a.	n.a.	n.a.	–55	n.a.	–45	n.a.
1961	–55	n.a.	–59	n.a.	–57	n.a.	–45	n.a.
1962	–49	n.a.	–50	n.a.	–50	n.a.	–47	n.a.
1963	–37	n.a.	–38	n.a.	–38	n.a.	–47	n.a.
1964	–44	69	–46	66	–50	62	–54	57
1965	–43	71	–49	64	–53	58	–54	57
1966	–48	65	–52	60	–50	62	–55	56
1967								
January–May	–59	51	–63	46	–55	56	–57	53
							(Round Cape voyage, by VLCC)	
June–December	+34	166	+33	165	n.a.		–72	31
1968	–2	110	–2	110	n.a.		–71	32
1969	–22	85	–23	85	n.a.		n.a.	36–39
1970								
January–May	n.a.	120	n.a.	119	n.a.		n.a.	40
June–December	n.a.	235	n.a.	230	n.a.		n.a.	(see text)

Sources and Notes:

Spot Rates. Mullion–Averages of weekly single-voyage averages computed by Mullion & Co. *Platou*–Average for single voyages from *R. S. Platou A/S*, Annual Reports. Figures for 1961 and 1962 are only approximate. Annual averages are biased downward because they are simple averages of weekly figures. The higher rates in the winter quarters correspond to more than a proportionate share of the oil moved. Therefore a properly weighted average would be somewhat higher. This bias is minor in 1960–62, but appreciable in 1963–66, when seasonal swings were large: there was practically no seasonal swing in the nine months before the June 1967 war.

Consecutive Voyage Rate: 1960–62–Harley Mullion & Co. (now Mullion & Co.), *Tanker Freight Statistics* gives average consecutive voyage rates (soft currency) for 6-months/3-year voyages. There is no allowance for seasons or size of ship. The rate for a 50,000-dwt tanker was estimated by extrapolation where no consecutive voyage fixtures were made. 1963–67: an average for fixtures is estimated from inspection of the weekly reports of Mullion & Co. and H. Clarkson & Co. The total range for a 50,000-dwt. tanker is not more than 10 percentage points.

Time Charters. See table IV-5 below.

[a]For Intascale rates see appendix IV-B. Worldscale was only promulgated in 1969; entries for earlier years are approximate equivalents to "old" Intascale, i.e., before sterling devaluation in November 1967.

n.a. = not available.

whole period the spot rate described a cycle, ending eight percent lower than it started; the consecutive-rate cycle was mild and ended where it started.

If my theory and calculations are acceptable, the time-charter rate declined, slowly but unmistakably, 12 percentage points in seven years; the reason will shortly be explained. Since one point of Intascale corresponded to 0.91 cents on the Persian Gulf–Rotterdam run, one would expect a decline of 11 cents per barrel of crude or of composite product at Rotterdam.

Incremental "Canal Ship" Determines Price of Freight Service. For the period 1960–67, transport of oil from the Persian Gulf to Europe required a transit of the Suez Canal. Because of the 38-foot draft limitation, tankers of more than about 60,000 deadweight tons could not transit the Canal fully loaded. However, it would not do to assign the lowest possible time-charter rate, i.e., that of the 60,000-ton ship, as the current long-term transport cost. In 1960 ships of 60,000 dwt. were relatively few, some 2 million tons in a world fleet of 65 million. Traffic through the Canal was mostly in "large" tankers of 30,000 to 45,000 dwt. But, as Table IV-3 shows, the situation was due for a very rapid change, and everyone was aware of it. Since better than half of all tankers on order at the end of 1961 were in the range between 40,000 and 65,000 tons, long-term charters were made at rates set with these ships in mind. By their expected availability soon, they depressed the current rates at which older tankers could be chartered.

Nevertheless, we have only narrowed the uncertainty, not removed it. Of the tankers old and new which could be relied on and planned for, it was neither the best nor the average tanker which was relevant to the determination of price, but rather the incremental tanker, or the highest cost new tanker which was needed in order to insure the flow of oil.

Companies preparing to bring oil from the Persian Gulf to Rotterdam encounter diminishing returns. Starting with the best tankers, the industry must use increasingly costly packages of draft, speed, deadweight tonnage, etc. If demand were low enough, the 70,000-ton ships would be sufficient for all Canal transit, at lowest rates. But as more freight service is needed, companies must draw in ships of 60,000, 50,000, and 40,000, at successive higher rates. Since our interest is in what determined the price at point of loading and destination, we must look not to the average or to the best available at any given moment, but rather to the most expensive which must be utilized. Older, smaller, and less efficient ships are extra-marginal and will fetch no higher rate because they can at best only be substituted for the marginal ship.

Table IV-3 shows the tonnage of ships on order, in the various size classes that may be regarded as "Suez Canal ships." The relatively wide spread of size calls for some explanation. Two vessels of the same deadweight tonnage may vary considerably in dimension and performance. During the early 1960s, efforts were made to build the largest size ship compatible with passage through the 38-foot draft limits of the Suez Canal. The "chubby" ship gradually raised the upper limit of tonnage to 60,000 and for a few ships

TABLE IV-3. "Canal Tankers" under Construction or on Order, December 31, 1960–1966

(*millions of deadweight tons*)

Size (000 *dwt*)	1960		1961	1962	1963		1964	1965	1966	
40–45	1.3*	(3.0)	1.1	0.5	0.4	(4.2)	0.0	0.0	0.0	(5.0)
45–50	4.3	(3.8)	3.3*	2.3	1.7	(7.5)	0.8	0.5	0.4	(8.8)
50–55	1.3	(0.6)	2.7	2.6*	2.1*	(3.2)	1.2	0.4	0.0	(5.8)
55–60	0.2		0.8	1.4	3.2	(0.6)	2.0*	0.6	0.2	(4.0)
60–65	0.0	(0.7)	0.2	0.2	1.6	(0.4)	1.5	1.0*	0.5	(3.0)
65–70	2.1		1.4	1.2	3.7	(1.1)	3.3	1.9	0.9*	(5.4)
Subtotal	9.2	(8.1)	9.5	8.2	12.7	(17.0)	8.8	4.4	2.0	(32.0)
Total, all sizes	15.4	(65.8)	15.7	14.0	19.2	(76.2)	17.7	20.6	27.4	(102.9)
Tonnage on order as percent of stock		23.4%				25.2%				26.6%

Source: Sun Oil Co., *Analysis of World Tank Ship Fleet*, respective years.

Note: Size classes are to be read: "between 40,000 and 44,999 tons, between 45,000 and 49,999 tons," etc. The columns in parentheses are the total fleet at that date.

*Smallest ship whose stock is being expanded.

even 70,000 tons. Moreover, it became common practice for ships of 70,000 and even 80,000 tons to be chartered only for the maximum load that could be carried through Suez, with additional tonnage paid for only when and as utilized. This was obviously a bet that in time the Suez draft would be increased enough to let these ships go through fully loaded. In time, these might have become our marginal ships.

Let us apply this theory to Table IV-3, and find the approximate marginal vessel, year by year. The size marked with a star is the smallest size ship whose stock is apparently still being expanded, by new orders exceeding scrapping. But by 1966, all ships below 65,000 tons were clearly a diminishing stock; probably also the 65,000-70,000 group, since new orders need not even have been positive.

Having identified the marginal ship year by year, it is possible to calculate or, more accurately, to identify its charter rate during 1960-67 in section A of Table IV-4. Ship, rate, and value in costs per barrel are summarized in section B of the table.

Occasionally independent published data can be found against which to check our estimates of freight rates.

1. For example, in mid-1964 a London broker calculated current voyage-ton equivalents in the approximate figures shown in Table IV-5. The Jacobs estimate for the lower end of the range is consistent with ours. The lower rates on the large ships were not available to anyone shipping from the Persian Gulf to Northwest Europe, unless they were willing to go around the Cape and bypass Suez. We shall consider this later.

2. In reporting on the state of the market in 1966, Platou cited a 52,000-ton motor ship which had been delivered in 1962 and was sold in 1966 with a time charter of $1.735. Appendix IV-A shows that this would be approximately Intascale less 55 percent, or 53 cents per barrel of oil carried including the Suez Canal charge. (The result is accurate only to within 2-3 cents per barrel, for there is room for minor error and change at a great many places.) Platou's comment is significant: the charter did not affect one way or the other the sales value of the ship. Therefore a charter equating to about minus 55 would be the current rate for such a ship. Reckoning the 1966 marginal ship in the 60,000-70,000 dwt class, its rate must have been at or slightly below Intascale less 55, and decreasing.

The fact that new orders for 52,000-tonners had disappeared shows that rates of Intascale less 55 were not profitable, could not attract new investment. Conversely, larger ships were profitable enough to draw in large new orders.

This freight cost is strictly appropriate to long-run investment decisions—to build or not to build a refinery, to make or not to make a major investment in crude production capacity. For such decisions, one must look to the whole productive life of the new facilities—though of course even considerable errors are tolerable since far-off mistakes have small present value. Over this period, the relevant variable is: the market price of transport service which can be *bought today* for use over a long time. This is the price of service by the marginal ship.

For shorter-term decisions, cheaper alternatives *may* be available. If one considers signing a three-year contract, for example, and if spot rates are at the moment well below the long-term supply price of the marginal ship and are not expected to revive immediately, a small premium above current spot rate may be enough to obtain a three-year charter; for the owner may prefer an assured lower return to the chance of a later higher return.

Working on the principle that doubts are always best settled in the direction of the long run, the rate in Table IV-4, section B, is the one used except when it is known that the f.o.b. or c.i.f. price refers to a shorter time period, and especially when it covers less

TABLE IV-4. Charter Summary and Incremental Ship and Transport Cost per Barrel, 1960-1967

A. CHARTER SUMMARY

Year	Charters (3 years or more)	Marginal charter rate (Intascale less %)
1960	One two-year time charter, 32,750 tons, IS -57, not close enough to be basis for estimate. I have extrapolated back the 1961 value, which happens also to be the one used in an earlier publication, and not challenged.	-45
1961	A group of 53,000-ton ships chartered at -45, a higher rate than for smaller ships.	-45
1962	All (but one) Canal ship charters in the 40,000-49,000-ton range; low is -50, high is -47; the one 56,000 tonner at -47.	-47
1963	Eight 50,000-59,000 tonners range from -46.4 to -54.2, all but one below -47; smaller ships from -44.7 to -50.2.	-47
1964	No charters for ships below 50,000. The 50,000-59,000 group range from -54 to -58; larger ships at lower rates.	-54
1965	Ships in the 40,000-49,000 group chartered at -56 to -53. No charters in the 50,000-59,000 group. Ships of 60,000-69,000 tons fixed between -58 and -60; larger ships at lower rates.	-54
1966	Smallest ships chartered are in the 50,000-59,000 group, one at -54.5, next at -56.6 and lower. Larger ships at lower rates.	-55
1967 (January-May)	Few term charters signed, rates seem lower than for corresponding ship sizes a year earlier.	-57

B. INCREMENTAL SHIP AND TRANSPORT COST PER BARREL,
Persian Gulf-Rotterdam Run via Suez, 1960-67

Year	Incremental ship (000 *dwt*)	Average T/C rate (Intascale less %)	Transport cost in ¢/*bbl* (incl. Suez toll)
1960	45	45	61
1961	45	45	61
1962	50	47	59
1963	50	47	50
1964	55	54	53
1965	60	54	53
1966	60	55	52
1967	65	57	50

Source: For section A, worksheets from Z. S. Zannetos, "Time Charter Rates," in Sun Oil Co., *Analysis of World Tank Ship Fleet*, 1966. Method explained in appendix IV-A.

TABLE IV-5. Large Tankers, Spot Equivalents, 1964

Size of tanker (000 *dwt*)	Duration of charter (*years*)	Intascale less (*percent*)
55-60	12	-55
70	5	-55
70	10	-58
95	...	-62½
97	...	-65

Source: John I. Jacobs & Co., *World Tanker Fleet Review*, June 1964.

than 12 months, or any number not close to a multiple of 12 and hence a period when seasonal fluctuations will not cancel each other.

The Average Freight Rate Assessment (AFRA). AFRA is the best known member of a family of computed rates; the weighted average of all charters in force during a given period.[12] This includes all spot charters fixed during that period, as well as all consecutive-voyage charters and time charters in effect, regardless of when they were signed. As a measure of the current flow of funds from the oil industry to the ship operating industry, AFRA may be useful.

But AFRA is no part of any current arm's-length transaction, and is therefore irrelevant to any market price or to any investment decision. The AFRA rate is not comparable to any f.o.b. prices or delivered prices for any long or short period.

1. AFRA mixes up rates for different services: spot, consecutive, and time charters. This is unfortunate but not necessarily fatal, since these three kinds of charters are imperfect substitutes for one another. An averaging together of all *new* charters signed during any given period might in practice be subject only to tolerable index number bias and imprecision.

2. AFRA includes clean products carriage (i.e., other than "dirty" crude and heavy fuel oil) which is inherently more expensive and is also in smaller more expensive ships. This is a less serious defect than previously, since AFRA is now calculated separately for three size classes, and clean products are carried in smaller sizes. If one stays with the larger ships, this is a fortuitous partial escape. But of course the lumping together of ship sizes, averaging together all charters in a class, is objectionable because it ignores the economies of scale and constraints like the Suez Canal.

3. Even if we overcame these other defects, the AFRA rate would still be not imprecise but irrelevant. To average in charters signed in the past at various times, under various supply and demand conditions at various rates, is to import confusion and irrelevance into the reckoning. Once the contract has been signed, bygones are bygones. If rates go up, an oil company which chartered previously at lower rates, or which built its own ship, has in effect secured an earning asset from which it will draw additional earnings so long as the contract lasts and prices stay high. If they no longer need the tanker, they can re-let the ship and take their profit. Contrariwise, if a company is stuck with charters made at higher-than-current rates, the damage has been done and the loss they incur is measured by the excess of what they must pay under the old contract over what they would pay now through new contracts. At any given moment, the opportunity cost of using the ship in their own operations is what it is worth *then*, either by saving the expense of a new charter, or by what it could earn on re-let. Whether we are operating our own ships, or chartering, or buying, or building, *the current long-term charter rate* is the market fact, the price relevant to a term contract for the sale of oil.

4. If tanker rates changed little over time, a bad theory might not have worked a great distortion. But in fact the trend of tanker investment costs and operating costs has been strongly downward. Hence, to average in the freight rates of previous years with this year

[12] A good summary of AFRA is in *Petroleum Press Service*, Aug. 1964, p. 292; there have been minor changes since. AFRA is announced every month for four size groups: 15,000–25,000, 25,000–45,000, 45,000–70,000, and 80,000–160,000. Time charters are converted to Intascale or Worldscale on the basis of the Aruba-London voyage. Spot charters are included in this manner: the Mar. 1 AFRA will include spot rates for Jan. 16–Feb. 14, etc.

produces a strong upward bias. Outside of Suez crises the AFRA rates are invariably much higher than current time-charter rates, which are the current cost of transport.[13] Applied to current f.o.b. prices and delivered prices, they would produce results absurdly high or low.

But AFRA rates did for a time have some importance as helping to explain the delivered price level in Europe and elsewhere. So long as governments could be persuaded that AFRA was a proper measure of freight costs, they would either permit or oversee the fixing of delivered prices of products covering AFRA costs. This has been particularly true in India, where the reports of two oil study committees are worth reading in this connection.[14]

In countries where refining was a private enterprise, the AFRA rates could be used as a kind of posted price. If every integrated producer-refiner used posted prices plus AFRA rates to calculate his landed cost of crude, the price of products would be determined within narrow limits. One need only add the refining margin including an adequate return on investment, and the resulting price would insure a decidedly high return on the whole integrated operation. But this could only be done if everybody adhered to the practice, and knew everyone else would do so too. The history of prices in the important consuming areas since 1956 is essentially a history of the breakdown of this informal understanding. By the mid-1960s, the understanding had been eroded everywhere but in the tightly regimented French market. "Based on the theory that a refiner must get a satisfactory return," AFRA rates are still used by the government to fix prices.[15] Moreover, AFRA may still be used by integrated refiners in reckoning for income tax purposes the delivered cost of crude to them. Legal rights and duties are not our affair. More generally, one could not claim that AFRA was of no use for any conceivable purpose. But AFRA is not a market fact.

Yet for sheer lack of any other tanker freight rate except spot rates, AFRA has long been and still is cited in respectable places. *Perspectives*[16] is probably the best governmental discussion of supply and price since World War II. Our criticism is a measure of its merits and its success—it set out its assumptions clearly enough to be understood, and therefore to be criticized. As late as 1964, it considered Intascale less 20 percent to be a current long-term tanker rate, and Intascale less 30 percent as a likely future long-term rate. Yet during 1961 a large number of ships, some afloat and some to be newly built, had already been chartered for long-term service at Intascale less 45 percent. The basic theoretical mistake was in ignoring the existence of a *current long-term rate*, which fitted the current price structure.

By early 1966, the sequel or supplement to *Perspectives* was issued. In the appendix, a rate of Intascale less 50 percent was projected for 1970 although it had already been in effect since 1963. Perhaps more important, in a later publication they could say no more

[13] Until July 1964 there was no AFRA for Canal-size ships. Thereafter the January and July figures for the 45,000–70,000 group was: 1964, –36; 1965, –39, –40; 1966, –42, –41; 1967, –46, –47.

[14] India, Ministry of Steel, Mines and Fuel. Department of Mines and Fuel, *Report of the Oil Price Enquiry Committee* (Damle Report), New Delhi, July 1961; Ministry of Petroleum and Chemicals, *Report of Working Party Group on Oil Prices* (Talukdar Report), New Delhi, 1965.

[15] *Petroleum Intelligence Weekly*, Feb. 1, 1966, p. 1; Apr. 8, 1968, p. 6. But see Farid W. Saad, "France and Oil: A Contemporary Economic Study" (Ph.D. diss., M.I.T., 1969).

[16] ECSC (European Coal and Steel Community), *Etudes sur les Perspectives Energétiques à Long Terme de la Communauté Européenne* (1962), Annexe 11 (1964).

about current marine transport than to note a slight firming of AFRA in the second half of 1966.[17]

The Relation of Long- to Short-Term Rates. In emphasizing the long-term rate, I may have done the reader a disservice by seeming to suggest that shorter-term rates are of little relevance to the problem of measuring transport costs. This is not true: the spot rate is an indicator—of sorts—of the current price of long-term service.

Single-voyage charters, consecutive voyage charters, short-term charters (say two-five years), and long-term charters form a chain of substitutes. Their prices cannot diverge by more than the net cost advantage of one over the other. In addition to brokerage, there are many costs involved in a large volume of frequent procurements. (a) It is inefficient, hence costly, to program production and refining while chasing after ships all over the world, waiting on the ship with the cargo, or vice versa. (b) There is the risk of not keeping a ship busy. Zannetos reckons a penalty for risk of unemployment if a ship is kept in the spot market. Since the charterer assumes the responsibility, the time-charter rate should be less than the spot, all else being equal.[18] It might be urged that the efficiency of using a given vessel on shuttle service more than compensates for the risk.[19] Obviously the balance of advantage will shift over time.[20]

One often hears that oil companies, especially large ones, must own or charter 85 percent or so of requirements, and are not interested in the spot market except for fill-ins. To describe a practice (not too accurately at that, since company practices differ) is not to explain it. Any time that the gap between two or more rates exceeds what the trade estimates as the net advantage, a movement out of one into the other will bring the two together again. As pointed out earlier, the need for new tonnage in an expanding industry, plus tonnage coming out of charter, plus the short-term market, means that charterers and owners are always in process of adjustment; movement from one form of charter into another is not something to wait upon but something in process all the time.

The long-term rate is set, as has been argued, by the cost (including acceptable return on investment) of buying and operating the incremental ship on a given voyage. As more lower-cost vessels are delivered, they increase the supply and force the higher-cost ships out of the time-charter market even before the contract is expired, since the charterer may re-let the ship at his discretion. Small ships have some advantages. Large ships are less flexible—can enter only a few ports—while the smaller the ship the more places it can enter. Smaller ships can go into clean products carriage, at higher rates, or into the grain trade.

Figure IV-1 shows the relationship. The rate curve whips up and down, above and below the long-term rate which serves as its anchor point. But the anchor itself slowly moves. While the delivery of new ships has no necessary perceptible effect on the current fluctuations of the spot or consecutive voyage charters, the increase in the supply of ships draining out of the term charter market into spot certainly has an influence on the average level of spot charters over a long enough period of time to average out random and seasonal fluctuations.

[17]ECSC, *Nouvelles Réflexions* (1966); and ECSC, Bulletin 67 (1967), p. 26.

[18]Zannetos, "Time Charter Rates."

[19]Or, as a broker put it in 1963: to the charterers, a spot rate of minus 55 to minus 60 means about the same thing as a five-year charter at minus 45.

[20]Hence Zannetos enters risk of unemployment and an efficiency premium as separate variables.

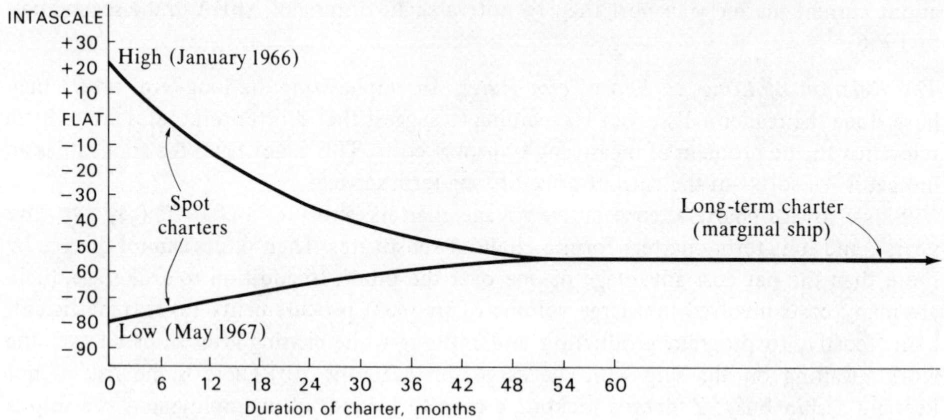

Figure IV-1. Tanker rate structure, Persian Gulf-Rotterdam, 1966-1967.

If the supply of tankers available for spot trading keeps increasing faster than the demand, rates fall to the point where some higher cost tankers are transferred into the grain trade, or laid up, or scrapped because not even operating costs can be covered. This is the way a competitive market works.

The single-voyage rate is therefore a genuine albeit inefficient and biased indicator of the long-term rate, which strongly affects but does not determine it. Table IV-2 measures its defects. But if there were nothing else available, then in a period free of massive shocks like a Suez Canal closure, a 12-month average of spot rates would for many purposes be a fair indicator of the long-term rate. Of course it depends on the purpose and on the degree of tolerable error; but, as a measure of the long-term rate, the average of spot rates is literally infinitely better than AFRA, which is no measure at all.

Reasons for Declining Time-Charter Rate

It remains to ask why tanker rates have trended downward so strongly. It was not because of any tanker "surplus." Obsolete high-cost ships were, of course, laid up and scrapped throughout the period. (Some went straight off the building ways into layup.) But even at the 1960 low, tankers on order were one-fourth of the operating fleet. Tanker owners were not merely replacing tonnage, but adding to it. Therefore tankers were not in surplus; there was a chronic shortage even at the May 1967 low.

Aside from seasonal troughs, there were no "low" or "depressed rates." In no sense has tanker transport ever been abnormally or artificially cheap. Tanker owners do not invest scarce capital funds, which could draw a return elsewhere, unless they hope to earn satisfactory profits. If tanker rates had not provided a satisfactory return, no new ships would have been ordered.[21] The stock of ships would have shrunk, and—especially in the face of rising traffic—the lower supply would have quickly forced up tanker rates and hence delivered oil prices as well. That is the only possible meaning of tanker "surplus" or of "depressed rates."

[21] An oilman in partial dissent: "The owners ordered ships because they counted on another Suez shutdown. And how right they were!"

But if tankers were actually in such demand that owners were increasing the fleet, then there was no surplus even in 1960, and no reason whatever to consider tanker rates temporarily low and due to rise.

In the years after 1960, tanker orders surged and the fleet grew more quickly. But, as Table IV-5 shows, the marginal ship grew larger, newer, and cheaper. Given a competitive tanker industry, the price of short- or long-term service declined along with the incremental cost of rendering the service.

But we need to question low ship prices, which may result from a temporary surplus of shipbuilding capacity which forces incremental cost of building new tankers well below the cost—including a return on new investment as the old facilities wear out. If so, once the surplus disappears, incremental cost will rise again, as will ship prices and tanker rates. We need to know: what prices can oil companies count on for future planning?

Table IV-6 presents a series compiled largely from Norwegian ship brokers' reports, confirmed over shorter periods by a number of other sources. The Japanese trade press, notably the monthly *Zosen* (formerly *Japan Shipping and Shipbuilding*) has in recent years carried the specifications, including price, of nearly every tanker sold for export. The French weekly, *Journal de la Marine Marchande* (*JMM*), reprints them together with less systematic bits of information about ship prices in Japan and elsewhere. The 1965 Annual Report of the Swedish shipbuilder Gotaverken noted as a commonplace that during the previous year Japanese prices had set the norm for world prices; and their data for prices paid by Norwegian shipowners are close to the Japanese. At an earlier date, the accepted rule of thumb had been that Swedish prices were about five percent higher, reflecting both higher costs to the builder and greater convenience and lower acquisition costs to the buyer. In late 1967, a Japanese shipbroker estimated Japanese prices to be 5-10 percent below European.[22]

Shipbuilding Innovation in Japan. Lower prices quoted by Japanese shipbuilders startled the trade in the spring of 1963.[23] Some felt the shock of pleasure, others of dismay. We shall need to look at the background.[24] In the late 1950s the Japanese government sponsored a consolidation of shipbuilding enterprises from about 50 to about 13. It was believed, probably rightly, that economies of scale required much larger yards and firms. The quick simulation of the competitive process, which forces out the less in favor of the more productive, accomplished under government auspices what might otherwise have taken years through private initiative. The new firms started operating in time partly to induce, partly to meet, the great upsurge in the building of new large tankers which began in 1963.[25] Since tanker operation is purely competitive, there was no barrier and, instead, a strong incentive to adopt the innovation. With remarkable speed, lower costs were

[22] *Journal de la Marine Marchande* (*JMM*), Sept. 28, 1967, p. 2161.

[23] *JMM* provides an excellent running account of the gradual realization in the trade.

[24] Ching C. Chen, "The Competitive Strength of the Japanese Shipbuilding Industry," unpublished paper, 1965.

[25] Japanese expenditures on shipbuilding equipment were as follows for the fiscal years ending in March (in millions of dollars at 360 yen per dollar): 1960, 28.6; 1961, 45.2; 1962, 55.5; 1963, 54.7; 1964, 65.6; 1965, 98.4; 1966, 98.7; 1967, 79.5; 1968, 117.1; 1969, 139.8. (*Zosen*, Aug. 1969, p. 29.) This is not a precise indicator of expenditure on tanker facilities, but it shows the trend. The 1968-69 fiscal year expenditures on tanker building equipment was $113 million (*Zosen*, Nov. 1969, p. 15). The low level in 1959-60 might be compatible with the idea of excess capacity, but not the strong rise through March 1963, accelerating thereafter.

TABLE IV-6. Ship Prices Paid by Norwegian Operators, 1951-1969

Year	Size (000 *dwt*)	$/*dwt*	Remarks
December 1951	19	220	Diesel
December 1952	20	182	Turbine
1953	27	135	Turbine
	18	140-148	
1954	46	150	Few firm prices
1955	46		No reported contracting
October 1956	46	114	Japanese prices above European, but were definite, not subject to escalation
End 1956	46	225	
Early 1957	46	200	
Spring 1957	46	230-240	Japan
End 1957	46	170-175-180	Prices definite; Japanese now at European level
1958	47	142-145	Europe, Japan
	19.5	187-200	
1959	47	120-125	
Mid-1960	47	120	
End 1961	80	122-126	Hypothetical "award"-type figure; no actual contracting
	51-58	127	
	47	135	
1962	53	125-130	Europe } Deep depression, few orders
		105-110	Japan } Deep depression, few orders
	47	115	
1963	87	96	Swedish prices match Japanese, allowance made for lower acquisition cost to buyer
	53-65	105-100	
Early 1964	70-75	90	Japanese set world prices
End 1964	87	94	
	70-75	97-95	
1965	150-160	75	
	120	80-78	
	90-95	90-88	
	80-85	94-92	
	70-75	97-95	
End 1965	200	65	
	87	98	
1st half 1966	210	63	
	175-200	75-65	
2nd half 1966	300	65	
	200	68	
	175-200	69-78	
	90-100	92-90	
	90-95	95-92	
	87	100	
1st half 1967	200	68	
	90-100	95-92	European prices only; no Japanese bids
2nd half 1967	240-250	71-70	
	210	70	
	200	70-72	Japan
		74-76	Europe
	90-100	100	
End 1968	210	79	
	200	83	
	90-100	105	
End 1969	280	89	
	250	92	
	200	95	

Sources: Compiled from R. S. Platou A/S (Oslo), Annual Reports; Fearnley & Egers (Oslo), Annual Reports; *Zosen*, Sept. 1968, p. 5.

translated into lower prices. Such few glimpses as we get of the earnings of shipowners show that those quicker to bring forth a new technology have had their reward. The rest complain.

The shipbuilding industry in Japan is highly competitive. A clear indication of this is in the complaint by a trade journal that prices could be raised with very little effect on total sales, were it not for the incorrigible independence of one of the largest firms.[26] That the "industry as a whole" would be better off is, of course, the monopoly rationale. Nor have shipbuilders and their governments failed to gather round and discuss moderating or throttling competition, by a halt on building new shipyards and a floor under prices, etc.[27] Nothing has come of this so far.

European Reaction to Lower Prices. The process in both tanker building and operation was summarized very well by the chairman of a large London shipbroker firm.

> If owners of more traditional size found it impossible to stand up against a totally new scale of economic costing, those wise or fortunate enough to be well up in the race by building larger and more efficient ships have done well. . . . The forces of supply and demand in a highly complex world market must be deemed to have worked outstandingly well, resulting in a steady reduction over the last nine years in the cost per ton of oil carried and, at the same time, offering sufficient incentive and also imposing compulsion upon individual owners to provide increasingly efficient and competitive ships.[28]

Plain good sense was also heard across the Channel, from the *Journal de la Marine Marchande*. To be sure, its editorial comment was resolutely protectionist. Low Japanese prices were a nuisance. The sensible policy was to live and let live, join hands, European and Japanese together, to keep prices reasonably high. But they soon ceased to question the permanence of Japanese prices. They reprinted all price information from published Japanese sources and added much of their own both for Japan and for Europe.

In addition to Japan, Swedish and Norwegian shipbuilders are ". . . a rationalized industry fully competitive price-wise."[29] This favorable picture may change in Sweden if the government insists that the shipyard companies take over the losing nationalized yard at Udevalla and absorb its deficit, on pain of losing the government financing they need to extend credit.[30] However, in mid-1968, Gotaverken had no difficulty in getting a loan of $6 million to enlarge a modern yard in order to build a series of 228,000-tonners.[31] It is difficult, however, to distinguish the particular financial stringency from the general shortage of capital funds.

The *Journal de la Marine Marchande* coverage of Swedish conditions has been particularly good, and in 1967, when some British observers were mourning over the financial difficulties of one Swedish firm, it analyzed the situation in detail to conclude that the firm was sound.[32] Like Mr. Glen, the *Journal* editors gave us lessons in how to tell positive from normative, the way things are from the way they "ought" to be.

[26] *Zosen*, Oct. 1964, p. 5.

[27] For a Japanese proposal, see *New York Times*, Apr. 16, 1968, p. 77. *JMM* has provided a good record of the many proposals and attempts.

[28] A. R. Glen (chairman, H. Clarkson & Co.), "Transport and Exports," *Institute of Transport Journal*, Jan. 1967, pp. 39–40.

[29] R. S. Platou A/S, Monthly Report, Mar. 1967, p. 2.

[30] *JMM*, Feb. 15, 1968, p. 359.

[31] Ibid., July 11, 1968, p. 1444.

[32] Ibid., Dec. 21, 1967, p. 2855; July 11, 1968, p. 1444.

But these voices did not dominate. There are few more melancholy spectacles than Europeans' bewilderment and frequent outrage, their dogged persistence in explaining tanker prices in every possible way except the simple and true one: intelligent application of some new methods and more old ones had given the Japanese and Scandinavians very low costs, translated by competition into lower prices. A simple tabulation of building time per ship, which fell by over 50 percent, would have suggested most of the truth. Instead: the Japanese shipbuilders were said to be subsidized by the government; they were selling at a loss in order to destroy European competition; they were subsidized or at least financed by the profits made elsewhere by the parent companies. In fact, some of the largest shipyards were unaffiliated, and there is no evidence that integration was a net advantage in finance or otherwise. Of the six largest Japanese shipbuilding firms, who account for nearly all the large tankers, only Mitsubishi does a major fraction in non-shipbuilding business: 71 percent in the fiscal half year October 1966-March 1967. The others were: IHI (61 percent shipbuilding), Hitachi (63), Mitsui (66), Kure (68), Sasebo (88). Moreover, the non-shipbuilding activities had tended to increase over time.[33] If anything, shipbuilding had been the "firm base" for a "cash flow" with which to "subsidize" other activities: just the contrary of the usual legend.

The Geddes Committee[34] was appointed to investigate the plight of British shipbuilding. They estimated "typical world shipbuilding prices" (p. 168): for tankers of 50,000-80,000 deadweight tons, $92-126 per ton; above 80,000 from $78 to $92. The size ranges are much too wide to be useful, and the prices are too high, but they chose well from "a wide divergence of views and experience" (p. 27).[35] They would have done much better to compile published data from *Zosen, JMM*, Platou, et al.

The Geddes group had an unstated industry model: a loose cartel, with prices at a level to keep all but the most incompetent in business, and not reduced except where necessary to gain volume (or prevent loss) for the trade as a whole. Any divergence from this model is considered strange, needing some special explanation. (Yet under competition a seller will expand output if it adds to his own profit, regardless of what it does to the industry as a whole.) They cannot understand how prices can be lowered despite the rising costs of

[33] *Zosen* June 1967, pp. 17-20, 48.

[34] *Shipbuilding Industry Committee Report*, Cmd. 2937 (London: HMSO, 1966). This is the best of the official and industry reports, including that of OECD.

[35] Incidentally, the Geddes Report regards the provision of long-term purchase credits as a subsidy (p. 29), which it is—but to buyers, not to builders. It does not explain the Japanese prices, but is an additional small reduction.

Shipbuilding Credit Facilities, mid-1960s

Country	*Maximum credit (percent)*	*Repayment period*	*Interest rate*
Belgium	85	10	n.a.
Japan	80	8-10	5.5 (e)
	80	15	4.5 (d)
Spain	80	10-15	4.5-5.5 (d)
	80	8-10	5.5 (e)
Sweden	80	8-10	5.5
United Kingdom	80	10	5.5
West Germany	80	8	5.5

Source: *Tanker and Bulk Carrier*.
Note: (e) export; (d) domestic order.

material and labor inputs. (Yet this is the characteristic trait of a progressive economy.) Along with the cartel model is a great vagueness. Swedish and Japanese profits are "said to be low." (Yet these "low" profits sufficed to bring forth a swift increase of capacity.) The investment of scarce capital and the strenuous recruitment of labor in Sweden and Japan are matters of public record. But they look for states of mind, not of competition: "The ambitions of Japan to dominate the world market or of Japanese firms to outdo each other's expansion may have led them to quote *unnecessarily* low prices . . . the fact is that prices seem to have been lower than they *needed* to be in order to equate supply and demand." (Page 43, emphasis added.) What they mean by "needed" or "necessary" is the price a monopolist or collusive group would charge. At lower prices, trade would largely be diverted to more efficient from less efficient concerns, not much increased in total. This, of course, was the argument made in the *Zosen* editorial noted earlier (see p. 119). Trade papers, themselves representing the industry as a whole, often carry advice to act like a monopolist. In an industry publication, it is special pleading, known to be such. But when this kind of thinking permeates the views of those responsible for analyzing an industry, the outcome is a mess of slogans.

The Post-1967 Long-Term Rate

The closure of the Suez Canal in mid-1967 heaped an acute tanker shortage atop the chronic shortage of the preceding seven years. It was aggravated by the Nigerian civil war and occasional shutdowns of pipelines from the Persian Gulf to the Mediterranean. Rates were not only high but often wildly fluctuating, and erratic contraseasonal movements often confounded the trade. It has been the most disturbed period since the end of World War II, and the end is not in sight as this book is completed.[36]

The intention now is to forecast a long-term rate under the new conditions. Previously the effort was to calculate the size and costs of the marginal Canal ship: the smallest (highest-cost) tanker within the 40,000-70,000 range whose supply was still being expanded, and whose term charter was governed by cost.

Table IV-7 shows that new orders placed in 1966 were about twice the average for the four preceding years, again proving that the slogan of "surplus" hid the fact of shortage. Then came the Suez crisis to speed up the process and create a wholly new dominant class.

Between 1968 and 1969, as Table IV-8 shows, new orders disappeared for 60,000-100,000 ton ships. As long-haul crude carriers, they are finished. The 100,000-150,000 class is still being expanded, and probably has a permanent place on the secondary crude runs to smaller ports. The 150,000-200,000 class is now seen, in hindsight, to be a mistake. However, the change was really less abrupt than it seems. Some ships ordered nominally of 175,000 dwt were really over 200,000; this is clearly evident if one tabulates the *Zosen* prices, which are fairly uniform by size class. A number

[36]The following account is based on the *JMM*, the annual reports of Platou and Fearnley & Egers, *Tanker and Bulk Carrier*, and the weekly reports of Clarkson. Analysis of the rates was made more confusing by the changes following the devaluation of sterling. In November 1967, when the pound was devalued from \$2.80 to \$2.40, the flat Intascale rate was maintained at the sterling equivalent. Consequently, in dollar terms the flat rates fell and a given fixture in dollars would appear higher on the new scheme. Furthermore, because some costs had increased in sterling terms a 3 percent increase was made on all flat rates from January 1968. The new flat rate therefore was 2.472/2.80 = .883 times the old. The new flat rates for Worldscale were not an exact proportion of Intascale rates. They averaged around 90 percent of new Intascale, about 80 percent of the old Intascale flat rates. (See appendix IV-B.)

TABLE IV-7. Changes in Tanker Fleet, 1962–1971

(millions of deadweight tons, except as noted)

Year	January 1 fleet	Deliveries	New orders	December 31 total on order	Scrappage	Price (\$/*dwt*), by ship size (000 *dwt*), at end of year: 210	87	96 (OBO)[a]
1962	62.1	5.1	3.2	n.a.	2.3	...	96	...
1963	65.1	5.8	10.8	n.a.	1.7	...	91	...
1964	69.2	8.5	7.7	n.a.	1.6	...	94	...
1965	76.0	9.5	10.9	n.a.	1.5	65	98	...
1966	84.9	10.3	16.2	n.a.	1.5	63	100	102
1967	94.4	8.0	24.2	42.4	1.3	70	103	107
1968	103.0	11.1	23.8	55.3	0.8	78	108	116
1969	114.1	16.4	23.5	59.3	1.7	90	127	128
1970	129.6	19.2	41.5	79.9	1.3	148	195	234
1971	147.5	...	...	...	...	...	...	...

Source: Fearnley & Egers, *Annual Review*, respective years. Note that total "on order" minus deliveries plus new orders do not exactly equal the next years "on order." The reason is partly rounding; moreover, the total "on order" is always a preliminary figure, with no later correction made. Other items are revised in later issues.

[a]Ore/bulk/oil carrier.

TABLE IV-8. Large Tankers Operating, on Order, Delivered, and Newly Ordered, 1968–1971

(all figures in thousands of deadweight tons)

Size range	Tonnage: January 1, 1968 Operating	On order	Delivered 1968	Newly ordered 1968	Tonnage: January 1, 1969 Operating	On order	Delivered 1969	Newly ordered 1969
60–80	15,901	1,616	1,055	561	16,956	1,404	392	...
80–100	9,795	3,135	2,933	2,446	12,728	2,648	1,347	366
100–150	6,725	4,520	2,785	1,363	9,510	3,098	1,517	1,819
150–200	1,327	3,907	1,468	2,027	2,795	4,466	2,669	0
200–250	409	23,562	2,910	7,049	3,319	27,703	7,437	6,848
250–	...	3,145	...	8,699	654	11,844	2,063	10,329
	34,157	39,885	11,151	22,145	45,962	51,163	12,970	19,363

Size range	Tonnage: January 1, 1970				Tonnage: January 1, 1971	
60–80	16,941	1,012	858		17,799	(1,935)
80–100	14,075	1,667	558	...	14,633	
100–150	11,027	3,400	1,927	3,436	12,954	4,909
150–200	5,464	1,797	694	...	6,158	1,124
200–250	11,410	27,114	12,988	13,590	24,298	27,616
250–	2,727	20,110	2,292	22,538	5,019	40,356
	61,644	55,099	19,217	39,564	80,861	75,940

Source: Computed from Fearnley & Egers (Oslo), *Large Tankers*, respective years. Scrapping assumed zero.

Method: Fearnley & Egers give ships operating and on order. If we assume zero scrapping (since most ships are very new), we may compute:

$$(\text{Deliveries})_{t-1}^{t} = (\text{Fleet})_t - (\text{Fleet})_{t-1}$$

$$(\text{New orders})_{t-1}^{t} = (\text{Fleet})_t - (\text{Fleet})_{t-1} + (\text{On order})_t - (\text{On order})_{t-1}$$

Obviously there was some scrapping in the 60–80 class during 1969; hence we reckon deliveries as (On order)$_{t-1}$ – (On order)$_t$. For the same reason, total computed newly ordered is less than the sum of the appropriate columns, which is too high by about one million tons.

of ships listed at 175,000 are inexplicably overpriced; but the mystery disappears if one adjusts for the difference in carrying capacity authorized in 1966, by changing the acceptable load line.

A more dramatic example of size understatement is given by the pathbreaking ships built for the Kuwait–Bantry Bay traffic. Even in 1968, Lloyds Register continued to list them as 276,000 tons, while they were widely referred to as 312,000; at the end of the year Gulf and IHI gave them as 326,000. But the draft was 81 feet 3-5/8 inches, which seemed to correspond to about 376,000.[37] In time this will doubtless be cleared up.

Whether the prevailing minimum size will keep going up, confounding the trade again as in the past, one cannot tell. But the marginal tanker for crude oil movements from the Persian Gulf to the main refining centers in Europe and Japan will be no smaller than 250,000 tons. All others "are intended for special trades requiring limitations in draft or other dimensions."[38]

What charter rate will suffice to procure the marginal ship? In the fall of 1967, at the height of the post-Suez crisis, about 30 new-building 200,000-tonners were chartered for 15–20 years at rates equating to between minus 73 and minus 75 (Worldscale 32 and 30). During 1968, there were some 10-year charters at about the same rate and a three-year charter at minus 66 (Worldscale 40).[39] Under 1967–68 conditions, therefore, the equilibrium long-term rate was around Worldscale 31. A 253,000-tonner in November 1968 went at what we would reckon (see Appendix IV–A(2)) as minus 75 (Worldscale 30). This would correspond to a 200,000-tonner at approximately minus 74 (Worldscale 31). A London broker's estimate is a little higher: Mullion reckons this charter at -69 to -74 (Worldscale 31 to 37).[40]

The year 1969 was one of the most disturbed experienced up to that time. Not only did short-term rates fluctuate wildly, but the long-term market was upset by rising ship prices (see Table IV-6), higher insurance rates, and a series of unprecedented disasters to large tankers. Hence, long-term charters were few. Late in the year, a three-year consecutive-voyage charter was signed for a 215,000-tonner at Worldscale 50, and a 10-year charter for a 263,000-tonner at a rate of about Worldscale 39.[41] These rates are mutually consistent, since a three-year charter must be strongly influenced by current high short-term charters, which have a disproportionately large share of its total present value, especially at a time of high interest rates. By the end of 1969, therefore, the best estimate of the long-term tanker rate, assuming the marginal ship to be a VLCC, was around Worldscale 40.[42]

The 1970 Cataclysm. "It is difficult to keep within bounds of sobriety when describing the tanker market for 1970."[43] Worldscale 40 was still expected as the long-term rate as

[37] *The Motorship*, quoted in *JMM*, Dec. 26, 1968, p. 2920.

[38] Fearnley & Egers, *Annual Review, 1968*; and see *Zosen*, Apr. 1969, p. 15.

[39] The sources are again the annual reviews of the Oslo firms and *JMM*. However, I have calculated the minus 73 by extrapolating from Platou estimates of other charters and checking with their table of time-charter to Intascale conversions. See also appendix IV-A. The article "Pipelines versus Tankers" in *Petroleum Press Service*, Feb. 1968, p. 59, gives the cost of a round-Cape voyage in a 200,000-tonner as 39 cents, or minus 73.6 (Worldscale 31).

[40] Ship Sale and Purchase Market Report: October-December, 1968.

[41] Fearnley & Egers, *Annual Review, 1969*; Clarkson, Dec. 12, 1969; Platou, *Annual Review, 1969*.

[42] Shell International and Esso expected a long-term rate of around Worldscale 37 late in the decade. Cabinet Task Force on Oil Import Control, *The Oil Import Question* (Washington, p. 19 and appendix F).

[43] Platou, *Annual Report, 1970*, p. 43.

late as the end of July,[44] despite the violent gyrations in the spot and consecutive voyage charters that followed the shutdown of Tapline and the cutbacks in Libya in May. Even in August, expectations did not seem to have changed. A VLCC was chartered for a single voyage from the Persian Gulf beginning January 1971 at Worldscale 225, and another was fixed to run January 1971–December 1973 inclusive, on terms equating to Worldscale 96.[45] The writer's basic principle is that the present value of the long-term rate is equal to the net value of the series of expected average short-term rates. We neglect the peculiar costs of spot operations and the risk of unemployment (see p. 115) and make two assumptions: that discounting is at 20 percent per year,[46] or 1.53 percent per month, and that the decline in spot charters is at a uniform percentage rate. If so, these two charters forecast a spot rate of about Worldscale 43 in mid-1973,[47] confirming the *Petroleum Intelligence Weekly* consensus. Over a relatively short period, the assumed uniform percentage decline does no great violence to nature, since it is reasonable to expect a moderating of the decline from month to month. Hence our assumed 20 percent discount rate may be not far from what the trade was actually using. It does not follow that the value of a dollar received 20 years hence is only 2.6 cents, unless one expects similar turbulence to be continued over 20 years.

After August, the spot charter rate fluctuated widely, but stayed in the high 200s. But there were few long-term charters. During this period the trade was rethinking its forecasts. Beginning in November, huge amounts of tonnage were fixed for long terms, at rates which continued for the rest of the year. *Petroleum Press Service* wrote of "established" or even "firmly established levels."[48] In three months there had obviously been a drastic change in expectations. Three-year charters, delivery 1972, were going for Worldscale 98; five-year charters starting 1973, for Worldscale 85; the same starting 1974, for Worldscale 71/72. We get some idea of a longer perspective by a transaction in late January, when eight VLCCs were chartered for 15 years, delivery 1973–75, at about

[44] "Most companies feel Worldscale 40 is in the cards for 1973." *Petroleum Intelligence Weekly*, July 27, 1970.

[45] Charters from H. Clarkson & Co.; conversion from Platou table. It should be noted that Clarkson frequently gives an estimate of Worldscale equivalents. There is usually no significant deviation from Platou, but where there is one I have allowed Clarkson to dominate, as being more timely.

[46] The discount rate applies to the net receipts of the tanker owner. He should be indifferent as between two streams which represent equal present values, after repayment of loans and interest thereon, payment of crew wages, and minor expenses. Therefore, the proper discount rate is not his average cost of capital, which may be quite low because most of his capital is borrowed, but a much higher rate, appropriate for a high risk. It is not pretended that the 20 percent is more than a rough guess, though it seems to be confirmed in this one instance.

[47] Let r = monthly discount rate, a = monthly decline of spot charters, t = time in months, and end-August 1970 = 0.

$$96 \int_4^{40} e^{-rt}\, dt = PV = 225 \int_4^{40} e^{-(r+a)t}\, dt$$

$$r + a = .0625, \text{ hence } a = .0472 \text{ per month}$$

$$e^{-35(.0472)} = .191, \text{ and } 225\ (.191) = 43.0\,.$$

[48] *Petroleum Press Service*, Jan. 1971, pp. 8–9, 37.

Worldscale 60.[49] Assuming mean time of delivery in 1974, one can make a rough present-value comparison which bypasses the rate of decline:

$$A \int_a^b e^{-rt}\,dt + B \int_b^c e^{-rt}\,dt = C \int_a^c e^{-rt}\,dt\,,$$

where A is the Worldscale rate between years a and b, B between b and c, and C over the whole period from a to c. If any two rates are known, the third can be derived. Assuming r alternatively at 10, 15, and 20 percent per year, B equals Worldscale 48.5, 44.8, and 41.1. Subject to the qualifications mentioned earlier, this means that the long-term rate expected by the turn of the year 1970-71 for mid-1973 was no longer about 40 but well over twice as high, at about 85. It was expected to decline, toward 1979, to the neighborhood of about Worldscale 45, and almost surely stay in the 40-50 range. Obviously this is no precise forecast. The risk factor should perhaps be increased, which would mean a higher expected long-term rate. Moreover, the result is sensitive to the estimated Worldscale equivalents, hence to conversion errors. But the least that may be said is that the expected long-term rate, having more than doubled for the near term, declines to somewhere near the old expected level by the end of the decade.

In mid-February several three-year charters were signed at separate rates for each year.[50] They are consistent with the levels mentioned by *Petroleum Press Service*. Taken together with the above calculations, they indicate the following profile of expected average annual tanker rates:

	1972	*1973*	*1974*	*1975*	*1976*	*1977*	*1978*	*1979*
Worldscale	110	96	84	73	64	56	48	43
Cents/bbl, P.G.-Europe	130	113	99	86	75	66	57	50

The nearer in time is any expected rate, the more uncertain its estimated value. The decade expectations may be substantially wrong for the second half, and amazingly wrong for the first half. In the years to come, there will be more examples of the short-term rate rising and falling drastically away from its long-term level, which is our chief concern.

Long-term charter rates are closely tied to estimated costs, but costs cannot explain the whole increase in 1970. Short-run expectations tell part of the story. Moreover, chartering large volumes of tonnage at rates above the expected spot average is a means of insurance. If turbulence continues, with cutbacks and embargoes (see Chapter VII, pp. 245ff.), there will be occasional shortages and panics. Similar fears during 1956-57 led to many ships being ordered or chartered for 1962 and even 1963 delivery at prices or rates which were absurdly high at delivery time. It is easy to see the mistake—in retrospect. But those who placed the orders and charters were not irrational, only trying to avoid being caught short on available shipping and so be forced to pay wildly inflated short-term rates. If one fears that during much of 1974-79 spot rates will again be at

[49]Clarkson, Jan. 25, 1971; my own conversion. See also ibid., May 28, 1971: Worldscale 60 was expected for a 15-year charter starting 1973-74.

[50]Clarkson, Feb. 12, 1971.

Worldscale 200 or 300, charter at 72 is a bargain. Probably the high rates contracted in 1970-71 reflected some panic mixed with prudence.

In 1970 price increases for tankers[51] were "staggering," especially in the last four months of the year, although difficult to measure exactly because of escalation clauses (mostly for wages) as well as changes in credit conditions. Because changes came so late in the year, they were difficult to estimate. Fearnley & Egers' estimate of up to 63 percent may be excessive; Platou estimates 48 percent for two important VLCC sizes. *Zosen* listings (adjusted by the writer for changes in credit conditions) would come to 53 percent. Prices increased, therefore, by about half. Assuming that operating costs have risen equally, which is probably somewhat exaggerated despite the rise in insurance, this would appear to forecast a long-term rate half again as large as the Worldscale 40 of late 1969, i.e., around 60. This would appear to conflict with the estimate of around 40 as the long-run normal. (See Table IV-9.)

However, the rise in tanker rates was cause as well as effect of the rise in ship prices. Ship prices ruling in early 1971 reflect short-run, not long-run, marginal costs. The yards

TABLE IV-9. Prices of Newly Contracted VLCCs, Japan, early 1971

Time	Size (000 *dwt*)	Unadj. price ($/*dwt*)	Terms	Adj. price[a] ($/*dwt*)	Delivery date
Aug.-Sept. . . .	261 (2)	100	cash	109	Dec. 1973, Mar. 1974
Sept.-Oct. . . .	269.5	108	80% in 8 yrs.	113	Dec. 1973
Oct.-Nov.	233	111.5[b]	80% in 8 yrs.	117	Dec. 1973
	262 (2)	111	80% in 8 yrs.	116	June 1972, Aug. 1974
	216	129.5[b]	70% in 8 yrs.	138	early 1974
	228	119	70% in 8 yrs.	127	Nov. 1973
	248	109[b]	80% in 8 yrs.	114	Mar. 1974
	276	109	65% in 6 yrs.	115	Dec. 1973
	477	95.4[b]	80% in 8 yrs.	100	May 1974
	220	100[b]	80% in 8 yrs.	105	Mar. 1973
Nov.-Dec.	218	105[b]	80% in 8 yrs.	110	Mar. 1973
	264 (2)	110	80% in 8 yrs.	115	Nov. 1973, Aug. 1974
	254	106[b]	80% in 8 yrs.	111	Mar. 1973
	261 (2)	115	60% in 8 yrs.	121	not stated
	231	113[b]	70% in 7 yrs.	118	Sept. 1973
	261 (2)	115	cash	126	mid-1974
Dec.-Jan.	226.5	129	70% in 7 yrs.	136	Mar. 1974
	261	115	60% in 6 yrs.	123	Nov. 1973
	235	115	80% in 8 yrs.	120	Feb. 1974

Source: Zosen, Oct.-Feb. 1970-71.

[a]Suppose that $100 is to be paid in eight annual installments with interest at 6%. Then each installment must be $16.10, by formula $i/1 - (1+i)^{-T}$. If the interest rate is increased to 7.5%, each installment must be $17.07, which is 6% higher. Applying this to the standard 80% payable in six years, the 1970 increase in the interest rate means an effective 4.8% increase in the price.

There are some sales for larger down payments, and some for cash. Assuming the cost of money to the buyer as 11%, the present value of one dollar per year to be paid out over eight years is $5.15. Multiplying this by $17.07, we have $88.00 as the present value cost of the obligation. If the buyer is now compelled to pay $100, the price increase is 100/88, 13.5% on the 80% formerly payable in installments, or 9.1%. If the buyer must pay 30% cash instead of the previous 20, then the increase is 1.35%; a 40% requirement means 2.7%, and so on.

[b]Payment in yen.

[51]This paragraph is based upon the annual reports of Platou and Fearnley & Egers, and on a comparison of ship prices in November-February 1969-70, with the same months, 1970-71, as given in *Zosen*.

are crammed with orders, working at top capacity on new designs, chronically short of labor, willing to pay anything for a little more speed. Some of the increase in construction cost may be permanent, i.e., for the labor which continues to be chronically short. Most or all of the cost increase will be reversed. New capacity is being provided on a large scale,[52] despite the fears of eventual excess.[53] If demand holds steady or declines and capacity grows, ship prices will decline, especially if efficiency improves. For a time, indeed, prices may decline to unremunerative levels. It makes perfectly good sense to invest in a new building yard in 1971 in order to earn high profits in 1972 and soon after, even if only variable costs are covered in later years, so long as the present value of expected profits, discounted back to the time of building, are greater than the needed investment.

Setting aside the changes of a cycle of boom-and-bust, which the rising demands for repair and modernization work will moderate in any event, improved technology will tend to lower real costs. Since the above-200,000-tonner is a relatively new ship, we should expect a rapid rate of progress in cost reduction as experience is gained. By the end of 1970, 151 such ships had been delivered and 272 were on order for delivery in 1975 or earlier. If the learning-curve rule of thumb were to hold, of a 20 percent cost reduction with every doubling of cumulative output, we would need to look forward to cost being reduced by about a fourth by 1975 (i.e., $151 \times 2^{1.485} = 423$, and $0.8^{1.485} = 0.72$). It is wrong to base a forecast on a rule of thumb, but even more wrong to doubt that there will be substantial cost reductions.

The world shipbuilding trade seems likely to remain competitive. An international agreement through the Organisation for Economic Co-operation and Development has mildly increased interest rates on credits to shipowners. The maximum credit period is to be eight years; the minimum interest rate is to be 6 percent and minimum down payment 20 percent, in effect putting the United Kingdom on the Japanese standard. If the persistent attempts at a world shipbuilding agreement were to succeed, the likely result would be vertical integration into shipping in order to capture the abnormal profits, with pressure on tanker rates.

The long-run equilibrium price of tanker service seems little different, after the cataclysm of 1970, from what it was earlier. But the importance of that equilibrium is less, because it is farther off. Such, at any rate, is the view from early 1971. "An increasing number of people have been engaged in research into the science of forecasting. The reliability of modern methods remains to be seen, but the industry is not likely to forget the mistakes made by the major oil companies in trying to assess their future requirements."[54]

Minor Factors: Suez, the OBOs. As early as the summer of 1964, some 3.5 million tons of crude were shipped around the Cape in tankers of over 60,000 tons; in 1965, 5 million; and in 1966, 19.2 million. About three-fourths of the voyages took place during the April-August low-freight period.[55] The respective average summer rates during those three years were Worldscale 48, 41, and 35, compared with annual averages of 66, 64, and 60.[56]

[52] *Zosen*, Oct. 1970, p. 15; Jan. 1971, pp. 5, 18; Feb. 1971, p. 7.

[53] Platou, *Annual Report, 1970*, p. 31; John I. Jacobs, *World Tanker Fleet Review* (Dec. 31, 1970).

[54] Platou, *Annual Report, 1970*, p. 32.

[55] Fearnley & Egers, *Large Tankers*, respective years.

[56] Platou, *Annual Report, 1970*, p. 32.

Probably few if any of the large tankers going around the Cape were on spot charter. But the true cost of the longer trip was the outlay on the additional tonnage that was needed to maintain the rate of deliveries. The spot or perhaps consecutive voyage rate was the opportunity cost of using the large tanker (under time charter) on the longer trip. This again shows how the spot and long-term charter markets are interdependent.

The decisive "bend" in the demand curve, where the benefit of using the Canal is no more than the dues, seems to be around Worldscale 35: that was where a contemporary estimate found the break-even point for a fully loaded 60,000-ton ship at Suez draft of 40 feet.[57] Larger ships had to go through part-loaded, with much higher break-even points.

Even before the 1967 war, therefore, the Canal was threatened with the loss of much of its traffic. The threat was masked by large unloaded ships returning to the Persian Gulf through the Canal. Indeed, the 175,000-200,000-ton "Shell-type ship . . . was designed" for the specific purpose of going round the Cape loaded, and returning in ballast through the Canal at the much lower ballast rate. This would have been much less profitable to the Canal authority, of course. Esso and Texaco ordered larger ships which cut out the Canal.[58]

But since the shift to VLCCs, the Suez Canal is no longer of any importance for the oil flow to Europe. The reopening would be the equivalent of a 5-10 percent increase in the world fleet.[59] The effect would be a temporary drop in single-voyage and short-term rates, which would quickly converge to the long-term level.

As Table IV-8 shows, the world fleet of the 1970s is composed of two specialized groups. Ships under 60,000 tons, which are the only ones able to use the present Canal both ways, cannot possibly compete with VLCCs twice rounding the Cape on the Persian Gulf-Europe run. Neither can the VLCCs compete with the smaller ships in carrying products, especially heavy fuel oil, to ports other than the big ones taking the VLCCs. A reopened Canal would not make it profitable to pull small ships out of a service where they have a comparative advantage in order to place them on voyages where their costs are much higher than VLCCs.

Ships of 150,000-200,000 tons could round the Cape loaded and return in ballast. But not all of this relatively small group are available for Persian Gulf-Europe runs, because they have a special advantage in shallower ports in Asia and South America. Moreover, experience with the first year's operation of 200,000-tonners built particularly to return through the Canal in ballast showed that return via the Cape was "only marginally more expensive."[60] Hence it is worth paying very little more for the right to traverse a reopened Canal in ballast, and very few ships would appear for even this marginal use.

The Suez Canal could in time be widened and deepened, and some turns could be rounded to accommodate 1,100-foot ships. But Canal economics are caught between the blades of a scissors. The larger the ship to be accommodated, the deeper must the Canal be dug, at greater expense, but the less is the value to the shipper of using the Canal. In Appendix IV-C the saving is estimated at about 38 percent of the cost of the round-Cape

[57] Fearnley & Egers, *Large Tankers*, 1967, p. 11.

[58] Platou, *Annual Report, 1968*, p. 18.

[59] John I. Jacobs & Co., *World Tanker Fleet Review*, Dec. 31, 1970, p. 10.

[60] J. H. Kirby, Managing Director, Shell International Marine Ltd., in speech to Royal Society of Arts, London, Mar. 3, 1969.

voyage. At Worldscale 60 and Worldscale 40 respectively, the saving on using the Canal is 25 cents and 17 cents per barrel. My earlier estimate of Worldscale 40 as a likely 1980 rate may, of course, be far wrong, but it would be imprudent, not to say reckless, for investors to make a large bet that it will be much higher all year round. Here again the prewar experience is instructive. The average 1966 rate was Worldscale 60. But during April-August 1966, with additional tonnage available in the low consumption months, the rate averaged about Worldscale 35. If a similar relation held, an average annual rate of 60 would imply a long stretch every year around 35, or 41 cents per barrel on the round-Cape run. A Canal charge exceeding 16 cents per barrel (38 percent of 41 cents) would mean a half-empty Canal during five months of the year.

Since there are no publicly available data on costs of digging a much deeper Suez Canal, it cannot be said that such rates are impossible. Probably nobody knows how much rock needs to be blasted out. But assuming that the job can be done, it is less attractive than the building of pipelines across the Suez isthmus—which can be provided in many increments in a much shorter time than a deepened Canal. In Appendix IV-C the cost of pipelining crude across the isthmus in a 42-inch line (and larger lines would be cheaper) is reckoned at around 14 cents per barrel to return 15 percent on the investment which seems necessary in 1971. This view of the pipeline plus VLCCs as being more economic is supported by Egypt's strenuous and finally successful attempt to obtain throughput commitments and financing for the Suez-Mediterranean (SUMED) pipeline. This effort would make no sense if a deepened Canal were a lower-cost alternative. The SUMED tolls will apparently be in the range of 18 to 20 cents per barrel.[61] The Trans-Israel pipeline (TIP) apparently charges a lower toll.

Moreover, the cost of using a pipeline is in part compensated by greater flexibility. At the Mediterranean terminal, tankers of all sizes can be used to load all types of products for all types of ports. A large stock of crude and products can be stored at the terminal for delivery anywhere in the Mediterranean within a few days. In effect, SUMED and TIP will be trans-shipment points, like the one at Bantry Bay or the Japanese Central Terminal Systems (CTS).

One interesting recent development is the growth of combined carriers: ore-oil and ore-bulk-oil (OBO).[62] At the beginning of 1970 there were 131 such ships (of over 60,000 tons), with a tonnage of 10.3 million tons. In 1969 they hauled nearly 10 percent of the seaborne oil trade, and their share is expected to reach nearly 15 percent by the mid-70s. For such ships, movement in and out of oil traffic will not be an expedient forced on the owners by very low rates, as is true of tankers hauling grain, but part of their normal operation. Two effects may be expected, though we cannot measure their importance. In the first place, the short-term market should be less volatile. Very low spot and consecutive rates will be refused by ore-oil and OBO owners, since the alternative will be something better than layup or grain. (Conversely, dull times in dry bulk cargo will weaken short-term oil rates.) Second, there will be some downward pressure on long-term rates because the ore-oil and OBO, although smaller ships and less efficient per mile than the specialized tanker, can engage in triangular or multiangular trades, such as Persian Gulf oil to Brazil, Brazilian iron ore to the U.S. East or West Coast, then metallurgical coal to Japan. A conventional tanker spends half its steaming time in ballast but a combined carrier much less, hence costs are reduced below those of the tanker. Oil freight

[61] *Petroleum Intelligence Weekly*, May 17, 1971.

[62] This paragraph is based on the Fearnley & Egers reports.

costs become a joint product with ore and ore-bulk freight costs, and there will be much mystification and complaints of "unfair competition" when combined carriers undercut tankers on some runs.

Summary and Conclusion

Where maritime transport is concerned, we are fortunate in having good published price data, which enable us to pass easily between f.o.b. and c.i.f. prices. Given an f.o.b. open market value, one can say what it is worth to a seller to have that oil delivered, for one knows what he would need to pay for the service if he were to hire it himself. Conversely, given a c.i.f. contract, one can make a tolerably precise calculation of the netback to the seller f.o.b. the Persian Gulf or other place of loading. One knows how much it costs him to provide that service, regardless of whether he hires it from someone else to supply it to the buyer, or uses his own facilities.

The basic cost figure at any given moment is the current long-term freight cost. It depends on the cost of building and operating the needed incremental ship.

An efficient competitive market mechanism senses these costs and reflects them in the long-term charter rate. This rate in turn determines, within limits, the average level of shorter-term rates, even including the highly volatile single-voyage rate. It is an error and needless complication to damn that rate as "marginal" and unimportant, or to suppose that its average level is beyond prediction.

The long-term charter rate decreased slowly in response to lower costs, and rose after 1967 only because of higher costs. Freight rates are not now and have not for a decade been abnormally low or temporarily depressed by a tonnage surplus. On the contrary, long-term rates have been usually and are now (in the first quarter of 1971) above long-term normal, because of a chronic and sometimes acute shortage of new efficient tankers. When the shortage eases, ship prices and long-term rates should ease. In addition, the incremental tanker may increase past 250,000 tons–but on this one can venture no guess.[63]

[63] For an earlier example of this kind of prediction, see "Oil Prices in the Long Run," *Journal of Business of the University of Chicago*, vol. 37, April 1964. For an opposing or at least a different view, see S. G. Sturmey, "Shipping in the 1960s," pp. 5, 14, and "On the Pricing of Tramp Ship Freight Service," p. 9 (lectures at the Institute for Shipping Research, Bergen, Mar. 25, 26, 1965.) For a useful discussion, see Walter L. Newton, "The Long-Term Development of the Tanker Market," *Journal of the Institute of Petroleum*, vol. 50 (1964).

CHAPTER V

Oil Prices, 1947–1957

The purpose of this chapter is to explain the evolution of crude oil prices from the end of World War II to the end of the 1956-57 Suez crisis.

Chapter III has analyzed the market structure of the oil trade outside North America. A high but slowly decreasing concentration leads one to expect high, stable, and perhaps slowly declining prices. But what we find is irregular change. The initial price, inherited from prewar and wartime, first rose, then fell by more than half during 1947-49, after which the evolution abruptly stopped and prices rose. This suggests that no simple hypothesis, either of competition or monopoly, will do in the oil market; more precisely, forces outside the market structure were powerful and not continuous.

We will therefore go step by step. The first task is to set up a simple model of competitive and noncompetitive action.[1] This is then checked against the actual data to see what is explained and what requires further evidence. The summary (p. 159) provides an outline of the conclusions reached.

A Simple Model of Competitive Adjustment

Delivered Pricing in General. Figure V-1 shows a producing center *A* from which oil (or for that matter any other commodity) is shipped to various points east. For the moment, point *F* should be disregarded. The cost of moving a barrel of oil from one location to the next (*A* to *B* to *C*, etc.) is shown by the slope of the line as 25 cents per barrel. The delivered price at the various points is as indicated, and the netback at the supply point *A* is the same $3.00 everywhere as the f.o.b. price at *A*. For, if the netback from shipping to any other point were higher, more oil would be shipped there from *A* to take advantage of the wider profit margin, driving down the price there until it was again equalized.

We assume now a new producer at point *F* who supplies only a part of the demand there. He charges a price equal to the price at *A* plus freight to *F*. He does not charge more because he cannot get more, and he does not charge less because he can sell all his oil at that price. Whether he calls $4.25 his "base price" or whether he charges "$3.00 plus $1.25 phantom freight" is semantics, not economics.

If this producer at *F* now expands his output to where he can supply more than the demand at *F*, he will sell eastward at *G* and *H* since he gets the same net return. He will not sell to the west of *F*, since he would have to accept a lower netback in order to sell in

[1] Economists will recognize that the model carries a number of simplifying assumptions, none to be accepted on faith. They are requested to suspend judgment until after reading the application of the model, in order to see whether the assumptions corresponded to fact.

those markets—e.g., e' from a sale at E. If we assume that he has the same production costs as the producers at A, his profits are higher, and the result of this premium is to attract capital into new production which will economize on the scarce resource of transport.

But "production costs at point F" is ambiguous. In general, average cost is not relevant to our problem, only incremental cost. Perhaps the oil deposit at F was shallow and permeable, but small. Average cost of a low production rate might be very low, but expanded output would soon impose sharply rising incremental costs. It would be profitable for the operator at F to keep expanding output so long as incremental (not average) cost was below price, which would be the cut-off point. If the incremental cost at F rises to \$4.25, the competitive equilibrium price stays there. With average cost below incremental, the producer at F will draw a geographical rent, but that is the end of the matter.

Now, however, let us suppose that even as output expands incremental cost remains lower than \$4.25, and production expands farther at F. What will be the effect on the price structure?

First possibility: production by a single (monopolist) operator or a collusive group. The price will not change. The operator will sell all he can at F, G, and H where his netback is \$4.25.[2] If there is no profitable market past H, he will sell in E at \$4.00 delivered, but will clear only \$3.75 (e') because he pays the freight. Thus there will be a two-price system at F, and higher revenues than if he cut prices on all of his sales. The segment $e'f$ shows the lower revenue per barrel on sales west.

Second possibility: production by a competitive group. The several members act in complete independence of one another. Each tries to sell in the most profitable markets, which are F, G, H. But so long as the price at F and points east exceeds even slightly the netback from shipping west to E, there will be an incentive to sell everything in F, G, H and stay out of E. Hence, under competition, if the output at F exceeds consumption in $F + G + H$, the price at F must keep falling as excess supply is pushed into it. When it falls to the point e', the \$3.75 netback from E, there is no longer any reason to cut prices, and the system is again at rest.

The new structure established E as the point of equalization where f.o.b. price plus freight is equal from either of the two sources of supply, A and F. If there is no change in either demand or supply, i.e., if \$3.75 is approximately the incremental cost at F, the structure is permanent. But if the new price is well above cost, it will keep opening the valve wider at F, for a further expansion of output. The price at F must fall again toward the netback from sales at D, which is \$3.25. The process will continue, depending on how much additional oil can be produced at F at what cost. It is assumed that after much expansion, incremental cost in the new area as it is developed reaches \$2.25. Then equalization is at point B, and the price at F has fallen to b', or \$2.25.

The evolution of the price registers the competitive force in the market. For if there were a single monopolist or a group of collusive sellers at F, they would maintain the price there at the old level and would be content to "absorb freight," i.e., grant discriminatory low prices in selling at points west of F. They would be much better off than

[2] Strictly speaking, there would probably be a very small price cut at F, G, and H to claim the market. Production at A would to some extent be curtailed, to some extent redirected to A, B, C, D, E, lowering the price there and hence at F, G, H. The result would be a slight lowering of the level, but no change in the structure.

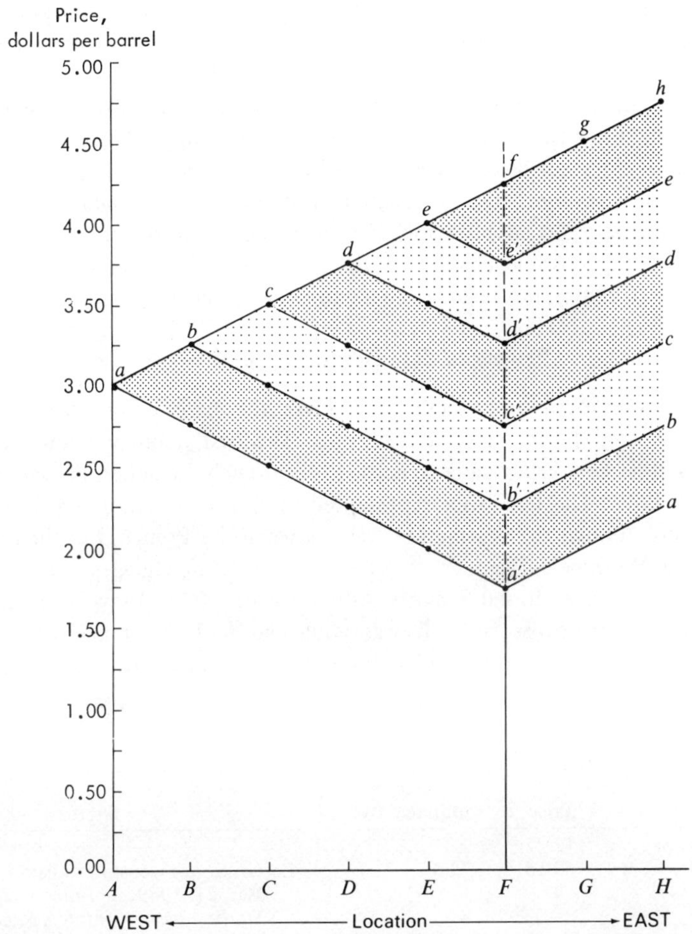

Figure V-1. **Evolution of a delivered price structure under competition.**

lowering prices equally to all. The sellers at *A* would also be better off with monopoly at *F*, for they could still continue to get the old prices in the old markets. Since customers would be indifferent to their source of supply, some buyers east of *D* might be ordering from *A* at the same time that buyers west of *D* might be ordering from *F*. This crosshauling benefits nobody and both groups of sellers would try to swap customers and save freight unless the law frowned on such a procedure. The crosshauling indicates imperfect monopoly or imperfect collusion.

Let us compare the old price structure, the straight line from *a* to *h*, with the new one, *abb'b*. Competition has done its work by offering a prize, in the form of abnormal profits, to expand output at *F*. The result is greater supply and a lower average price paid by consumers. The shaded area, *bb'bh*, shows the savings to them, and the difference between discriminating monopoly and competitive prices.

Savings are actually understated, for up to this point we have assumed that the price at *A* was constant under any conditions. But if there is a normal competitive market at *A*, producers need to cut their delivered price at *F* and eastward to stay competitive. As product backs up at *A* and the price level falls, high-cost inefficient output at *A* is shut out of the market. This secondary effect is not shown in the diagram, but the reader can visualize it as the whole price structure moving down, while keeping the same form.

Another possibility of great practical importance is that during the adjustment period the overall demand has been rising so that the increasing supply is just balanced, and the price at *A* and *B* does not change. *Figure V-1 takes the level of oil prices as given and analyzes only the structure.* It is deficient in not showing the change in level which is generated by the increased supply which forces the change in structure. The figure is not deficient in omitting possible changes in the level which have nothing to do with the evolution of the structure, but result from a worldwide price inflation.

Postwar Revolution, 1948-1949

Changed Level and Structure. Table V-1 gives the bare chronology of the price reductions following the increases in 1945-48. The first publicly known arm's-length purchase of Middle East crude came in early 1948, though the action probably was taken at the end of 1947. The new price represented both a reduction and a departure from delivered pricing, though the destination was known to be French. But the contract tied the price to the Venezuela-West Texas price so far as future changes were concerned. In June 1948 Mobil, as the Mutual Security Administration (MSA) saw it, inaugurated the Eastern Hemisphere price as such. It was equalized to U.S. Gulf-Caribbean supplies delivered in London. In its April 1949 reduction, Esso's contracts provided that the first

TABLE V-1. Price Chronology, 1947-1953

Date	Price	Change initiated by:	Source
1947	$2.22	...	...
First half 1948	2.15	Gulf	Letter, S. A. Swensrud to Paul G. Hoffman, Mar. 21, 1949, reprinted in *Petroleum Study, Hearings before a Subcommittee on Interstate and Foreign Commerce*, H.R. 81st Cong., 1st sess. 1950, p. 421.
June 15, 1948	2.08	Mobil	*Statement of MSA (Mutual Security Administration) Concerning Its Relations with International Oil Companies with Respect to Petroleum Prices*, Aug. 15, 1952, mimeo., attachments 2-8.
July 7, 1948	2.07	Esso	*Statement of MSA*
Sept. 15, 1948	2.03	Esso	*Statement of MSA*
Sometime before Mar. 25, 1949	1.99	Unknown	Correspondence among ECA (MSA), consultants, and Esso. See *Petroleum Study*, pp. 421 ff.
Apr. 1, 1949	1.88	Esso	*Statement of MSA*
July 15, 1949........	1.85[a]	Gulf	*Statement of MSA*
Sept. 7, 1949	1.75	Caltex, Esso	*Statement of MSA*
Sept.-Dec. 1949	1.65	Gulf	Letter, S. A. Swensrud to Paul G. Hoffman, Dec. 19, 1949; reprinted in *Effect of Oil Imports on Domestic Producers, Hearings Before the Select Committee on Small Business*, H.R. 81st Cong., 1st sess., 1950 p. 374.
Unknown	1.75	Unknown	
April 1, 1953	1.60	BP	*Platt's Oilgram Price Service* Apr. 9, 1953.

[a]Allowing 10 cents quality differential on Kuwait crude.

15 cents reduction in East and West Texas crude would not affect its Ras Tanura price. In September 1949 this formal link with the Western Hemisphere was dropped and the contracts provided simply for the seller's posted price at the time of lifting.

More than two and one-half years were to pass before there was another reduction at the Persian Gulf, but this time there was no response at New York and the U.S. Gulf. Indeed, two months later U.S. prices actually were increased. But this is to get ahead of the story. Before looking at the 1950-53 period, we need to show the significance of the previous two years. Table V-1 explains nothing of what it records, and it understates the decline because of the rapid worldwide price inflation during the period. By late 1949 the price level in the United States and elsewhere had risen so swiftly that the Gulf-plus price at the Persian Gulf would have been \$3.81. In order to allow for inflation, Table V-2 has been stated in terms of the level of prices and freight rates of September 1949, and shows three price structures under those conditions. The reader is advised to look at Figure V-1, and consider that Points *A-F* are respectively the U.S. Gulf, New York, London (and Northwest Europe generally), the Mediterranean, and the Persian Gulf.

In Table V-2, column I represents the price structure of U.S. Gulf-plus. It corresponds to the straight line *ah* in Figure V-1. Actual prices at New York, the United Kingdom, and the Persian Gulf are all equal to the U.S. Gulf price plus eastbound freight, points *b*, *c*, and *f*. Netbacks are respectively b', c', and *f*. Freight rates on westbound shipments are not connected with prices. For example, the price at the Persian Gulf (line 3, line 4), plus freight to the United Kingdom, obviously does not correspond to the U.K. price. The producers at *F* "absorb freight" in successively larger amounts on shipments to points *e*, *d*, and *c*. They meet competition from *A* by discriminating and taking netbacks e', d', and c', e.g., \$3.37 less \$0.66 from delivering to the United Kingdom.

Column II represents the price structure $acc'c$. The price at the United Kingdom (*c*) still corresponds to U.S. Gulf plus freight. But it now also corresponds to the Persian Gulf

TABLE V-2. Crude Prices, September 1949 Level, Three Price Structures

	Structure Type		
	I	II	III
Component of price	U.S. Gulf-plus	Equalization at U.K.	F.O.B. Persian Gulf
1. Actual U.S. Gulf f.o.b. price[a]	\$2.71	\$2.71	\$2.71
2. Freight, U.S. Gulf to Persian Gulf	1.10	1.10	1.10
3. U.S. Gulf-plus price at Persian Gulf (lines 1 + 2)	3.81	3.81	3.81
4. Actual Persian Gulf price	3.81	2.43	1.75
5. Freight, U.S. Gulf to U.K.[b]	0.66	0.66	0.66
6. U.S. Gulf-plus price at U.K. (lines 1 + 5)	3.37	3.37	3.37
7. Freight, Persian Gulf to U.K.[b]	0.94	0.94	0.94
8. Actual price at U.K.	3.37	3.37	2.69
9. Freight, U.S. Gulf to N.Y.[b]	0.25	0.25	0.25
10. Gulf-plus price at N.Y. (lines 1 + 9)	2.96	2.96	2.96
11. Freight, Persian Gulf to New York[b, c]	1.21	1.21	1.21
12. Actual price at New York	2.96	2.96	2.96

Source: Statement of MSA (See Table V-1). See as a useful check: Federal Trade Commission, *The International Petroleum Cartel*, chap. X, whose detail is very similar but harder to follow.

[a] Includes minus 4 cents product yield differential.

[b] USMC less 35.5%. (For explanation of USMC, see chap. IV, p. 104, n. 5.)

[c] Includes 10.5 cents U.S. tariff rounded to 11 cents.

price plus freight eastward (line 4 plus line 5), because of the lower Persian Gulf price. Hence the U.K. price is no lower, but prices in the Mediterranean are lower. The Persian Gulf producers no longer "absorb freight" on shipments to the United Kingdom, since the f.o.b. price plus freight add to the actual price. But they do still absorb freight on shipments to the United States.

Column III represents the price structure *abb'b*. The price at the United Kingdom no longer corresponds to U.S. Gulf plus freight, as in columns I and II, but it still corresponds, as in column II, to the Persian Gulf price plus freight eastward. This is because the Persian Gulf price has again been lowered, and the U.K. price with it.

New York in column III is in the same position as was London in column II. The price corresponds both to the U.S. Gulf plus freight and to the Persian Gulf plus freight. It is no lower than previously. But just as equalization at London benefited not customers there but those eastward, so equalization at New York benefits customers at London and points east.

By September 1949, the price at the Persian Gulf was considerably less than half of Gulf-plus, and differences elsewhere corresponded. The Gulf-plus price for oil delivered in Northwest Europe would have been $3.37, while under Persian Gulf f.o.b. it was $2.69. The lower total revenues from European sales were 68 cents per barrel, $550,000 daily, and this sum was, of course, to multiply tenfold with the growth in European refining.

Competitive Process or "Basing-Point System"? Table V-2 is a close fit to our model of a competitive evolution, but only for the period it covers. The theory developed here obviously runs counter to the commonly accepted one, and it is also an important point of departure for later periods. We had better pause to consider a rival theory presented in the Federal Trade Commission report.[3]

The report says often that the oil companies have devised a system whereby prices at any given market are identical "regardless of point of origin." If prices in any given market were not identical, that would really be something to explain. If a given commodity sells at more than one price at one time in one market, this means either that an apparently single market is really two or more markets, or that the market is controlled by a monopoly which can increase profits by separating customers and charging different prices to different classes according to their ability to pay.

As our model showed, and price history confirmed, "identical delivered prices" are compatible either with competition or with an extreme case of discriminating monopoly. Since they are compatible with either, they are characteristic of neither. What matters is the *structure* of prices and its relation to transport costs; and the *level* of prices, and their relation to incremental cost.

A little reflection will show that the U.S. Gulf-plus price structure under pre-World War II conditions did not, in and of itself, necessarily indicate any departure from purely competitive conditions. For exporting oil companies the U.S. Gulf price was an external fact. The Gulf-plus price structure was a competitive adaptation to it. The U.S. Gulf and Venezuela, with practically the same distance to all consuming points, were the only

[3]FTC, *The International Petroleum Cartel*, Staff Report to the Federal Trade Commission submitted to the Subcommittee on Monopoly of the Select Committee on Small Business, U.S. Senate, 82d Cong., 2d sess. (Committee Print No. 6, 1952), especially chap. X. Chaps. I-IX contain much useful material also, especially that pertaining to the period before World War II. While weak in analysis, the actual price data are largely accurate, or at least agree with this writer's. That makes it possible to compare the two theories.

exporters to the Eastern Hemisphere. Romania, the Persian Gulf, and Indonesia were all in the position of point F in our original simple model, for they did no more than supplement the consumption of the areas they served.

To demonstrate control of supply, or price higher than competitive, one needs (and finds) independent evidence. Before World War II there was a cartel. The post-World War II evidence on cost (Chapter II) can be coordinated with price structure data to permit a much more detailed and complex history.

American Background: "Basing-Point Price Systems." However wrong or meaningless, the FTC formula is highly persuasive to anyone with an American antitrust background, and this includes far more people today than it did in 1952.

Under "basing-point price systems" involved in some celebrated U.S. litigation,[4] price was quoted only for the point of delivery, and it equalled the price at the nearest base plus the all-rail freight to destination. In terms of Figure V-1, a producer met or matched prices b, c, d, and e, and realized the respective netback b', c', d', and e', so that there was great discrimination at the point of production. The steel and cement associations compiled freight rates, and even when errors crept into the books, as they were bound to, members adhered to the erroneous rates until the revisions were published. There were complaints to the trade association and sometimes punitive reprisals when some producers by accident or design quoted a lower freight rate and hence a lower price. Sellers refused to consider cheaper freight rates by truck or water in order to lower delivered prices. This did not arise from a churlish distaste for letting a customer have—or provide—cheaper transit. The keystone of the whole system was a refusal by every seller to quote f.o.b. or to permit diversion in transit. The system collapses if buyers can buy freely f.o.b. For the more distant buyers paying the lower f.o.b. price (e.g., a buyer at b paying b') and able to take delivery at the point of shipment, would turn a quick profit by immediately reselling to others at an intermediate price (e.g., selling at the f.o.b. price c' to a buyer at d). The system would break down and over time approach a uniform f.o.b. system, from something resembling the line ah to the line $abcc'c$.

The purpose of the collusive "basing-point system" was to head off any possible independent action, prevent any divergence among sellers' offers, and therefore preclude competitive pressure to push all sellers down in the direction of the lowest offer. In a free and well-informed market, price would gravitate to a uniform lower level, as it did in the oil trade. The purpose of basing point and other delivered price systems was precisely to block that force of gravitation and enforce a predetermined identity at a higher level. In the U.K. example, every seller, if he could, would have adhered to the $3.37 price because it was much more profitable. Under a collusive "basing point" regime, it would be there to this day.[5] The FTC can hide the elementary fact that Persian Gulf oil prices went on an f.o.b. basis by mid-1949.

Machlup, in his criticism of the basing point system, proposed:

> *A Test of Non-Discrimination* . . . A seller with no intention of discriminating among different buyers will be indifferent to their personal, occupational, functional,

[4] Samuel H. Loescher, *Imperfect Collusion in the Cement Industry* (Harvard University Press, 1959).

[5] Matters have not been improved by the acrimonious and voluminous controversy, in and out of American courts, about the legal merits of "delivered price systems." Defenders have argued that since competition would produce identical prices, it followed that any given set of identical prices could only result from competition. The FTC turns this completely around, substituting "collusion" for "competition." It is fallacious pointing in either direction.

> geographical, social, national, racial or any other characteristics. He will not ask them before quoting his price where they wish to use the product. . . . If a buyer chooses to take delivery at the producer's gate, he may do so and will not be charged more than any other buyer pays—explicitly or implicitly—at the mill gate. . . . By this test sellers under a basing point system are quickly revealed as being intent upon discrimination. They are not indifferent as to the destination to which their products are to go. They believe that a buyer should not even interest himself in the prices f.o.b. place of shipment and they insist on quoting delivered prices for the actual destination points of their products.[6]

This test may be applied to the statement in December 1949 of the then president of Gulf:

> Sales of Kuwait crude are made f.o.b. Mena al Ahmadi and transportation is for the account of the purchaser. The cost to the purchaser is the going market price at the time of sale irrespective of whether the oil is to be shipped to Europe or to the United States. We have sold Kuwait crude oil to purchasers who transported it to France and to other purchasers who transported it to Philadelphia. The transportation of the oil was arranged and paid for by the purchaser.[7]

Since nobody has disputed either Mr. Swensrud's knowledge or his veracity, Machlup's test is obviously passed.[8]

Reasons for the 1948-49 Change. Why the price structure was transformed has not yet been explained. This is a serious deficiency, for explanation of past events is always an implicit forecast. As our simple model showed, the force of competition puts prices down until they are somewhere in the neighborhood of incremental costs. Why did the 1948-49 events happen at all? And why did the price decline stop so soon?

The ECA-MSA (Economic Cooperation Administration, later known as the Mutual Security Administration) during 1948-52 financed very large purchases of crude oil and products imported into Western Europe under the Marshall Plan.[9] The authorizing act had required that purchases be at the "lowest competitive price." Gulf-plus prices and very great discrimination did not look competitive. The ECA administrator asked the advice of a group of consultants.[10] In two brief communications still worth reading today—admirable in saying much in brief and yet saying no more than absolutely necessary to dispose of a difficult problem—they gave the administrator just the understanding he needed.[11] As the bargaining agent for roughly half of Persian Gulf output and with a legal compulsion not to overpay, MSA could make itself heard when it insisted

[6] Fritz Machlup, *The Basing Point System* (Philadelphia: Blakiston, 1949), pp. 137-38.

[7] See Table V-1, Swensrud-Hoffman letter of Dec. 19, 1949.

[8] There is an amusing sidelight on the semantics of the "basing point system" in the 1969 agreement signed by Ente Nazionale Idrocarburi to import natural gas from the Soviet Union. (*New York Times*, Dec. 12, 1969, p. 93.) The ENI-Soviet announcement "said the unit price for the gas—which was undisclosed—represented a breakthrough in the 'basing point' price system in effect in world markets. . . . The Italian-Soviet accord is based on the principle that buyers whose transport costs are high should be given unit price concessions in compensation." This is, of course, nothing more nor less than the discriminatory system which was eliminated in 1947-49 and succeeded by f.o.b. pricing. It is surely as much of a "breakthrough" in the reverse direction.

[9] This account is based on the documentation in *Petroleum Study, Hearings before a Subcommittee on Interstate and Foreign Commerce*, H.R., 81st Cong., 1st sess., 1950.

[10] Max Ball, Walter J. Levy, Edward S. Mason, Sumner T. Pike, George W. Stocking, L. S. Wescoat.

[11] Later attempts by MSA to obtain refunds on alleged overpayments foundered in a confusion it is fortunately not necessary to unravel. See U.S. v. Standard Oil Co. of California et al., 155 F. Suppl. 121 (S.D.N.Y. 1957), 210 F. 2d 50 (2d Cir. 1959).

that Persian Gulf prices be a uniform f.o.b., equal to what was being realized on a modest but not negligible volume of sales to the U.S. East Coast.

Can it, however, be said that the intervention of MSA was decisive in bringing about so swift a revolution in the structure, halving the Persian Gulf price in less than two years? Professor Frank may be right in thinking it would have come about anyway, though not as soon.[12] But we must consider an opposing view.

Dr. Penrose's view. Edith Penrose, stressing the integration of the Persian Gulf producers, says that "the market *prices* of crude oil could not have determined the 'competitiveness' of oil from the Middle East . . . because they [the Companies] were the buyer-refiners as well as the seller-producers of oil. . . . [A company] decision [on 'the sources from which it would lift its oil'] can sensibly be made only with reference to *costs*."[13] This is certainly true. Dr. Penrose suggests that the price cut was "a deliberate decision of the major Companies (taken, to be sure, under government pressure) [rather] than the consequence of competition. . . . It was as *buyers* of crude that the Companies decided when and how far prices should fall, and surely one of the most important factors in their decision was the desirability of using low-cost Middle East crude in their own refining operations in the Eastern Hemisphere, particularly in Europe."[14]

But the companies were not "buyers of crude"—all of them were substantial net sellers (Appendix Table III-B). And since Dr. Penrose (rightly) denies the importance of the "selling" within an integrated company, she must deny the importance of "buying." There was no conflict between using low-cost crude and charging high prices on sales to outsiders.

Dr. Penrose points out that the posted prices of crudes landed at European refineries would be the basis for calculating product prices.[15] Hence, higher crude prices meant higher product prices also. Thus the companies lost on lower crude prices directly, and they lost much more indirectly. She also points out that high crude prices saved taxes and made entry into refining more difficult for unintegrated rivals.[16]

Thus Dr. Penrose offers no reason why the price structure became (for a time) nondiscriminatory, and at a much lower level. I share her reluctance to say to what extent the reaction was generated within the market and to what extent imposed by the largest customer who was also the government of most of the companies. But uncertainty over the relative importance of possible causal factors should not make us deny the result. It would be a mistake to argue: the companies are few and not fully independent; therefore a given limited price change cannot result from competition. This kind of stultifying nonreason is clearly seen in the UN-ECE document,[17] and even more clearly in the FTC report. Dr. Penrose will have none of this sophistry, but neither has she a theory.

The Stop to Competitive Evolution

By late 1949, as we have seen, the U.S. East Coast became the farthest market for Persian Gulf crude oil. The f.o.b. price fell low enough for it to be delivered, given current

[12] Helmut J. Frank, *Crude Oil Prices in the Middle East* (Praeger, 1966), pp. 45, 60.

[13] Edith T. Penrose, *The Large International Firms in Developing Countries: the International Petroleum Industry* (London: Allen & Unwin, 1968), p. 185. Emphasis in original.

[14] Ibid., emphasis in original.

[15] Ibid., p. 186.

[16] Ibid., p. 190.

[17] UN, Economic Commission for Europe, *The Price of Oil in Western Europe*, E/ECE/205 (Geneva, 1955).

transport costs at the time, at a delivered price equal to U.S. domestic or Venezuelan crude.

But the parity relationship did not last; by 1951, Persian Gulf crude was "overpriced" in relation to prices at New York. More precisely, there was again systematic discrimination, this time against European buyers, though on a scale not comparable with that of 1947.

Moreover, if the competitive process had continued in effect, the Persian Gulf f.o.b. price would not have continued at parity with the East Coast but would have gone well *below* it, though one cannot say precisely how far below. Finally, Persian Gulf exports to the United States would have increased far more than they actually did.

Taken together, these three developments brought about an abrupt stop to the competitive evolution of 1947-49.

Parity Prices, 1948-52. As Table V-3 shows, Persian Gulf prices at parity with New York rose in the period 1948-49, while the quoted f.o.b. price fell. As that market strengthened with increased demand and lower transport costs, sales rose and the evolving Persian Gulf f.o.b. posting came down to meet them, until the two converged at the end of 1949.

As noted in Chapter IV, the relevant tanker rate would lie somewhere between the Two-Twelve rate (a two-year charter) and the five-year time charter rate, its precise location depending on the proportion of crude oil and products sold on short- and long-term contracts. One cannot expect any precise correspondence between the Persian Gulf price and the netback[18] from New York, so long as the market is subject to changes in tanker rates and so long as bargains are made with imperfect knowledge of the future. Contracts must be fulfilled; an upward move one month or year may be cancelled by a decline the next; each bargain contains a host of special terms and considerations which if they were expressed in money terms might just offset any apparent difference. What we need to look for are substantial and continuing differences between the quoted price of $1.75 and the realized price, which in the last quarter of 1949 was in the range $1.78-$1.84.

We must first dispose of the idea that the spot netback from sales to the United States (column 11) is *the* competitively determined price to which the f.o.b. Persian Gulf price should have tended.[19] For months on end the netback was negative, or at least below producing cost. Moreover, taxes and royalties were not waived or reduced. Such sales do not take place under competition.

But the Ras Tanura[20] netback from the Two-Twelve refutes a competitive explanation for the behavior of prices during the Korean War. Our model applies as well in one direction as another. Had producers at the Persian Gulf been independent and free to compete on price, the increase in tanker rates in late 1950 would have cut off exports to the United States because each producer would have confined himself to the more remunerative European market, where delivered prices could reflect the higher tanker rates. This would have only slightly increased European supply and lowered the European

[18] A netback is defined as the price realized at a distant point of sale, less the transport cost, or less the sum of transport and refining cost, where the sale is of refined products.

[19] This was the prosecution's theory in the SoCal case, supra p. 138, n. 11. For a summary, see Wayne A. Leeman, *The Price of Middle East Oil* (Cornell University Press, 1962), chap. V, especially pp. 137-40; and Frank, *Crude Oil Prices . . .*, pp. 57-60.

[20] Loading port in the Persian Gulf.

TABLE V–3. F.o.b. Prices, Delivered Prices, Netbacks, by Months, 1948–1952

Year	U.S. Gulf f.o.b.	Freight, U.S. Gulf/N.Y. Spot	Freight, U.S. Gulf/N.Y. Two-twelve	Freight, U.S. Gulf/N.Y. 5-yr. TC	N.Y. delivered price basis Spot	N.Y. delivered price basis Two-twelve	N.Y. delivered price basis 5-yr. TC	Freight Ras Tanura-N.Y. Spot	Freight Ras Tanura-N.Y. Two-twelve	Freight Ras Tanura-N.Y. 5-yr. TC	Ras Tanura netback Col. 5 less col. 8	Ras Tanura netback Col. 6 less col. 9	Ras Tanura netback Col. 7 less col. 10
	(1)	(2)	(3)	(4)	(5)	(6)	(7)	(8)	(9)	(10)	(11)	(12)	(13)
1948													
J	$2.75	$1.00	n.a.	$0.36	$3.75	n.a.	$3.11	$5.32	n.a.	$1.49	–$1.57	n.a.	$1.62
F		0.88	n.a.	0.36	3.63	n.a.	3.11	5.54	n.a.	1.49	–1.91	n.a.	1.62
M		0.57	n.a.	0.36	3.32	n.a.	3.11	3.74	n.a.	1.49	–0.42	n.a.	1.62
A		0.57	n.a.	0.36	3.32	n.a.	3.11	2.89	n.a.	1.49	0.43	n.a.	1.62
M		0.49	n.a.	0.35	3.24	n.a.	3.10	2.63	n.a.	1.46	0.61	n.a.	1.64
J		0.38	n.a.	0.35	3.13	n.a.	3.10	1.52	n.a.	1.46	1.61	n.a.	1.64
J		0.31	n.a.	0.35	3.06	n.a.	3.10	1.35	n.a.	1.46	1.71	n.a.	1.64
A		0.29	n.a.	0.39	3.04	n.a.	3.14	1.34	n.a.	1.63	1.70	n.a.	1.51
S		0.29	n.a.	0.38	3.04	n.a.	3.13	1.27	n.a.	1.59	1.77	n.a.	1.54
O		0.32	n.a.	0.37	3.07	n.a.	3.12	1.35	n.a.	1.56	1.72	n.a.	1.56
N		0.40	n.a.	0.37	3.15	n.a.	3.12	1.69	n.a.	1.54	1.46	n.a.	1.58
D		0.36	n.a.	0.37	3.11	n.a.	3.12	1.59	n.a.	1.52	1.52	n.a.	1.60
1949													
J		$0.32	n.a.	$0.36	$3.07	n.a.	$3.11	$1.57	n.a.	$1.49	$1.50	n.a.	$1.62
F		0.30	n.a.	0.35	3.05	n.a.	3.10	1.41	n.a.	1.47	1.64	n.a.	1.63
M		0.28	n.a.	0.34	3.03	n.a.	3.09	1.75	n.a.	1.42	1.68	n.a.	1.67
A		0.25	0.44	0.33	3.00	3.19	3.08	1.19	1.70	1.39	1.81	1.49	1.69
M		0.23	0.44	0.32	2.98	3.19	3.07	1.02	1.70	1.35	1.96	1.49	1.72
J		0.18	0.44	0.32	2.93	3.19	3.07	0.85	1.70	1.35	2.08	1.49	1.72
J		0.18	0.44	0.31	2.93	3.19	3.06	0.76	1.70	1.32	2.17	1.49	1.74
A		0.17	0.44	0.31	2.92	3.19	3.06	0.80	1.70	1.30	2.12	1.49	1.76
S		0.21	0.44	0.30	2.96	3.19	3.05	0.93	1.70	1.27	2.03	1.49	1.78
O		0.25	0.33	0.30	3.00	3.08	3.05	1.02	1.24	1.27	1.98	1.84	1.78
N		0.27	0.33	0.29	3.00	3.08	3.04	1.10	1.24	1.25	1.96	1.84	1.79
D		0.32	0.33	0.29	3.07	3.08	3.04	1.27	1.24	1.22	1.80	1.84	1.82

TABLE V–3. (Continued)

Year	U.S. Gulf f.o.b.	Freight, U.S. Gulf/N.Y.			N.Y. delivered price basis			Freight Ras Tanura-N.Y.			Ras Tanura netback		
		Spot	Two-twelve	5-yr. TC	Spot	Two-twelve	5-yr. TC	Spot	Two-twelve	5-yr. TC	Col. 5 less col. 8	Col. 6 less col. 9	Col. 7 less col. 10
1950 (cont.)	(1)	(2)	(3)	(4)	(5)	(6)	(7)	(8)	(9)	(10)	(11)	(12)	(13)
J		$0.27	0.33	$0.29	$3.02	3.08	$3.04	$1.13	1.24	$1.22	$1.89	1.84	$1.82
F		0.21	0.33	0.20	2.46	3.08	3.01	0.43	1.24	1.16	2.03	1.84	1.89
M		0.25	0.33	0.20	3.00	3.08	3.01	1.00	1.24	1.16	2.00	1.84	1.89
A		0.23	0.33	0.20	2.98	3.08	3.01	0.93	1.22	1.16	2.05	1.86	1.89
M		0.24	0.33	0.20	2.99	3.08	3.01	0.99	1.22	1.16	2.00	1.86	1.89
J		0.24	0.33	0.20	2.99	3.08	3.01	1.00	1.22	1.16	1.99	1.86	1.89
J		0.26	0.33	0.26	3.01	3.08	3.01	1.08	1.22	1.16	1.93	1.86	1.89
A		0.32	0.33	0.27	3.01	3.08	3.02	1.34	1.22	1.19	1.73	1.86	1.83
S		0.47	0.33	0.28	3.22	3.08	3.03	2.09	1.22	1.20	1.13	1.86	1.83
O		0.57	0.48	0.29	3.32	3.23	3.04	2.42	1.84	1.22	0.90	1.39	1.82
N		0.66	0.48	0.29	3.41	3.23	3.04	2.79	1.84	1.22	0.67	1.39	1.82
D		0.88	0.48	0.29	3.63	3.23	3.04	3.22	1.84	1.22	0.41	1.39	1.82
1951													
J		$1.05	0.48	$0.35	$3.80	3.23	$3.10	$4.33	1.84	$1.46	–$0.53	1.39	$1.64
F		1.14	0.48	0.37	3.89	3.23	3.12	4.79	1.84	1.52	–0.90	1.39	1.60
M		1.06	0.61	0.35	3.81	3.23	3.10	4.37	1.84	1.47	–0.46	1.39	1.63
A		0.83	0.61	0.35	3.88	3.36	3.10	3.61	2.23	1.46	–0.03	1.13	1.64
M		0.48	0.61	0.34	3.23	3.36	3.09	2.07	2.23	1.46	1.16	1.13	1.67

J	0.50	0.61	0.33	3.25	3.36	3.08	2.07	2.23	1.39	1.18	1.13	1.69
J	0.45	0.61	0.34	3.20	3.36	3.09	1.95	2.23	1.46	1.25	1.13	1.67
A	0.48	0.61	0.34	3.23	3.36	3.09	2.13	2.23	1.44	1.10	1.13	1.65
S	0.50	0.61	0.35	3.25	3.36	3.10	2.40	2.23	1.46	0.85	1.13	1.64
O	0.71	0.72	0.35	3.46	3.47	3.10	3.23	2.62	1.46	0.23	0.85	1.64
N	0.94	0.72	0.37	3.69	3.47	3.16	4.40	2.62	1.54	–0.71	0.85	1.58
D	1.13	0.72	0.39	3.88	3.47	3.14	4.84	2.62	1.63	–0.96	0.85	1.51
1952												
J	$1.14	0.72	$0.39	$3.89	3.47	$3.14	$4.72	2.62	$1.64	–$0.83	0.85	$1.50
F	1.14	0.72	0.42	3.89	3.47	3.17	4.49	2.62	1.73	–0.60	0.85	1.44
M	1.10	0.72	0.42	3.85	3.47	3.17	4.30	2.62	1.73	–0.45	0.85	1.44
A	0.65	0.78	0.42	3.40	3.53	3.17	2.78	2.85	1.73	0.62	0.68	1.44
M	0.40	0.78	0.42	3.15	3.53	3.17	1.91	2.85	1.73	1.24	0.68	1.44
J	0.42	0.78	0.42	3.17	3.53	3.17	1.86	2.85	1.73	1.31	0.68	1.44
J	0.35	0.78	0.42	3.13	3.53	3.17	1.59	2.85	1.73	1.54	0.68	1.44
A	0.39	0.78	0.42	3.14	3.53	3.17	1.63	2.85	1.73	1.51	0.68	1.44
S	0.48	0.78	0.42	3.23	3.53	3.17	1.76	2.85	1.73	1.47	0.68	1.44
O	0.43	0.65	0.42	3.18	3.40	3.17	1.52	2.40	1.73	1.66	1.00	1.44
N	0.46	0.65	0.36	3.21	3.40	3.11	1.54	2.40	1.56	1.67	1.00	1.55
D	0.44	0.65	0.33	3.19	3.40	3.08	1.51	2.40	1.44	1.68	1.00	1.64

Sources, by column:
(1) is a mixture of East and West Texas crude, the basis for Venezuelan prices, not yet posted. Prices from *Platt's Oilgram Price Service*.
(2): Conrad Boe Ltd., A/S.
(3): *Platt's Oilgram News Service* or *Platt's Oilgram Price Service*.
Note: See preceding chapter for definition of tanker rates. For f.o.b. price comparable with cols. 12 and 13, see table V–1.
n.a. = not available.

price. In continuing exports to the United States at much lower realizations the exporting companies preferred to accept a thin or zero profit margin on part of their output rather than precipitate the general price decline that would have reduced profits on total output. Each knew the others would refrain from any attempt to take the more profitable European business away from their rivals; each one, therefore, acted for the interests of the group.[21]

A seller operating ships under long-term charter at a low rate might still have been realizing profits on sales to the Western Hemisphere. (See column 13.) But putting the same oil into the same ship and directing it to Europe would have meant a higher return. Or, what comes to the same thing, the true cost of shipment is not the accounting figure of tanker hire but rather the sacrifice of revenue on a trip to the more renumerative markets. Hence, when importers shipped their own output to their refineries in the United States in 1948-49, and above all in 1950-52, they were accepting a much lower netback than was available in the Eastern Hemisphere. The only explanation for this is that it was more profitable to dump surplus oil in the United States.

Some U.S. companies did not produce, but purchased at the Persian Gulf (at prices not publicly known) and shipped to the United States. Presumably, they earned a profit by shipping it to the United States for refining, but they would have earned considerably more on a much smaller investment and with much less risk by using U.S. crude and reselling their Persian Gulf crude in Europe. Presumably, these buyers were obligated not to follow this procedure, which reinforces the conclusion reached above that dumping by some or all the producers kept the excess supplies from lowering the Eastern Hemisphere price.

By the end of 1952, a relatively small reduction in the f.o.b. Persian Gulf price would have equated it again to the East Coast price and permitted larger sales, which were especially desirable because production was actually being cut back and there was some excess capacity. Under these circumstances the stable price is evidence of a block to competition. The competitive parity established in late 1949 was becoming ever looser (though there was no reversion to Gulf-plus). It is time to examine its limitations.

The Parity Theory Incomplete. Earlier, I expressed admiration for the consultants' two memoranda, which covered the immediate problem so well. The competitive parity theory was that the f.o.b. Middle East price could never go any lower than the price that would just permit exports to the United States, even under the competition of many suppliers. If, therefore, it went to and stayed at such an equalizing point, the price would be "defensible" because it would be the same as the price resulting from vigorous (or pure) competition. This idea was in accord with the view of Esso's president and was elaborated by Walter J. Levy.[22]

The theory is correct within limits, and wrong outside of them. (Particularly in the next chapter we shall see what massive blunders it leads to.) The consultants (and Mr. Holman) respected the limits, but did not state them. The events just after their pronouncement force us to do so.

[21] Had the exporters channeled all their shipments to Europe, the demand for transport would have fallen, the tanker rate risen less, and netbacks on U.S. shipments would have been higher. But since the amount of shipping saved was a very small percentage of the world total, the effect on tanker rates and netbacks would have been correspondingly small and probably imperceptible.

[22] Letter from Eugene Holman to Paul G. Hoffman, Feb. 23, 1949, reprinted in *Petroleum Study*, p. 415; W. J. Levy, "The Past, Present and Future Price Structure for the International Oil Trade" (Third World Petroleum Congress paper, 1951), paras. 18, 48, 50.

If the reader looks back at Figure V-1, he will see that the parity theory sets forth two conditions: that a seller receives the same net revenue from all customers, regardless of location or other characteristics; and that at any destination there are "identical delivered prices, regardless of origin." But the parity theory *ignores comparative production costs* which, as was concluded in Chapter II, are largely development investment costs little affected by short-run output changes. A parity price is not stable where the difference between price and the cost of providing new capacity is far greater in one production center than in another. In that case, the competitive process causes output in the more profitable center to rise so much more quickly that it lowers the price and closes the price-cost gap. The process was underway in 1947-49, but not in 1950-53. To understand why the structure stopped evolving during 1950-53, we need to consider comparative production costs.

Comparative Laid-Down Costs and Prices. Chapter II gives us a basis for estimating Persian Gulf development costs in the early 1950s as perhaps half again as high as in the mid-1960s, say 18 cents per barrel. Taking Saudi Arab royalty—the highest at 21 cents per barrel—incremental tax-plus-cost (the sum of highest cost and highest royalty) was thus 39 cents.[23] At any higher netback, therefore, a Persian Gulf producer would be making extremely high profits, a more than sufficient inducement to expand output manyfold at only a moderate increase in cost and at an enormously greater total profit.

The 1950-53 Persian Gulf price, though far below the price of two years earlier, was far above incremental cost of greatly expanded output. The competitive process would therefore lead to a large increase to be shipped to the U.S. East Coast. Hence the additional 10-cent reduction to $1.65 noted by Mr. Swensrud was to be expected, with more to follow.

But it did not happen. Indeed, the reduction was quietly cancelled. The price remained at $1.75 for some years, and then actually increased. Nor did imports to the United States increase greatly. The price structure and the trade stream froze together.

Up to this point, price and production have evolved according to the competitive scenario set forth above. (Whether the competition has itself been explained is another story.) But the competitive hypothesis cannot explain what happened in 1950-53. Imports to the United States hit some new barrier.

Hypothesis: Tacit Collusion. An obvious hypothesis is that the Persian Gulf producers had reached an understanding amongst themselves to limit exports to the United States to a very low figure. They considered it more profitable to let production expand no faster than Eastern Hemisphere market demand—i.e., to plan on selling a smaller amount at a higher price.

The concentrated market structure shown in Chapter III was surely no obstacle to such an understanding, which could have been reached without any word put to paper. The objection that the oil companies did not aim to maximize profits will not be seriously advanced by anyone. (Yet how long and seriously and hyper-ingeniously it is argued about big business generally—so long as it need not be tested against the behavior of companies in a real-world market.) The American companies who were Persian Gulf

[23]Cf. *Investigation of the National Defense Program. Hearings before a Special Committee* . . . U.S. Senate, 80th Cong., 1st sess., 1948, part 41: "Petroleum Arrangements with Saudi Arabia," pp. 24756-7, 24923, 25364, where cost-plus-tax is reckoned at 25 to 40 cents. The royalty would have been more than 50 percent of profits; hence, there would have been no income tax to add.

producers could have run the crude into their East Coast refineries. In addition, they could have sold large additional amounts of crude. They did not do so—this is the anomaly to be explained.

A monopoly would be guided by its marginal net revenue. Consider even a drastic reduction, say 50 cents per barrel. This would have offset the cost of shipment from the U.S. Gulf Coast to the East Coast, and permitted a shipment by pipeline almost anywhere in the United States even under 1950 conditions. In 1950, Persian Gulf production was about 1,750 thousand barrels daily (TBD), and a reasonable expectation for 1955 would have been considerably less than the 3,200 actually reached; but let us assume perfect foresight anyway. A 50-cent reduction would then mean a 1955 loss of $1.58 million daily. Against this must be set the hoped-for increase in revenues from sales in the United States. On each barrel sold at $1.25 and costing them 40 cents, they would clear 85 cents. In 1950, U.S. refineries used 5,750 TBD, and by 1955 took 7,500 TBD. Of this they needed only to capture another 1,860 TBD and the loss in revenue would be offset. On a larger outlet they would have been better off. Therefore the only obstacle to a complete capture of the U.S. market would have been the ability of U.S. and Venezuelan crude to compete there, at prices 50 cents less.

To judge from what is written, the universal opinion is that on the basis of comparative cost Persian Gulf crudes would have swept the whole Western Hemisphere off the board. If that were true, it would utterly discredit the monopoly hypothesis. But it is far from the truth; it springs from the pernicious habit of looking at averages. Even in 1968, as seen in Chapter II, there was plenty of Venezuelan crude to compete with Persian Gulf on the basis of comparative costs. In the 1950s, with tanker rates far higher, the proportion must have been much larger. For the United States, we have less basis for comparison because the U.S. cost data are so far inferior. But surely one is entitled, by analogy with Venezuela, and considering the wide dispersion among states and also within Texas for one year (see Chapter II, pp. 48ff), to conclude that some large fraction could compete even in distant areas, and certainly in its own vicinity.

Obviously, any attempt at calculation would be a pretense. But if even an incremental fourth of U.S. refinery runs had been captured by the Persian Gulf producers, they would have been compensated for their price cut. Capture of even a third of the U.S. market would have brought enormous profits, far more than by staying on the fringes as they did.

Hence, the simple monopoly theory breaks down in its turn. Let us now consider a more complex monopoly theory.

Hypothesis: Tacit Collusion and American Producer-Refiner Integration. We now consider the fact that about half of Persian Gulf output was produced by six concerns (Esso, Shell, SoCal, Texaco, Mobil, Gulf) with very large interests in U.S. producing and refining. A reduction in Persian Gulf prices and large imports into the United States would mean a reduction in domestic U.S. prices, and since most of their production was in that country, on balance it would not have paid them.

Like the simple monopoly theory, this one is internally consistent, and makes good business sense. It also makes us see that the six large producers did not have the same interests. Esso made no secret of its desire to keep Persian Gulf crude out of the United States. Its policy was: Western Hemisphere crude for the Western Hemisphere.[24] Esso and

[24]*Petroleum Study*, p. 417.

Shell had very large interests in Venezuela. Had the old pricing system continued, either Gulf-plus or at least Europe-minus (equalization in Northwest Europe), the reduction to $1.75 would never have come about and they would have been better off supplying the East Coast from Venezuela. There were also political elements in their reckoning, to be explored later. Of the four others, Gulf's Venezuelan interests had in effect been signed over to Esso and Shell control,[25] with Gulf retaining largely a beneficial interest. This counted for very little against the huge potential output from Kuwait. (Indeed Gulf was to be the most oppressed of the four companies by the impasse over U.S. imports.)

Moreover, British Petroleum (BP) and Compagnie Française des Pétroles (CFP) had no United States interest at all, and no reason to refrain from exporting. They had no common pocketbook with the other six, and some kind of side-payment would have been necessary to compensate them for this self-denial. This would have been difficult and would have left a trail of documentary evidence. The American companies were and are highly sensitive to antitrust possibilities.

But even if one assumes that Gulf, BP, and CFP were somehow paid off, an even worse weakness in the theory serves to remind us not to let "integration" mask the need for analysis. The six companies accounted, in round percentages, for 45 percent of U.S. refining and 30 percent of U.S. crude output.[26] Every one of them was in deficit on U.S. crude supply, even adding Esso and Shell Venezuela output. Had they done no more than substitute Persian Gulf crude for U.S. crude costing over twice as much, their profits would have been very great and the impact on U.S. domestic prices small, possibly negligible. For only very high-cost crude would have been backed out of the U.S. market.

In general, the six companies would have reaped all of the gain on cheaper crude, and not much of the loss—nothing like 30 percent. First, they had a disproportionately small amount of extremely high-cost and stripper production, and hence would have borne less than 30 percent of the reduction. Even more important was the automatic run-down of their domestic crude production because of natural decline. In 1950, the six companies produced about 1,725 TBD. By 1955, an 8 percent decline rate would have brought this down to 1,170 TBD, only 15 percent of the national crude production. On sales to unintegrated refiners, they would, of course, have gained much in addition. And with every succeeding year, as they let their domestic output run down, the net gain to them would become all the greater.

Or let us imagine a really drastic reduction in the Persian Gulf price, say 50 percent or 87 cents per barrel, and an equal drop at the U.S. East Coast. Then the net benefit to an integrated producer-refiner, from substituting a barrel of Persian Gulf crude for a barrel of purchased U.S., would be $2.96–0.87–($0.40 + 1.21) or 48 cents.[27] Then the gains and losses to the integrated American companies are crudely summed up in Table V-4. It

[25]Cf. FTC, *The International Petroleum Cartel*, pp. 171-90, and *Prospectus* of Gulf Oil Corporation, May 6, 1963, p. 5.

[26]Crude production (gross) from annual reports 1947-52; refining throughput cannot be directly calculated and I have therefore taken percent of capacity from the semiannual publication of the U.S. Bureau of Mines, *Petroleum Refineries in the United States* (1948 to 1956).

[27]Let A represent the crude oil price at New York, which a refiner would need to pay for purchased crude. Let C be the production cost plus tax-royalty payments for Persian Gulf crude, and let D be the Persian Gulf-New York freight rate. The gain realized by a refiner by substituting his own Persian Gulf crude for a barrel of purchased crude is the difference between A and the sum of $(C+D)$. But if this substitution were on a large enough scale to provoke a price reduction, of amount B, the refiner's gain would be $A-B-(C+D)$. In the text, the values for A, C, and D are taken from table V-2, and B is assumed to be one-half of A.

will be noted that the profits reckoned on Persian Gulf output are over and above the 20 percent return on investment assumed as a cost, and they assume that cost to be nearly twice as high as in the late 1960s. Moreover, in time they could have supplied the whole country, and the offset in line 6 of the table would have dwindled.

These gains may look unduly large, but they are not. Oil production and refining is not a "zero-sum-game" where the winners get only what they take from the losers. Here the winners get much more because there has been a shift from costlier to cheaper methods of producing oil. The lower price level for all sellers is for the most part a mere reflection of lower costs.

In any case, the idea of the six companies restricting exports to the United States because they feared that lower prices would cost them too much in this country cannot stand examination. Because the figures used are perforce rough, rather generous allowances have had to be made; but because better data would only make the conclusion stronger, it hardly seems necessary to pursue the matter.

To summarize, the abrupt 1950 stop to the price-export evolution cannot be explained as the industry reaching the competitive equilibrium. Nor does it make sense as a tacit understanding, nor as the protection of a vested interest in the American market. At this point, therefore, one reaches an important though limited conclusion: there was *some external block* in the way of normal profit-seeking behavior. Having at least brushed away some false solutions and shown the general direction of the real one, we turn now to look outside the market, at U.S. regulatory policy.

U.S. Production Controls. For about 15 years before 1950, the U.S. domestic price had not been free to move in any direction except occasionally upward. Supply was and is controlled in an uncoordinated but effective manner by state regulatory commissions. The method and its result have been fully analyzed elsewhere.[28] Very briefly: in each of the most important producing states, the producing companies state their requirements for the forthcoming month to the state regulatory agency, of which the best known is the Texas Railroad Commission. The commission determines the most likely total, and then allocates it to every single well in the state as a proportion of the well's total "basic allowable." The basic allowable is not producing capacity; it is determined in a slapdash manner with the general object of compensating wells for their higher costs, e.g., their depth. Compensation is waste. For under competition, the higher costs of deeper wells would shut them out of the market, except as offset by cost-reducing factors such as greater productivity. Furthermore, stripper wells—broadly speaking, those incapable of making over 10 barrels daily—are exempted from regulation. Thus high-capacity, low-cost wells are held back in order to preempt the market for high-cost production. The same exception is made for pressure maintenance projects ("secondary recovery"). Obviously, a price high enough to make high-cost production profitable makes lower-cost production vastly profitable, and has encouraged the drilling of many thousands of unnecessary wells. This is even more encouraged by the system of granting allowables to wells rather than to pools or even leases. In order to get more production out of a given reservoir, more and more wells have been drilled into it even if they merely raised costs and added little or nothing to capacity—which was in excess anyway.

[28] Wallace F. Lovejoy and Paul T. Homan, *Economic Aspects of Oil Conservation Regulation* (Johns Hopkins University Press for Resources for the Future, 1967); M. A. Adelman, "Efficiency of Resource Use in Crude Petroleum," *Southern Economic Journal*, vol. 31 (1964), pp. 103–9, 116–22.

TABLE V–4. Gains and Losses to Six Integrated American Persian Gulf Producers

1. Refinery output, 1955, total U.S.	7,500	TBD
2. Share of six companies (Esso, Mobil, Gulf, SoCal, Texaco, Shell)	3,375	TBD
3. Domestic 1955 production of six companies (allowing 1950–55 runoff at 8% annually)	1,170	TBD
4. Crude deficit (line 2 minus line 3)	2,205	TBD
5. Gain on substitution of Persian Gulf crude (line 4 × $0.48)	$1.06	million daily
6. Loss on own output because of price decline (line 3 × $0.87)	$1.02	million daily
7. Maximum additional market (line 1 minus line 2)	4,125	TBD
8. Maximum additional profit (line 7 × $0.48)	$1.98	million daily

Sources: See text, p. 147, fn. 26.

But look as one will through analysis and history of the system, one will find no villains. It was devised in great haste, under the pressure of gross physical waste—oil running over temporary open-air pits, befouling creeks and the like. An irrational system of subsoil rights, plus the recent discovery of two great oil fields, plus the onset of the Great Depression, had all contributed to the urgent necessity for quick action to repress supply. But the temporary survives. The longer the system lasted, and the greater the excess capacity it generated, the greater the fear of even a mild degree of competition. Only by first-hand contact with American oilmen can one understand how firmly the system has gripped their minds.

Those outside the United States have been even more misled, not only as to the nature of oil production economics, but about the politics of oil in the United States. Nobody would deny that the oil industry is very influential in the states and in Washington. No industry that size could fail to be. But since the big companies are so well known, and since they seem to bestride the industry like a colossus, it seems self-evident that theirs is the predominant voice. This is the contrary of the truth. State regulation has gone to the point where the industry as a whole is hurt by it. But small producers, landowners, local supplying industries, the labor force and local communities they support—these benefit greatly; they have the votes, and their support keeps the system in power today as it did over 30 years ago.

Any change in the system is extremely difficult because of the balance of interests that has been built up around it. Any change is bound to hurt somebody and threatens that "chaos is come again."

This system is powerful politically, but fragile and vulnerable to additional uncontrolled output. The ill will in Texas over a slightly more permissive policy in Louisiana is eloquent testimony. So is a report issued by the Antitrust Division of the Justice Department, which is worried that "even moderate liberalization of controls in one area might swamp stabilization efforts in others . . . and wipe out precisely those independent elements . . . which are so necessary to competition."[29] Not for the first nor the one-hundredth time, solicitude for small business is the bulwark of wasteful monopoly. Others have seriously proposed to treat production from the offshore United States as "foreign."

However, this last modest proposal has got us ahead of the story. Given the ultra-tight control of output existing in the post-World War II industry, the price of oil was not free to move. Hence there was no market mechanism to cut back high-cost output. Greater

[29]Ibid., Adelman.

imports would therefore have produced no effect on the price, but merely a barrel-for-barrel curtailment of domestic output. Low-cost output would be cut back most. For stripper wells had (and have) a preemptive right to the market, while more productive prorated wells are only permitted to make up the deficit—which would keep shrinking. The thermostat had been disconnected, and there was and is nothing to keep the furnace from getting hotter and hotter. Indeed, as the more productive wells' output was cut back, unit costs would keep rising, to make foreign oil look even more attractive.

Of course it would be a mistake simply to extrapolate the process until the last barrel of U.S. crude oil was produced at a Gulf Coast price still \$2.96. Long before that happened, the whole prorationing system would have crumbled, and the price collapsed even farther than if competition of imports had been allowed to compete in the first place.

U.S. Import Controls, 1948-57: Congress. The fragility of the system to even small amounts of uncontrolled increment explains the strong reaction in the producing states and in the 81st Congress, with three investigations during the 1949-50 term.[30] During this time also, the Federal Trade Commission decided to investigate an alleged international petroleum cartel. For them and the Senate Small Business Committee (not to be confused with the House Committee) the "most important" question was whether oil imports were "causing injury to independent American oil producers." All four groups failed to uncover any respectable evidence that imports were hurting anyone. This failure was not for lack of effort.

The strong reaction was out of all proportion to the mild stimulus.[31] But the oil state congressmen knew what they were doing. Had the oil industry been competitive, the impact of imports would have been roughly proportionate to their percentage of total output. But here the real threat was: faced with eventual collapse of the prorationing system, the states would bolt out—every state for itself. The congressional investigations were at once a reassurance to the states and a threat (not a bluff) to the companies that if they did not keep imports down to a very low figure, legislative action would be taken. And so long as the Congress did their share, the states could make would-be importers think twice; for against the chance of Congress allowing them to import in the future, they had to weigh the chance of having wells shut in on one pretext or another, on the bulk of their current production.

All parties showed a commendable reluctance to take measures any more sweeping than needed. The Small Business Committee of the House of Representatives suggested a cartel: "voluntary agreements, approved by the President, to control the importation of foreign-produced crude petroleum and petroleum products in the U.S." They rebuked the importing companies for failing to recognize "the equitable necessity for mutuality of flexibility." And they pointed with justified pride to their accomplishment: "The activities of the importers have been brought into the open for public scrutiny with the result that we now know who is importing what from where and in what quantities; no longer are individual activities submerged in totals."[32]

[30]*Effects of Foreign Oil Imports on Independent Domestic Producers*, Report of the Subcommittee on Oil Imports to the Select Committee on Small Business, H.R., 81st Cong., 2d sess. 1950; *Causes of Unemployment in the Coal and Other Specified Industries*, Report of the Committee on Labor and Public Welfare, U.S. Senate, 81st Cong., 2d sess. 1950; *Petroleum Study: Petroleum Imports*, Progress Report of the Committee on Interstate and Foreign Commerce, H.R., 81st Cong., 2d sess. 1950.

[31]These hearings were my own primary education in oil economics. Economists are like geologists: their trade is finding anomalies and explaining them.

[32]*Effects of Foreign Oil Imports*, p. 2.

TABLE V-5. U.S. East Coast Crude Oil Imports, by Importers of Record, 1946-1957
(*imports in percent of total*)

Company	1946	1947	1948	1949	1950	1951	1952	1953	1954	1955	1956	Programmed second half 1957
Esso	36	33	31	22	16	19	19	15	13	12	11	8
Mobil	7	13	16	16	15	15	16	16	14	11	9	7
Texaco	a	a	3	12	14	10	9	9	9	10	9	6
SoCal	2	3	4	4	7	8	11	12	12	11	10	7
Gulf	28	26	22	23	23	19	19	20	20	18	18	12
Subtotal	73	74	76	77	74	69	74	71	68	61	57	40
Six others[b]	27	26	24	23	26	31	26	29	29	32	34	41
All others	...	...	...	...	...	...	...	...	3	7	8	19
Total (in TBD)[c]	235	265	343	405	467	469	524	552	601	678	718	916
Number of importers[d]	8	9	11	11	11	11	11	11	16	18	19	19
Canadian crude (in TBD)	...	...	...	...	...	...	...	...	4	10	37	n.a.

Sources: 1946-53, Independent Petroleum Association of America and Texas Railroad Commission, with some minor inaccuracies.

1954-56, programmed second half 1957; "Petroleum Imports, Report of the President's Special Committee to Investigate Crude Oil Imports, July 29, 1957" (Memorandum from the White House, mimeo.).

Notes: Detail may not add to total because of rounding. There are small discrepancies between the totals shown here and the U.S. Census-Bureau of Mines import figures. We cannot use the latter because it would exaggerate the share of "all others."

[a]Unknown but insignificant.

[b]Atlantic, Cities Service, Indiana Standard, Sinclair, Sun, Tidewater.

[c]Excluding Canadian crude.

[d]Twenty-eight importers received allocations for the final quarter of 1958, excluding importers of Canadian crude.

n.a. = not available.

The importance of pinpointing the importers, and of knowing where to apply the threats, is obvious. (See Table V-5) But by the time the report was issued, the urgency was gone. For one thing, the mild 1949 recession was definitely over and domestic consumption had risen; for another, Venezuelan imports were practically unchanged during the first half of 1950. It would seem that the countermeasures had succeeded. But we cannot really be sure. In July 1950 came the Korean War, with its strain on crude supply and tanker space. In April 1951, Iran seized the BP concession, and production shut down almost totally. Despite the remarkable surge in output on the Arab side of the Persian Gulf, crude was still short through the middle of 1952, and 1951-52 imports were below the 1950 level.

After the 1950-52 Respite: The TRC. But by the end of 1952 it was apparent that imports were again on the increase, and now it was the Texas Railroad Commission that applied the counterpressure. For January 1953, and subsequent months they demanded and secured from all producers a highly detailed statement of import forecasts for the next five months, month by month, and by gravity numbers.[33] The large importers

[33]The best published source is *Platt's Oilgram News Service* or the New York *Journal of Commerce*, around the middle of each month. The first entry, to my knowledge, is on Feb. 13, 1953, covering January to May 1953. The *Oil and Gas Journal* summaries are less detailed and less useful.

TABLE V-6. Imports Related to Crude Refined in United States, 1947-1960

Year	Crude refined: Million barrels daily	Percent imported: Total	Percent imported: Canadian	Percent imported: All other
1947	4.7	5.8	...	5.8
1948	5.1	7.0	...	7.0
1949	5.5	7.6	...	7.6
1950	5.3	9.1	...	9.1
1951	5.7	8.6	...	8.6
1952	6.5	8.8	...	8.8
1953	6.7	9.7	0.1	9.6
1954	7.0	9.4	0.1	9.3
1955	7.0	11.2	0.6	10.6
1956	7.5	12.5	1.6	10.9
1957	7.9	12.9	1.9	11.0
1958	7.9	12.0	1.0	11.0
1959	7.6	12.6	1.2	11.4
1960	8.0	12.7	1.4	11.3

Source: U.S. Bureau of Mines, *Mineral Industry Surveys, Petroleum Statement, Annual.*

divided forecasts into eight different groups. Thereby each party made its intentions known to all the others in advance in great detail. These reports constituted an agreement between the commission and each importer to limit imports, which amounted also to an agreement among the importers themselves. The subsequent record of actual imports constituted a compliance report.

Any such action by private parties would have been a criminal violation of the Sherman Anti-trust Act. But however legal, it was a highly imperfect agreement because the parties could never meet to work out the market shares. Essentially, it held the line—for a limited time, as will be seen.

The TRC import agreements are a useful documentary record in giving details of imports nowhere else available. A danger was apparent at once. Kuwait importation, in 1949-52 around 65 TBD, was up to 125 TBD by the first half of 1953. What was more threatening, a third of it was being imported not by the producer Gulf, but by customers. Imports by an integrated producer-refiner are limited to his own needs over and above his own production and his own purchase commitments. They could affect price only by reducing demand for American domestic crude very slightly. But imports to be sold to third parties contain price terms which others can see and act upon. The proportion of Kuwait crude was negligible, but if Gulf could expand sales quickly and make very large profits for themselves and their customers, the hold over all others would break down rapidly. If the large East Coast refiners who did not produce abroad were to start buying crude from the overseas producers, the pressures to cut price to gain additional sales would be irresistible. Not only would crude oil be backed into the Southwest and other producing areas, the large buyers could offer very large blocks of additional business, for quick delivery, on which profits would be enormous even at sharply cut prices. Our previous calculation, however sketchy, leaves no doubt on that score.

Hence the trickle through the dike was cause for some concern. Yet the domestic producers felt strong enough in their new protection against imports that the price of U.S. crude was increased in June 1953. Whoever raised the price was counting on the Texas and other commissions to restrict output to any point needed to keep the price at the new higher level. We will return to this price increase later.

The success of the Texas Railroad Commission as pivot man in the import-limitation agreement is, as usual, hard to assess. Imports (other than Canadian) increased by only 0.8 percentage points during 1953, but this was partly due to a slowdown in consumption and the fulfillment of contracts made in 1952 and already covering most of 1953. Imports in 1954 were actually cut back. But the opposition to oil imports was not lulled, and called on the national Administration, which now first enters the scene. It seems a fair inference that the Texas Railroad Commission could not have held the line permanently.

Further Escalation of the Struggle Against Imports, 1954. A special Cabinet committee was named in July 1954 and reported in February 1955.[34] On the basis of memoranda submitted by both importing and domestic interests to its operating subcommittee, the committee concluded that imports were a possible threat to national security[35] and should be held at the 1954 proportion to domestic output by voluntary action. As a result, the work of the Texas Railroad Commission was taken over by the Office of Defense Mobilization. The Trade Agreements Extension Act of 1955, Section (7)(b), empowered the President, after investigation, to limit crude oil imports.

By mid-1954, there were signs of future trouble. A large international company let it be known that it was investigating loss of trade to a centrally controlled rival.[36] This would indicate price reductions, with headquarters approval, which the lower echelons had not the power to grant. A large East Coast refiner was ready "to take a plunge and buy Middle East oil if price is right."[37] Later reports spoke of tanker charters being arranged for a long-term processing deal.[38] These portended independent action which had to be repressed.

"The policy of voluntary restrictions worked with reasonable success until the middle of 1956."[39] At that time, however, danger threatened. The seven "established importers," as they were officially called (Atlantic, Gulf, Sinclair, Mobil, SoCal, Esso, and Texaco), had brought into the East Coast 536 TBD in 1954, 551 in 1955, and only 556 in 1956. But the "new importers," starting with 64 TBD, went to 128 and then to 186 in 1956. As this became known, the larger importers became restive, fearing with reason that the more the newcomers brought in, the less would be left for them. In order to head off a collapse of the agreement, the government held public hearings in October 1956; but then suspended action because of the Suez crisis. Shortly thereafter, the Director of Defense Mobilization said he had acted on the 1955 report but had "never obtained an agreement from all the companies, only from some of them."[40] It is a revealing fact. Even when lawful and encouraged by government, cooperative behavior by industry was limited and insecure because not every company would benefit. As the situation returned to normal in early 1957, the established importers were committed to bringing in

[34] *White House Report on Energy Supplies and Resources Policy*, Feb. 26, 1955.

[35] This was never defined or explained. Nothing in the official literature has any relevance to security before publication of *Cabinet Task Force on Oil Import Control, The Oil Import Question, A Report on the Relationship of Oil Imports to the National Security* (Washington, Feb. 1970).

[36] *Platt's Oilgram Price Service*, Aug. 12, 1954.

[37] *Platt's Oilgram Price Service*, Aug. 17, 1954.

[38] *Platt's Oilgram Price Service*, Nov. 10, 1954.

[39] "Petroleum Imports: Report of the President's Special Committee to Investigate Crude Oil Imports, July 29, 1957." (Memorandum from the White House, mimeo.), p. 2.

[40] *Emerging Oil Lift Program.* Hearings Before the Subcommittee of the Committee on the Judiciary and Committee on Interior and Insular Affairs, U.S. Senate, 85th Cong., 1st sess. (1957), p. 24.

630 TBD, a much higher rate of increase, and even this was less important than the fact that some planned a larger increase, which others would soon match. Worse yet, the new importers were programming 341 TBD![41] The committee therefore recommended one last try at voluntary compliance, failing which the President should impose compulsory quotas.

But even as this new federal force was mobilized on the anti-import side, their flank was threatened by the buildup of productive capacity in Canada. Although costs there were perhaps no lower on the whole than in the United States, the prorationing system was less rigid and hence less disabling. Moreover, since Canadian crude had a locational advantage in the western states, the high American price could be relied on to draw in a considerable supply. More important, a substantial part of this output was produced by companies, American or Canadian, with small or no interests in Texas or Louisiana, and therefore not subject to pressure or persuasion. There was, of course, no shadow of a "security" reason for excluding Canadian oil, but the situation posed a considerable danger. Crude production, backed out because of Canadian competition, would not overhang the market and threaten prices so long as the commissions cut back output. But a crude-buying refiner who saw his rivals obtaining cheaper crude from their own production in Canada was likely to try compensating himself by buying more overseas crude. He may have resisted the temptation earlier, as part of a general understanding that would benefit him, but to see others fudging on the agreement was reason for him to feel released from it.

From 1954 to 1956, Canadian imports accounted for half the import increase, which was both proportionately and absolutely the largest increase since 1950.

Suez Crisis, "Voluntary" and Mandatory Controls. In late 1956 came the Suez crisis. Middle East imports were stopped and Venezuelan imports reduced. On January 1, 1957, the U.S. price was increased by about 35 cents per barrel, thus providing imports with an irresistible lure. Shortly after the resumption of Suez Canal traffic in April 1957, it was clear that the old pattern was being followed a little too strongly for comfort. Imports increased rapidly, particularly those from Kuwait, with substantial amounts brought in by companies new to the scene. The seven "established importers" were soon joined by fifteen others. Under the voluntary plan it was necessary to make room for them. They were small, and had no producing interests, and could not be pressured by producing states or Congress. Whether they bought Kuwait crude from Gulf or BP was immaterial; if they could not be stopped, the trickle would become a stream and then a flood. In mid-1957, therefore, following the July 29 Special Committee report[42] the Administration again threw fresh force against imports by setting up a formal "voluntary" program of detailed import quotas. In the aggregate, it seemed to have some success; Canadian imports were sharply down for reasons not altogether clear, and from all other sources they decreased—except Kuwait again, which was now a fifth of the total. Table V-5, above, shows that as late as 1953 the five Persian Gulf producers had accounted for 71 percent of imports, and their six customers for the rest. By the second half of 1957, the producers were down to 40 percent, the six large customers had actually surpassed them, and eight small companies had nearly a fifth. In a word, the Persian Gulf producers no longer had imports under control, and could not hold a collective line, although much of most of what "others" brought in was their output. In late 1957, the administrator of the

[41] *"Petroleum Imports Report . . . 1957,"* p. 2, app. table II.

[42] *"Petroleum Imports, Report . . . 1957."*

"voluntary" plan proposed that the seven cut imports 10 percent below the 1956 rate, and 22 percent below the level of the second half of 1957, in order to permit the newer fifteen to raise their imports by 40 percent. Thereupon Gulf threatened to "reevaluate" its earlier decision to abide by the program.[43]

But three of the newcomers held out for larger increases, and while their imports ran only moderately above quota, they would not promise to stay inside it. Tidewater was intransigent because the plan amounted to "confiscation" of the value of its new plant built at Delaware City to process foreign crude. To the smaller concerns the price was right, there was money to be made, and statesmanship was for the big.

Throughout 1957 and 1958 there were constant representations from all sides. The Special Committee made a *Supplementary Report* (March 24, 1958), accusing "three substantial importers [as] not in compliance," and recommending import quotas for no less than 33 firms. They again reported in June, and in December; but their last report in March 1959 said that "excessive imports by companies who have not complied . . . [and] the likelihood of increased noncompliance among companies . . . asked to cut back imports [to make room] for newcomers to the program" made mandatory quotas necessary. The legal power had already been provided by Congress, and quotas became effective on April 1, 1959.

This quick résumé of U.S. import policy suffices for at least one conclusion. Imports into the United States have never been free since World War II. No company was free to consult its own interests and its own profit. To suppose that the mandatory oil import program of 1959 forced the world oil industry into surplus, or did anything but hold the line, ignores the documented record.

European Protectionism Reinforces American. There might still have been a considerable incentive for the Persian Gulf producers to cut prices in order to take business away not from competing oil producers but from fuels competing with oil. But, like U.S. domestic oil, European coal was not something to be touched by profane hands seeking higher profits. By mid-April 1953, MSA had expressed its concern over the restrictions on the petroleum trade imposed by various governments of Europe. The trouble had started in the uncontrolled tanker market as it relaxed from the Korean peak.

> Delivered costs to Europe have fallen 16-27 percent in recent months. . . . France provides one example of the type of action MSA finds disturbing: on May 1 the French government plans to impose a new tax system on fuel oils aimed at cancelling the effects of an expected price cut resulting from a drop in freight costs. . . . The object is to prevent lower fuel oil prices which would cut into the market held by higher priced coal now piling up in French collieries.[44]

The protection of high-cost coal production in Europe, like the protection of high-cost oil producers in the United States removed the strongest incentives for the oil producers to lower prices and expand markets.

[43] *Wall Street Journal*, Sept. 18, 1957, p. 37.

[44] New York *Journal of Commerce*, Apr. 13, 1953, p. 16; see also May 4, 1953, p. 9, col. 6, reporting an interview with Cornelius J. Dwyer. See Richard L. Gordon, *Energy Policy in Western Europe: The Reluctant Retreat from Coal* (Praeger, 1970); and "Energy Policy in the European Community," *Journal of Industrial Economics*, vol. 13 (1965), pp. 219-34.

As early as 1959, before the fuel oil excise taxes were levied in Britain, Germany, or Belgium, the Organisation for European Economic Co-operation noted that all member countries had policies inhibiting the most efficient use of energy. OEEC, *Toward a New Energy Pattern in Europe*, 1960; since it was issued in January, it must have been completed in middle or late 1959.

The Line Reformed, 1950–53. Thus the two impulses behind the 1948–49 revolution ceased to exist. The market could no longer be widened into and past the U.S. East Coast. The MSA believed itself to be forced by its mandate to insist on equality and non-discriminatory treatment at any given moment. Whatever its legal merits, which led to prolonged litigation, this doctrine acted to chill the incentive to shade the price. With no competitive incentives to chisel or cut, the Persian Gulf and Venezuelan oil producers left price alone. Presumably they did not communicate on the subject of controlling output or price. For if the New England conscience is deeply rooted in fear of the police, noncommunication is rooted in the knowledge that several Big Brothers are watching and can examine witnesses and records. But it was reasonable conduct to resist chiseling. A producing-refining company had every reason to believe that rivals were aware of the same facts and would react the same way to serve the group interest. But this expectation might not have been sufficient, had it not been confirmed by the knowledge, through the joint venture operations (Chapter III), that rivals had no plans for expanding output to upset market shares.

Rising Prices, 1950–1957

We now try to see what, if anything, happened to oil prices after the end of 1949, when the Persian Gulf quotation for Arabian settled out at $1.75. But there are two stories here, not one. The oil companies did some persistent mild chiseling on some sales, as Appendix V–A shows. But on two occasions there was a sharp reverse movement: the two American price increases of 1953 and 1957 were promptly though not completely copied in the world oil trade, and hence prices were actually higher in 1957 than they had been nine years earlier.

The 1953 Price Increase. As we saw earlier, the mounting waste of market-demand prorationing had imposed higher expenses on the U.S. industry, and the entry of the Texas Railroad Commission had instilled confidence that a higher price would not draw in additional imports. Consequently, in June 1953 some bolder spirit increased the price and all others followed. Soon thereafter prices were raised outside the United States, though by a lesser amount.[45] Neither price change had anything to do with "meeting competition"; neither was a response to competitive supply and demand, and both were continued in the face of oversupply. But it will be necessary to look at each separately.

In the states, the long-run or potential excess capacity resulting from a cartel had already become short-run excess capacity. In District III[46] excess capacity was already 32 percent of current production.[47] Carrying this excess body fat of course explains the industry's higher costs. Without the rigid production controls, the price would have been falling until the excess disappeared.

Outside the United States, the price rise served as the visible signal to which all concerned could respond. Each expected everyone else to increase and to abide by the higher price. Given few producers overlapping in joint ventures, the reliability of each toward the others had already been confirmed. The market structure was a necessary but

[45] See *Petroleum Press Service*, July 1953.

[46] District III, one of the five geographical areas set up by the Petroleum Administration for Defense, consists of New Mexico, Texas, Arkansas, Mississippi, and Alabama.

[47] Capacity is as estimated by the Independent Petroleum Association of America's Committee on Productive Capacity for Jan. 1, 1954. The estimates cannot be precise, and IPAA makes no claim that they are.

not sufficient condition for making water flow up hill and price to rise. But the outside signal was needed in order to avoid overt collusion.

Had oil producers in 1953 been free to ship as much oil as they liked to the United States, then the price to other customers could have been no less than the netback from shipments to the United States. On that basis, then, the response was a simple competitive one, and there is nothing to explain. But if the reader agrees that imports were not free in 1953, then the competitive theory cannot explain the worldwide response to the price increase.

The 1957 Price Increase: In the United States.[48] At the height of the Suez crisis, the international companies requested higher production, which the Texas Railroad Commission refused. Louisiana was somewhat more permissive, but not substantially so. Although capacity was in greater excess than ever, Gulf Coast crude stocks were depleted. With supply thus choked off and demand surging as a result of the Suez Canal shutdown and the IPC pipeline cut, it was a foregone conclusion that prices would rise—they had, indeed, already risen in Venezuela. The records of the Texas Railroad Commission hearings and the statements of some of its members make interesting reading. Commissioner Thompson several times said that Europe could have more crude oil if they would only take gasoline and clean out the U.S. product surplus that way. To take superfluous gasoline was a waste of tanker capacity and European storage; it was also a veiled price increase—though the veil was as thin as Salome's. The price increase registered what had already happened, or at least begun to happen.

A picture of the commission as coldly exploiting the necessities of Europe is overdrawn. It was an article of faith, sincerely believed then and now, that the industry "needed" and "deserved" a higher price because of the "higher costs" which market-demand prorationing had forced upon them.

Moreover, the basic rationale of market-demand prorationing is *fair shares for all.* To permit increased output throughout the state would have meant ever-larger inventories in West Texas, whose pipeline connections to the Gulf Coast were insufficient to take off larger production. To permit higher output at the Gulf Coast but not elsewhere would have been "discrimination," unfairness to those in the interior. Until West Texas inventories were drawn down there would be no increase in production anywhere. There was none until February, when the price had gone up.

The commissions cannot be faulted for carrying out laws they had sworn to uphold. They did not seize or usurp power; it was thrust into their unprepared hands. The national Administration, having declared that the national interest required higher oil production, did nothing whatever to bring it about. The tentative side-glances of sub-Cabinet officials toward the South, the timid hope that "our good friends in Texas" would "do what they felt would be helpful . . . in . . . carrying out the President's wishes" are beyond sober analysis.[49] One Louisiana commissioner remarked, truly enough, that if the Administration really wanted higher output they could easily say so directly to the commissions instead of making vague statements directed to nobody in particular.

[48]The following section is based on the published *Hearings* underlying: *Petroleum Survey, Preliminary Report of the Committee on Interstate and Foreign Commerce*, H.R. Report No. 314, 85th Cong., 1st sess., 1957; *Petroleum, the Antitrust Laws, and Government Policies, Report of the Committee* on *the Judiciary*, U.S. Senate, 85th Cong. 1st sess., 1957. *Platt's Oilgram News Service*, *Platt's Oilgram Price Service*, and *Oil and Gas Journal* have also been helpful.

[49]*Emergency Oil Lift Program and Related Oil Problems* (1957), part 1, p. 137.

Why the federal government thus delegated the conduct of foreign policy to the Texas Railroad Commission will never be known. The conduct of Secretary of State Dulles during the earlier phases of the Suez crisis makes it plausible that he wanted to restrict the flow of oil to teach the French and British a lesson.[50] But plausibility is not proof. A simpler and therefore better hypothesis is that with Mr. Dulles hospitalized during November and a desk piled high with pressing matters during December, the matter was allowed to drift until the price had been raised and the oil would be forthcoming, after which there was no need for action.[51]

One curious echo of the Suez affair may be noted. It would be tempting to say that the Administration, having brought on the price increase by allowing restrictions of supply, tried to salve its bad conscience by bringing suit against a score of oil companies for collusive price fixing.[52] But the temptation should be resisted. The documents of that litigation show plainly how little the Department of Justice understood oil prices or any prices. The defendant companies, as might be expected, pointed to the shortage that had been forced by the Texas Railroad Commission. The government retorted in its brief that there had been no shortage because the Texas Railroad Commission had said there was none. The guardians of monopoly were to set the standards for competition.

The 1957 Price Increase: In the World Market. We see the peculiar and sporadic but effective mechanism of price control which had evolved under certain constraints. As a result of oligopoly plus government block to competition plus price-raising regulation in the United States, supply was held as close as possible to anticipated consumption. If demand outran supply, prices were allowed to react upward; whereupon nobody sold at less than the new price. This ratchet effect needed no intercompany communication (which may have occurred, but we do not know). It did require a good understanding of what was possible.

Looking back now from 1957 to 1949, the result is surely remarkable: a price increase despite a price-cost gap of several hundred percent. The average Venezuelan posting was up by 28 percent and the (declared) crude price by 30 percent. The posting on Arabian 34° was up by over 20 percent.[53] Realized prices must have gone up by little less, for higher postings meant higher taxes per barrel of crude oil produced. Hence, postings would not be increased without the expectation that realized prices would also go up on signal. Another piece of evidence is the realized price of heavy fuel oil, which as Appendix VI-A shows, was higher in 1957 than just before Korea.[54]

[50]See the admiring but not uncritical biography by Richard Goold-Adams, *John Foster Dulles* (Appleton Century, 1962).

[51]Against this hypothesis one might note the deliberate refusal in November to reconvene the Middle East Emergency Committee (of oil companies). See *Emergency Oil Lift Program and Related Oil Problems* (1957), part 4, pp. 2401, 2410, 2834. Moreover, at a press conference of Feb. 6, 1957, he said "Europe's oil shortage isn't bad enough to require an appeal by President Eisenhower for more U.S. petroleum production. . . . Even as Mr. Dulles spoke . . . Defense Mobilization Chief Flemming asserted that curtailed shipments to Europe were below his expectation and resulting in a severe crisis for that area." *Wall Street Journal*, Feb. 7, 1957, p. 3. The target set was 500 TBD. Since actual shipments of 475 TBD contained much useless gasoline, the shortfall was substantial.

[52]U.S. v. Arkansas Fuel Oil Corp. et al. F. Supp. (N.D. Okla., 1960) CCH par. 69, 619.

[53]Arabian 36° crude was first posted at $1.75 on Nov. 1, 1950 and raised to $1.97 in July 1953, then discontinued. It was replaced by 34° posted at $1.93, which was increased to $2.08 in June 1957. Since each degree of gravity was valued at two cents, we may speak of an increase in 34° Arabian as being from $1.71 to $2.08, or 21.5 percent.

[54]In a forthcoming work, Professor Zuhayr Mikdashi calculates the real price of Kuwait crude as rising from $1.22 (1951) to $1.32 (1953), declining to $1.22 (1955), then rising to $1.72 (1958)–an increase of 40 percent.

SUMMARY

1. To explain price structures and changes therein it is necessary, though not sufficient, to consider costs of production. The parity theory of the ECA (MSA) consultants is correct inside narrow limits, and was correctly used, but is wrong outside of them.

2. The 1948-49 price changes were a substantial but incomplete competitive adaptation to new cost conditions.

What weight should be assigned to spontaneous competition among the companies and how much to the competition induced by the U.S. government and the ECA (MSA) consultants, one cannot say.

3. The price evolution stopped abruptly in 1950. Under competition, it would have continued to a far lower price level. A simple theory of tacit collusion breaks down in the face of contrary evidence. A more complex monopoly theory, based on vested interests of U.S. based companies in their home market, breaks down also. All three theories predict much larger imports, at lower prices than actually happened. This suggests that some outside force blocked the buyers and sellers from doing what comes profitably.

4. There is ample evidence that imports into the United States have always been severely limited. The widely held belief that they were only restricted in 1959 is wrong by at least ten years. Moreover, even with the approval and supervision of the United States government, the "voluntary" import limitation was so threatened by nonconformist behavior that it had to be made mandatory.

5. The block to competition stopped the price decline. There was no longer any inducement to chisel or shade price in the hope of expanding outlet. Anybody contemplating a price reduction knew that it would be met by all other sellers, could not expand sales, and would benefit nobody. The few producers were able to avoid any price reductions, or any market-expanding rivalry which might have led to price reductions. They could not initiate any price increase. But in 1953 and 1957 they were able to use higher prices in the United States as a signal. Under competitive conditions no price response would have been made outside the United States.

By twice following the U.S. example, oil prices were raised by about 25 percent during 1949-57, despite the facts that costs were only a fraction of price and that huge amounts of oil were available for making into producible reserves in a short time.

The price increases of 1957 marked the zenith of postwar control of the market. After 1957 it was all downhill until 1971, but even that is a lower peak.

CHAPTER VI

Oil Prices, 1957–1969

In this chapter we look at the behavior of Middle East and North African crude oil prices and European product prices in the period between the two Suez crises of 1957 and 1967, with an afterword on more recent years. The areas covered were—with due respect to Venezuela, Japan, and other important centers—the principal foci of the world petroleum market. The United States had cut itself off from that market a decade earlier—the 1959 quota systems bolted a door long shut. The inter-Suez decade is a natural unit of study, for at each end a major supply crisis threw the usual arrangements into disarray.

Our concern is with the structure of real prices that emerged in the late 1950s: a shape dimly seen in the mosaic of hundreds of items, their precision ranging from excellent to very poor. Strictly speaking, there is no "world price" of oil, but a structure of prices. Variations in quality, quantity, location, and length of contract—all stand in rational relationship to each other, and if the estimates I have arrived at are wrong, the failure cannot be attributed to the structure. No greater mistake could be made than to suppose, for example, that long- and short-term contracts are so "different" that they cannot be compared. They are substitutes for each other. In practice, a small discount or premium equates them.

The price of heavy fuel oil (termed "residual" in the United States) is given much attention. It can be shown to be a rough but valid indicator of all product prices and crude oil prices, both as to level and as to trend. Moreover, the price of heavy fuel oil has been badly misunderstood, above all in Europe. A costly mistake has been long maintained. As will be seen in Chapter VII, the resulting wasted investment in coal and nuclear power has important effects on the supply and price of crude oil.

The 1967–69 interlude is a good time to pause and look back. The Suez closure crisis has so disturbed the price structure that some time must pass before we can see the appearance of any new norm or equilibrium. Moreover, the coming of very large tankers will give us new points of origin for price quotation. The terminals in Ireland and Japan to be served by tankers of three to four hundred thousand tons and eventually perhaps half a million tons, will permit considerable economies in the transport of crude oil.[1] So will the Suez Isthmus ports. Crude oil may soon be quoted f.o.b. the superterminals. This price will be lower than the Persian Gulf f.o.b. plus transport in smaller ships. It will thus become a competitive advantage for those firms having much crude to sell and wishing to quote a lower price in Europe or Japan but not to other customers, or for those firms

[1] As might be expected, the pioneer in these efforts was Gulf Oil Corporation, a very large producer but a relatively small refiner-marketer.

lifting crude oil into their own tankers at the Persian Gulf. But the complication of superterminals fits the scheme of this study, since estimates of Persian Gulf prices include the Persian Gulf equivalent of oil sold as delivered. This requires that we reckon the price of tanker service, calculated in Chapter IV.

Analysis of Arm's-Length Crude and Product Prices

Posted Prices–Hail and Farewell. In 1957 posted prices were raised all over the world in response to a U.S. increase. There was no scarcity or higher pressure of demand on supply. But with higher American prices, unchanged foreign prices would have supplied an even larger reward for importing into the United States by independent refiners, or by anyone who could buy in the cheaper market and sell in the dearer.

Higher posted prices meant higher payments to governments. It only made sense to raise them if higher prices would actually be received. In mid-1957, therefore, posted prices were real prices. Discounts for some particular areas or customers must have been relatively very small. Moreover, although the postings referred to crude, the great bulk of the companies' receipts was from products, and they would not have undertaken to pay more if they had not expected to receive more. To the companies, the posted price was the accepted foundation for product prices. To us, it may serve as an index number: 1957 crude and product prices were higher than they had been eight years earlier, after the postwar revolution.

But not for long: it was soon apparent that the higher prices were being so widely discounted or so poorly translated into product prices that the higher postings meant simply more money paid out. For the outside observer, a piece of information was lost. To the industry, the 1957 increase soon appeared as a serious mistake. In 1959 posted prices were reduced, and the governments resented it. In 1960 another reduction provoked the formation of the Organization of the Petroleum Exporting Countries (OPEC). The cuts were partly restored, but much more important was the clear implied warning: posted prices must never again be reduced. Since then, postings have ceased to be market facts and have been merely a way of calculating taxes and royalties due to host governments.

A Purge of Non-Prices. Posted crude oil prices are merely a means of roundabout calculation of what comes close to a per-barrel tax on production. Transfer prices from one to another division of the same corporate entity are simply bookkeeping notations to permit the corporation to minimize its total tax bill. There is no market without independence or freedom to bargain. No market, no price! Hence, posted "prices" and intracorporate "prices" need detain us only the few moments necessary to suggest why anyone takes them seriously. First, posted prices were for years treated as approximate norms, to which temporarily "depressed" prices would tend to converge in the long run. Second, no little has been said in defense of intracorporate prices as being "fair and reasonable." True or not, this does not make then relevant to investment decisions.

Does the same strict logic that purged a mass of pseudo-prices now also threaten the one I propose to use? We need to consider some objections to the meaning or relevance or importance of crude prices (or product prices converted to crude equivalent). In finally rejecting these attacks, we will see that they were not made in vain. A hypothesis may be false, but the act of considering it makes us aware of new facets of reality and leads to a deeper insight. Hence we owe much to those holding ideas with which we disagree. Francis Bacon said that a little learning was conducive to atheism but deeper knowledge to belief. To which Santayana added: surely Bacon did not mean that one reverted to the original hodge-podge of ideas, but rather to the same conclusion purified of error and ambiguity.

Small Percentage of Market. The first objection to considering arm's-length prices is that they cover only a small percentage of the market. No precise estimate is possible. Table III-3 (p. 90) tends somewhat to underestimate arm's-length because (1) it assumes that all crude production outside the main centers is kept out of the market by ownership or government regulation; (2) subtraction of all U.S. seaborne imports exaggerates because the trading in U.S. quotas gives all the seaborne imports (nonexempt) a very substantial market component; (3) the producing and refining of the group of companies are treated as though they were one company. As will be seen later, there is considerable buying and selling among the larger as well as the smaller companies. The long-term supply contracts among major producers are not somehow qualitatively different and noncomparable. While the conditions vary a great deal, all of the crude sold among the producers is paid for, and is subject to bargaining and the alternatives open to both parties.

Coverage, Independence, Fringe Benefits. But one cannot tell how well arm's-length transactions are covered by our sample of arm's-length prices, and this is a serious weakness. It is unimportant that the proportion may be small. What matters is not the percentage sampled, but, rather, the absolute number of independent observations.

We must distinguish two kinds of independence. Transactions close together in time are *causally* interdependent in the sense that the contracting parties will be strongly influenced by recent transactions and will only move a limited way from them. A given offer or bid at a certain price is determined essentially by the price reached at the last similar transaction, plus or minus whatever change is warranted by changing market conditions since then. *Statistical* independence concerns the various sources of data. For example, a figure derived from one public bid or offer is independent of a figure derived from another one, in the sense that whatever chance errors creep into the one have no influence on the second. Yet if both offers are reported by the same news-gathering agency or source, omissions and ambiguities in one may also be in the other.

Our greatest single problem lies in the "fringe benefits" of quality differentials, credit terms, buy-back provisions, etc., that can give rise to the appearance of fluctuations or of uniformity, which really is not there. For these reasons, it has seemed necessary to set out in painstaking detail the specifications of every usable transaction the writer has observed. They are subject to appreciable error, whose importance must not be exaggerated. Few indeed of any of the sources of variation exceed 5 cents per barrel, which is to say under recent conditions about 3 to 4 percent, and these errors tend not to be self-reinforcing.

Alleged Bias of Arm's-Length Crude–"Excess Capacity." The next criticism pertains not to the validity of the crude oil prices as measurements of the crude oil market, but rather to their wider relevance. The assertion is not that crude oil open-market transactions are "too small" a segment of the whole oil flow to be meaningful. Instead, specific reasons are given why arm's-length transactions, far from being a representative sample of all transactions, are necessarily biased downward, especially in recent years. Roughly since the ending of the Suez crisis in late 1957, there is alleged to have been excess capacity in crude oil production in the Middle East and Venezuela, of the order of 20 percent. Since only about 10 percent of the crude oil output enters the open market,[2]

[2]This is exaggerated, as Table III-3 shows, but faithful to the argument as usually stated.

the excess 20 percent has all been focused upon or channeled into the 10 percent open market, resulting in chronic oversupply and a price weakness which is not characteristic of the great bulk of the market in which crude oil is eventually sold, i.e., as refined products.

This argument is logically sound, for the conditions assumed do lead to the conclusions. But it does not fit the facts. The evidence cannot possibly be reconciled with 20 percent excess capacity, or indeed with any excess capacity whatsoever after 1959, or 1960 at the latest. Every industry and every firm must have some unused capacity in order to meet the seasonal and chance peaks and valleys in sales output. There is a trade-off between the cost of more producing capacity and more storage—each must stand idle some of the year, and an oil company seeks the cheapest combination. But this brings no downward pressure on prices. Capacity that is only occasionally and unpredictably available to fill other orders is not excess at all. For the incremental cost of using it is not merely the variable expenses, but, much more important, the lost opportunity to produce for another account if need be. Insurance used is no longer available.

Unused capacity in summer is most illusory because storage is not available. Sales made at lower prices during the summer to a buyer with storage are no indication of a lower price level; they are equivalent to paying a rental to the storage owner.

In the oil-producing industry outside the United States something on the order of 15 percent is needed as an insurance against irregular and seasonal peaks.[3] Unused capacity at the end of 1963 was reckoned by the chairman of British Petroleum at 1.8 million barrels daily, 13 percent of non-Communist non-U.S. production.[4]

Strong additional proof came during the second Middle East crisis in mid-1967, when oil was in urgent demand. In Venezuela, which was closer to Europe and therefore economized the most scarce factor—tankers—output in July utilized "virtually all the spare productive capacity."[5] Yet the July peak (later months were lower) was only 5.7 percent above the last previous high in January and February. Because Venezuela and the United States contributed relatively little additional crude oil, all tankers available in the Persian Gulf could be directed to Iran, which did not join the Arab boycott of Britain. Before the boycott, Iran shipments were 28 percent of total Persian Gulf tanker liftings. Hence if, say, half of normal tanker service had been available, Iran liftings could have doubled before tankers were any constraint. After the confusion in early June, tankers were no constraint.[6] Yet August output was only 5.9 percent over the January high. As an oilman said in a personal letter: "Normal operational planning in each major producing country must allow for the constantly increasing requirements and for operational fluctuations in off-take, which two factors demand more of a cushion than the relatively modest increase experienced in Iran."

In Libya the increase was larger, 14.5 percent, but it was mostly in new areas and projects coming into production a little earlier than planned. Comparing December 1966

[3]Zeb Mayhew, *International Oil and Gas Development* (Dallas: American Institute of Mining, Metallurgical and Petroleum Engineers, jointly with International Oil Scouts Association, 1961).

[4]Sir Maurice Bridgman, Board Chairman of BP, addressing American Petroleum Institute, November 1963. What does matter is what he calls the ability to expand capacity "enormously" in 12 to 18 months.

[5]*Petroleum Intelligence Weekly*, Aug. 28, 1967, p. 3.

[6]"While Iranian oil has been hurt by tanker shortage, it is now being given priority call on tankers available in order to help fill the otherwise serious gap in oil supplies flowing to Britain." (*Petroleum Intelligence Weekly*, July 10, 1967.)

with August 1967, the largest producer, Oasis (Continental, Marathon, and Amerada-Shell), showed no change.

Such small increases do not *precisely* measure unused capacity, but they are incompatible with the idea of any excess that is available for commitment and might therefore increase offerings and press down on price. Firms attentive to profit try to minimize stocks of goods *and* services. Single companies are often temporarily in excess or in deficit because they cannot predict sales and availabilities perfectly. But they can trade, and so long as the system as a whole is not in surplus there is no pressure on prices.

But let us now suppose that the preceding argument is in some way completely wrong, and that there *was* excess capacity in 1957. Middle East output grew during 1956–66 at about 10.5 percent per year. Hence, a 20 percent excess would have disappeared within 20 months by simply not expanding capacity and letting demand catch up. The fact that price weakness persisted is the strongest kind of evidence that no surplus of producing facilities was at work.

Four-Fifths Unimportance. We come now to the most important criticism of the use of crude oil market prices, and we will be compelled to accept it. Since over four-fifths of oil output moves through integrated channels, there is no *necessary* relationship between the arm's-length crude price and the crude equivalent of what is finally realized from the sale of refined products. If price is the reward of production and the inducement to invest, then surely the net receipts from the 80 percent are four times as important as the receipts from the 20 percent. Investment decisions will be based largely on the reported integrated profit. Hence the open-market crude price is not meaningless, but it is relatively unimportant. (Even if it were meaningless, intracompany transfer "prices" would still be not information, but noise.)

If the difference or value-added between f.o.b. price of crude and realized price of product were relatively small, then the problem would not be important. The product price, less "downstream" cost, would have to be very close to the f.o.b. price; and the latter could then be used with little inaccuracy as a representative sample of the total. But clearly this is not the case, since realized prices are more than twice as much as f.o.b. prices, and therefore there is room for much divergence. Hence, to make sense of prices we need to consider the operation of the vertically integrated structure.

Vertical integration, even when as great as in international oil, is not the kind of general idea from which we can draw any given conclusion. Even as description, the stereotype of a seamless single conduit, "integration from well-head to pump," is badly oversimplified. We need further particulars and an explicit argument.

Competition-Monopoly Hypothesis. Let us first suppose pure competition throughout the transport-refining-marketing sectors. That is, no seller has any influence on the price, and the rate of return on investment is just enough to draw in the resources needed. Then realizations from products would be strictly related to the crude price. It would be a matter of chance that oil companies sold crude f.o.b. or c.i.f., or integrated all the way to the consumer. The open-market price would in strict logic be representative of the total of crude oil produced and sold in whatever form. Now to the opposite extreme: if the oil companies sold as monopolists in their final product markets, they would in each case charge what the traffic would bear. The margin above crude price would have no particular relationship to the transport-refining-marketing costs involved. If so, even a precisely known open-market crude price could not serve as a general index of the real market value of crude oil, for most of that value would be determined outside the market.

At this point, we have no reason to accept or reject either of the opposing hypotheses. Each will have to be examined in turn to find out how far we get in applying it to the market. This may seem a pedantic exercise. Obviously the companies do not work under pure competition. But the exercise is a necessary one, because the extent of divergence from pure competition at each level of operation and the relative importance of each level are what really decide our problem. Competition is not used here as an encomium or means of pronouncing judgment, but simply as a tool of analysis. Those unable to suppress for the moment their urge to pronounce sentence will never reach a tenable verdict.

Fallacy of Return on Total Integrated Investment. But before looking at the various markets, let us define the point of view: prices and investment as seen by a producing company with reservoirs available for immediate development, i.e., able to make regular deliveries in one or two years. The company's alternatives are to sell the oil as crude f.o.b., *or* sell it plus transport services as crude c.i.f., *or* sell it plus refining services at wholesale, *or* sell it plus marketing service at retail. Each choice beyond bare f.o.b. sale involves added costs and added investment, which must be weighed against the expected added profits. Oil companies do elaborate calculations to carry each investment to the point where the present value of the expected additional receipts will just exceed the necessary additional investment. Corporate planning tries to push each opportunity toward the point where incremental rates of return are equalized at the margin, so that the company could not improve its current position or present value by making any change in its investment plans. And the incremental rates of return in production, transport, refining, marketing, etc. need not be anywhere near the average rate of return.

Investments in each of these four activities include a risk, both commercial and political, not at all equal to that of the others.[7] An additional investment in any one of these activities does not require investment in any other. Integration is a problem in management choice, not imposed by nature. Hence it is a serious mistake, regrettably often committed, to compare total investment and total profit in one global figure, except as an accounting after the fact for the use of the shareholders' money as a whole. Even here, conventional accounting can give some seriously misleading results. Profits and rates of return in published financial statements have no clear or close resemblance to the rates of return that guide management decisions. But that is quite a separate problem. Even if our accounting system satisfied any or all of our theoretical requirements, it still would be wrong to disregard the options open to management each year, above all because the particular choices made in one year are not necessarily those that it will make in any other.

Therefore, just as it has been shown that oil companies are not necessarily profit maximizers with respect to the f.o.b. open market price, neither can they be called profit maximizers with respect to any particular investment pattern or product sales pattern. We must analyze separately the elements of the margin from crude to products.

The Refining Margin

A market does exist in processing services, and contracts are drawn up whereby refiners process for a fee. Unfortunately, they are few, and none of them is in the public domain. Hence one is forced to rely altogether on the cost calculations of the refining

[7] Compare the treatment of risk in the Appendix VI-C with that in Chapter II, pp. 53–56.

margin. Appendix VI-C presents the costs of refining an average barrel of the crude used in Western Europe, given the investment in new refinery capacity installed in each of the years 1960-69. I have tried to make clear the many sources of error and imprecision, but these estimates are at least correct in theory. They are based on present-value prospective cash inflows and outflows. The sources of error seem to be mutually independent. The possible understatement of investment requirements per daily barrel, which would make the estimates too low, is unrelated to the possible understatement of depreciation allowances granted in Europe, which tends to make them too high. Capital cost is reckoned to be about 20 cents; operating cost, 15 cents; and hence total refining costs, 35 cents. But a company paying no income tax, either because it earned no profit or because it costed in its crude oil at a fictitiously high delivered price, would save over 5 cents.

"Downstream Losses." If we follow one popular line of argument, the requirement of any positive profit whatever may seem to be a drastic overestimate of refining supply price. For refining-marketing operations are generally considered as profitless or unsatisfactory, and we have heard much of the "downstream losses" that "must be incurred" if a company is to market its crude. But downstream "losses" reckoned by deducting from receipts an intracompany accounting figure, rather than arm's-length prices for crude and for tanker service, have no meaning. These fictitious prices are a legal and open method whereby taxes are minimized. Yet how seriously people take the artifacts of their own making. There is wisdom in the Biblical warning not to worship graven images. One must never treat what he has himself fashioned and can change at his convenience as if it were either an external fact or a general rule. "Downstream losses" are an acceptable way of beginning the analysis if one subtracts the "loss" per barrel from the arm's-length crude price to see how much the crude will really fetch, net, in comparison with an arm's-length transaction. If, after making due allowance for the cost of the additional capital employed, the downstream investment allows a better return than could be had by selling the crude, there is a real downstream profit. If not, the downstream investment should not have been made. We have read countless times that a given market in refined products was profitless or woefully unprofitable; yet the companies continued to pour in heavy investments year after year without fear of a stockholder's suit. In refreshing contrast is the announcement by Continental that they expected a 4- to 5-year payout (or possibly 3.5 to 4.5) on a British refinery starting in 1969.[8] Yet how easily could one convert a 20-25 percent rate of return to a "downstream loss" by using posted prices and AFRA rates. Let us follow good management practice rather than bad economic theory.

Prices of Products at European Refineries

The value of a barrel of refined products sold in Western Europe in 1960-70 is shown in Figure VI-1. The composite price, minus the necessary cost of transport and refining, yields a derived Persian Gulf or North African realization, or netback, comparable to the open market price. It is the single most relevant figure for an investment decision in crude oil production. Consequently, the manner in which it is derived and the limitations of the data will be explained in some detail before discussing their significance.

Final product prices—i.e., to ultimate consumers—are not available and probably not needed, for they depend on a variety of local marketing conditions and are subject to great and unpredictable variation. For example, retail margins were extremely fat in the

[8] *Wall Street Journal* (WSJ), Dec. 13, 1968, p. 12.

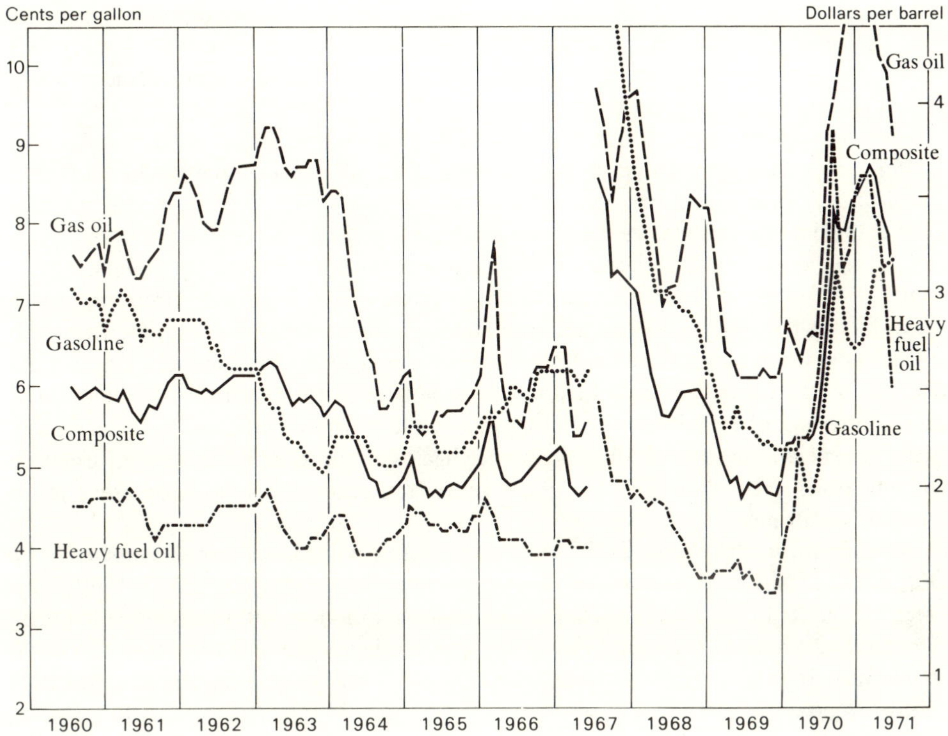

Figure VI-1. Rotterdam product prices and realization per barrel of crude charge, by months, 1960-1971. (Source: Appendix Table VI-B-1.)

United Kingdom for some time, and this led to strong rivalry with sellers willing to give up a good deal of the large profit in order to be sure of holding the rest.

We stop instead at the wholesale product market, partly whole tanker loads but more frequently barge loads. The Western European open market has centered in Rotterdam, with many tributaries at Antwerp, Hamburg, and southern Italy. These prices are all interrelated, and the differentials correspond approximately to differences in transport cost.

Sources of Price Information. (See Appendix VI-B.) Since March 1960, there have been at least two continuous price series. One is from *Platt's Oilgram Price Service* (*POPS*). Originally, they presented a "high"—posted Caribbean prices plus fictional AFRA rates to Europe—and a "low"—Caribbean market prices, mostly single cargoes, plus spot freight rates.

By mid-1960, Western European prices had greatly diverged from the Western Hemisphere pattern since gasoline was relatively cheap, selling for less than gasoil (light oil and diesel oil), while by Western Hemisphere standards heavy fuel oil was relatively expensive, being only a little below the price of crude oil delivered in Western Europe. This is in strong contrast to the United States and Venezuela, where a 34° crude sold for well over a dollar per barrel more than the heavy oil taken from it. At the end of 1962, the "high" was dropped as not useful. We may guess that by early 1960, there were few if any transactions at the "high," but that it had for so long been regarded as a "normal" to which prices would some day gravitate that it was worth keeping for purposes of comparison. By 1962, however, its irrelevance was plain. When later that year the *Perspectives* report was published[9] its skepticism about the relevance of Caribbean prices and notional freight rates had thus been justified.

POPS has always shown a noteworthy zeal to improve and widen the data series, and in 1965-66 it took several steps in this direction. First, all connection with the Caribbean was dropped. This was more important in theory than in practice. For so long as Caribbean shipments to Europe were an important part of total output, the Caribbean price had to stay in the neighborhood of the European price less freight. In other words, it would be a mistake to suppose that Caribbean prices ever determined European, or vice versa. In fact there was a mutual accommodation between them. But with the European market now much larger than Caribbean output, the latter had become the tail, not the dog. *POPS* now avowedly based its quotations, both at Rotterdam and in southern Italy, on the offerings of refiners and importers there, and in 1966 these were put on a weekly basis. In general, one may say that the service had two sources of Rotterdam information and published each once a week. These prices are close enough to each other to be mutually confirming.

In addition to the regular *POPS* report, there are also those of the *Aussenhandel für Mineralöl* (*AFM*), an association of German oil importers, dating from early 1960. It has been alleged that they were reporting prices somewhat too low, but this is not convincing. Persistent understatement would lose credibility, and hence the series would not be worth publishing. It is plain from the publications of *AFM* that they mean the so-called Price Mirror (*Preisspiegel*) to be taken seriously and devote some time and expense to preparing it. The *AFM* series quotes barge lots, while the Platt series has in recent years quoted both tanker lots and barge lots; the spread is usually 25-50 cents per metric ton, 4-7 cents per barrel.

The third regular source has been *Europa Öl-Telegram* (*EOT*), published since early 1963 in Hamburg, which specializes in the German scene and also gives prices for all the main products including heavy fuel oil (which *AFM* reported only for a short while). This report used to be made in somewhat casual fashion, but by 1965 it was being put out weekly in regular tabular form. Price reporting again became sporadic after May 1967.

These series provide strong mutual support. A valuable check is in occasional press reports of particular transactions or of the range of bargaining, as well as in official statistics for some product prices, particularly of heavy fuel oil. Supplementary data, though not nearly as much as for heavy fuel oil, are available for naphtha, whose price is that of regular gasoline minus about ½ cent per gallon. Similar data for gasoil were not used because both series (*POPS* and *AFM*) were complete and checked well with each other.

The problem of whether a set of reported transaction prices of this type can be taken to represent a much larger volume of oil is not new. In his study of the Gulf Coast market, Hamilton[10] showed in detail how *POPS* was a useful though imperfect indicator of the level and movement of product prices. We are more fortunate in having not one but three series, as well as supplementary information as a further check.

Meaning and Limitation of Published Transaction Prices. We do not know how much oil is involved in the particular transactions recorded in the various sources covering the

[9]European Coal and Steel Community (ECSC), *Etudes sur les Perspectives Energétiques à Long Terme de la Communauté Européenne* (1962), Annexe 11 (1964).

[10]Daniel C. Hamilton, *Competition in Oil* (Harvard University Press, 1958), pp. 21-28.

Western European product market. But the total volume of oil sold ex-refinery in cargo lots or barge lots is very large. Heavy fuel oil, which is about 40 percent of the volume of European output and nearly 20 percent of the value, is nearly all sold in cargo or barge lots. Where it is not, the integrated company as a final distributor faces the competition of a considerable number of rivals. Much light fuel oil is also sold to independent distributors—perhaps the greater part of it, as in the United States. A substantial though minor fraction of integrated gasoline and diesel oil is also sold through independent distributors (about 10 percent of Esso A.G., for example), and some of the rest is sold directly to large users such as companies' automobile fleets.

Indeed, the emergence of price reporting is in itself an indication of the development of a wide market. The reports carried in *POPS* and other publications are needed by those *not* making the transactions. Price reports are wanted by people who are bargaining on their own and want to see what others are paying or getting as a point of departure for making their own particular bargains.

Relation to Final (Consumer) Markets. The wholesale market is not only a large part of the total, it is very closely related to the final markets, which embody the wholesale product plus additional marketing services. Suppose the wholesale open market price rises, and the marketing margin is thereby so compressed that the profit on the marketing investment is reduced below the cost of capital. The independent marketers will contract operations by letting inventories run down and not buying any additional products. In general, the compression of a marketing margin may not have much immediate effect because much of the cost of doing business is fixed in the short run. The margin must be severely compressed before it begins to reduce the supply of distributive services. (In the long run, there may well be an overreaction, and so we may be started in a cycle of endless fluctuation, damped or explosive—the well-known cobweb effect.) But in an industry constantly growing, compression of any margin operates in the short run according to the rules governing the long run. An unacceptably low distributive margin will keep independent distributors from expanding. Thus supply is quickly restrained below the level where it would otherwise be, and the rising price effect is felt soon.

Contrariwise, if the wholesale market price falls, it profits independent suppliers to expand their operation. Some on the margin of decision will enter the market, thereby bringing down prices to consumers.

Or suppose there is a surge of demand in final markets and prices rise. Both integrated and independent firms try to sell more product so long as the additional receipts more than cover the cost of incremental capital and labor. Even some integrated companies, who would profit more by expanding their own crude oil production but cannot do so on short notice, especially because of offtake limitations, occasionally find it profitable to buy on the open market. But the same effect is felt if integrated firms merely withhold part of their usual supply from the open wholesale market and divert it to their final consumer market. The wholesale price level rises.

The integrated producer-refiners are active both as buyers and sellers in the open wholesale market, which is not, and cannot be, the domain only of the few independent refiners. *POPS*, *Petroleum Intelligence Weekly* (*PIW*), and other sources leave us in no doubt on that score. Thus the open product market has strong links to what those integrated companies are realizing and expect to realize in their final arm's-length transactions. True, integrated companies buy and sell only small amounts, aside from heavy fuel oil. There is a healthy repugnance at the idea of marginal transactions making the market—the tail wagging the dog. But this is nothing new. Alfred Marshall long ago denied that prices were determined by marginal supplies and demands. It was rather, he said, with an emphasis unusual in him, that we "*must go to the margin*" to study the interaction of total supply and total demand.[11]

Myth of "Dumping" Surplus Product. Above all, the wholesale market is not a dumping ground for "surplus" material at "marginal" bargain prices. The stereotype of a "lunatic fringe" of refiners constantly burdened with too much product and constantly dumping it, while staying in business, makes lively reading but is not a tenable theory. If there is an extra shipload that the owner must get rid of because he has no storage space, others will snap up the offering as soon as it is attractive–lest still others do so. Similarly with a marketer, integrated or independent, who must have a tanker load or barge load and is willing to pay a high price for blending stock or to retain a valuable customer and his reputation for dependability. Just as the seller is rarely forced to part with his goods at the lowest price he could be forced to take, so the buyer is rarely forced to part with his money at the highest possible price he could afford to pay. As with the single-voyage (spot) tanker market, the price may fluctuate, now above and now below an average that does not greatly differ from what can be obtained for the product in Western Europe generally, including the net realization on product sent farther through an integrated channel. An integrated company selling or buying small amounts through the open market is not necessarily charging or paying a higher or lower price than on the rest of its output.

Stable Wholesale Markets Despite Fluctuating Short-Term Tanker Rates. The distinction between spot and contract prices deserves a closer look. For as with tankers, so with products: they are imperfect substitutes and hence a small differential may well exist. But it is limited. Suppose that the value of a barrel of products at Rotterdam, say $2.00, was a modest 10 percent below the "true" or nondumped price of $2.20. Assume further that within a month the barrel could be resold at the "true" price. Product storage is expensive, say $5.00 per barrel, and the product would represent another $2.00, for a total investment of $7.00. An additional product movement and maintenance of the storage facilities would cost about 10 cents per barrel (barging at Rotterdam costs between 25 and 50 cents per metric ton). Hence the net profit would be 10 cents per barrel. Twelve such transactions per year would bring $1.20 on an investment of $7.00, or a profit of 17 percent with very little risk. Hence most of the investment could be with borrowed funds, affording a handsome return on the company's own funds–indeed, a much better return than would be available even in a lush Persian Gulf producing property. An integrated firm would be better off curtailing its output there by a few barrels in order to buy from the "lunatic fringe." Of course this is a rough comparison, but it proves that there are powerful forces holding the spot price within at most 10 or 15 percent of a normal price free of irregular swings.

Comparison with the spot tanker market shows the supposedly thin and unreliable wholesale market to be remarkably steady. Prices frequently change, but slightly and continuously. This is not a market of very small availabilities, where small increments can send the price appreciably up or down in a day or a week. In noncrisis years, few weekly changes are as much as 5 percent. Where spot tanker rates may fluctuate in a range of 200–300 percent (again, in noncrisis years), the greatest such change in the Rotterdam

[11] Alfred Marshall, *Principles of Economics* (London: Macmillan, 1961 ed.), Book V, p. 410. Emphasis in original.

composite was the 1964 decline of 20.3 percent, extending over seven months. (Gasoil fell by 35.4 percent over eight months.)

But if the comparison with tanker rates is sound, then it cuts both ways. As has been shown (Table IV-2, p. 109), the average of spot rates in a given year is a valid but imprecise proxy for the time-charter rate, because spot tanker service and time-charter service are imperfect substitutes for each other. In any given noncrisis year, the error involved in using spot to approximate time charters has ranged from 4.6 to 21.9 percent of the time-charter rate. Furthermore, in the short run the direction of movement may not be the same. In 1960-63, spot rates rose 13.7 percent, while time-charter rates fell by 4.0 percent. There can be no doubt that Rotterdam spot product prices at any given time diverge from contract prices to some substantial extent. We will later see whether the difference is systematic.

The 1967 crisis saw the European free market under a very unusual strain. On September 18 *PIW* remarked that "so few spot sales are being transacted that many prices are not especially meaningful," implying that other prices were meaningful. Thus heavy fuel oil was stable at $14 per metric ton with even "sizable quantities" available with "no difficulty." Even at that time the composite Rotterdam price did not reflect the extremes of the current spot tanker rate. This is a market reflecting current supply and demand over a planning period, not small amounts dumped haphazardly.

Real Defects of the European Product Price Series. Our composite product price includes only the three main products: regular gasoline, gasoil, and heavy fuel oil. The minor products are credited with the same average value as regular gasoline and light fuel oil, although naphtha sells for less and premium gasoline and liquid petroleum gas (LPG) for more. This does not necessarily bias the estimate of crude oil price delivered in Europe. Since refining cost is greater for premium gasoline, a wider refining margin would be subtracted from a higher price. But this certainly brings in a possible error in either direction.

The weighting scheme used, which is based on the various national statistics, cannot be precise either, because usage is not consistent from country to country. Furthermore, at least within limits, some products can be sold for more than one purpose, and the prices may vary. The heavier naphthas and lighter kerosenes on the borderline between naphtha and light fuel oil may be classified differently in two different countries. Some may be dumped into middle distillates and sold as heating oil, while some may eventually go into jet fuel. Any nonvolatile product may be added to residual fuel oil. Naphtha may be sold at a higher price when destined for gasoline rather than for town gas or petrochemical use. The more competitive the market, the less discrimination of this type we would find; but information is lacking on its past or current importance.

Furthermore, the prices themselves are never precise because a transaction may have significant "fringe benefits." It would be impossible, or at best impossibly expensive, for the price reports to be very detailed about these.

All these elements of error may be quite substantial at any given time and place; for more precise work we need better data. But there is one other reason why the Rotterdam composite price should not be considered as an average of arm's-length prices. It is essentially a trade price, in two senses. "The trade," i.e., refiners and distributors of oil products, trade among themselves, at prices set by knowledgeable people who know how much they can realize and how much they need to pay. There is no general reason why outsiders should pay any more, but one cannot ignore specific evidence that some and perhaps many of them do.

Anticipated Higher Prices and Term Contracts. If buyers expect prices to rise in the future, then contract prices must exceed spot. During 1957-67 many and perhaps most buyers generally expected higher prices, even much higher ones. We may disregard the official optimism of the oil companies, but not the report of a prominent oil consultant in 1961,[12] nor warnings at a UN seminar,[13] by a committee of experts appointed by the German government in 1962,[14] by the European Coal and Steel Community's working party in 1962 and 1964 (see footnote 9), and by the advisory committee of the Alliance for Progress in 1963.[15]

At the end of 1963, the official truth on both sides of the Atlantic was well summed up in a report published by the U.S. government. Prospects of steam coal exports were considered very good if the price were only slightly lowered. This conclusion was firmly based on an expected price of over $18 per ton for heavy fuel oil. Crude oil prices were considered depressed and due to increase. In addition, refinery margins were considered abnormally low, especially on heavy fuel oil, which was being sold at "marginal cost"–a term not explained. Therefore: "The long-term supply price of fuel oil in Europe will certainly be higher than the depressed prices recently prevailing at the margin of the market in industrialized countries."[16] The emphatic "certainly" reflected the emphatic consensus in Europe and America, as did the scorn for "the margin of the market," i.e., the European open market. (The writer had occasion to warn, on the basis of 1963 data, that either U.S. coal prices would be brought down to equate to about $12 per metric ton–those "depressed prices" which then declined further–or else that steam coal exports to Europe would disappear.[17] Between 1963 and 1968, exports of U.S. steam coal to Europe fell 62 percent[18] despite EEC energy consumption rising 20 percent.[19])

In November 1965, a spokesman for the United Kingdom Gas Council, a large naphtha buyer, considered naphtha available at "7.9 cents per gallon while there is a surplus. The picture might well change in the future since naphtha will shortly cease to be a surplus product."[20] At the time he spoke, regular gasoline had been rising for two years, and sold at Rotterdam for 5.2 cents (and naphtha for roughly 4.8). On the eve of war, it was under 6.5 cents and even in March 1968, with the Suez closed, good quality naphtha went for only seven cents.[21] Moreover, the Gas Council contracts required escalation when freight rates soared after May 1967. In mid-1969, "majors would be glad to sign long-term

[12] Walter J. Levy, *Current and Prospective Developments in World Oil: Implications for Western European Energy Policy* (Bundersverband den Deutschen Industrie), pp. II-29-30, III-30.

[13] C. A. Heller, in UN, *Techniques of Petroleum Development* (64, II.B.2), p. 196.

[14] *Untersuchung über die Entwicklung der gegenwärtigen und zukünftigen Struktur von Angebot und Nachfrage in der Energiewirtschaft der Bundesrepublik unter besonderer Berücksichtigung des Steinkohlenbergbaus* (Berlin, 1962), usually abbreviated to either *Energie Gutachten* or *Friedensburg-Baade Report*.

[15] Alianza para el Progresso, Comité de los Nueve, *Evaluación del Plan de la Nación 1963-1966 de Venezuela*, Sept. 1963, pp. 100-4.

[16] U.S. Department of the Interior, Office of Coal Research, *The Foreign Market Potential for U.S. Coal* (1963), vol. 2, Annex, p. C-35.

[17] "Le Charbon Américain en Europe Occidentale," paper at Colloque Européenne d'Economie de l'Energie, Grenoble May 7-8, 1965, abridged as "American Coal in Western Europe," *Journal of Industrial Economics*, vol. 14, Apr. 1966.

[18] *International Coal Trade*, June 1968, p. 6.

[19] ECSC, 16e Rapport Général, "Annexe Statisque–Energie."

[20] Institute of Petroleum, *The Petroleum Industry in the United Kingdom* (London: IP, 1966).

[21] *Petroleum Intelligence Weekly*, Sept. 25, Oct. 9, 1967; also Mar. 4, 1968.

contracts" at 5.4 cents,[22] but it is a safe guess that the Gas Council was paying half again as much. They had the worst of both worlds.

In late 1966, heavy fuel oil was said to cost "large users" in Britain $22.14 per metric ton, or $16.78 free of tax.[23] Doubtless, some of this was distributive markup and local transport. In March 1968 "a leading oil company" and the National Coal Board were asked to quote for delivery (but not for keeps) in central London. "The price came out within pennies of each other at just over $13.80 per ton of coal equivalent."[24] Assuming a long ton of 29.5 million Btu, this would be about $21.75 per metric ton of oil equivalent. At this time, Rotterdam barge lots were selling at $12.50[25] and the barging distance inside Europoort is not much less than from the Thames estuary to central London. The excise tax on oil accounted for only $4.60 of the difference of $9.25 per ton. What this really shows is the noncompetitive British market, where sellers of heavy fuel oil can still price at coal parity, which is best for the coal and oil trades as a whole.

The Rotterdam Composite, 1960–70: A Summary. (See Figure VI–1, p. 167, and Table VI–1 below.) For the whole 11-year period, the trend is broadly downward, but no more summary statement would be justified. It is not merely the usual index number problem, of a composite representing essentially independent demand-supply factors that bear on each component. To anticipate the argument of the next section: because petroleum products are in joint supply, an increase in the price of any one product encourages higher refinery output. Thereby it tends to lower the prices of the others. But where the proportions are variable, this inverse relationship is moderated by the ability to convert one product into the other. For example, the surging demand for fuel oil caused more crude to be charged and more gasoline refined, depressing its price. Yet some gasoline (naphtha) could be sold as the lighter fractions of gasoil, while the heavier gasoil fraction could be dumped in heavy fuel oil. But in any case, the price of any one product is

TABLE VI–1. Rotterdam Product Prices, 1960–1970

Year	Regular gasoline	Gas-diesel oil	Heavy fuel oil	Value per barrel of crude charge
	(cents per gallon)		*(dollars per barrel)*	
1960 (June–Dec.)	7.0	7.5	1.91	2.47
1961	6.8	7.8	1.86	2.45
1962	6.4	8.4	1.78	2.54
1963	5.4	8.8	1.79	2.49
1964	5.2	6.7	1.74	2.13
1965	5.3	5.7	1.81	2.02
1966	5.9	6.2	1.72	2.12
1967 (Jan.–May)	6.1	5.8	1.70	2.05
1967 (July–Dec.)	11.1	9.2	2.10	3.24
1968	7.2	8.1	1.76	2.53
1969	5.5	7.4	1.51	2.04
1970 (Jan.–May)	5.0	6.6	2.10	2.24
1970 (June–Dec.)	6.5	9.2	3.33	3.19

Source: Appendix Table VI–B–1.

[22] *Petroleum Intelligence Weekly*, July 7, 1969.
[23] *Wall Street Journal*, Dec. 21, 1966, p. 5.
[24] *The Economist*, Apr. 6, 1968, p. 54.
[25] *Platt's Oilgram Price Service*, Mar. 15, 1968.

subject to two separate factors: (1) the overall demand for products (and hence for crude) that determines the height of the structure, and (2) the respective demands and costs of transforming the crude into the various products. Indeed, over the 126-month period, excluding 14 months of tanker crises, we see only 7 months where all three products moved in agreement up or down.

From July 1960 through October 1960, prices were stable. From October through February 1961, all three series tended to decline. There followed a period of divergent movements to April 1962, for gasoline-naphtha and heavy fuel oil were fairly stable while gasoil described a cycle of rising and falling, as did the composite. Again there was divergence from April 1962 through February 1963, and hindsight showed gasoline to be in a fairly long decline from the beginning of 1960. But heavy fuel oil was almost a mirror image of naphtha throughout this whole period, and it strengthened while naphtha was weakening and gasoil rose strongly again. The composite therefore rose to the all-time peak of not quite $2.65 in February 1963.

There followed a long decline to a low of $1.94 in mid-1965. Gasoil declined spectacularly, from 9.2 to 5.7 cents. Gasoline fell almost as sharply through late 1963, then staged two cycles with a constant rising trend, so that by mid-1965 it was clear that there had been a bottoming-out, and there were well-founded hopes of a continuing increase as petrochemical and town gas demand began to take hold. Heavy fuel oil described two and a half cycles during this period, with three declines and two increases, and stayed in the middle of the range.

In mid-1965, the three principal products came closer together than ever before, then diverged again. The volatile gasoil price described two sharp cycles through mid-1967, with no particular tendency. Naphtha-gasoline and heavy fuel again diverged, naphtha rising strongly but then perhaps topping out at the end of 1966, while heavy fuel oil declined very slowly, and ceased to fall when naphtha ceased to rise. The composite was stronger in 1966 than in 1965, but appeared to be weaker in 1967 on the eve of the June war.

Price data compiled by W. Cipa of Gelsenberg, a large German independent, are set forth in Table VI-2. The level of prices is not comparable with the prices used here, for the products are from a Libyan crude and the refinery locations have higher transport costs and prices than does Rotterdam. However, the trend is comparable. From 1964 to

TABLE VI-2. Refinery Realizations, Western Europe, 1964-1967: Sales Value per Barrel of Products Refined from Libyan Crude

Country	1964	1965	1966	1967[a]
France	$3.10	$2.96	$2.90	$2.77
	...	(-.045%)	(-.022%)	(-.045%)
Britain	3.03	2.73	2.66	2.50
	...	(-.099)	(-.026)	(-.060)
Italy	2.24	2.17	2.50	2.47
	...	(-.029)	(+.154)	(-.014)
West Germany	2.74	2.60	2.50	2.09
	...	(-.051)	(-.038)	(-.164)
Average	2.79	2.44	2.65	2.45
	...	(-.125)	(+.086)	(-.075)

Source: W. Cipa, *Realizations Obtained by German Refineries, and Consumer Prices for Petroleum Products in Germany Compared with the Situations in Other Member Countries of the European Common Market and in Great Britain* (Essen: Gelsenberg, A. C., May 1967). The average is weighted by respective refinery output of the four countries; 1966 weights are used for 1967.

[a]First third of year.

1965 the drop is most severe in Britain, where competition was breaking into a cozy market; mildest in Italy, where the margin was most compressed; overall, more severe than at Rotterdam, which was a competitive market. From 1965 to 1966, the decline was much milder in France, Britain, and Germany, with a 15 percent increase in Italy. This would constitute an aggregate increase of 1 or 2 percent; the Rotterdam composite shows a stronger rise. From 1966 to the first third of 1967 there was a fall everywhere, as in the Rotterdam composite. The weighted average yield is closer to the Rotterdam composite than any one country. In one sense, it is a statistical fluke, but it may also register the effect of the Italian fluctuation.

The jump in tanker rates in mid-1967 (see Table IV-2, p. 109) sent prices up sharply, though it is not possible to calculate a precise relation. By 1969, prices were about at the January-May 1967 level, although short-term tanker rates were still well above. Hence f.o.b. prices must have declined. During the first five months of 1970, tanker rates and prices rose together; during the second half they soared together, and it is safe to predict that they will plummet together, but with significant divergences. A more complete discussion is given in the Epilogue at the end of this chapter.

The Refined Products Price Structure

The structure of prices in the wholesale market can be used to check the validity and precision of the estimates just cited. If the structure is in accord with refinery economics, then the odds are much higher that the composite is accurate, and any particular product price also. It will be shown that the structure of prices is internally consistent, and further that the price of heavy fuel oil is consistent with independent evidence. Heavy fuel prices will then be reviewed to confirm that the general price level of oil, crude, and products together was rising up to Suez I and declining thereafter. The relation between coal and heavy fuel oil is particularly revealing of the political influence on the oil market, which remains of first rate importance. Knowledge of its past will give us a basis for prediction.

Basic Economics of Joint Supply in Refining. A given price or margin may at any given time be temporarily below (or above) the level necessary to evoke a permanent supply of the product at the current rate of output. If costs (including the necessary return on investment) are not covered, supply shrinks and price rises. It would, of course, be wasteful to let the market act in this clumsy way, eliminating productive facilities now in order to bring them back later. But if the suppliers can foresee the price rise, they reduce their offers but do not leave the market so long as prices stay above current variable costs; they keep their capacities, and are not ready to sell off their assets at any price as the start of their exit.

So much for the total refining margin. But individual petroleum products are joint products, and it is altogether useless to seek or to pretend to have found the costs of the individual products—costs that do not exist. It is a great pity that so many people wonder that heavy fuel oil is priced at less than the crude oil from which it is extracted. (A ton of river sand containing gold nuggets would fetch more than the sand sold for concrete mixing after the gold had been removed.) One sometimes hears that heavy fuel oil "ought to make a fair and equitable contribution to amortizing the investment," an idea that is a barrier to serious discussion. It is a waste of time to try to impute this or that part of refining costs to this or that product. Refining a ton of crude oil is profitable, and will be done if the receipts cover the *total* costs of the operation, including the necessary profit; while no *individual* product will be forthcoming if the market price is above its incremental cost.

But this idea of "incremental cost" is too often misunderstood. The notion that incremental refining cost is necessarily below average cost, or below the price, must be rejected. When and if a refinery is working at less than optimum percentage of capacity,[26] an additional barrel can be produced at a rather low cost, below the average cost; then the price of the refined products is under downward pressure. On the other hand, when a refinery is being pushed toward maximum output, the cost of the additional barrel becomes higher and higher because of storage cost, product deterioration, lack of normal downtime for maintenance, cleaning and repair, etc. In general, incremental cost rising with higher output expresses the resistance of output to expansion, and gives the signal that production is pressing against the limits of capacity. When this situation is expected, or when it arrives unexpectedly, it is time to plan for expanding capacity. Thus the reckoning of short-term incremental cost—that of making the best choice given the present capacity—is replaced by considering long-run incremental cost—that of adding the best type and amount of additional capacity, either in a new refinery or in an addition to one already on stream. In general, when a branch of the economy is rapidly expanding, its incremental cost is a mixture of short and long term, and is higher than the current or future incremental cost at optimum levels of output.

In the United States, a slow growth of demand, the availability of natural gas liquids, and visbreaking (viscosity breaking) all reduced the need for distillation; hence refinery capacity was practically static during 1956-65. There was much well-founded complaint about that famous or infamous "incremental barrel" that had always to be denounced at every trade meeting and in every trade magazine. By 1965, production had grown into existing capacity, incremental cost had risen to the neighborhood of price, and prices tended upward.

In Western Europe and Japan the case was different. Aside from some small and localized surpluses, the incremental cost of the petroleum products taken together was not below the average cost, the incremental barrel was not cheaper, and prices were therefore not under any pressure from that source—from others, to be sure, but not from that one. It is a pity that this view of the incremental barrel as being ultra-cheap—justified by the facts, for a time, in the United States—should have been mechanically transferred to markets where it corresponded to nothing. Perhaps excess refining capacity will come later, as producers of crude oil try to realize their profits at the product level, but it does not yet exist.

But even with no pressure on the price of the whole output, there can surely be a surplus of any particular product available at a very low incremental cost. For when products are joint in variable proportions, the incremental cost of a single joint product does exist within certain limits even if its average cost does not. This type of incremental cost is perhaps too complex for precise statement except in mathematical terms, because two phenomena must be taken into account: (1) outlays that vary with the refinery pattern and result from the decision to produce more of one and less of another product, and (2) the opportunity cost, or the loss of revenue from the products being sacrificed. Rising price and output of any one product puts downward pressure on the price of every other product. Hence the tendency, already noted on page 174, for prices to move in opposing directions.

The Price of Heavy Fuel Oil (HFO). Another way of stating the problem is more elegant. Heavy fuel oil is a simple combustible, which is worth no more than any other source of

[26] There may be almost as many optima as refiners. See *Petro Chem Engineer*, Oct. 1968, p. 23.

heat, allowance being made for handling costs a little lower than coal and a little higher than gas. All other refined products have higher value uses with no near substitutes. Therefore the HFO price cannot go appreciably above the price of crude oil, for if it did, consumers would burn the crude entire.[27] But the HFO price can go to zero or even below.[28]

But the prices of the other products cannot go below the price of crude, for then it would not pay to refine them out. They would have only fuel value. Therefore the basic rule is: the more are prices of other products above crude oil, the more is heavy fuel oil below it. Conversely, the nearer the prices of other products are to their lower limit (the price of crude oil), the closer is heavy fuel oil to its upper limit, which is also the price of crude oil. At this limit, there is no longer any refining.

So far, however, we have assumed that the refining margins were competitively determined, so that incremental processing costs and prices stayed close together. It is competition, therefore, and not the good or bad policy of oil companies, which keeps the price of heavy fuel oil below that of crude—permanently, and in long-run equilibrium. Let us test this thesis further by looking to the one place where heavy fuel oil does sell for more than crude.

A Significant Exception—The Persian Gulf. In the Persian Gulf fuel oil has sold above crude since the early 1960s, at least.[29] Let us examine the reasons, first from the point of view of the customers, then from that of the producer-refiners.

The bunkering trade is relatively important at the Persian Gulf; in 1965 it was 40 percent of all offshore sales ("bunkers" plus "exports").[30] But ship operators have not the choice of burning crude oil because, until 1970 at least, this has been dangerous or too expensive.[31] The nearest alternative to buying heavy fuel oil at the Persian Gulf is to buy it elsewhere, which is obviously much more expensive. Hence, if all Persian Gulf suppliers abide by an understanding to price just enough under bunker oil at the nearest large supply source, they can collect that price for bunkers regardless of its relation to cost.

Let us now consider the other 60 percent of Persian Gulf heavy fuel output, "exports" for onshore use east of Suez. Customers for onshore use could drive down the price of heavy fuel oil to that of crude by buying crude instead. But the consumption of heavy fuel oil per customer is usually very low. Crude oil for direct burning must either be carried in small ships or stored and transshipped. This would raise its price by more than the crude-HFO differential, since the volatile light ends make storage and reshipment dangerous or expensive. Only the few largest customers find it economic to take larger ships and install storage, and they can get the lower prices. HFO prices would be lower if

[27] The technical problems, particularly flash point, are solved cheaply enough in onshore facilities, as Japanese experiments showed before 1963. (*Petroleum Press Service*, July 1963, p. 255.) It is usually better, however, to take off the lightest fractions first. The topped crude is then worth more than the original untopped, but only for so long as it takes to find a market for the lighter fractions.

[28] If the demand for heavy fuel oil disappeared (say if natural gas were available at very low prices) its value would be negative since there would be a cost of disposal. The incremental cost of more severe cracking would then be in part offset by the saving in disposal cost, in addition to the value of lighter products secured. In most of the United States this is not far from the actual case.

Low-sulfur fuel oil may of course sell for more than high-sulfur crude oil.

[29] In March 1969, it went below the crude price. If it stays there, the exception of course becomes historical, but is no less worth analyzing.

[30] U.S. Bureau of Mines, *International Petroleum Annual* and its more frequent, detailed, and useful predecessors.

[31] In 1969 plans were announced for experimental burning of crude oil in tankers.

electric power plants throughout non-Communist Asia were free (or under some compulsion) to obtain heavy fuel oil from the cheapest source, and free to substitute crude. Few of them are, and hence no active market can develop.

Let us look now more closely at the suppliers. It was assumed earlier that the profitability of refining a given product depends on the cost of crude (plus incremental refining cost) as compared with the revenues from that product. For an independent refiner, the cost of crude is its price. In recent years an independent Persian Gulf refiner would have been delighted to buy crude at \$1.20–\$1.40, and resell the fuel oil at \$1.55, not to mention an even larger profit on the more valuable light ends. As independents did this, increasing the supply of fuel oil, and as alert buyers saw what was in the wind, the price of fuel oil would necessarily sink below the price of crude. It would not need a very large independent refining segment to accomplish this, for the existing independents would find it profitable to expand.

The competitive opportunity cost of a barrel of crude must exceed that of a barrel of heavy fuel oil, because crude can profitably be made into more valuable products, while fuel oil cannot. To charge an equal or even higher price for fuel oil is to discriminate in price against fuel oil buyers. Under competition this cannot be done.

But at the Persian Gulf, there is no independent refining fringe. For the integrated producer-refiner, the cost of crude is not the price but the much lower cost-plus-tax. Nor is there an opportunity cost. The producer gives up no crude sales. He can supply as much crude oil as needed, and the price of crude need not affect the price of HFO. Because the price is far above the cost of bringing forth more crude oil *or* more fuel oil, there is no interaction between the two. The Persian Gulf producer-refiners are not forced to compete on price; and their discrimination in charging a higher price to some customers than to others, taking into account cost differentials, is commonplace in any monopoly. (If anyone considers the term too harsh, and calls it a noncompetitive price determination of Persian Gulf fuel oil, I will not object. The great bulk of Persian Gulf crude output is sold under more competitive conditions.)

Thus we have not only seen a limitation on our rule, but gained some useful insights both into the European products market and into a state of mind. The European market, served both by integrated and independent refiners, has a price structure characteristic of competition. We will see shortly that it is precisely the competitive nature of European refining that so baffled observers inside and outside the industry. Their implicit model, universally taken for granted, was monopoly, and they were right in feeling that actual prices were not conforming to that model.[32] The Persian Gulf exception not only probes our rule—it makes it more specific and useful.

Heavy Fuel Oil Prices, 1950–67. (See Appendix VI-A.) Heavy fuel oil is a large fraction of total output outside the United States—35 to 60 percent by volume and about 15 to 30 percent by value. Prices are set almost entirely by arm's-length bargaining with large resellers or electric power companies and other large industrial consumers. The product is relatively homogeneous, and therefore a series of prices is a good representation of market trends.

As has been pointed out, under competition the price of heavy fuel oil cannot exceed that of the crude oil from which it is extracted except under very unusual conditions and for a limited length of time. Furthermore, the higher the percentage of crude processed into heavy fuel oil, the less the price of the fuel oil can fall below the price of crude. The

[32]Cf. chap. IV, discussion of tankship prices.

percentage of crude processed into heavy fuel has been high (35-40 percent) and steady in Europe over the period that interests us. We can therefore treat the price of heavy fuel oil as the low end of the range within which a market price of crude oil would fall. Hence a series of HFO prices goes far to give us a rough indicator of crude oil prices. Because it is rough, its value is limited. A rising trend in fuel oil prices relative to crude could for a time mask a falling tendency for crude oil and products generally, and vice versa. Yet a fairly good heavy oil price series tells us something we could not know from even excellent series of gasoline (naphtha) or middle distillates, which can fluctuate much more sharply, and which need not assume any particular value in relation to crude oil.

Figure VI-2 presents a very approximate and smoothed-out picture of heavy fuel oil prices in Europe from 1950 to 1967. As Appendix VI-A shows, there is not much doubt of the general trends. Putting aside the Korea and Suez crises, the price was rising from 1950 through 1956. After Suez, it dropped to an all-time low, and continued to decline very slowly.

Why did HFO prices rise during 1950-56, up to the eve of the Suez I crisis? It is not really an answer to say that crude oil prices were rising during that period (cf. Chapter V). For the crude oil market was too small, and the price data too meager, to let us treat crude oil prices as an independent causal factor. And the crude price rise was to some extent a reflection of the HFO price rise. How do we account for it?

One explanation would be: a slow, circumspect monopoly process. Prices would be cautiously raised. The trade would look around to see if there were repercussions. Then another signal, such as an increase by a customary leader, would start it again. But this theory is supported by no evidence. Furthermore, the HFO price increases were often

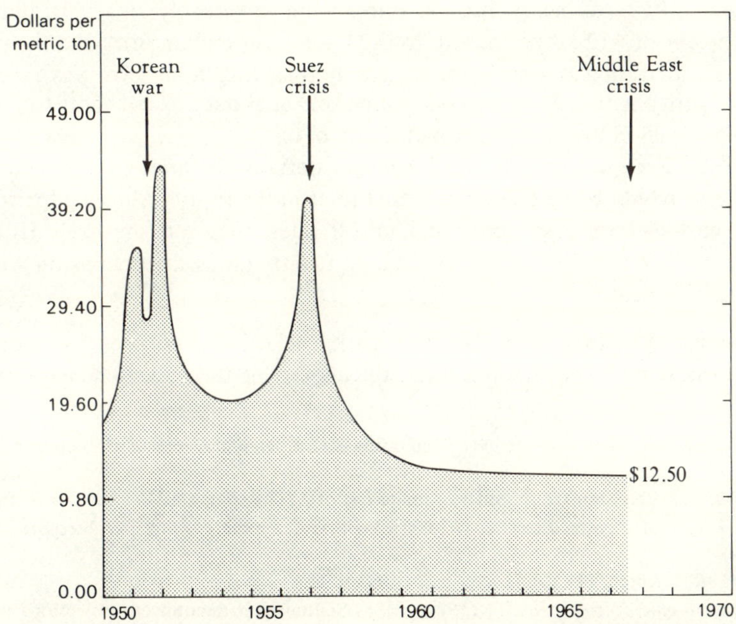

Figure VI-2. Sweden: Heavy fuel oil price development, 1950-1967. (Source: Swedish State Power Board, *Elforsorjningen under 1970-talet* [Supply of Electricity during the 1970s], 1967, p. 43.)

accompanied or preceded by reports of shortages, of buyers bidding higher to obtain supply. Above all, there is no example of prices rising or even holding in the face of excess capacity. When demand slacked in 1953-54, fuel oil prices temporarily fell. Inability to hold the price line—the easiest of all small-group policies—during even a mild stagnation of demand discredits the idea of tacit understanding to maintain prices.

In Appendix Tables VI-A-7 and VI-A-8, we look more closely at the market where fuel oil competes with coal. A comparison of absolute prices is impossible with available data. But the Danish series (see Appendix Table VI-A-5) indicates that coal and fuel oil must have been equal within a narrow range. The ECSC data show that French and German mine-mouth coal prices rose strongly in 1950-53, paused in 1954, and then diverged, German prices rising strongly through 1958. British coal prices as delivered to electric power plants rose without interruption from 29 cents per million Btu in 1950-51 to 48 cents in 1957-58.[33] American coal import prices are the truly "marginal" tail of the dog, rising by 50 percent in one year, falling by 40 percent the next three, then repeating the movement.

During 1951-57, European-Caribbean fuel oil was only a minor part of the total market in steam fuel and could not for long sell at much more or much less. As coal grew more expensive, largely because of real increasing scarcity, fuel oil prices rose with coal. This simple parity theory worked quite well. It broke down altogether after Suez I. As the ECSC rather belatedly reported in early 1961: "Whereas calorie for calorie the [price] ratio had stood for years at about 1 : 1, since the latter half of 1959 it has been shifting quite definitely to the advantage of fuel oil."[34] Since then, European coal prices have been relatively stable, rising slightly. In Great Britain the cost of fuel to the electric power plants was fairly stable from 1957 to 1965, then rose to 55 cents per therm (predevaluation equivalent) in 1966-67.[35]

Throughout 1957-69, heavy fuel oil prices were remarkably steady within plus or minus 10 percent of $12.50 per metric ton (31 cents per million Btu). Yet the price was considered an aberration or was ignored. The normal long-term level was taken to be approximate parity with U.S. steam coal, which sold at about $13 per short ton, equating to 50 cents per million Btu or $20 per metric ton of oil.

Etude sur les Perspectives,[36] by far the most careful of the price studies, assumed that U.S. steam coal would be forthcoming at that price in the future. Allowing for what they thought to be long-term cost trends and tanker rates, they put the likely HFO parity figure between $17 and $19. Even this muted realism aroused reactions on which it is distasteful to comment.[37]

The Breakdown of Parity. Why did the parity relation, confirmed by econometric studies,[38] break down? The reason is brief: the underlying theory was incomplete. Within

[33] United Kingdom, Central Electricity Generating Board, *Annual Report and Accounts*, 1959-60, p. 84.

[34] European Coal and Steel Community, *Annual Report, 1960*, paras. 124-25.

[35] United Kingdom, Central Electricity Generating Board, *Annual Report and Accounts*, 1966-67, p. 10.

[36] ECSC (1962), Annex 11 (1964).

[37] In the *Economic Journal*, vol. 74 (1964), E. F. Schumacher denounced projections to 1975 as "bordering on mendacity." Mr. Schumacher's affiliation with the National Coal Board was not stated. In several places, including the April 11, 1963 *Times* (London), he had predicted not 13 but 37 years ahead, to A.D. 2000.

[38] R. Regul, "Energy Studies," *Institute of Petroleum Review*, vol. 12 (1958), pp. 299, 303.

limits the degree of competition in heavy fuel oil did not matter; beyond those limits, the higher coal price had to drive *down* the fuel oil price.

Before the first Suez crisis, coal-oil parity was a simple competitive adjustment–provided that one neglects the very low production cost of oil. Had that cost been in the neighborhood of price, growing coal shortages and rising coal and oil prices would have forced more intensive oil development, raising costs. But the higher were current coal prices–and, more important, expected coal prices–the greater the profit inducement to expand heavy fuel oil output. After Suez I, the force exceeded the resistance. The more buyers were attracted to oil and shopped around, the wider the market, the greater the inducements to shade prices. The price began to gravitate toward cost.

But the parity theory and the "mere by-product" slogan prevented recognition. The ECSC-EEC working party never left off "parity" until the 1966 publication of *Nouvelles Réflexions*. Others reacted with growing bafflement and frustration. How *could* it be? The large international companies must be to blame, "waging economic war" and exercising and "abusing" their huge size and "power."

If "power" means monopoly, this was the contrary of the truth. The oil companies wanted to make money and avoid trouble. The best strategy for money-making was to keep fuel oil prices moderately below coal. A small but permanent price differential could attract a large amount of new business. However, the best way to avoid political dangers was to keep oil prices at least a little higher than coal. To reconcile these objectives was difficult. What actually happened was the worst of both worlds for the industry as a whole–i.e., for a rational politically aware monopolist. Oil prices fell far more than necessary to get the business–indeed, low enough to convulse the European coal industry and its numerous defenders. In the German parliament, both sides displayed "growing resentment against international oil companies." Social Democrats attacked them for "the abuse of economic power," injuring German consumers by charging them too little. From the other side of the floor, an expert of the government party spoke for them and nearly all Europe. "International companies can digest losses caused by the murderous price war in [the] German energy market more easily than German companies can. Present rock-bottom prices won't last forever, but the German companies in [the] meantime will go to the wall and international companies will profit."[39]

The UN Economic Commission for Europe was glad to confirm in 1967, looking back over two years, "the recent declarations of the international petroleum companies which foresee a stabilizing, indeed an increase, of the prices of fuel oil in the Western European market."[40]

Had the oil companies been smaller and more numerous, prices would have started dropping even sooner, and come down more. Like all sellers, they gave ground in spite of themselves. They tried to be statesmen, but competition kept breaking through.

The stronger the demand for heavy fuel oil, the weaker the price of naphtha, whose surplus was notorious and constantly discussed. But there was much more than talk; petrochemical plants were redesigned (or newly built) to use naphtha as charging stock, and in 1963 there came onto the market two commercial processes for cheaply converting naphtha into town gas. Naphtha prices stabilized, and after late 1964 began to increase strongly–regular gasoline being a slightly distorted record of naphtha. And as this happened, fuel oil, which had since early 1960 been coming ever closer to the

[39] *Platt's Oilgram Price Service*, Dec. 3, 1964, pp. 3–4.

[40] UN, ECE, *La Situation du Marché Charbonnier en Europe et ses Perspectives*, 1967, p. 7.

composite value of the product barrel, slowly sank away from it. After the summer of 1966, heavy fuel oil seemed to be strengthening. How far this movement would have gone we cannot tell, for the 1967 war intervened. Because Libyan oil was so much cheaper delivered to Europe and its output could be rapidly expanded, it displaced much Persian Gulf crude oil. Libyan crude was lighter and might therefore be expected to furnish more naphtha and gasoil, weakening their price and strengthening that of heavy fuel oil. But the high pour point of the Libyan crudes made them unsuitable for middle distillates. These were largely dumped into the residuum, the price of which was back to prewar levels as early as September 1968 and sank to an all-time low over the next year. The composite and naphtha had about resumed prewar levels by late 1969, and middle distillates had reached a somewhat higher level.

At this time a new force became important enough to change price relationships. The Libyan crude is very low in sulfur and was in such oversupply that the low-sulfur premium went effectively to zero. It gradually became apparent that it was worth shipping Libyan crude from the Mediterranean refineries to the East Coast of the United States, where municipalities were beginning to enforce low-sulfur requirements that recently had been enacted.[41] The rising demand and premium price, particularly anticipating the 1969-70 winter, raised HFO prices considerably from a little over $9 to $12 per metric ton. The longer-term effects of American demand on the price structure depend also on the size of Libyan oil deposits and the rate and cost at which they can be expanded, and also on whether cheap dewaxing for middle distillates can be devised.

Summary. The European product price structure since early 1960 has seemed altogether in accord with the basic economics of oil refining under competition. Because of competition in the European market, the integrated producer-refiners had to meet the prices of the small minority of independents. Changes of individual prices are consistent with what is known about real scarcities of the respective products. The precision of the publicly known price series leaves much to be desired, but criticism of them as marginal or irrelevant is logically and factually wrong.

Crude Oil Prices as Derived and as Observed

Having set forth the elements of the product price structure and calculated the downstream elements, we now turn to crude oil liftings. Prices will be estimated from two directions: first, from the Western European product prices just discussed we derive the netback at the Persian Gulf; second, actual arm's-length sale prices will be estimated from press reports and from customs figures. Then, an attempt will be made to summarize and integrate the data from the various sources.

Realizations from Western Europe. Table VI-3 presents the annual averages of the Rotterdam composite, the refining margin, the long-term tanker rate, and the Persian Gulf netback, or realization. It should be read not only from left to right but also from right to left. That is, the Persian Gulf netback was no mere mechanical reflection of the changing Rotterdam composite, for the latter was affected by changes in crude prices tabulated in Appendix Table VI-E-1, and also by the slow decline in tanker rates.

[41] In mid-1969, the U.S. East Coast delivered price after trade discounts was $1.65 a barrel for residual oil with no sulfur limits. Cf. my "L'Offre et le Prix du Fuel Lourd aux USA," *Revue Française de l'Energie*, June 1965; and A. D. Little, *Future Market for Utility Coal in New England*, Report to Office of Coal Research, U.S. Department of the Interior, Nov. 1968, p. 176, p. 176n.

TABLE VI–3. Rotterdam Composite and Computed Persian Gulf Realized Price, Annual Averages, 1960–1967

(dollars per barrel)

Year	Rotterdam composite	Refining margin	New time charter freight rates of marginal Canal ships: INTA less	New time charter freight rates of marginal Canal ships: Per barrel	Derived Persian Gulf price	General price level
	(1)	(2)	(3)	(4)	(5)	(6)
1960 (Jul.–Dec.)	2.47	0.36	–45	0.61	1.50	100
1961	2.45	0.34	–45	0.61	1.50	101
1962	2.54	0.34	–47	0.59	1.61	102
1963	2.49	0.31	–47	0.59	1.59	103
1964	2.13	0.31	–54	0.53	1.29	105
1965	2.02	0.32	–54	0.53	1.17	107
1966	2.12	0.33	–55	0.52	1.27	110
1967 (Jan.–May)	2.05	0.33	–57	0.49	1.23	111

Sources and Notes by column:

(1) Table VI–B–1. There is no deduction for barging cost. Rotterdam refineries, like nearly all others, sell mostly barge lots; Italian coastal refineries, particularly in Sicily and Sardinia, sell mostly tanker loads. A steady stream of barge lots costs more to deliver, per unit, than the same amount into a tanker. But against this we must reckon a storage cost if tanker offtake is not continuous. Which cost is lower depends on the timing of deliveries; there is no general answer.

(2) Capital plus operating cost, average new capacity installed (table VI–C–8).

(3) Highest rates on time charters three years or over, converted to P.G.–Rotterdam voyage-ton equivalent (chap. IV, table IV–2).

(4) Flat Intascale taken as 91 cents per barrel plus 12 cents Suez Toll.

(5) Col. (1) less sum of cols. (2) and (4).

(6) GNP deflator, United States, calculated by U.S. Department of Commerce, Office of Business Economics.

The mild 1960–61 decline in the Rotterdam composite could be accounted for by the decline in refining costs, so that the netback to the integrated refiner was not affected. In 1962 Rotterdam receipts rose appreciably, and the netback was further improved by cost reductions in refining and transport. In 1963 the receipts deteriorated with the availability of light crude from North Africa, new low-cost refining capacity, and lower long-term tanker rates. Although independent refiners were in difficulty in 1963 and loudly said so, the integrated companies were only mildly affected until late in the year. But the downward movement clearly overran itself in 1964 and 1965, since prices dropped considerably more than could be accounted for by the continuing slow reduction in costs. The netback therefore fell from February 1963 to June 1965, and quite sharply. Nineteen hundred and sixty-five was a very difficult year and, as noted earlier, in April the margin seemed insupportably low for the independent refiner; the Rotterdam price level was due to improve.

Nineteen hundred and sixty-six saw a recovery to approximate 1964 average levels of price and netback. In early 1967 prices again started to fall and netbacks deteriorated, but perhaps this was merely the gasoil price going through a familiar gyration, or perhaps it was a more important movement. It is true that gasoline-naphtha prices fell in February-March, but they recovered altogether, while heavy fuel oil declined very slightly and then stabilized.

Over the seven years from July 1960 to May 1967, prices at Rotterdam fell by about 18 percent, from $2.47 to $2.02. The impact on the Persian Gulf netback was about 31 cents, or 21 percent. Exact percentages give an impression of more regularity and precision than is justified. Let us say that prices tended to fall by something between 18 and 20 percent in seven years, or between 2.5 percent and 3 percent per year. This mild price movement is best and most simply explained as the low costs of developing, producing, transporting, and refining crude oil, slowly transmitted through a very restrained competitive process to the market prices.

"Perverse Netback Movements." Demand is higher in the winter months and tankers more intensively used, thus raising the incremental transport cost and (in a competitive industry) the price as well. This sometimes confuses the discussion of netbacks. If high demand increases freight rates, which are subtracted from the c.i.f. price, the netback decreases, other things being equal. Higher demand and lower prices seem to make no sense. But the mistake is in focusing on the arithmetic–delivered price minus freight equals netback–and forgetting that other things are not equal. With higher demand for freight, the oil is more expensive delivered, but the price received for the oil exclusive of freight service–the netback–is not affected.

Crude Oil Prices: Trade Reports. (See Appendix VI-E.) The trade press occasionally reports public bids, chiefly to South American countries, and on prices set on sales of crude supply rights for independent refineries, in the public or private sector. Unfortunately, various terms of the bargain may or may not be identified. The most important are: quality, delivery, credit terms, buy-back rights, and the currency of payment. Lighter crudes are generally, but by no means always, more valuable; the reckoning used is 1.5 cents per degree API, which, as explained in Appendix VI-H, is probably better than the industry rule of thumb of 2 cents per degree API. Appendix VI-H also explains why sulfur might account for as much as 6 cents per barrel.

In any case, these terms are viewed differently by different people, and so one must expect much variability–prices may look the same that are not, and may look different when they are the same. Transport can be accounted for when all special conditions, if any, are revealed–chiefly the depth of the unloading port.

Credit terms are converted into cash-equivalents. Our calculations of interest cost assume the seller to pay the long-term bond rate of a large credit-worthy firm in Western Europe, where rates are well publicized. The cost to the seller is greater than this, for there is a risk of default or delay. One would be closer to the truth using the bond rate paid by the buyer, which registers this risk; but there are no data publicly available. Furthermore, if the seller must consider the funds that he extends on credit as part of his permanent capital, renewed or newly extended as fast as it is repaid, then its cost is his general cost of capital, which is not 6-8 percent but well over twice that (including an allowance for taxes). Since the cost of extending credit is understated, prices tend to be overstated. This overstatement is a small but not negligible amount, perhaps 5-10 percent.

One is compelled to discard some observations. Buy-back is very difficult to handle unless the market value of the product taken in return is known. Currency terms are sometimes not revealed.

Customs Reports. The second source of information is in the reported customs statistics, particularly for Italy and Germany (see Appendix VI-F). The bulk of these figures is quite useless. First, they are mostly not arm's-length transactions but mere entries on the books of an integrated company. Second, the shipments and import values are only identified by exporting country, with sometimes an additional subdivision such as currency of the transaction or nominal place of incorporation of the supplying company. Hence the imports under any given heading may be a mixture of real sales and irrelevant transfers. Third, both real sales and intrafirm transfers may be pursuant to new contracts or old ones. When prices are on the decline, the level of average values declared is much too high. But as old contracts are replaced by new at lower prices, then even if the price had ceased to decline, the average unit values still would continue to fall as they approached the new level.

In order to avoid these pitfalls, with few exceptional cases all average figures have been disregarded. Wherever a series appears of "supplies from Country X" or "supplies shipped from Port Y," the concentration has been on prices that are substantially and continuously below the average. Where the amount of oil shipped is given, and it is obviously a single tanker load, it has merely been noted as such. But a succession of single tanker loads, apparently on some kind of continuing basis, indicates a commitment for a longer period of time; and if for more than a year, the time charter rate, rather than the spot or consecutive voyage rate, has been subtracted from the reported c.i.f. value.

The customs valuations f.o.b. are only occasionally worth a glance. The c.i.f. value is considered as possibly an arm's-length transaction. When one finally arrives at what seems to be an arm's-length price, it is subject to the same uncertainty about quality of the crude and the nature of transaction as bids reported in the press.

However, published customs figures can be used in tabulating the imports from Soviet Russia. (See Appendix VI-G.) These are real arm's-length transactions. The Russians have always sold for as good a price as they could, although this was at first below the price that oil companies could obtain. The discrepancy has tended to narrow, and in recent years Soviet prices on new transactions are not always the lowest. We will not waste time refuting arguments that the Soviet Union sells at less than market prices for "political" reasons. (See Chapter VII.) These make no more sense than the arguments that oil companies sell heavy fuel oil at less than the "natural" price—coal equivalent—for some other discreditable reason. But although the Soviet Union is a seller like all others, it is not exactly like them. Soyuzneftexport has seemed more willing to make barter deals or term contracts. As a newcomer whose capability and reliability were not well known, it had to offer some compensating advantage, which might be a lower price.

It is not easy to summarize this miscellaneous collection of data, but one can state a few conclusions. For comparison with the netback from European product prices and with each other, all prices are adjusted to the basis of a Gach Saran type crude of 31° f.o.b. the Persian Gulf. There have been prices equivalent to around $1.00 as far back as 1960. These were for the Italian contracts with the U.S.S.R. and Esso. Arm's-length prices generally seem to have risen in 1961 and 1962, fallen through the end of 1965, risen slightly from 1965, and seem to have lowered again in 1967. Thus the movements correspond in direction with the Rotterdam composite price.

In 1960 Soviet crudes were the most prominent low-priced sellers. From 1961 their export prices rose, and by 1965 they were no longer among the lowest. In 1964 and 1965 North African, particularly Libyan, crudes were the most notable low-priced sellers. Libyan crudes were said to be offered in the $1.30s, or equivalent to 93 cents and up. (See Appendix Table VI-E-1, and see German imports in Appendix Tables VI-F-2 and 3.) In early 1967 low prices seem to be originating in Persian Gulf crudes. New sellers, or those determined to increase output, are at least the most visible low price sellers. The case of Libya is discussed further below.

Although the lowest prices changed very little over the period, there was clearly an increase in the number of buyers at these lowest prices. At the same time captive buyers, such as affiliates, were gradually paying less than the full posted price that many still paid in the early 1960s. These trends cannot be specified any more precisely, but the German average monthly import figures are shown in Table VI–F–1.

Delivered prices in the Latin American market were remarkably steady from 1960 to 1965 ranging ten cents above and below \$2.00. However, freight rates during that period fell so that f.o.b. netbacks rose (an opposite example of perverse netback movement from the one discussed above). There are three possible explanations of the difference in trend between Europe and Latin American netback prices and all may be partly true. The first is that Brazil's Petrobras, the most aggressive buyer, did not make its transactions public after 1964, so that the nature of the data changed. Another possibility is that Latin America became a less desirable market as the countries' financial position deteriorated. Or, what comes to the same thing, the value of the credit increased and hence the cash-equivalent price really decreased. The third is that North African crudes, which were especially competitive in Europe because of their location, had a much smaller advantage in South America (14 cents against 32 cents at Intascale less 55).

In the 18 months before the second Suez closing, price equivalents ranged from about \$1.05 to \$1.25. (See Appendix Table VI–E–1.) Open market prices for Iranian crudes were around 40–45 cents discount off posted, or \$1.18–\$1.23 for Gach Saran. Our estimates of open-market prices for 34° Iranian in 1966–67 also form a ladder or structure. The least-favored, such as India, paid around \$1.40. Publicly bid crude seems between \$1.20 and \$1.35; good-sized European refiners paid perhaps \$1.20. Kuwait crude was selling to German and Italian refiners around \$1.20, or about \$1.25 Gach Saran equivalent.

Some German and Italian import prices netback as low as \$1.00, but some of these would reflect contracts made earlier. An interesting contract for new business was for Arabian crude to the Hess refinery in the Virgin Islands, equivalent to a Gach Saran price of \$1.06.

Interpretation of Libyan Crude Price Data. Libyan production increased very rapidly during the period 1960–67. Much of it was and is in the hands of newcomers and has gone into both integrated and independent refineries serving the Northwest European market. Hence it merits some discussion.

Libyan prices tabulated in Appendix VI–E seem to range around a central value of \$1.60 f.o.b. Another way of estimating Libyan selling price in 1965 is from the size of the tax increase, which yields \$1.62. (See Table VI–4.) But the range in 1965 was great.

In order to calculate equivalent f.o.b. values in the Persian Gulf, we have assumed a freight differential around Intascale less 55 percent, on which reckoning an equivalent barrel is worth 32 cents more in Libya. The lighter crudes have been evaluated at 1.5 cents per degree gravity and sulfur premium at 5 cents per barrel. (See Appendix H.) Hence a barrel of 39° Libyan crude selling at no more than \$1.60 f.o.b. Libya equated to a 31° Iranian crude at Kharg at \$1.10.[42]

I am inclined to accept the lower Libyan netbacks in 1964–65 as real, but to waive any precision in comparing Libyan and Persian Gulf prices. First, arm's-length sales were largely by newcomers, meaning chiefly the Oasis group—Continental, Marathon, and

[42] Because of the wax content of Libyan crude oil, the 5 cents is too high and the allowance should perhaps be zero, which would make the Iranian equivalent \$1.15.

TABLE VI-4. Approximate Calculation of 1965 Oasis Average Price, Libya

(dollars per barrel, except as stated)

1. Posted Price 40°, Es Sider	$ 2.220
2. *Less* 7.5 percent allowed discount	0.167
3. *Less* gravity allowance (40° – 27°, at $0.001325 per degree difference)	0.017
4. *Less* marketing allowance	0.005
5. *Equals* posted price, adjusted on new tax basis	2.031
6. Oasis production, 1965 (million barrels)	184.5
7. Additional taxes due under OPEC settlement (millions of dollars)	64.5
8. Additional taxes per barrel (line 7/line 6)	0.346
9. Royalties, 12.5 percent of $2.22	0.278
10. Tax saving due to expensing of royalties (one-half of line 9)	0.139
11. Additional tax based on higher accounting price (line 8 – line 10)	0.207
12. Increase in accounting price (double above)	0.414
13. Previous accounting price (line 5 – line 12)	1.617

Sources: Lines 2, 3, 4, 9–OPEC settlement terms, applied to Libya. Lines 6, 7–Continental Oil Co., *Annual Report, 1965*, p. 2.

Amerada, in equal shares. Much has been said about their "unstatesmanlike," "short-sighted," or "foolish" conduct. There have also been more temperate remarks about newcomers with "weak hands"–poor silly fellows who must get their money back soon and hence must dump oil. Nothing is more untrue. These companies are well financed and managed, under no pressure to dump, and will sell at whatever the traffic will bear. Far from being under pressure, they were, at least in 1963-66, free of one business constraint. An established company would be reluctant to accept a lower-than-current price. It would tend to weaken prices everywhere and would cost some appreciable fraction of current revenues–indeed, perhaps more than the additional sales. Hence a new contract, profitable in itself, might make a company produce more oil, incur higher costs, yet receive lower total receipts. The net revenue on the additional output (the incremental or marginal revenue) would be too low and might even be negative.[43]

But newcomers need only compare the prospective costs of developing a given amount of production with the price offered or the netback expected from integrating downstream. In 1963-66, Oasis companies did not lightheartedly invest development money, then look around in fright for customers and offer oil for any price above short-run operating costs plus taxes.

But although this accounts for the willingness of the new producers to quote lower prices if they must, it does not account for the ability of competing producers to maintain–if, as the data suggest, they did maintain–their own prices at a higher level. It would be consistent with the rapidly changing supply sources of German and Italian refiners, where Libyan crude took over much and perhaps most of the independent segment.

When the Libyan independents were forced to pay 35 cents per barrel more in taxes in January 1966, there was widespread satisfaction and expectation of higher prices, not only in Libya but everywhere.[44] In fact, there was apparently a very slight increase, but it

[43]To take the simplest example, if one has been selling 100 units at $1, total receipts have been $100. To reduce the price to 90 cents in order to sell another 10 units, while also accepting the lower price on the previous units, means that the sale of the additional 10 units will actually reduce total revenues by 10 cents, despite the higher volume. Hence, even if the cost of the additional output were zero, to sell it would reduce profits.

[44]The writer predicted no price increase: see *Platt's Oilgram Price Service*, May 25, 1966, p. 1.

needs no tax to explain it.[45] Incremental cost must have risen sharply, perhaps temporarily. In retrospect, we can tell that the Oasis reservoirs (as then known) were fully developed, with incremental costs-plus-tax of more output rising sharply toward price. Output rose 28 percent in 1966, but nearly all this must already have been placed; over the next two years, the increase was only 4 percent, and the big Dahra and Waha fields even declined. Consequently, Oasis no longer had any new crude to offer, and no expanded markets to gain at lower prices.

Another reason for stable to firm Libyan prices would be the better netback from Rotterdam sales in 1966 than in 1965. Some and perhaps most of the contracts with independents were written as a margin below the current netback.

Comparison of Rotterdam Netback and Reported Crude Prices. Given strong competition in transport and refining, which is proved by independent information, f.o.b. open-market prices and derived f.o.b. netbacks should be equal. I estimate that the netback on a 31° crude for an integrated producer-refiner was approximately $1.20-$1.25 during January 1966-May 1967, while Gach Saran 31° (or equivalent) fetched $1.18-$1.25 f.o.b. The apparently precise agreement is fortuitous, but better observations would at least be fairly close. Since these two price series are independent as to their sources, each serves to reinforce the validity of the other.

Discrepancies are due to chance and to imperfect observations, but some seem due to better purchasing; as between Argentina and Brazil, for example, about 10 cents a barrel. Again, if we look at the difference in price between large and well-financed West-of-Suez and East-of-Suez buyers—the latter of which basically means Japan—the difference was at one time over 10 cents per barrel and later down to perhaps 5 cents.

We ought not to assume that the f.o.b. price series and the netback series should be close together because each is the opportunity cost of the other. Under purely competitive conditions, the cost of crude oil to a producer contemplating an integrated transfer would be the price at which he could sell it in the open market. Conversely, the cost of an open-market sale would be the sacrificed netback from a sale of products in the downstream markets. But the purely competitive assumption does not hold. When a company makes a sale in the f.o.b. market, it is not forced to withdraw a corresponding amount from the integrated channels, or vice versa, except where it has been caught unprepared. Indeed, one reason for carrying some spare capacity is precisely to avoid this kind of short-run embarrassment. Anything that can be sold at current prices can be produced, and profitably. Hence, there is no powerful competitive force reaching inside the crude producing firm to keep these two prices close together. What has been demonstrated here is a more limited fact: there is enough competition to keep each price within the other's margin for error.

Contracts Between Major Companies. Before summarizing the results of this inquiry, let us pause to estimate price of crude sold among the largest eight companies under long- or short-term contract. This should be considered as market oil. There is no warrant for treating it as within a closed circuit. To assume that the majors act as a unit is to beg the whole question of what in fact they do. They have no common purse or profit. Oil is only bought by an integrated company if it is worth buying in preference to own-production

[45] Older established companies in Libya who, according to press reports, encouraged the Libyan government to raise tax rates, did not serve their own interests well. For the government had unilaterally changed the terms of the concession; the precedent may not have been needed, but it will be cited from now on.

or buying elsewhere. Hence the price must lie between the lower limit of production cost and the upper limit of the lowest price available elsewhere.

The prices set on these long-term contracts are a well-kept secret,[46] yet they are not too difficult to guess within small margins for error. The long-term contracts cover mostly Kuwait and Iran output, the latter considered higher cost. As shown in Chapter II, the real cost of Iranian Consortium output in 1966-68 was about 10 cents per barrel, equally divided between capital charges and current expenses. Table VI-5 shows the computation of taxes and royalties on light and heavy crude, into which 1966 output was about equally divided. Cost-plus-tax on Iranian Light was a convenient $1.00 per barrel. The favorite pattern for prices has been a partway price, most likely halfway, between cost-plus-tax and the market price (which was once the posted price).[47] If the lowest prices paid for Iranian Light are about $1.20, then the halfway price paid on long-term contracts ought to be $1.10.

TABLE VI-5. Iran Consortium, 1967: Calculation of Difference in Tax-Plus-Cost Between Light and Heavy Crudes

(*cents per barrel*)

1. Cost per barrel, Iranian Consortium, 1967, all crudes	12
2. Tax per barrel, all crudes	83
3. Difference in posted price, Iranian Light less Iranian Heavy ($1.79-$1.63)	16
4. Difference, adjusted for 6.5% marketing allowance (16¢ × 0.935)	15
5. Royalty at 12.5% difference (15¢ × 0.125)	2
6. Net posted price difference (15¢ less 2¢)	13
7. Tax (at 50%) corresponding to posted price difference	6.5
8. Total difference in tax-plus-royalty (line 5 plus line 7)	8.5
9. Tax on Iranian Heavy (83.3 – 4.3)	79
10. Tax on Iranian Light (83.3 + 4.3)	88
11. Total cost-plus-tax, Iranian Light (line 1 plus line 10)	100
12. Total cost-plus-tax, Iranian Heavy (line 1 plus line 9)	91

Sources: Chap. II; IOOC Annual Reports.

Note: It is assumed that 12¢ is a convenient midpoint between estimated real cost of 10¢ and accounting cost (text below, n. 48) of 14¢.

Obviously, every one of these figures is approximate. But a little reflection and experimentation with better figures (which some readers will have) will show, I believe, how insensitive is this $1.10 to errors. The accounting cost will not agree except by accident, but even a large percentage error will make little difference.[48] Taxes depend on well-defined rules, and they are the bulk of the tax-plus-cost lower limit. As for the upper limit, when we know of the Argentines paying in the range $1.30-$1.40, and the Japanese about 10 cents less, we cannot get it much lower than our assumed $1.20. Moreover, the two errors can only be significant if they are both in the same direction: if one is positive and the other negative, the halfway limit will be little affected. Our seller will not sell below the lower limit, and our buyer will not buy above the upper limit. Since it is senseless for a large, integrated producer-buyer to pay appreciably over $1.20, or for a seller to accept less than $1.00, the maximum chance of the price being below

[46] I thought at one time that they were known to other international oil companies, but am now inclined to think they are not. If they were, it would strengthen the inference of the paragraph.

[47] *Petroleum Intelligence Weekly*, Dec. 7, 1964.

[48] In early 1968, a much quoted and no doubt deliberately leaked set of costs were: Iran 14 cents, Saudi Arabia 9, Kuwait 6. (See *New York Times*, Apr. 14, 1968, p. F-1). One cannot tell how they are reckoned, or what they mean, but the result of using them is close to what has been reckoned here.

$1.05 or above $1.15 seems small. Hence the estimate is correct to within 5 cents, because the contracts can only be profitable to both parties within a certain market framework.

Summary. 1. During 1957-67, the trend of prices was obviously down, but the rate of decline was mild and the variation great.

2. During 1965-67, the arm's-length price of 31° or equivalent crude at the Persian Gulf varied from $1.00 to $1.25 per barrel, and Libyan crudes were in that range, allowing for freight and quality difference.

3. The netback on a barrel refined into products sold at Rotterdam approximated $1.20-$1.25 a barrel in 1966-67, close to the higher range of Persian Gulf or Libyan prices. The netback is the most relevant figure for determining prices to the consumer and also investment decisions in production.

4. These prices were quite stable; they were above costs-plus-taxes and far above costs. It is far-fetched to speak of them as "depressed."

Epilogue: 1967-1970

Since May 1967, European product price data have been better than ever before. *Platt's Oilgram Price Service* has had considerably more detail on the bargaining moves and countermoves of buyers and sellers, and on the forces of supply and demand at the moment. It has been abundantly clear that the Rotterdam and Italian price quotations are linked to sales by integrated and nonintegrated firms throughout Continental Europe, and to a lesser extent Great Britain. But one cannot fit together the pieces to see whether the old structure has changed in some ascertainable way. The binding element is the appropriate tanker rate. Under stable conditions, this approaches the long-term rate needed to pay for building the marginal ship, which is now the 250,000-ton tanker. But this is a new size, which is still very scarce, and whose long-run equilibrium rate is still uncertain. (See pp. 123ff.) Since May 1967, the tanker market has been continually, and often wildly, out of equilibrium.

Table VI-6 shows how the calculation assuming a stable tanker market becomes impossible under instability. The first line in the table, for January-May 1967, works out

Table VI-6. Rotterdam Composite and Computed Persian Gulf Realized Price, Annual Averages, 1967-1970

(dollars per barrel)

Period	Rotterdam composite	Cost of moving a barrel of crude: Single-voyage (spot)	Cost of moving a barrel of crude: Long term	Refining margin	Derived Persian Gulf price		General price level
	(1)	(2)	(3)	(4)	(5a)	(5b)	(6)
1967 (Jan.-May) . . .	2.05	0.44	0.49	0.33	1.28	1.23	111
1967 (July-Dec.). . .	3.24	1.95	0.37	0.33	0.96	2.54	115
1968	2.53	1.30	0.37	0.33	0.90	1.83	118
1969	2.04	1.00	0.44	0.33	0.71	1.27	124
1970 (Jan.-May) . . .	2.24	1.40	0.47	0.33	0.51	1.44	129
1970 (June-Dec.). . .	3.19	2.72	0.85	0.33	0.14	2.01	132

Sources by column:

(1), (4), (6)–Table VI-3; col. (4) assumed constant after May 1967.

(2), (3)–Table IV-2 and Appendix IV-B.

(5a), (5b)–Col. (1) less col. (4), less cols. (2) or (3), respectively.

(6)–GNP deflator, United States, calculated by U.S. Department of Commerce, Office of Business Economics, 1960 (Jul.-Dec.) = 100.

to a Persian Gulf f.o.b. of $1.28 using average spot rate, and of $1.23 using long-term charter rates for marginal Suez Canal ships. The true rate is somewhere between, nearer the long- than the short-term rate, but its precise location makes little difference because the market was in equilibrium.

Since then, the gap between long- and short-term rates has always been large and sometimes enormous, and f.o.b. prices are absurdly high or low using the one or the other extreme. A safe statement can be made for 1969: since the Rotterdam composite was about the same as January-May 1967, and since refining margins could not have changed appreciably in a short time, and since the relevant tanker rate was almost certainly above that of early 1967, *the 1969 Persian Gulf f.o.b. must have been somewhat lower. This is confirmed by the observations of f.o.b. crude prices being somewhat lower (see Appendix VI-I).* But nothing more precise can be said. Consecutive-voyage rates in 1969 for the period between one year and three years were in the range Worldscale 50-60,[49] or about 60 to 70 cents per barrel. This would suggest an f.o.b. value of between $1.00 and $1.10 per barrel, which is compatible with observations in the range of some 10 to 20 cents higher.

The downward price movement begun in 1957 carried all the way through Spring, 1970, with only occasional and temporary reversals.

[49] R. S. Platou, *Annual Report, 1969*, pp. 10-11.

PART THREE

SUMMARY AND CONCLUSIONS

CHAPTER VII

The World Oil Market Today and Tomorrow

The Three Price Levels

It is time to bring together the various strands of the analysis, to extract from past experience an idea of the future price of oil. But "price" in economic theory is not one but three.

1. The market price, explored in Chapters IV-VI, means the current rate of exchange of money against goods: in particular, what is being paid today for a large-volume stream over an appreciable time ahead.

2. The competitive supply price, analyzed and measured in Chapter II, indicates the tribute due to nature: the least that needs to be paid to bring forth the stream of output. This can only be found out by analyzing the outlay for an investment made at one point of time, and the rate of output it will generate.

3. Since the investment is only made in hope of profit, the decision to invest will only be taken if the expected revenues will exceed or at least equal the supply price. Somewhere in the calculations there must be a reference price,[1] which measures the expected benefit of the investment, as against its cost.

Under perfect competition, which is as unreal but also as useful as absolute zero in physics, these three prices become one. The price-production system is in complete equilibrium, and there is no incentive to increase or decrease the rate of output. The study of real markets is a study of how these prices diverge and of the forces generated thereby.

Human error makes them diverge all the time, as do changes in desires and in technology. Demand may be greater than expected, raising market price. If the change is expected to persist, the reference price moves up too, additional investment expands output, thereby raising supply price, pushing the three toward equality. Before they ever get there, some new shock is applied to the system, and so on as long as the industry exists.

The second basic reason for divergence is market control. A single-firm or group monopoly will not necessarily expand output even if reference price stands above supply price. It will make the additional calculation of incremental net revenues. Only if the reference price is sufficiently higher than supply price to offset the drop in market price will it expand output.

But the discrepancy is still there. The automatic valve, so to speak, remains opened. A manual control has been shut, which another hand can open. Unless the monopoly retains

[1] This has no connection with "tax-reference prices" in some of the producing countries.

its cohesion, and permits nobody to touch the valve but its own agent, it will be opened because it is profitable to exploit the price-cost gap.

This was the reason for studying the market structure in Chapter III. There the reader was warned that it was only the start of the inquiry, and that the exercise of control over price could only be understood by looking at prices, their level, structure, and evolution. But the job is incomplete until we look at the expected or reference prices that guide companies and governments in their investment decisions.

The Reasons for the 1957–1970 Price Decline

The Divergence of Market Price and Supply Price: A Factor of 10. In Chapter II the 1962–68 crude oil supply price (reckoning 20 percent return on investment) was estimated to be about 10 cents per barrel in the most expensive Persian Gulf area (Iran) whose output was needed for current or even much higher consumption rates. Costs in Libya were shown to be lower when account is taken of the transport differential, though smaller reservoirs might make the future marginal cost higher.

In order to estimate future costs, it was assumed that reserves in existing fields could be expanded at a much lower rate than has been feasible in the United States. Further assumptions were: no new discoveries and no improvements in technology.

The greater the output, the greater is the pressure upon reserves. Hence it was necessary to estimate cumulative consumption during 1969–85, in order to estimate 1985 costs.[2] Here again extreme assumptions were made: zero natural gas, zero nuclear power, and a rapid exit of non-U.S. coal. Thus one allowed for the most drastic possible reductions in the price of oil—not that I expect it—in order to get a maximum estimate of production, the strain on reserves, and therefore on long-run incremental cost. The concept of Maximum Economic Finding Cost (MEFC) made it unnecessary to estimate finding costs separately.

The result was an estimate of 20 cents per barrel—not the most likely value for cost in 1984 but, rather, a maximum value. Even major corrections cannot touch the conclusion that if the price of oil were set by competitive supply and demand, it would now, and for at least 15 years, be only a small fraction of its current value.

But one may have proved too much. If the price has been so much higher than the competitive level, why should it not go higher still, or at least stay where it is? A satellite can be launched into space to revolve almost indefinitely—not in defiance of gravity, but rather because a sufficient counterforce has been provided. What we must now analyze is the nature of this counterforce: the barriers to competition.

The Barriers to Competition: Cartel or Commodity Agreement? The market described in Chapter III was concentrated and integrated enough to permit some large departures from competition. But the market structure was only an enabling condition, not a sufficient cause for the wide gap between price and cost. The price history showed to what extent government policies overrode the interest of the eight-company group at some times, helped it at others. The historical question—to what extent does group action explain what happened?—is a forecasting problem. One thing seems well settled: the international

[2]On the basis of 1962 data, I estimated in "Oil Production Costs in Four Areas," in 1966 Proceedings of the Council on Economics, AIME, p. 56, that scarcity could not affect prices through 1980. With better knowledge—and the passage of time—the horizon has been pushed back another five years.

oil companies do not get together to fix prices. There have been a Federal Trade Commission Report[3] and three suits by the U.S. Department of Justice,[4] not to mention six congressional investigations. Whatever else they came up with, consultation on prices and output was something they did not find.

Because everybody knows this, it is elaborately discounted and dismissed. Even a hardened skeptic, insistent on facts and logic, will often collapse altogether in the face of: "Now, my dear fellow, let's not be naive! They don't actually need to get together. I mean–oligopoly and all that." But the mystery lies precisely in the "all that," much as in the "etc." in the title of Queen Elizabeth I, when all Europe wanted to know if she did or did not proclaim herself the head of the Church of England, and that great and prudent lady "et-ceterated herself."

As it happens, there was a cartel between the wars, and the companies did have to get together, repeatedly.[5] They started with an agreement on principles and a sound distaste for detailed regulation: settle strategy and let tactics be flexible. But with a product mixture as varied as petroleum, and with many product dimensions offering as many new channels for competition to find when it was dammed up in other channels, the companies found themselves writing more and more detailed specifications. And therefore, in a postwar period even longer and more turbulent,[6] the impossibility of getting together has meant at the very least a substantial limit on their power to act. "Oligopoly" was a fact in world oil in 1947, and it is today, but the degree and effects are far less.

The postwar evolution of crude oil prices, particularly up to the end of 1949, has been analyzed in Chapter V. The FTC Report's repetitive rhythmic beat of "identical delivered price, regardless of point of origin" was meant to lull the critical sense of the judge in a great antitrust suit that never came off. There are few more definite signs of a block to competition than delivered prices *not* the same, regardless of point of origin. Putting aside the subliminal persuader and looking only at the facts, it is instructive to read them in the light of a complaint made some eight years later by Sheikh Abdullah Tariki.[7] In his view, the oil companies should have equalized the price at London,[8] perhaps just barely undercutting Venezuelan oil, instead of equating with it at New York. If the Middle East producers had been able, they would have equated at London and increased profit. Like all good things, this would have taken some self-denial. Since the net return on shipments to London was much higher than on shipments to New York, the normal competitive response would have been for each individual seller to channel as much output as possible toward London, and to keep doing so as long as any such differential remained, driving it toward zero. Each company had to resist this pressure, and equate the price to a London sale in the confident expectation that everybody else would: a united front and refusal to

[3]FTC, *The International Petroleum Cartel, Staff Report to the Federal Trade Commission submitted to the Subcommittee on Monopoly of the Select Committee on Small Business*, U.S. Senate, 82d Cong. 2d sess. (Committee Print No. 6, 1952), hereinafter referred to as FTC Report.

[4]*U.S. v. Standard Oil Company (New Jersey), et al.*, D.C. Civil Action No. 1779-53, hereinafter referred to as World Oil Complaint; and D.C. S.D. N.Y. *Final Judgment*, Nov. 14, 1960; hereinafter referred to as World Oil Consent Decree. *U.S. v. Standard Oil Co. of Calif., et al.*, 155 F. Supp. 121 and 210 F. 2d 50 (2d Circ. 1959); hereinafter referred to as MSA Case. *U.S. v. Arkansas Fuel Corp., et al.*, CCH Par. 69.619 (N.D. Okla., 1960); hereinafter referred to as Tulsa case.

[5]FTC Report, chaps. VIII and IX, pp. 275-79, 282-86, 288-304, 311-48.

[6]From the Achnacarry agreement to war was 11 years; World War II has been over for more than 25.

[7]Sheikh Abdullah Tariki, "The Pricing of Crude Oil and Refined Products," *Proceedings, Second Arab Petroleum Congress* (Beirut, 1960), pp. 13, 15, 20. It should be compared with the FTC Report, pp. 366-67.

[8]Chap. V, table V-2, col. II.

supply at any lower price. The structure that Mr. Tariki wanted was that of a more effective Middle Eastern monopoly. Instead, there was literally a revolution in the price structure: up changed to down. Gulf-plus became a single f.o.b. price that just equated with the netback in the most distant market—the U.S. East Coast.

The fact that prices evolved as they would under competition in 1947–49 only sharpens the question of why the price evolution ceased and was in part reversed in 1950–57. Chapter V then explored the failure to import into the United States on a scale that would have been profitable for each individual producer. This destroyed the hypothesis of competition. Considered next was the most simple theory of group action, then a more complex monopoly theory. Both kinds of agreement could have been reached simply by everybody waiting for somebody else to start something, and hoping he wouldn't. Yet both theories had to be rejected. The group as a whole would have been money ahead even at much lower U.S. prices and output. A wholly independent support for this conclusion is the 11-year escalating struggle over imports in the United States, which shows that the companies were never free to import. Even a cartel administered by government could not restrain imports—control had to be compulsory. In Europe, coal protectionism was an obvious fact of life, and during the mid-1950s restraint on output in the face of increasing coal scarcity helped push up the price of heavy fuel oil.

The companies were few. Because of sales contracts, joint ventures, and integration into final consumer markets, every company knew every other company's development plans well enough to know that nobody else was planning to reduce prices in order to take sales away from any other. But they also needed the restrictions on imports into consuming countries as additional assurance against individualistic conduct: the less to gain by chiseling—shading the price—the less the chance of it happening.

The 1953 and 1957 price increases in the United States gave a signal they gladly followed. Otherwise, the eight-company group was able, without any known consultation, to stand fast, try to gain business by non-price methods, and leave the price alone.

To do nothing was indeed to do much, and only the Middle East-Caribbean oligopoly permitted it. But by rejecting a stable "oligopoly" as a sufficient explanation of changing prices, and by focusing on what actually happened and the need to explain it, we rid ourselves of the notion that the companies did or can in the future do any more. To be sure, since 1960 they have been permitted, by special leave of the U.S. Department of Justice, to get together to fix prices, divide markets, and limit output.[9] But this, however useful a tool, is not a shield.

If forced to summarize, one might defend the following. Since World War II, in effect if not in form or intent, there has been and still exists an informal but effective commodity agreement in crude oil. The governments in the consuming countries restrained competition at home, and the international oil companies refrained from competing on price. Many would say that the commodity agreement worked very well. Large amounts of oil were forthcoming at a cost far below that of other sources of energy. The high price provided an incentive to establish reserves for a long period ahead. Be that as it may, we live with the consequences.

[9] World Oil Consent Decree, sec. V, subsecs. (A) and (C). The original World Oil Complaint charged both restricting imports *and* excessive dumping into the United States. See IV-8-b and V-11-(2), (3), and (4).

Some may find it odd that antitrust proceedings should have this result, but the guiding genius of the law has always been Mr. Facing-Both-Ways. Others will maintain with some heat that the consent decree could not possibly permit any such thing; they should read it.

New Elements in the Market. The problem is why the price line could not be held after Suez 1956-57 as it had earlier held and even advanced.

Chapter III showed the slow change in the market structure. Concentration of output decreased, the number of firm-equivalents rising from 5 to 10. Yet this would in itself probably overstate the degree of change, since the share of the great Persian Gulf joint ventures stayed about the same. (See Chapter III, Tables III-1 and III-2.) The slow growth of an arm's-length crude oil market permitted increased rivalry in the sale of crude oil, and by giving nonintegrated refiners a source of supply permitted them to compete at the product level.

In 1957, market crude (i.e. sold at spot or contract by a producer to a refiner) was about one million barrels daily, of which the eight large companies must have furnished nearly all. By 1968, market crude was perhaps 4 million barrels daily, of which newcomers furnished perhaps one million. This would be only 7 percent of the world market, as has been reckoned here. Perhaps, therefore, the newcomers really had little to do with the price decline after 1957, since they had so little of the total world productive capacity and markets. The old internationals were already cutting prices before the impact of the newcomers was felt. There was price cutting in sales to Japan in 1955, which involved continuing relations and deliveries: these are more important than a one-shot public bid. The rise of Caltex' joint venture (Nippon Oil) in the Japanese market seems to have involved price cutting. After Suez, British Petroleum (BP) and Compagnie Française des Pétroles (CFP) especially were believed to be leaders in price reductions, but the majors "started it."[10]

In fact, the data in Chapter VI seem to indicate that the rate of decline was steepest in the late 1950s, before newcomers had any noticeable amounts to sell. The period of most rapid price decline seems to have been after the Suez reopening in early 1957. But then after the lowered postings in mid-1960, prices stabilized with even some tendency to rise through early 1963, after which they declined again through mid-1965. This second decline coincided with the first big Libyan exports. Yet gasoline, which would be particularly important for the light Libyan crude, had clearly been declining since the second quarter of 1961. Heavy fuel oil had been stable or decreasing and, like light fuel oil, it started to drop in March 1963. To the extent that Libyan feedstocks were replacing Middle Eastern, one would rather expect heavy fuel to hold or strengthen.

After the first half of 1965, strength was all in the light products and weakness in heavy fuel. By the end of 1968, 18 months after the closing of the Suez Canal, heavy fuel oil had reverted to the previous May 1966 low. The data can be reconciled with Libyan exports having some substantial effect in 1963, though even this is not overwhelmingly probable. But from the middle of 1964, no independent influence can be ascribed to them.

The Libyan tax increase in January 1966 was widely hailed as a force to raise prices.[11] But by April 1967, the expectation was that "the new Libyan tax system [will] clobber the independent price cutting refiners *in a couple of years' time.*"[12] By June 1967, "prices . . . had fallen somewhat farther than even pessimistic forecasts would have suggested."[13]

[10] *Petroleum Press Service*, vol. 26 (1959), p. 87; *Oil and Gas Journal*, Aug. 15, 1960, p. 83.

[11] The writer ventured to disagree. See *Platt's Oilgram Price Service*, May 27, 1966.

[12] *The Economist*, Apr. 27, 1967, p. 390. Emphasis added.

[13] M. E. Spaght (president Royal Dutch/Shell), presentation to Financial Analysts Federation, Boston, May 6, 1968 (multilithed), p. 10.

We may never know whether old-timers or newcomers were more important, partly because some facts will never become public and partly because the two are interdependent. Had the old internationals been able to arrive at a firm understanding, they could have followed the most profitable road: disregard the price cutters, who had little crude. Thereby they would have sacrificed only a little sales volume and nothing at all in prices. But they were not free to do this. Forced occasionally to follow their particular rather than the general interest, they have been occasional competitors rather than constant statesmen, and have met price cuts at least part way at crude and product levels.

The Newcomers in Production. It was (and is) to be expected that in the longer run the newcomers would account for a larger share, as Chapter III showed. It is an axiom in industry economics that the larger the market, the more room there is for competition and the smaller is the relative size of a firm large enough to survive. Moreover, the larger the market, the larger and more available is the complex of independent service and supply companies. It was clear by the mid-fifties in Venezuela and the Middle East that relatively small companies could find and sell oil profitably. Governments therefore have made a point of promoting competition among bidders for leases, getting as many independent firms as possible to explore. Today, a single lease covering the whole country is unthinkable. Libya, the greatest of the new countries, had welcomed 28 concessionaires by 1968, breaking all precedent in speed of growth. The lesson has not been lost on other countries: the more the exploring companies, the faster the development. Reasonable men may disagree on whether oil was ever a "natural oligopoly," but it clearly is not one today.

The Arabian Oil Company (AOC) was formed by a group of Japanese refiners, electric power companies, and other large buyers. It was not "access" or "assured supply" they were after. They could have signed up all the oil they wanted. Plainly, they considered even discounted prices to be far above the cost of finding-plus-development-plus-production, even for newcomers who cannot do it as cheaply as the old concerns. They had no producing experience and conducted a pure checkbook operation. With $10 million, a rather small sum in world oil, they hired the talent.[14] Once oil was found, sums for development came in with little difficulty.

The most spectacular success story is that of Occidental Petroleum, a once-small "second generation" Libya company who signed a concession agreement in March 1966, made the first discovery in December, and 11 months later was estimated by DeGolyer and MacNaughton to have reserves of 3 billion barrels on the basis of 20 percent development—with only a small part of the concession tested thus far.[15] In the second half of 1969 they produced at 650 thousand barrels daily (TBD).[16] As with AOC, Occidental's progress was fueled by the hope of future gain, not by past profits. The manpower and facilities which did so much so quickly were available for hire. The entry barrier of risk is still great. The barrier of money has dwindled, and so has the insider's know-how.

[14] *Petroleum Press Service*, vol. 28 (1961), p. 224; *Oil and Gas International*, vol. 2 (Jan. 1962), p.29.

[15] Occidental Prospectus, June 1968.

[16] At a 3 percent decline rate, 650 TBD initially amounts to 4.2 billion barrels in 25 years, which is as far ahead as one need look. Probably the decline rate is lower than 3 percent at 650 TBD, *or* more can be produced with no greater a decline; either way the cumulative expected output, i.e., reserves, seem much larger than the estimate published. The two figures are not inconsistent, however, since a year of intensive development intervenes.

A worthwhile study would be the role of the contracting firm. The offshore drilling industry is perhaps the biggest single part of the complex; in mid-1968, its trade journal identified 320 mobile units, tenders, and self-contained platform rigs. Oil companies, including the Soviet Union and other national companies, owned 85; the other three-fourths were available for hire.[17] The analogy with oil tankers is striking and, in my opinion, accurate. As for geophysical work, contractors do about 60 percent in the United States, but 90–95 percent in the Middle East and about 80 percent in Africa.[18]

The Soviet Union. The Soviet oil trust is a large, old-established company which considers itself entitled to the share (19 percent) of world trade it had in 1930–33.[19] The Russian incursion into world markets has been well and extensively covered.[20] There is no evidence that the Russians sell at prices unrelated to costs, or are dumping in order to disrupt world markets for political ends. Nothing is so easy and cheap as to accuse the price-cutter of price "warfare," selling below cost in order to drive all others out of business, leaving him in a position to exploit without mercy. To the student of industry economics, it is the most dreary familiar complaint to be heard at all times in all countries. Soviet prices have been set low enough to make the sale, and no lower. The supposed shocking low prices were only moderately below capitalist sellers' real arm's-length prices. But they soon entered a very inelastic range of their particular demand curve. Since the Russian market in the non-Soviet world was (and is) a politically limited share of the government and independent refiners, they could not, past a certain point, increase their sales by further price-cutting. Hence, if profit were their aim they would not cut further, nor did they. Their attempts, starting late in 1961, to raise prices—and their occasional losses of contracts—are only consistent with a seller trying to make the most of his every bargain and now and then overreaching himself. As early as 1962, their export company was repeatedly underbid. In 1963 they complained that the international companies were practicing "artificial competition" and meanly underselling. When they offered to take British ships for Soviet oil, whose importation is forbidden by Labor and Conservative governments alike, they warned that the deal was not to be sweetened by "cut prices" for the oil. Later they said "cutthroat" prices by U.S. oil companies were damaging all sellers.[21]

Prices to members of the Soviet bloc are much higher than prices to buyers outside, even after allowing for the other provisions of the bilateral trade deal of which the oil shipment is part.[22] A rational monopolist meets competitive offers when and as he must, as may be seen in Figure V-1. These high prices do not "subsidize" or make possible the

[17]*Offshore*, vol. 28 (July 1968), pp. 75-98.

[18]See Annual Reports of the Committee on Geophysical Activity of the Society of Exploration Geophysicists, published annually in the December issue of *Geophysics*.

[19]*Oil and Gas Journal*, Oct. 31, 1960; and *Petroleum Press Service*, vol. 27 (1960), p. 406.

[20]Robert E. Ebel, *The Petroleum Industry of the Soviet Union* (American Petroleum Institute, 1962); National Petroleum Council, *Impact of Oil Exports from the Soviet Bloc* (Washington, 1962), and *Supplement* (1964). Robert W. Campbell, *The Economics of Soviet Oil and Gas* (Johns Hopkins University Press for Resources for the Future, 1968). *Petroleum Press Service* has given the matter good, unsensational coverage.

[21]*Platt's Oilgram News Service*, Feb. 15, 28; Mar. 7, 11, 1963; *Platt's Oilgram Price Service*, Feb. 12, 1963.

[22]Horst Mendershausen, "The Terms of Soviet Trade," *Review of Economics and Statistics*, vol. 42 (1960), pp. 152–63; and discussion by Franklyn D. Holtzman. See also Campbell, *The Economics of Soviet . . .*, pp. 233–34.

low prices to non-Soviet buyers. The sales to non-bloc buyers have been profitable, and the sales to Soviet-bloc buyers even more profitable.

However, we know all too little of the costs of Soviet oil, which cannot be understood except as part of the whole finding and developing process. The Russians were late in developing petroleum, and paid for their neglect.[23]

They found a disproportionate amount of gas, and it pushed corresponding amounts of oil (much cheaper to ship) out of the producing region, precisely as had occurred in the American Southwest after World War II.[24] Moreover, gas transmission is a voracious consumer of steel line pipe, which the Russians could not produce in sufficient amounts. To get the pipe they needed foreign exchange and therefore exported oil—acting entirely in accord with comparative advantage, however much or little they were aware of it.

At prices ruling in the 1960s, oil finding, development, production, and export were a rational, profitable activity.[25] But we know very little about the costs of their oil and gas. The estimates of 64–68 cents per barrel and 2.5 cents per thousand cubic feet are not explained and hence not acceptable.[26] Professor Campbell estimates average lifting costs in 1963 of about 40 cents per barrel, ranging from a low of 19 cents.[27] He warns against using these figures, which neglect an interest charge, as a basis for investment planning. At another point (p. 78), he draws attention to the fact that the return to the investment in exploration and development drilling seemed to be diminishing after 1951–55. This was a shrewd hit. The incremental cost of Soviet oil must have climbed rapidly in the late 1960s, and have become better recognized; but full recognition lagged because of the absence of an interest rate in their cost calculations. By mid-1968, Russian difficulties in supplying export oil were well known. By the end of 1969, the Soviet Union was urging the European Communist countries to find new sources of oil in Africa and the Middle East. Russian aid policy was said to exacerbate the problem because technicians and oil equipment were being sent to countries such as Egypt, Iraq, and Syria. "They really can't afford it," one Western oilman said.[28]

Professor Campbell noted in May 1970 that the emphasis on developing Western Siberian oil and gas resources, under very unfavorable cost conditions, "looks like a desperate gamble rather than sensible planning. It seems significant that the initiative for this approach has come almost exclusively from the Party, and the oil industry people are pointedly unenthusiastic about it."[29] In any case, both economic and political interests of the Soviet Union are best served by high prices, which they can undercut slightly. Thereby they profit as exporters, and gain the good will of both producing and importing countries—whose only standard of comparison is in what the Western countries are offering.

[23] Campbell, *The Economics of Soviet* . . . , pp. 8–13; Arthur W. Wright, "Soviet Investment Planning in the Fuel Industries" (Ph.D. thesis, M.I.T., 1969).

[24] M. A. Adelman, *The Supply and Price of Natural Gas* (Oxford: Blackwell, 1962), pp. 94–110.

[25] Cf. Campbell, *The Economics of Soviet* . . . . , p. 239, and Adelman, *The Supply and Price* . . . , p. 100.

[26] National Petroleum Council, *Impact of Oil* . . . (1962), vol. I, p. 83.

[27] Campbell, *The Economics of Soviet* . . . , pp. 136–40.

[28] *Petroleum Intelligence Weekly*, June 10, 1968; *New York Times*, Nov. 24, 1969, p. 6; Dec. 4, 1969, p. 4; *Oil and Gas Journal*, Dec. 29, p. 90; *Petroleum Intelligence Weekly*, Dec. 1, 1969.

[29] Testimony of Robert W. Campbell, *Hearings before the Subcommittee on Strategic Arms Limitation Talks of the Committee on Armed Services*, Part 2, U.S. Senate, 91st Cong., 2d sess. (1970), p. 73.

The Independent Refiners. Some of the independent refiners are private, like Petrofina, which in 1954-60 tripled its sales and its crude deficit alike.[30] In 1967, Petrofina made a new 10-year agreement with BP, at the same time as Esso and Mobil.[31] On the basis of data in Chapter VI, I would expect that Esso and Mobil paid not more than $1.15 and Petrofina not more than $1.25 for 34° Iranian, and about $1.10 and $1.20 respectively for 31°.

Some independent refiners are government enterprises, and some quasi-government, notably Italy's Ente Nazionale Idrocarburi (ENI). In 1962, 68 percent of Soviet crude and 21 percent of product exports went to governments; private nonintegrated refiners took 32 and 56 percent; "others," zero and 29 percent.[32] This is an indication of where any independent oil producer would also hope to sell. What all these refiners had in common was the ability to do arm's-length bargaining. They bought crude oil at market prices and transported oil at market rates, but sold finished products at prices near those of integrated companies which reflected posted prices and AFRA rates. The results varied with the response of the older internationals. Where it was too sluggish, the independent refiners were able to profit and expand market shares. Eventually they brought product prices down, not because they wanted to, and still less because they were ignorant or poorly financed, but because they looked to price, not incremental net revenue. Having small market shares, they did not fear to spoil a market which was not theirs to spoil.

In Chapter VI special attention was paid to fuel oil. Where price is not much higher than incremental cost, prices should rise in the face of a brisk demand. But where price is much higher, the expansion of the market puts the price under the greatest downward strain. *Nouvelles Réflexions*[33] pointed out how the more rapidly expanding EEC markets were the first to "translate the declining tendencies of the world price level of petroleum products." An earlier report had neatly summed up conflicting economic and political pressures:

> The large oil companies are understandably reluctant to make cuts in prices which, if they tended to push up oil sales, would almost certainly lead to demands for further protection for the coal industry. On the other hand, depending on the degree of competition from "independent" suppliers, they may have to concede reductions to some consumers simply to avoid the latter taking cheaper supplies from elsewhere.

Because none of the established concerns can be sure of what the others are doing, and because rewards for success in getting and keeping the market share—and penalties for losing it—are the greatest, it is hardest to maintain a given price. Furthermore, since fuel oil distribution, unlike gasoline, was relatively easy to enter, independents were soon playing a significant if minor part. And their uncontrolled capacity was soon playing its traditional role. Hence the paradox that the rise in the price of coal in the mid-50s, increasing the competitive advantage of oil, caused the price of oil not to firm but to collapse.

Gasoline was for a long time a safer product, but the rising demand for fuel oil brought a great deal of gasoline into the market (which we might well remember the next time we

[30] *Oil and Gas International*, Mar. 1961, pp. 18-21.

[31] *Petroleum Intelligence Weekly*, Apr. 24, 1967.

[32] National Petroleum Council, *Impact of Oil... Supplement* (1964), p. 32.

[33] European Coal and Steel Community, *Nouvelles Réflexions sur les Perspectives Energétiques* (Luxembourg, 1966), pp. 14, 60. Council of Europe, Consultative Assembly, *European Energy Problems*, Doc. 1463 (Strasbourg, 1962), pp. 70-71.

hear fuel oil called "a mere by-product"), at the same time that independent refiner-marketers were also coming in.[34] A decade of experience received a classic summing up:

> Gasoline has over a period of years been available at European refineries at prices below the British wholesale level. It was therefore possible for owners of filling stations to buy and import gasoline into the UK and to undercut the established level of retail prices without any loss of margin. In this way the leading suppliers became vulnerable to competition from dealers turned wholesalers or from new wholesalers building up their own retail chains. *They persisted for too long in regarding this as a temporary phenomenon of distress for marginal suppliers, even while it continued through a decade in which the refinery capacity of Western Europe was doubled.*[35]

The new wholesalers also suggested to dealers that by taking a lower margin they could increase sales and profits. British retailers were at first incredulous. "But they did learn–as may be seen from the growth in the number of filling stations in the hands of newcomers and the decline in those controlled by the established suppliers."[36] In March 1967 Esso began cutting prices at three points and by mid-1968 prices were firm at the lower level.

Summary. The 1957-70 history may be summed up: The modest growth of oil sales outside the vertically integrated companies, by producers with a small share of total sales (integrated plus nonintegrated), reacted back upon those companies. Some independence in buying and selling mobilized, though it did not cause, some of the excess of potential capacity in the international oil trade and made it affect the price. Had the U.S. crude oil market been open to substantial imports around 1950, large refiners independent of Middle East and Venezuelan production would have been able to shop around for the lowest possible prices, and to offer large and profitable outlets. The price would have come down then, earlier and faster. To the extent that such outlets open up here and there in the world, it becomes more difficult to support the current price level.

The Crude Oil Reference Price

The Crude Oil Reference Price and Continued Exploration. Naturally, there is little publicly available data on prices expected in the long run. Many have reason to be silent. Others consider the matter as so obvious as to need no discussion at all. Still others express themselves in a rather misleading way. The consensus may be less strong and confident than in 1962, but it still exists. Yet one rarely sees the rationale or explanation. A document from the European Economic Community Commission is an exception in stating a coherent theory: There is a risk of oil resources getting more scarce, as oil companies shrink from the growing burden of exploration.[37] It is not clear whether the authors hold this view themselves, but there seems no reason to doubt that the EEC Commission and Council of Ministers do.

The consequence of this belief is greater exploration-development investment, therefore even greater abundance of oil and gas resources and even lower supply prices. Although in Chapter II I disclaimed the need or ability to reckon finding costs and

[34] *Petroleum Press Service*, vol. 28 (1961) p. 163; *Oil and Gas International*, Nov. 1961, p. 38; July 1962, p. 89.

[35] *Petroleum Press Service*, June 1968, p. 209. Emphasis in direct quotation has been added.

[36] Ibid.

[37] M. Albert, G. Brondel, G. Ascari, "Situation Financière des Sociétés Pétrolières dans la Communauté Européenne," EEC No. 1948/II/68-F.

assumed new discoveries at zero, the exploration effort is worth a glance to indicate how radical was this understatement.

Exploration Effort has not tended to decrease Eastern Hemisphere geological-geophysical work;[38] it touched a new high in 1968 and was half again as large as it had been 10 years earlier. During 1962–67, directly estimated expenditures on geophysical work seemed to increase; crew months were stable or possibly slowly increasing. Estimation is extremely difficult, however, and subject to substantial error.[39] Let us refrain from guessing what this continued activity means for finding new fields. The effect of chance is too great. Furthermore, the shift to marine exploration may compound the constant improvements in efficiency: "Fast marine operations, where 24-hour days are often the rule, will produce 1,000 line miles and more a month. [Compare this] to a land crew where 50 miles is a good month's work. . . ."[40] There is no question that offshore exploration is in the midst of a constant and growing boom, because it is generally so much more efficient. It also promises lower development costs. First, wells are close to tidewater ports. The per-mile transport cost may be horrendous, but the total per-barrel cost is likely to be small. Second, as L. G. Weeks has pointed out, offshore sediments are younger, softer, and more quickly drilled.[41]

As in the past, large amounts of money will be spent by public and private companies to no purpose but, as in the past, there will be more fields and great reserve potential. As explained in Chapter II, a company that has already more reserves than it can hope to produce may be better off with still more crude to lower costs. BP, the most crude-rich company, was well advised to search in Abu Dhabi and Libya, for by taking part of the load off Kuwait it slowed down the process of increasing cost through lower pressure and more water cut. As was seen in Chapter II, Esso profited by the Libya search and would have done so even if there had been no Canal closure in 1967; the higher cost of the crude was more than overborne by better quality and transport cost savings. Diversification of supplies, as noted earlier, is a reduction of political risk. A given amount of capacity in a second country is preferable to the same amount of capacity under identical physical cost in the first country, for it allows a lower risk premium in both countries, hence lowers the cost of capital, r, and cost per present-barrel-equivalent (PBE).

We often hear that the continuing search for oil "will tend to raise costs." Taken one way, this is the precise contrary of the truth. As new field competition slows the rate of growth in the old fields, the strain on old-field resources is lessened, and costs are held down. But the argument that exploration raises the cost of oil involves a hidden assumption: that somehow the price will be kept high enough to keep many new, less efficient fields in operation as well as the old low-cost ones. Then of course the arithmetic average of oil costs actually incurred may be higher. It is an indirect way of saying that the reference price will somehow be kept as high as the current market price, or higher.

Crude-rich companies may also search for crude which will not save costs, if they cannot otherwise get into a protected market. The best examples of exploration at high cost are in the United States, Canada, and Australia, where the higher price is supported by the government.

[38] Estimated by the Chase Manhattan Bank and unfortunately also including lease rentals.

[39] Annual Reports of the Committee on Geophysical Activity of the Society of Exploration Geophysicists, published annually in the December issue of *Geophysics*.

[40] Ibid., 1966.

[41] See Appendix II–E, p. 297.

But for old-timers and newcomers alike, an additional reason for continuing to spend large sums on discovery is a reference price much in excess of cost. Or, what comes to the same thing, the price is so far above cost in the best deposits that even a mediocre field may soon pay out.

Some widely accepted reasons for oil exploration are either mistaken or else distorted reflections of the real, and hence are worth a brief glance. A crude-deficit company does not hunt for its own oil because vertical integration is "normal" and "natural." This is like explaining historical events by "national character," which someone called the last refuge of the baffled historian. A company searches when there is a good enough chance that (1) the total of finding costs in various places plus (2) the development-operating cost in the one or few places where it is found will be considerably less than (3) the price he must pay some other producer for the oil.

This principle is usually misstated by saying that the producer "needs" high profits in one area to enable him to search in others. But past profits have nothing to do with present expenditures for future production. Where money can be made, money will be found; but where there is none to be made, one can stuff a firm with profits but one can't make them explore.[42]

The Tensions of Price Maintenance: Output Shares. Whenever price is kept far above cost, there is not only a general surplus; perhaps even more important, the market-sharing mechanism has been suppressed. Were price in the neighborhood of incremental cost, the operator's decision would be simple in principle though complex in practice: carry output everywhere to the point where higher production rates would not pay. But the current price level is an incentive for every operator to produce and sell far more than he now does. He must restrain his offerings in order not to spoil the market. But the incentive remains, repressed but felt. The urge for additional sales and profits puts the market under a severe continuing strain. In order to reconcile the objectives of more sales volume and no price cutting, producers give away some of the excess of price over cost by special deals for some customers. They may operate marketing-refining facilities at little or no additional profit, which is a disguised and (if possible) localized price cut. A variant is to tie up long-term outlets by loans or other inducements that equate to lower net prices. Or one may try to gain business at the final customer level by non-price rivalry. All this drive for more volume is correctly cited as proof of "how competitive" is the industry, just as one might point out that a cup is one-third empty or two-thirds full.

As more and more companies appear, the circle of hands becomes wider and looser, and breaks more easily. Small producers can do increasing damage by shading prices, because others may follow suit. The companies' strength always lay in their being slow to react. But they cannot let markers be taken away from them, and in the round of chiseling, retaliation, and further chiseling the whole price structure threatens to crumple.

But, as shown in Chapter V, a threat to a price level calls forth political action to defend it. As opposing forces escalate, prediction of the final outcome becomes more difficult. Let us now examine the effect of the producing-country governments on the price level—as it has been and as it promises to be.

[42]The only qualification is that current expensing of exploration work and of dry holes makes their after-tax cost so low as to bias management decisions in the direction of exploration. But the limited force of the qualification is to be seen in the drop-off in U.S. exploration as prospects worsened after the early 1950s. (M. A. Adelman, "Trends in the Cost of Finding and Developing Oil and Gas Reserves 1946–1966," in S. Gardner and S. Hanke, eds., *Essays in Petroleum Economics* (Colorado School of Mines, 1967).)

The Producing Countries and Oil Prices

Posted Prices and Per-Barrel Payments: The Floor to Prices. As pointed out in Chapter I, it is axiomatic that a good discovery means a dissatisfied landlord who wishes he had held out for more. He will look around for any possible way of getting more. Once this basic relation is understood, a great deal of history can be compressed into a page.

The original concession agreements in Venezuela and the Persian Gulf provided for a fixed money payment per barrel. Later, in Venezuela the producing companies agreed to pay enough in addition to effect an equal or 50 : 50 division of profits. This change was soon applied to the Middle East; the refusal to apply it in Iran led to the 1951 seizure, which was not restored until 1954.

Thus by the early 1950s the tax was on profits, but much of the burden was shared with home governments. For the payments were considered as income taxes, and hence were deductible dollar for dollar from income tax to the home country.

A tax on profits meant that the companies had to account for receipts and outlays. In order to put a value on oil exports, they needed to set prices, and to make them public so that each government and each joint venture would know what the others did. The self-interest of the companies required them to post a price no higher than what they actually realized from its sale as crude or ultimately as products. Hence, as explained in Chapters V and VI, for a short time there were publicly known prices whose level and movements were in the neighborhood of real transaction prices: good indicators of what they expected to realize from crude oil in the near future. They were, in short, a reference price—which explains to some extent why some still regard them as such.

During the late 1950s government revenues rose greatly, both in total and per barrel. But the per-barrel figure for any one year is far from precise. Payments in any one year may lag production or contain nonrecurring items.[43] Even Table VII-1, which is confined to Saudi Arabia and Iran and shows companies separately, has some irrational per-barrel movements. However, the salient fact is that Aramco per-barrel payment went from 17 cents in 1946 to 80 cents in 1956-57 (one should probably disregard 1955). Price reductions, both posted and third-party—for Aramco sales to third parties were until 1966 entered at real not posted prices—account for the decline through 1960.

The governments' resentment over the 1958-60 price reductions is a time-worn tale. It led in 1960 to the formation of the Organization of the Petroleum Exporting Countries (OPEC). The effectiveness of this body cannot be appraised, for one does not know what would have happened had it never existed. Moreover, very little OPEC activity is public. But publication and statements have been marked by sober good sense and devotion to the one big objective of increasing the governments' revenues.

Whoever gets the credit, 1960 marks a long step away from income taxes and back to per-barrel payments, which governments have gradually raised. The negotiations of 1962-63 changed the treatment of royalties: instead of being treated as income tax, they were treated as a deduction from revenues.[44]

[43] The data published by the United Nations, and the unpublished UN data used by Issawi and Yegeneh (which the UN did not make available to this writer) are obviously in agreement as to trend, but differ greatly and erratically from table VII-1. (See sources beneath table.)

[44] For example, if receipts over and above costs (however defined) were 100, and royalties 14, companies had previously been liable for another 36 in order to make the total tax 50. Under the new system, gross profits were reckoned at 86 (=100 - 14), tax due 43. Hence total payments to governments became 57 (=43 + 14). The companies received some offsetting concessions because certain discounts (much smaller than what actually prevailed) were to be deducted from receipts.

TABLE VII-1. Payments per Barrel: Saudi Arabia and Iran, Selected Years

(*cents*)

Year	Aramco	Getty	AOC	Iran Consortium
1946	17	...	...	...
1947	17	...	...	...
1948	n.a.	...	...	...
1949	n.a.	...	...	...
1950	28	...	...	...
1951	28	...	...	...
1952	n.a.	...	...	...
1953	n.a.	...	...	...
1954	67	50	...	...
1955	96	59	...	...
1956	80	59	...	...
1957	80	84	...	...
1958	77	69	...	...
1959	74	72	...	...
1960	69	74	...	...
1961	70	80	67	
1962	69	74	27	74
1963	70[a]	70	56	80
1964	77	69	54	82
1965	78[a]	71	62	83
1966	83[a]	68	48	83
1967	88[a]	71	63	84
1968	84	59	62	87
1969	82	67	63	86
1970	89	n.a.	n.a.	82

Sources: Annual Reports of Saudi Arabian Monetary Agency; Annual Reports of Iran Consortium.

Note: Sometimes (e.g., AOC 1965–66) payments are overstated for one year, understated for another. The missing years cannot be supplied from UN, *Economic Development in the Middle East* because for those years where I have both series, the discrepancies are very large and are not explained. Charles Issawi and Mohammed Yeganeh, *The Economics of Middle East Oil* (Praeger, 1962), appendix table 6, also utilize the UN series.

[a]Excluding nonrecurring payments.

n.a. = not available.

TABLE VII-2. Payments per Barrel, Seven Countries, 1957–1970

(*cents, U.S.*)

Year	Kuwait[a]	Saudi Arabia[a]	Iran	Iraq	Others	Total P. Gulf	Libya	Venezuela
1957	79.6	88.2	86.8	93.1	91.3	85.7	...	103.0
1958	81.7	81.7	89.0	88.9	92.4	84.8	...	111.6
1959	77.8	75.8	83.6	82.4	89.1	79.8	...	98.4
1960	76.4	75.0	80.1	78.6	88.7	77.7	...	89.2
1961	74.4	75.5	75.8	76.5	87.2	75.8	62.7	92.9
1962	74.8	76.5	74.5	76.7	83.4	75.9	64.7	97.2
1963	74.3	78.7	79.7	80.7	79.2	77.9	65.0	98.6
1964	76.9	82.0	81.8	80.1	58.4	78.4	62.9	95.4
1965	78.9	83.2	82.9	81.7	57.7	79.5	83.8	95.6
1966	78.4	83.4	83.2	81.3	81.0	81.1	87.0	95.8
1967	79.3	84.8	82.5	85.2	81.5	82.6	101.6	102.2
1968	80.5	87.8	83.7	90.7	84.1	85.0	100.7	101.4
1969	80.8	87.1	80.9	91.4	87.0	84.3	100.0	103.6
1970	82.9	88.3	80.8	94.2	88.7	85.8	109.0	109.2

Source: *Petroleum Press Service*, Sept. 1971, p. 327.

[a]Including half Neutral Zone.

In 1966-67 another round of negotiation provided that the allowable discounts be phased out through 1973. The expectation was that this would add about 7 (± 2) cents to the tax due per barrel. In round numbers, therefore, payments to governments in Iran and Saudi Arabia and the other big Persian Gulf concessions would as a consequence have averaged around 90 cents per barrel. Hence the floor to price, which is tax plus cost, would be near or above $1.00 per barrel.

Table VII-3 shows the method of calculation, into which the reader may fit any given posted price and nominal tax rate on the fictitious profits.

TABLE VII-3. Tax on Two Arabian Crudes Under 1967 and 1971 Conditions

(dollars)

	Arabian light 34°		Arabian heavy 27°	
	1967	1971	1967	1971
1. Posted price	1.800	1.800	1.470	1.470
2. *less* marketing allowance	0.005	...	0.005	...
3. discount (.065)	0.117	...	0.096	...
4. gravity allowance	0.019	...	...	...
5. cost[a]	0.100	0.100	0.100	0.100
6. royalty (.125)	0.225	0.225	0.184	0.184
7. Pro-forma profit (line 1 *less* sum of lines 2 through 6)	1.334	1.475	1.085	1.186
8. Pro-forma income tax	0.667	0.737	0.542	0.593
9. *Payment due* (lines 6 plus 8)	0.892	0.972	0.726	0.777
10. Add back: cost (line 5)	0.100	0.100	0.100	0.100
11. *Tax-plus-cost*	0.992	1.072	0.826	0.970

[a]Assuming accounting cost is about the same as is calculated in chap. II.

Table VII-3 confirms the price estimates in Chapter VI in one direction: there was a more than adequate return on investment in selling even at the $1.15 calculated as the bottom of the range—the price on long-term contracts among the large eight companies. There is still room for some moderate reduction.

One can also make a rough estimate of the division of profits among companies and governments. Assuming that average f.o.b. receipts from crude sales and netbacks at the Persian Gulf were about $1.20 in mid-1969, taxes 85 cents, and operating costs 5 cents, the companies received on their investment about 30 cents, or 26 percent of the total gross profit. It would be more accurate, in my opinion, to consider total costs as about 10 cents and profits of $1.10, split about 83 : 17. This is a long way from 50 : 50. But the proportions have not much importance for price determination.

The "OPEC terms" were accepted by all member countries, although Kuwait, taking longer to ratify, made counter-proposals: that it reserve the right to raise its tax rates in the future, to assert claims for back taxes based on prices as posted before the 1960 reductions, and to litigate the claims in the Kuwait courts.[45] The Kuwait position had been adopted after mature consideration. Mr. al-Mazidi, their representative on the Kuwait Oil Company (KOC) board, had urged in a memorandum that if KOC did not come to terms, Kuwait could impose them unilaterally. If KOC thereupon left the country, Kuwait could operate the concession itself. "We can seek new buyers for our oil as we can offer relatively favorable terms." (See below, p. 214.) Eventually, Kuwait did accept

[45]*Middle East Economic Survey*, Feb. 18, 1966, p. 1.

the OPEC terms, with some modifications not made public. One reason may have been a shortage of cash.[46]

Before 1966, three producing countries (Venezuela, Libya, Indonesia) still taxed the companies according to actual receipts from the sale of oil.[47] In 1967 the companies agreed to adopt a set of "tax reference prices" somewhat below the Venezuelan posted prices, on the basis of which profits would be reckoned and taxes assessed. Perhaps the companies' willingness to agree had been influenced by the events in Libya. Here the government, encouraged by some of the older internationals operating in the country, asked all companies to go from real prices to posted prices as a basis for tax. The Oasis companies demurred at the higher tax burden and made counter-proposals, whereupon the Libyan government unilaterally imposed the posted prices as tax-reference prices. The Indonesian case is more complex, but in the context of the discussion need not concern us.

The complications of the various agreements should not hide their essence: the income tax has been transformed into an almost pure per-barrel tax. Lower costs can slightly increase the amount; higher accounting costs can slightly decrease it, but the room for variation is small.

The Floor to Current Prices. The per-barrel tax is now the basic support of the current price level, and many times more important than cost or company profits. For even if there were pure competition in oil production, none could sell at less than cost-plus-tax.

If my cost estimates are anywhere near correct, the 1957-70 cost-plus-tax floor never came under strain. No company in the big Persian Gulf concessions has had to make painful decisions about whether they could or could not afford to sell at the current price. The floor has been effective at a distance, because the oil industry lives mostly in the future. No prudent management will contemplate receiving bare cost-plus-tax under any long-term contract, since there is good reason to expect new demands. Sometimes the buyer may agree to escalate for higher taxes.[48] Furthermore, companies with small market shares, who would be disposed to act competitively and never mind spoiling the market, have generally higher costs. Hence their floor is higher, and some of them may be near it right now.

The producing nations have become a cartel that sells a license to produce. In general, a cartel exists to keep the price above cost. Each member is always tempted to chisel and sell at a somewhat lower price to increase volume and profit: do your friend before he does you. Hence the cartel must have a machinery to detect cheating or register non-cheating, and to assure each member that all the others are observing the price.

Cost to the producing nations of supplying the license is zero. But the system of fictitious posted prices, fictitious income taxes, and real excise taxes, is simple and strong. The tax is a public record, putting each under the scrutiny of all. A persistent and substantial down-drift in any nation's tax, not explained by a trend from higher- to lower-taxed crude, is evidence of cheating. Furthermore, tax changes are difficult or impossible to keep secret. Hence the OPEC nations need follow only the simplest strategy like the companies before them: do nothing.

[46] *Petroleum Intelligence Weekly*, June 7, 1965, Aug. 19, 1968.

[47] For Venezuela, this is not quite accurate. A special commission examined prices and refused to recognize any they regarded as unduly low.

[48] *Petroleum Intelligence Weekly*, Oct. 28, 1968.

But we must consider what forces might breach even a strong floor. Even more important, we need to consider the urge of the producing nations, against their own better judgment, to do away with the excise tax system that is their shield and buckler, and lose all by claiming all.

Continued Appropriation by Government. Host governments have claimed the whole price-cost gap; attest a 1968 OPEC resolution, the gist of which follows.[49]

Governments should themselves explore and develop oil to the maximum extent possible. As a second best, they may enter into contracts whose "terms and conditions . . . shall be open to revision at predetermined intervals."

Where the governments have no "participation in the ownership of the concession-holding company . . . the Government may acquire a reasonable participation, on the grounds of the principle of changing circumstances." Relinquishment of acreage of the current concessions shall be "progressive and more accelerated."[50]

The operator's income shall be based on posted or reference prices, "as determined by the Government" and shall change "in such a manner as to prevent any deterioration in its relationship to the prices of manufactured goods traded internationally."

A government may at its discretion guarantee that a tax rate will not be changed, but "the operator shall not have the right to obtain excessively high net earnings after taxes." If the government cannot persuade the operator through negotiations, it will unilaterally estimate and collect the excess. Earnings shall not exceed a level "the reasonable expectation of which would have been sufficient to induce the operator to take the entrepreneurial risks necessary."

Finally, except as provided by the government itself, all disputes with the operator "shall fall exclusively within the jurisdiction of the national courts."

The general principle is not very different from a 1961 resolution of the United Nations General Assembly which urged member nations: "To pursue policies designed to ensure to the developing countries an equitable share of earnings from the extraction and marketing of their natural resources by foreign capital in accordance with the generally accepted reasonable earnings on invested capital;"[51]

However, the OPEC resolution is much more specific in defining "reasonable earnings": the amount just sufficient to make it worth the operator's while to stay and operate.

Soon after the April 1968 OPEC meeting, the Saudi Arabian petroleum minister, Yamani, gave a noteworthy speech and interview in Beirut.[52] It is worth a close examination, at several places.

Mr. Yamani recognized the limit: "So long as they [the companies] are able to continue making a decent profit, they should be only too glad to remain with us, particularly in the sort of stable investment climate we have in Saudi Arabia." The irreducible "decent profit" is the barely adequate return, which is cost as reckoned in Chapter II of this study.

[49] No. XVI-90, June 25, 1968.

[50] The status of relinquishment at present is as follows: in Iraq, all except the actual producing area of IPC; in Iran, well over half, with more scheduled; in Kuwait, about one-third, and the offshore; in Libya, relinquishment of most of the concession is written into each agreement.

[51] See A/Res/1710 (XVI), Jan. 5, 1962, p. 2. See also UN Resolution 2158, Nov. 25, 1966, "Permanent Sovereignty Over Natural Resources." Useful background material is in A/5056, Dec. 18, 1961, pp. 20, 43, 56.

[52] *Middle East Economic Survey*, May 17, 1968.

But Mr. Yamani's other demand was for participation: Petromin, the Saudi Arabian national concern, must have a half-share in Aramco. Later in the year, the Kuwait Minister of Finance and Oil made a similar demand.[53]

Neither minister gave the terms of participation. My guess would be: about what was suggested in Chapter III as the zero-effect model (pp. 83-84). That is, the government would make an investment in new producing facilities proportional to the amount of oil it wanted to offtake. Moreover, it might "pay" for the share by debt obligations to be cancelled by profits on the production and sale of crude oil. If the estimates of Chapter VI are correct, the option to buy at a quarterway price will be rejected, because Mr. Yamani and his colleagues know that this is only about as good a price as private companies are paying now on long-term contracts.

The governments know that by taking up relinquished acreage and letting out new concessions as partnerships between the national oil company and the foreign operator, they will some day have large amounts of oil to sell. But this is a slow uncertain process. They want to be able to sell large amounts soon.

Mr. Yamani set no limits to his government's freedom of action. "If they [the companies] now choose to withdraw from the cooperative spirit, the result could be sad." The new agreements with France (discussed below) and others gave the national company an equal or preponderant share, and it was

> . . . absolutely essential for the [older companies] . . . to follow suit if they wish to continue operating peacefully in the area. . . . Any delay will be paid for by the oil companies concerned. . . . If it isn't done in due time, then things could turn out badly for all of us. . . . Control over all oil operations is our objective. . . . The major part of our long-term plan will be achieved within ten years . . . [but] it may take up to 25 years to attain full control.[54]

The most radical Arab nations, Iraq and Algeria, were "using ways and means which we don't consider reasonable," but the goals were the same.

In a later statement, Mr. Yamani said:

> The government of course has sovereignty and it can legislate; this is indisputable. But in our case it is against our philosophy to take unilateral legislative action against foreign investors in our country. In other words, we have this weapon but we don't want to use it. We are doing our utmost to avoid trouble but it may be inevitable.[55]

It is my opinion that these two objectives—appropriation of all surplus, and participation—are really one and the same. But first one may well ask to what extent the producing countries have the power to "legislate" the takeover of some or all of a concession.

Of course they have the physical force on the spot, and can expel anyone or take anything, without regard to "inky blots and rotten parchment bonds." But this hardly matters. Indeed, there has been too much talk of expropriation, nationalization, confiscation, just compensation, and the like, without considering the economic issues. The parties are negotiating about profits, and ownership is only the right to future profits.[56] As the Kuwaiti member of the KOC board noted, taxes could be increased to where development was unattractive because profits were low. Then there could be nationaliza-

[53] *Wall Street Journal*, Oct. 8, 1968, p. 15.

[54] *Middle East Economic Survey*, May 17, 1968.

[55] Ibid., June 7, 1968.

[56] A significant exception: when the concessionaire is another government. See below, pp. 215ff.

tion, but it would not cost much to compensate the companies because their low profits would mean their properties and producing rights were not worth much anyway.[57]

Expulsion of an oil company is the government's maximum sanction just as withdrawal is the company's. In practice, it may be impossible to tell one from the other; all we know is that the parties have been unable to make or renew a contract. It is a false distinction to differentiate between negotiation and expulsion. The power to expel is the power to make (or change) an agreement; actual expulsion is a sign of nonagreement.

Confiscations, like wars, strikes, and lawsuits, are the breakdown of normal relations. They happen when both parties think they can win, or when the weaker party feels that he must ultimately lose a prolonged struggle, and would rather risk all on one throw of the dice.

One must examine the inhibitions on the producing countries. It is well to start with the worst possible case from their point of view: total stoppage of oil production. This forces one to recognize how varied are these countries in their dependence on oil.

In five producing nations the petroleum sector[58] is less than a fourth of the total gross national product. Their production in million barrels daily in the first half of 1970 was: Algeria (0.9), Indonesia (0.9), Iran (3.8), Nigeria (1.0), and Venezuela (3.7), totalling 10.3 million barrels daily. Iraq oil income was roughly a fourth of GNP, and it produced 1.5 million barrels daily. In all others, oil income is well over half of GNP, approaching 100 percent where the sparse local population abandons its traditional livelihoods. Production in million barrels daily in the highly dependent countries in the first half of 1970 was: Abu Dhabi (0.6), Muscat and Oman (0.4), Kuwait (2.9), Saudi Arabia (3.6), Libya (3.7), others (0.1), totalling 11.3.[59] (Hence–if I may anticipate–a defection of this group would be fatal to any united action by the whole group of producing countries.)

A different view is given by the U.S. Department of the Interior. When asked how long producing countries could hold out without oil revenues the response was that since they were less developed countries, "all of them would probably be able to do without oil revenues for a long time."[60] Apparently the theory is that the poorer is a country, the lower the per capita income, the larger a proportion is its oil income, the more easily they can do without it. Perhaps the basic syllogism is: Iran is an "underdeveloped" country, and held out without oil revenues for two years (1951-53) before negotiating; ergo, "underdeveloped" countries can hold out "a long time." But in 1950 oil revenues to Iran were $45 million,[61] as compared with $940 million in 1970,[62] and similar proportions hold for other old producers.

[57] *Platt's Oilgram News Service*, Apr. 2, 1965.

[58] "The petroleum sector" means direct payments to governments plus the other local income generated by the oil operations: wages, and earnings from sale and contract services rendered to the producing company and the local refinery, if any. (This excludes any multiplier effect, i.e., income generated by the spending of oil-based incomes. Hence it greatly understates the importance of oil income, since the public sector of the economy depends entirely on oil payments to governments.) Inspection of the Aramco, IOOC, and KOC annual reports and the reports of the Bank of Libya seem to indicate, as a very rough guess, about 15 percent of direct revenues. In the case of Iraq, I reckon: 550 million barrels, $465 million in oil revenues, plus another $45 million in respect of local services (only 10 percent, because of very little local activity), or $510 million petroleum income. This is about 22.5 percent of total Iraq GNP.

[59] Population refers to 1968, per capita GNP to 1966; multiplying them together is a crude expedient (*World Bank Atlas*, 1970). These are significantly lower for Iran and Saudi Arabia than estimates made by those governments. Nigeria output is for June 1970; Libya, April 1970.

[60] *Platt's Oilgram News Service*, July 28, 1970.

[61] Zuhayr Mikdashi, *A Financial History of Middle East Oil Concessions* (Praeger, 1966), p. 23.

[62] *Petroleum Press Service*, Sept. 1970, p. 324.

Yet it would be wrong to equate expulsion of any individual company with stoppage of oil revenues. When the companies had "the power of disposal," the oil could not be sold except through the concessionaire. The power of disposal is no longer an absolute sanction; it still is an important risk. The host country must assume that the expelled concessionaire has transport, refining, and marketing facilities, and will seek crude oil supplies elsewhere, depriving his ex-host government of revenues, by benefiting another government. This may be the moment of truth for the producing countries as a group, for the second government that permits the ex-concessionaire to be supplied is, of course, undermining the position of the first government. If he is then cut off by all the producing countries, the company has no choice but to submit or to cease operations. And had the company been confronted with a prior refusal by the second government, it would have submitted in the first place, and met the terms of the first government. Expulsion of a concessionaire is therefore a test of solidarity, which the governments are naturally not anxious to undergo.

In the short run—say within three years—they cannot replace the refining-marketing apparatus of the integrated companies. If others do it for them, it must be done quickly. In countries heavily dependent on oil, before the people starved the government would be seized by any group who could borrow to buy arms on the security of the billions in annual oil revenues. And they would be supported, or tolerated, by those who had lost jobs or incomes or who saw that in time, as their neighbors' production expanded, they would have the worst of all worlds—loss of oil revenues while the consuming nations would be supplied anyway.

Today, unlike the 1950s, the expelled concessionaire may be replaced by another foreign company. But the host government cannot expect terms as good as from the old company. The risk premium that a new private company must in prudence assign to the return on its investment must be considerably higher than the old company's premium. First, the very fact that an old contract has been repudiated makes the new concessionaire less secure: it can happen to him too. Second, he cannot sell the crude oil as profitably, though he surely can sell some of it. The crude market is narrow, and a sizable increment would depress the price in the short run. A refining-marketing system takes time and money to build. Moreover, and especially if the home government of the expelled company gives it diplomatic-legal support, this may restrict sales by the new company. Courts or administrators in various large consuming countries may hold that the crude oil and products made therefrom are still the property of the old company. In a country like France, where a decision in favor of an old company is least likely, the new company will not be helped much, because the market is nearly all preempted.

For these reasons, the new concessionaire's minimum rate of return will be higher than the old company's. This does not make expulsion and replacement impossible. The old company may have been pigheaded and asked more than its true minimum, more than what it would have held out for if it had taken the government's threat seriously. There is an array of probabilities that dictate the higher risk premium and insure that, on the average, the new company will not be able to offer as good a deal as did the old. Perhaps this, rather than the temporary cash shortage, was why Kuwait drew back from such a decision in 1965-68.

For the producing company, the penalties for not replacing the crude are severe: idle ships and refineries. Some companies can replace the crude from producing properties in other countries, presumably at somewhat higher cost. Others will have a share in one or more joint ventures in another host country. There may be problems in overlifting past production shares previously established, which would require the consent of their partners. But more important, as just explained, the second host government would have to decide whether to pocket the additional revenues from the additional output, or to renounce them for the sake of solidarity with the first host government.

The companies would have some analogous problems. They might set up contingency plans, whereby any of them cut off or forced out of any concession could buy from the others at a price which would not penalize him too much, perhaps cost plus tax plus some low royalty. Home governments might well try to promote them or even require them; their refusal to accept crude oil produced in a confiscated concession would be an important support for an insurance scheme.

But this is not a zero-sum game where one party gains what the other loses. As Mr. Yamani argued in 1968, the integrated companies should not be deprived of their producing function to become mere buyers of crude, for that would bring about a "drastic reduction in oil prices, and the only beneficiaries would be the consuming countries." A year later he warned it would be "disastrous" to make buyers out of producers.[63] However, Mr. Yamani takes no notice of a favorable factor: the reluctance of any integrated oil company to give up own-produced oil for purchased oil even when the price is identical with cost-plus-tax. For the price must be entered as the crude oil cost to the refining affiliate in a consuming country, and this price is surely lower than the price previously recorded on the interaffiliate transfer. (See below, pp. 231ff.) Hence the integrated company will be willing to pay more in cost-plus-tax to the producing country, since there is an offsetting gain in lower income taxes in the consuming country.

This aside, Sheik Yamani's warning is correct and of utmost permanent importance. If the oil companies become buyers of crude oil (even if their producing subsidiaries remain in place as contractors) and the producing nations sell, there is no longer a means to detect cheating. (See above, p. 210.) The known agreed-upon price floor has disappeared, and far below it there is only the ground of cost. The height is enough to give even the most hardened seller an acute fit of vertigo. Hence, while any individual company is expendable to the producing nations, the companies as a whole are not.

Thus each company and each government has reason to fear the losses the other can inflict, and each knows that *all companies and governments together will lose if an expelled company becomes a buyer.* For these reasons, we expect few cataclysms. But while a showdown, with each party refusing to give way, is very unlikely at any given time, the odds are much higher that it can happen at some time.

Each party will call on its fellows to help. If the host governments appear to be losing out, they may well play their last card—a concerted shutdown to force the consuming-country governments to force the oil companies to give in. The consuming-country governments, for reasons explored below, may be more than a little inclined to help the producing nations. If they will not help, production will soon resume. Most of the producing countries cannot hold out, and when even a few defect the rest must follow. Nobody will gain from the confrontation.

The writer's one policy suggestion would be to the immediate and ultimate benefit of all—a stockpiling program to render Europe, Japan, and the rest of the OECD nations safe against a concerted shutdown.[64] The cost of stockpiling would be repaid several times over by ending wasteful coal production and by heading off wasteful nuclear power

[63] *Petroleum Intelligence Weekly*, June 16, 1969.

[64] See Appendix to Part Three.

generation. The producing nations would profit by the greater outlet, even at lower prices. Both sides would gain from such an agreement. The chances of its adoption are low.

Government Influence over Crude Oil Price: Current Status of Production Control. There is a striking omission from OPEC's Resolution XVI.90 (above, n. 49). Nothing is said about regulation of total output or of member countries' shares. Yet at the original 1960 Baghdad meeting, Resolution I.1 provided that members were to "formulate a system to ensure the stabilization of prices by, among other means, the regulation of production." The OPEC members and staff have therefore been aware of the problem from the start. OPEC made two attempts to set up output quotas: for 1965-66, and for 1966-67. But actual 1965-66 output patterns diverged much more widely than can be accounted for by chance and unforeseen events. Member governments were openly doubtful. The 1966-67 program contained a significant change of emphasis: instead of being agreed maxima, like the shares of joint venture offtakers, they became near-minima, for it was stressed that every member nation had a right to at least its programmed output increase. Thus even the halfhearted first program had aroused the conflict of interests among member nations which must be overcome if any price level is to be held.

Even the bare mechanics of prorationing by OPEC or a similar body would be complex and perhaps impossible. The Texas Railroad Commission needs to make monthly determinations of permissible output, and watches inventory figures very carefully.[65] It is hard to imagine OPEC doing anything of the kind; nor would the organization be in a position to make estimates with sufficient accuracy for a much larger worldwide system where much of the stock is in transit or in unknown hands. It would be to the interests of the oil companies to sell more than their quotas in the earlier months of the year, and work back to their compliance level by the end of the year. Otherwise an oil company might find itself with unsold stocks and no place to put them, which means there could be false signals given out by inventory changes. A forecast too restrictive and quotas too low could mean physical shortages and discomfort or worse in the consuming countries during a severe winter, which would be greatly resented. To forestall this, larger inventories would have to be carried, which would not only be expensive but, unless controlled, could be a source of price weakness. They would have to be impounded in some way—kept out of normal commercial channels and somehow policed.

On the eve of the 1968 meeting a chief item on the agenda was production quotas, or at least a start toward guidelines. There had been publicly stated disagreement on the guiding principle: predictably, Iran argued for population, Saudi Arabia for reserves, and so on. An additional problem was how to phase in the growing production of the state-owned oil companies without disrupting prices.[66] Neither issue was faced.

The omission of quotas was not for lack of time, since we are officially told: "The Resolution is the result of many months of effort, and of long discussions among representatives of Member Countries, during which many different possibilities were explored." Shortly thereafter, the Secretary-General of OPEC said that there would be "renewed erosion in prices" unless there were a production-control agreement.[67] In the

[65] See current issues of *Oil and Gas Journal* or *Platt's Oilgram News Service*, and the work of Wallace F. Lovejoy and Paul T. Homan, *Economic Aspects of Oil Conservation Regulation* (Johns Hopkins University Press for Resources for the Future, 1967), pp. 136-41, 219-30.

[66] *New York Times*, June 23, 1968, p. F-1.

[67] *Wall Street Journal*, Sept. 9, 1968, p. 9.

Spring of 1970, Venezuela was reported as exerting a maximum effort to have production controls discussed at the June 1970 OPEC meetings.[68]

If the market-sharing mechanism of incremental cost does not regulate output, then some kind of agreement must substitute for it. Whether population or reserves or cost or alphabetical order be chosen as the criterion, each government must be freed from the temptation to produce and sell a little more, very profitably, by chiseling just a little on the price.

The pressure upon any member of a cartel to break away is proportional to the difference between price and cost, multiplied by the amount of additional output available at that cost. Offsetting this is the unknown of how much the price will be depressed by his own action, over and above what his fellows will do in any case.

In Chapter II it was calculated that if output were increased to where reserves were 20 times production, costs might double: from 10 to 20 cents. This is still a tiny fraction of the available price. Noted also was the fact that 15 years' production was considered a suitable time limit for reserves by Mr. Berkin of Shell and Mr. Bridgeman of BP. Reserves are an inventory, whose carrying costs need to be minimized. In the United States, end-of-1969 reserves were 9.3 years, with no observed rise in cost; in Venezuela, the number is about 12, and little Bahrein has perhaps served as a laboratory example, its ratio dropping to 6.2 in 1968, then restored to 12.4 in 1969. This suggests that the Persian Gulf–North African optimum lies somewhere between 10 and 15, but it is safer to make it 15 to 20.

Under 1969 conditions, given time, Saudi Arabia alone, or Iran plus Kuwait, could have supplied all of Persian Gulf–North African demand at a small rise in cost. Available output was from 3.2 to 4.3 times actual.

TABLE VII-4. Output Rates Attainable from Current Reservoirs, Persian Gulf–North Africa, 1969 Conditions

	Output (*mill. BD*)	Reserves (*billion barrels*)	Available output (*mill. BD*) assuming reserve, production ratio of: 20:1	15:1
Iran	3.3	54	7.5	10.1
Iraq	1.5	28	3.8	5.1
Kuwait	2.5	68	8.4	11.2
Saudi Arabia	2.9	140	17.5	23.4
Other Persian Gulf	1.8	41	8.8	11.8
Libya	3.1	35	4.8	6.4
Other North Africa	1.2	13	1.7	2.3
Total	16.3	380	52.5	70.3

Source: *Oil and Gas Journal*, Dec. 28, 1969.

How fast output can be expanded is suggested by Libya, where it went from 1,748 thousand barrels daily in May 1967 (the last pre-crisis month) to 2,660 thousand a year later—a growth of 52 percent. A large Persian Gulf country could probably do as well after a six-month warmup. The Libyan reserve : production ratio in 1967 was less than half of that in the Persian Gulf. Hence, even one defector from an agreement can do several of the parties heavy damage, since he can expand output and replace them.

The Basic Problem: Maintaining a Cartel. If a government aims to take everything over and above the minimum necessary return on investment, posted prices are irrelevant. For

[68] *Petroleum Intelligence Weekly*, Apr. 6, 1970.

if a posted price, less allowed cost, yields a tax which leaves the operator more than the necessary minimum, the government is committed to tax it away. On the other hand, if a proposed tax is such that the operator cannot pay it, the government must decide how much to give up in order to keep the rest. Posted prices are no help in making either adjustment. In fact, they are an impediment. The structure of differentials among posted prices is based on API gravity numbers, which are related to market value loosely, and to cost not at all.

The producing countries are reluctant to give up the reliance on posted prices. But if they want to take as much as the traffic will bear, they must recognize costs and actual prices, which vary among companies and locations. This means, in effect though not intent, to move toward an income tax. But if the governments move to take the surplus over cost for each barrel of crude oil, the stabilizing effect of per-barrel payments will be lost. If the tax obtainable depends on real prices, and must be adjusted to them, the tax is no longer a price floor. Countries must consider the risk of taxing themselves out of the market. They will watch, and be watched, for opportunities to gain production at the expense of those demanding too much. They must appraise the trustworthiness of their neighbors. (See p. 214 and pp. 263ff.) The closer the governments get to their objective of taking all the surplus, the more they take on the job of price-making decisions.

Where the government has taken the whole surplus over cost, the operator will find life simpler. He cannot obtain more than his allowed return, and the government cannot force him to take less. In effect he is a hired contractor and gets his contractual fee whatever the price received by the owner—the government. Of course, the owner may use him as a sales agent and reward him accordingly: the higher the price secured, the larger the fee. This would be common prudence and would tend to slow any price decline. But with no sale, no fee, and the sales agent will shade the price to get the business. He is free of the inhibition of a fixed tax-per-barrel floor.

The Producing-Country Governments and Price-Tax Shading. A private company, running its own "concession," but taxed to or near the economic limit, is nothing but a hired contractor-operator for the government-owner. The difference between such private companies and national oil companies is a difference more of style than substance. In either case, the governments have the authority and responsibility for price-output decisions.

To forecast how governments will act when they have the power generally, let us review such few examples as exist of where they have had such decisions to make. So far, they have been only marginal decisions.

1. One piece of evidence is in the type of concession[69] awarded to Entreprise de Recherches et d'Activité Pétrolières (ERAP), the French national oil company, by Iran, Iraq, Saudi Arabia, and Libya.[70] (Similar agreements have been signed with others.) The agreements are extremely complex, but plainly ERAP is acting as a contractor, to be rewarded by a share of profits realized, not for a fixed payment per barrel. Realized prices are specified as a tax basis in the NIOC-ERAP agreement. Moreover, ERAP has

[69]The term "concession" is used loosely to mean any agreement providing for exploration and/or development in any given area, under any type of terms. For the purposes of this discussion one need not ask whether in any legal sense the agreement ought to be called a "concession" or a "contract."

[70]Reprinted in full as special supplement to *Platt's Oilgram News Service* and *Petroleum Intelligence Weekly*, Jan. 2, 1967. See Thomas R. Stauffer, "The Erap Agreement: A Study in Marginal Tax Pricing," paper delivered at Arab Oil Congress, Baghdad, Mar. 1967.

itself publicly noted that price reductions are not to be at its expense.[71] This removes the floor to prices. Realized prices are to be estimated by a committee of experts "ascertaining by documentation, testimony and all other means of acquiring accurate information, the f.o.b. selling prices of crude oil delivered and sold to third parties in the Persian Gulf." A look at Chapter VI and its appendices shows the task to be laborious but surely feasible.

Another line of evidence is in the proposed and actual sales of crude and products by the national oil companies.

2. In 1965 KOC agreed to sell crude to the Kuwait National Petroleum Company (KNPC) refinery at $1.10 per barrel. (The refinery was to do the stabilizing and desalting; the cost of these operations is about one cent per barrel.) In itself, a price of $1.10 for Kuwait crude was nothing unusual, probably even a little higher than the price under long-term contracts among the Persian Gulf producers. Much more important is the provision that on incremental sales of its refined products, which in practice means all sales, KNPC pays KOC only 60 cents per barrel. So low a price is possible only because on these sales Kuwait waives four-fifths of the tax due it.[72] Thus government take as a floor to price has been lowered. And if the KNPC refinery is as uneconomic as it seems—an untried refining process and a very disadvantageous location—the necessary rebates may even be greater. (Similarly, the Libyan national oil company is exempted from taxes for 10 years.)

3. In 1966 the Iranian Consortium agreed to sell to the National Iranian Oil Company (NIOC) incremental oil over a five-year period, starting with 40 TBD and rising to 120 TBD. The oil was not to be sold in the international market, but shipped only to East Bloc countries in barter deals. The price to NIOC was 50 cents per barrel, made possible because, as in Kuwait, much or most of the royalty was waived by the government.

4. Shortly thereafter Saudi Arabia demanded and received similar incremental purchase rights from Aramco, and in November 1967, they announced a barter deal with Romania.

The amounts are small and the sales are insulated as far as possible from the world market. But the insulation cannot be complete and can sometimes be more apparent than real. At the same time as the Romanians were negotiating with Iran, they were also negotiating a sale of products to Shell for shipment into Britain if the prohibition on East Bloc imports were to be relaxed.[73] Such a sale might not disturb prices but it points a general moral: once the East Bloc nation gets the oil, it can resell it even without receiving and reshipping it. Blurring and hiding the origin and destination of oil shipments has become a fine art. Moreover, the influence of these shipments will be felt wherever an East Bloc nation bargains with an oil supplier. For it will ask terms at least as good as it can get from the producing countries on the special incremental oil.

5. Late in 1966, Caltex proposed to Indonesia that they cut the export price from the current semiofficial $1.77 (Indonesia has only one posted price) in order to get additional arm's-length sales. Indonesia refused, but then approved a 17-cent reduction for a firm in which it was an equal partner, which marketed its royalty crude oil in Japan. Late in 1967, Caltex began selling at 15 cents less per barrel, apparently with tacit government approval. The government took 9 cents lower tax per barrel (since they received 60

[71] *Petroleum Intelligence Weekly*, Jan. 1, 1968.

[72] *Petroleum Intelligence Weekly*, July 19, 1965; Mar. 13, 1967.

[73] *Petroleum Intelligence Weekly*, Dec. 19, 1966; Nov. 6, 1967.

percent of profits) in order to get higher sales. The Indonesian director general for oil later said he believed in "getting the best price available while remaining flexible to meet competition." He looked forward to a big jump in output. The government owned the oil, the companies being only contractors, and set prices, which had to have "flexibility to meet the always changing competitive situation."[74] This flexibility is the complete antithesis of the rigid posted-price floor which has served so effectively in the Persian Gulf and the Caribbean.

6. Decisions on incremental business may also involve decreased sales. When threatened by the loss of Venezuelan heavy fuel oil sales on the U.S. East Coast because of high sulfur content, the Venezuelan government promptly made special arrangements for Creole and Shell to build desulfurization facilities at their refineries. "Both government and company officials were tight-lipped as to exactly what this entailed."[75] But there was to be a special low reference price on all exported heavy fuel oil, desulfurized or not, as well as quick depreciation and exemption from import duties on equipment purchased abroad.[76] These are merely three ways of lowering taxes per barrel. Faced with the choice of large sales at a lower tax per barrel, versus no sales at the existing tax rate, the government chose the additional sales at the lower tax per barrel to get higher total revenues.

National Oil Companies. The preceding evidence indicates that when and as the national oil companies have large amounts of oil to sell, there will be more competition and price cutting. The national companies are not inhibited by tax payments and can go as far toward cost as need be to get the sale. Private companies cannot do this while they run into the cost-plus-tax floor. Moreover, national companies are not mere commercial ventures; they embody national pride, yearning for respect, desire to show the world that they too can run an oil business. National oil companies *must* sell large amounts of crude and/or product, cost what it may.

But incremental crude sales are not a loss operation for the short run. They will yield a handsome return on a small or zero investment. For the long run they may yield a loss or a gain. It depends on the chances of prices holding if national companies play the game and do not cut prices; and on the use of the money gained today and used in the interim. The producing countries are by no means alike in the productivity of their investment in economic development. What is good for countries like Iran and Iraq, with unused land, water, and population resources, may be a mistake for small barren areas like Kuwait or Abu Dhabi. The national companies will send prices down, but whether a particular company harms its own nations is another question.

How large are the amounts that the national oil companies are likely to have available for sale in the near future? Until a short time ago, the outlook appeared to be dim.

In an interview in January 1968, the then secretary-general of OPEC first buried the production quotas as "a transitory measure . . . largely experimental . . . ."[77] Thereafter the colloquy:

[74] *Petroleum Intelligence Weekly*, Oct. 24, 1966; Jan. 9, 1967; Apr. 17, 1967; Jan. 8, 1968; May 13, 1968; May 20, 1968.

[75] *Platt's Oilgram News Service*, Jan. 9, 1968.

[76] Venezuela, Ministry of Mines and Hydrocarbons, press release of Dec. 25, 1967 (Creole); of May 4, 1968 (Shell).

[77] *Platt's Oilgram News Service*, Jan. 27, 1968.

Q. It is apparent that national companies also have had to offer oil at substantially below posted prices in order to break into world markets. How is this justified within OPEC? Are any specific norms being established for national companies?

A. As far as I know, none of the national companies is leading prices downwards because none has the very large volumes that would enable it to do so. . . . It is too early yet to talk of specific norms being established. We are still in the exploratory stage.

But as stressed earlier, a national oil company selling incremental output, and the local concessionaire selling more are only two ways of getting at the same result. Late in 1965, it became known that the Aramco offtake agreement was being revised to make it "more attractive for partners to buy more from Aramco rather than increase purchases elsewhere." It was expected that there would be "substantial" increases in Aramco production, especially in heavy crudes.[78] In fact there was a larger increase in Saudi Arabian output during 1965-66 (just under 400 TBD) than anywhere else; even the percentage gain was the largest in the Persian Gulf outside the newly developing Abu Dhabi fields. The nature of the 1966 change can be guessed (see below for the writer's reasons) as follows. Previously, the offtaker had been permitted to buy at a halfway price between tax-plus-cost and posted price. For example, with posted price $1.80 and cost-plus-tax 95 cents, the halfway price would be $1.37, and this would not be worthwhile under 1965 conditions. The new deal was probably a quarterway price, about $1.16. Perhaps it was further reduced by the temporary OPEC discount.

The repercussions in Iran were strong. For reasons explained in Chapter III the Iranian Consortium found overlifting a particularly dangerous business, since it gave incremental oil to concerns with small market shares, who would find it profitable to sell even at much lower prices because they were not bothered by the reaction upon existing receipts, of which they had too little. A "top executive" of one of the older large companies wrote an "internal" memorandum which he obligingly sent to a trade publication: "Unlimited volumes of very low-cost oil would be put into the hands of non-integrated companies which have no investments at stake, thus magnifying geometrically the downward pressures on the market—far more than the pressures from other new crudes being peddled by newcomers who at least have investments to recover."[79]

"Investments to recover" is the standard industry irrelevance. Past investments are bygones. But the essential point is well taken: the shareholders producing the great bulk of Consortium oil would lose on higher output, which might displace some of their sales directly and lead to lower revenues on all the rest. Nonintegrated companies are feared because of their future investments, production, and sales.

As a partial concession to Iran, the additional crude was made available to the National Iranian Oil Company, as described earlier. But the government was not appeased. Particularly after the 1967 war, it pressed for either more crude to NIOC, or else a guarantee of higher output. It felt entitled to specially favorable treatment as a specially reliable source: not being an Arab country, it had never halted or limited output in the summer of 1967. It demanded a 20 percent increase per year for five years, which would have had Iran producing 2.5 times as much at the end of the period as the beginning; say from the 2.5-MBD average of 1967 to 6.2 MBD in 1972. (A "normal" 10 percent growth rate

[78] *Platt's Oilgram News Service*, Nov. 8, 1965.

[79] *Petroleum Intelligence Weekly*, July 8, 1968.

would bring them to 1.6 times the initial amount, i.e., from 2.5 to 4.0.) The government of Iran was apparently not satisfied with the offer of a quarterway price for overlifting.[80]

The North Rumaila Field in Iraq. Discovered in 1953, Rumaila pools are highly prolific, with output per well around 11-12 TBD.[81] The northern portion was not developed, however. In 1961 the Iraq government ordered exploration to stop. The famous expropriation Law 80 confined Iraq Petroleum Company (IPC) and subsidiaries to acreage currently producing. The rest of the country, previously all IPC concession area, reverted to Iraq. IPC was particularly anxious to recover North Rumaila; negotiations in 1965 nearly restored it, but eventually broke down—an incident well worth research. In late 1967, when France had gained much credit with the Arab nations during the Israeli war, Entreprise de Recherches et d'Activités Pétrolières (ERAP) was awarded a large concession in Iraq. It was "ratified with great pomp in Baghdad [and hailed throughout] the Middle East as a great victory over Anglo-American imperialism,"[82] but did not include North Rumaila. Instead, Compagnie Française des Pétroles (CFP), a partner in IPC, performed the curious exercise of negotiating for new rights in a territory which it still claimed as its own. Furthermore, CFP let it be known that it would include its IPC partners in any exploitation rights it obtained. The cloud of dispute over North Rumaila should not hide the economic issue: Iraq government insistence that production be increased. Seizure of the field was only a means to put more pressure on IPC. The objective was the same as for Iran; only the methods differed. It serves to remind us again that the rate of production, which rarely gets into public notice, is yet one of the most important company-government issues. Iraq has been the least successful of the large producers in expanding output, which has stagnated since February 1966.

Political complications over North Rumaila were many. In December 1967, France lifted an arms embargo on weapons shipments to Iraq and other Arab countries, in the hope that this would "strengthen the French hand in the oil and sulfur negotiations." Iraq wanted a high payment per barrel of crude produced, and reversion of all rights after five years. Perhaps this was why later it was discovered that there would be a delay in shipping the promised French fighter planes to Iraq.[83]

Early in 1968, Iraq was reported to be demanding about \$1.20-\$1.25 per barrel,[84] indirectly confirmed by a prediction by INOC of profits of \$9.00 per ton,[85] which equates to \$1.20 per barrel. This gives a clue to what the government thought was obtainable. But if my calculations are anywhere near correct, it would be reckless for a private company to pay any more than \$1.00 at the outside. Perhaps Iraq wanted CFP to agree on a partnership deal for Rumaila of the kind they had with ERAP, who had in the meantime secured one in Libya, and which the French had been operating in Algeria since the independence of that country. "Algerian oil technicians are known to be active as advisors in Iraq."[86] Also, the government's hand was strengthened by an agreement for

[80] *Petroleum Intelligence Weekly*, Mar. 4, 1968. The writer's guess is that the offer was presented as matching the current Aramco arrangement, hence that the Aramco change was to a quarterway price.

[81] The published data are erratic and not always consistent; the potential may be higher.

[82] *Le Monde*, Nov. 23-29, 1967.

[83] *New York Times*, Dec. 8, 1967, pp. 1-24; Dec. 18, 1967, p. 8; Dec. 22, 1967, p. 8.

[84] *Petroleum Intelligence Weekly*, Apr. 8, 1968.

[85] *New York Times*, Apr. 11, 1968, p. 63.

[86] *New York Times*, Apr. 21, 1968, p. E5.

Soviet technical assistance for INOC, which could cover North Rumaila if Iraq so decided.[87]

Apparently unable to find a private contractor to pay $1.20, Iraq finally decided to have INOC develop North Rumaila. Its announcement had some cost calculations, probably taken from one of the plans presented to them, hence probably overstating costs. The outlay expected was $16.8 million, for an expected 105 TBD in three years.[88] The necessary investment would seem at first sight to be $164 per daily barrel. Since the potential of North Rumaila is much larger, and since half the original outlay was for a pipeline to the port of Fao, a steady-state investment per barrel would obviously be less than $164. If one assumes a 2 percent decline rate (since it is only a single field), and a 10 percent discount (since INOC has no political risk), the present-barrel-equivalent factor is 7.63, and development investment cost is 5.05 cents per barrel. For a private foreign company needing to use 20 percent, or even higher in view of the turbulent history, development investment cost would be about 10 cents, total cost about 20 cents, and $1.20 tax would bring cost-plus-tax to about $1.40. This would be unprofitably high.

The announcement was greeted with ill-concealed annoyance in France, where it was asked how Iraq discovered overnight it could do the job itself. Nobody in Paris believed it could be done without technical and financial help.[89] This remains to be seen. There was a later announcement that North Rumaila reserves were 5 billion barrels, and that a production of a million barrels daily was planned, with Soviet equipment and assistance.[90] However, there have been no reports on drilling before the end of 1969. In mid-1970, a contract was awarded to build the pipeline from Rumaila to Fao on the Persian Gulf.[91]

As just seen, a national oil company and a private foreign company earning "a decent profit," are only two ways of the government acting as owner (rather than landlord), taking the whole profit and thereby taking responsibility for pricing. All the government decisions reviewed here went the way of lower price to gain outlet. None should treat this small sample as proving that governments will always, or even usually, choose to sell more at a lower price. The purpose of our review is to lay bare the problem of choice. Additional sales bring clear and immediate gain, but at the risk later of lower prices. There must always be a comparison, a weighing of advantages.

The crucial comparison is between two present values: sales gained today versus revenue lost tomorrow. If a barrel of national company sales simply displaced a barrel of tax-paid sales, the gain would be relatively small and soon negative as prices on all sales declined. But if a barrel of national company sales can be tucked into one of the many corners of the market at the untraceable expense of all other competitors, without displacing one's own current sales, then the higher profits on the incremental output might well more than offset the higher later losses. Any general answer is useless. For the particular oil producing nation, its choice depends also on the discount rate. Twenty percent might be a good approximation to the real rate in Iran, in view of the rapid growth of the national product; 7 percent seems to be an official discount rate, since it

[87] *Petroleum Intelligence Weekly*, Jan. 8, 1968.

[88] *New York Times*, Apr. 11, 1968, p. 63.

[89] *Le Monde*, Apr. 11–17, 1968.

[90] *World Oil*, Aug. 15, 1969, p. 188.

[91] *Petroleum Intelligence Weekly*, June 8, 1970.

has been used for the very purpose of measuring the present value of payments deferred to the Libyan government.[92]

Conclusion: Producing-Country Governments Undermine Prices. Control of the market has passed from the hands of the companies, who only had it in a special and qualified sense, but through 1970, at least, it found no identifiable permanent group of holders. OPEC has apparently been a useful instrument–and in the writer's own judgment, has served as an admirable center for negotiations–but it wields no power, not even a delegated one. National oil companies will by their nature be price cutters. The governments are now demanding large ready reserves for their national companies, to have them enter the business soon and on a large scale. Every country can very profitably produce much more than any likely quota, and there is no automatic price-cost mechanism to regulate market shares.

The governments are less able to operate a successful cartel than the companies. Not only do they lack the companies' experience, but they also lack the intercompany contacts at two levels: crude production and sales (the joint ventures) and the refined product markets. These contacts are necessary for sound decisions on when to meet competition, when to beat it, when to disregard it. There is nothing to prevent the governments from going into the refining-marketing business, but the large investments needed and the mediocre returns will slow down their entry and minimize their contact. Furthermore, their entry will increase the number of competitors at both levels.

Thus the increasing role of the governments in the market will tend to increase competition and reduce prices. This is the contrary of what they want. It should surprise nobody who keeps in mind either economics or history. Adam Smith emphasized how sellers produce market results they never intended. History is sprinkled with examples of groups who exerted great power to bring about their discomfiture or even ruin.

Yet the argument is not complete and we may yet be forced to reverse the conclusion of declining prices–once we have examined the role of the consuming-country governments.

Consuming-Country Governments: Reference Prices and Energy Policies

As with the industry, so with the consuming-country governments: the reference price of oil is everywhere in the air, rarely on paper. The oral tradition is strong and clear. First, market prices have been temporarily low (for over a decade now) and must increase. Second, as consumption rises it will dry up the "surplus," giving an additional boost to prices. Third, the demands and pressures of the oil producing nations for higher revenues will send up the price even more. But estimated prices committed to writing, on the basis of which decisions are to be made, are rare.

Oil Reference Prices. The voluminous OECD literature never touches on specific prices, only availability and "reasonableness." *Perspectives* is, of course, the great honorable exception in stating its assumptions and trying to calculate from them. As we saw earlier, its forecast in 1962-64 was heavy fuel oil at $18.00 per metric ton; but in *Nouvelles Réflexions* (1966)[93] it was reduced to $12.50. This was about 20 percent above the market price as of the middle of 1967; yet it was much lower than what governments

[92] Bank of Libya, *Thirteenth Annual Report, 1968-1969*, p. 137.

[93] See above, n. 33.

were calculating. Bechtel Corporation[94] noted that the fuel oil reference price was about 40 cents per million Btu, or $16.60 per metric ton (and coal at 50 cents, or $20.75 oil equivalent).[95] When the report appeared in October 1967, heavy fuel oil at Rotterdam had averaged about $13.00 over a decade, then reached a 10-year peak of $15.00 (36 cents per million Btu), after which it bumped down to a new low of about $9.30 in October 1969, with the Suez Canal still closed, before the spasms of the tanker market, which have not ceased as this book is completed, put rates above June 1967. These price comparisons, and the ones below, must all be made with due allowance for the worldwide inflation. (See Table VI-3 above, p. 183.) All else being equal, a ton of heavy fuel oil at $12.50 in 1971 prices is about as good a buy as a ton at $10.50 in January-May 1967.

Great Britain: Nuclear Cost and Oil Reference Prices. Such information as is available on the British reference price may indicate a significantly lower one. Table VII-5 shows that the estimated costs at the new nuclear power plants in Britain would be matched by oil costing about $16.10 in 1965. Two years later, if one allows both for devaluation and for revised (increased) costs, the net change is negligible: to $16.30. Yet if one considers the Ministry's estimates of oil-fired power costs, the implied price of oil was definitely lower, around $14. (It would not be proper to recompute for devaluation, since the price is of an imported material, and the estimate was made before the pound was devalued.) Either oil-fired plants are ruled out from the start, except in certain unusual circumstances, or the price of oil is expected to increase. Perhaps the British government follows both these rules, since they are mutually compatible. It is also significant that one must laboriously calculate the reference prices from the documents, as is done in Table VII-5. Nowhere is there a straightforward statement of the reference price, still less any reason or any evidence for it.

British power plants pay much more, even ex tax, for fuel than Danish plants. Yet Britain is closer to crude oil sources, and some though not all harbors are deeper. Moreover, British fuel oil is usually produced at local refineries; hence, transport to Britain is in larger carriers as crude oil, rather than in small ships as fuel oil. Finally, British refineries are nearly always located close to the power plants—which will not otherwise burn oil—consequently, handling and storage costs are minimized. The difference in price must be due to the more competitive Danish market. The way in which competition was repressed in Britain was to make the actual oil price charged, including tax, about equal to that of coal. The oil companies met the price of coal almost exactly, absorbing the fuel tax. The limited market permitted to them was available if they just undercut the price of coal; there were no additional sales to be had by bidding lower. Once again the principle was confirmed: noncompetition meant higher prices. The British fuel market in 1968 was where the Rotterdam market had been in 1957-58.

It probably is not worth while to translate these expected heavy fuel oil prices into expected crude oil prices at the Persian Gulf. That would require assumptions about freight rates and refining costs expected by the respective government. There is no evidence that any such calculations were made.

The use of higher reference prices for nuclear power than for fossil fuel plants implies higher future oil prices. This is consistent with what little we know about other governments, though in contrast to the drastic downward revision made by the EEC Commis-

[94] W. K. Davis and F. V. Karlson, *Nuclear Energy in the United States and Western Europe* (Bechtel Corp., Oct. 1967), p. 23.

[95] *Platt's Oilgram Price Service*, Dec. 29, 1969.

TABLE VII–5. British Reference Prices of Heavy Fuel Oil, 1965 and 1967

(all items in mills [tenths of a cent] per kwh, except as stated)

	1965		1967
	Original estimate	"Optimistic" variant	Revision
	(1)	(2)	(3)
A. Fuel Oil Price Equating to Nuclear Power Cost			
1. Total power cost, Dungeness B	5.37	4.43	6.07
2. Oil-fired plant: capital cost	1.19	1.07	1.19
3. Oil-fired plant: operating and maintenance	0.30	0.30	0.30
4. Oil-fired plant: subtotal	1.49	1.37	1.49
5. Fuel cost of barely competitive oil-fired plant (line 1 less line 4)	3.88	3.06	4.58
6. Equalizing fuel oil price in cents/million Btu (for heat rate assumed, see note)	38.8	30.6	45.8
7. Equalizing fuel oil prices, dollars/metric ton	16.10	12.70	19.00
8. Post-devaluation (line 8 × (2.40/2.80))	...	...	16.30
B. Fuel Oil Price Implied in Oil-Fired Plant Cost Estimate			
9. Ministry of Power estimate, oil-fired plants under construction in 1965	4.79	4.79	4.90
10. Implied fuel cost (line 9 less line 4)	3.30	3.42	3.41
11. Implied fuel oil price, excluding tax: cents/million Btu	33.0	34.2	34.1
12. Implied fuel oil price, excluding tax: dollars/metric ton	13.70	14.20	14.15

Sources and Notes: Except as stated, U.K., Ministry of Power, *Fuel Policy, 1965* (Cmnd. 2798) and *Fuel Policy, 1967* (Cmnd. 3438).

Line 2, capital cost $95 as revealed in *The Economist*, May 29, 1965, p. 1060. Discount rate 7.5%, from both *Fuel Policy* editions. Fossil plant service life 25 years, plant factor 80, assumed in cols. (1) and (3). In col. (2), 25 years and 85% following "optimistic" variant set out in *Fuel Policy, 1965*.

Line 3, from Philip Sporn, "Nuclear Power Economics" in *Nuclear Power Economics 1962 through 1967*, Report, Joint Committee on Atomic Energy, 90th Cong., 2d sess. (1968), pp. 25–36.

Line 6 implies heat rate of 10,000 Btu/kwh for United Kingdom, as against 8,650 for United States (Sporn, 1964). Factor calculated from *Fuel Policy, 1965*, where tax of 2 pence per imperial gallon (462/metric ton) is equal to 0.11 pence/kwh, which comes to 4,200 kwh per ton of about 42 million Btu. This is partially confirmed in F. P. R. Brechling and A. J. Surrey (*National Institute Economic Review*, May 1966, p. 32), who show *average* national heat rates of 12,300 in the United Kingdom and 10,400 in the United States. If one assumes that *new-plant* heat rates are in the same proportion, Sporn's 8,650 for the United States would correspond to 10,400 in the United Kingdom.

sion. The explicit calculations of *Perspectives*[96] contrast with the brief unexplained conclusory statements of *Fuel Policy.*[97] Yet it is known that the Ministry of Power has made a very elaborate linear programming model of the fuel industries, which contains prices (especially of coal and gas) as explanatory variables.[98] There is no awareness that the "prices" are political artifacts having no relation to social costs. They result from barring coal and certain oil imports; taxing oil to protect coal; reserving the electric power and steel industries to coal; and reserving electric power to nuclear generation planned with the coal "price" as a standard of reference for cost. The direct effects of government actions are amplified by the support they give to the oil industry's reluctance to compete.

[96] European Coal and Steel Community, *Étude sur les Perspectives Energétiques à long Terme de la Communauté Européene* (1962).

[97] U.K., Ministry of Power, *Fuel Policy, 1965* (Cmnd. 2798) and *Fuel Power Policy, 1967* (Cmnd. 3438).

[98] C. I. K. Forster, "The Statistical Basis of National Fuel Policy," paper published by the Institute of Actuaries, 1969.

In time, the price constellation will change drastically and falsify the calculations—just as other very competent econometric studies have gone astray.

Energy Policies: Coal Protection. There is a basic difference between private and public reference prices. Private companies who make mistaken investments and bring about excess supply must bear the consequences in lower prices and perhaps losses. But sovereign states, within their own borders and possibly across them as well, can prevent competition or minimize it. Hence their mistaken forecasts may in the short run be self-fulfilling. We must therefore look at fuel and energy policy in the consuming countries.

The literature on the subject tells us little of what the governments do. Above all, there is avoidance of evidence or data or even assumptions about specific prices and costs. Governments have lived from crisis to crisis, or election to election, trying to find some resolution of the political forces pressing upon them. It is a durable, comforting illusion that governments listen to someone in the back room who has calculated the balance of private and social costs, national security, miners' retraining, etc. There is no evidence of it.[99]

Coal protectionism is of course the oldest and most important policy, and has tended to keep oil prices higher than they would otherwise be, by weakening inducements to aggressive oil competition. Coal protection has been most effective in sectors owned or much influenced by government, particularly electric power and steel. To shut off a large part of the market from stern bargaining and search for low prices has made an important contribution to keeping prices high. Electric power production is owned by the government in Britain, France, and Italy; and in Germany and Japan the government has not permitted private companies to buy fuel freely. Of the Western European power plants going into operation in 1967-71, over half were designed to burn coal.[100]

We have an extraordinarily valuable piece of evidence to prove that *no* European or Japanese coal would be produced now if it were exposed even to the limited competition now at work in oil. British coal is cheaper than Japanese and probably cheaper than Continental coal. Lord Robens, of the National Coal Board, said early in 1967 that if given time and, of course, much money he could produce *some* coal at 3 pence per therm, which equates post-devaluation to 30 cents per million Btu, i.e., oil at $12.45 per metric ton.[101] If this is the best that a near desperate[102] partisan can promise—let alone per-

[99] For an excellent critical review, see Richard L. Gordon, "The Goals of European Energy Policy," *Economia Internationale delle Fonti di Energia* (Milan), Sept.-Oct. 1968; and "Forecasting and Policy Making—the Case of European Energy," *ibid.*, Jan.-Feb. 1969 (both in Italian translation). It is most unfortunate not to have them available in English. For an example of backroom analysis which (despite its narrow terms of reference) deserved the attention it did not receive, see Commissariat Général du Plan d'Equipement et de la Productivité, *Rapport Général de la Commission de l'Energie: IV Plan* (Paris, 1961), pp. 71-73, 253-57; ibid., *V Plan* (Paris, 1965), pp. 53-77.

[100] *International Coal Trade*, July 1968, p. 8.

[101] See Appendix to Part Three. In 1966-67 only 1.2 percent of the coal sold to the Central Electric Generating Board (U.K.) cost less than 35 cents per million Btu (post-devaluation). (Memorandum submitted April 24, 1962 by CEGB to Select Committee on Nationalized Industries on "The Implications of North Sea Gas.") These prices include transport charges. But the "wide variation in pithead prices" of which the CEGB speaks indicates substantial price discrimination. The writer believes that the cheapest coal was sold below cost; and since the National Coal Board is still making heavy net investments, the true social cost includes capital as well as labor consumed. The coal contracted for large aluminum works, at a secret price, is a similar case. This can neither be proved nor disproved until the facts are made public.

[102] This word is used advisedly. Lord Robens stated that if the Seaton Carew (or Hartlespool) power station were not coal-fired, no future power station ever would be. The decision went against him. *Times* (London), Aug. 22, 1968, p. 17.

form—then no coal can possibly compete. Oil has been selling at lower prices for years, and can be produced at a fraction of that.

Given a lag of about 25 years between the fact and its realization by government, European coal would disappear by about 1990. In the interim, the coal labor force will shrink as will its voting strength. The lessening opposition to free oil imports will tend toward increasing competition and lower prices. The outcome is predictable only in the long run, however.

Nuclear Power. As coal slowly passes from the picture, much or most of it will be replaced by nuclear power. With current British technology, nuclear power is far more costly than oil, as has been noted earlier in this chapter. In France, an attempt was made to develop a reactor which would run on natural uranium, obviating the need to buy enriched uranium (or enrichment services) from the United States. The reactor was a technical success, but far too high-cost; consequently, a special committee recommended in early 1968 that the program be ended, but that France do enough building and operating of nuclear power plants to gain operating experience and readiness for the next more economic generation of plants.[103] They were reluctant to use American technology, and hence were disposed to consider favorably any alternative reactor. But the British Advanced Gas-Cooled Reactor (AGR) received short shrift because it was as costly as the method they had decided to scrap.[104] The dismissal was no surprise to those noting the failure of British manufacturers to win any export orders. AGR plants are the new British coal mines. It is sad to see the stubborn delusion that the AGR is as good as the American reactor, only not as well sold.[105]

If U.S. technology is used, one can apply the estimates for a plant the size of Commonwealth Edison's Dresden II, near Chicago. There have been astounding increases since 1966 in equipment prices and plant construction costs, and no end was clearly in sight in 1970, but the increases have been greater in nuclear plants.[106] Hence the comparison shown in Table VII-6, which refers to 1967 conditions, is too favorable to nuclear plants. The calculations are based on an 800-megawatt plant, one-third larger than Dungeness B. (General Electric has stated that it costs 30 percent more to build a nuclear plant abroad.)[107]

The comparison is obviously rough. (See, however, Table VII-7 below.) Capital costs in Europe are usually assumed (however falsely) as lower than the 12.5 used by the sources used here. But attempts at finer adjustment would be pedantic in view of the fact that before October 1969 no commercial nuclear reactor was in operation. All the costs so widely discussed over the last few years are estimated by scaling up by a factor of three or more. One must also note the long and costly delays in the building of the first commercial plant, Oyster Creek. The cost of delay probably swamps the increased manufacturers' prices since announced.[108]

[103] Commission Consultative pour la Production d'Electricité d'Origine Nucléaire, *Les Perspectives de Développement des Centrales Nucléaires en France*, Apr. 1968, pp. 5-8.

[104] Ibid., p. 15.

[105] *The Economist*, July 25, 1970, p. 62.

[106] Philip Sporn, "Developments in Nuclear Power Economics, January 1968-December 1969," in *Nuclear Power and Related Energy Problems—1968 through 1970*, report for the Joint Committee on Atomic Energy, 92d Cong., 1st sess., December 1971. In Table 1, he estimates capital cost of a 1,100-megawatt reactor plant at $204 per kw.

[107] *Wall Street Journal*, Sept. 15, 1966, p. 1.

[108] Dungeness B will probably not operate until 1974.

TABLE VII-6. Europe: Nuclear and Fuel Oil Power Costs, Dresden II Modified, 1967

Element of cost	Nuclear	Oil
Capital	2.86	2.00
Fuel	1.82[a]	2.19[b]
Operation and maintenance	0.30	0.30
Nuclear insurance	0.10	...
Total	5.08	4.49

Source: Philip Sporn, "Nuclear Power Economics," in *Nuclear Power Economics 1962 through 1967*, Report, Joint Committee on Atomic Energy, 90th Cong., 2d sess. (1968), pp. 25-36. An earlier version of this paper was the source for W. K. Davis and F. V. Karlson, *Nuclear Energy in the United States and Western Europe* (Bechtel Corp., Oct. 1967), p. 15. Adjustment described in text.

[a]Carrying cost of nuclear fuel inventory.

[b]Price 25 cents per million Btu, $10.40 per metric ton.

TABLE VII-7. Sweden: Thermal and Nuclear Power, 1967

	Nuclear	Oil	
Size of plant (mw)	750, two units	750, two units	
Capital cost ($/kw)	126-155	105	
Fuel cost (cents/million Btu)	...	24.4	34
Total cost (mills/kwh)	3.86-4.25	3.97	6.59

Source: Swedish State Power Board, *Elförsöorjningen Under 1970-Talet* [Supply of Electricity During the 1970s] (1967), pp. 40, 44, 79. Kroner exchange, $0.1933; large calorie, 3,968 Btu; load factor, 68.5%; discount rate, 7%. At a 20-year life, this would equate to an annual capital charge of 9.1%. If we used 12.5% instead, a nuclear cost comparable with the previous table would be 5.3-5.8.

If, therefore, the building of nuclear power stations were governed by comparative costs, very few stations would be built outside the United States. It is significant that the Scandinavian nations, which combine distance from fuel sources with a high level of electricity use and of technical competence, have been slow to enter: the Swedes are now committed to their first nuclear power plant (see Table VII-7) and the Danes postponed theirs after a careful survey.[109] Swedish 1967 costs seem higher than U.S. costs, despite their using the same technology, though the capital cost in late 1969 in the United States would be at least half again as high.

The Swedes, having no oil or coal interests to cloud their judgment, and making no attempt to forecast oil prices, simply used the latest price, which had been quite steady since 1960, rounded to the next higher unit: $13.15 per metric ton.[110] (The Danes buy somewhat more cheaply; their report looks to a maximum around $12.00.) At this point, the choice will almost surely go to nuclear power, since there is no cost of insuring security of supply.

But for some, nuclear power may be more than an economic choice, more even than a shiny new toy. It may heal the egotisms bruised by disappointing growth rates or the supposed "technological gap" behind the United States. It proves that one is modern and

[109]Braendselsgruppen, *Elsam-Kraftimport Underso/Gelson–en Analyse of Kraftvaerksudbygningen i 1970 erne* [Fuels Group, Elsam-Electric Power Import Study–An Analysis of Electric Power Plant Construction in the 1970s], 2 vols. (Copenhagen, 1965). An English summary is available, and the tables are all easily read.

[110]The cost of shipment Rotterdam-Stockholm, assuming a spot rate of Worldscale 50 (old Intrascale −45) for small ships, would be about 80 cents per ton. Hence the Rotterdam equivalent price would be $12.35, which is near the middle of the $11.50-$13.50 range of *Nouvelles Réflexions.*

up to date. A national resource—highly educated manpower—is wasted by having it do useless things instead of learning useful ones or pursuing knowledge, and one justifies this in the name of education. M. Jourdain was more to the point. "Est-ce que les gens de qualité en ont? . . . J'en aurai donc."

In Germany, much of the reluctance to sign a nuclear nonproliferation treaty has stemmed from a fear that it would slow down or prevent the development of power reactors; reactions have ranged from fury against "another Versailles diktat"[111] to the ingenious suggestion that the "true" American motive for sponsoring the treaty was to keep its monopoly of nuclear fuel.[112] It recalls another ingenious complaint that the German coal industry was in trouble because "after World War II the victorious allies demanded that the chemical industry shift from a coal to a petroleum base."[113] The German Minister of Science has noted (without necessarily agreeing) that Germany opposes the treaty because it would stifle peaceful nuclear programs.[114]

In India, the late Dr. Bhabha expected exhaustion of world fuel reserves and considered nuclear power cheaper for India than coal. Oil was not considered.[115] Since the reactors are of only 180 mw each and scale economies are very strong, costs are very high. Yet in Bhabha's September 1964 paper, nuclear costs were estimated to be lower than those estimated by Sporn in July 1964 for a unit over three times as large.[116] We will not ask how. The Indian plant is at Tarapur, north of Bombay, less than four days' sail across the Arabian Sea from the Persian Gulf.

André Malraux has recorded conversations with Nehru, which vividly evoke his gallant refusal to be crushed to earth by the poverty and ignorance whose burden he had assumed. One cannot begrudge a dying man a last illusion: "India is an underdeveloped country only in a special sense, since she is building atomic reactors."[117] Nuclear power is a heavy burden to India, particularly because it sterilizes part of its technical brainpower, who learn an uneconomic technology and produce nothing worth its cost.

Nuclear reactors will be built, therefore, regardless of market prices for oil, not only because reference prices are so much higher than market but also for noneconomic gratification. But a commitment to nuclear power means a barrier to oil and the loss of an incentive to compete in order to gain sales. Hence nuclear power is a force for higher oil prices.

Consuming-Country Government Oil Policies: Local Refining. We must take as given a strong dislike of the large international oil companies, and a feeling of irksome "dependence" on them. In most underdeveloped countries the local refinery is in the public sector; elsewhere the pattern varies greatly. But in all countries there is some degree of public commitment to local ownership of refining.

[111] See the roundup of German press opinion, *Der Spiegel*, Feb. 27, 1967.

[112] Uwe Nerlich, "Defense Policy and Technological Relations," paper at Conference on Technological and Scientific Problems in the Relations between Europe and the United States, convened by Twentieth Century Fund and Agnelli Foundation, Turin, Nov. 1967.

[113] *New York Times*, Apr. 17, 1967, p. 57.

[114] *Nucleonics*, May 1967, p. 86.

[115] H. J. Bhabha, "Need for Atomic Energy in the Underdeveloped Regions," lecture, Second UN Conference on Peaceful Uses of Atomic Energy, Geneva, 1958; (with N. B. Prasad), "A Study of the Contribution of World Energy Requirements and the Economics of Nuclear Power with Special Reference to Underdeveloped Countries," Third UN Conference, Sept. 1964.

[116] Philip Sporn, "Nuclear Power Economics," in Joint Committee on Atomic Energy, 90th Cong., 2d sess. (1968), p. 49; lower interest rates are a negligible factor.

[117] André Malraux, *Antimémoires* (Gallimard, 1967), p. 350.

These locally owned refineries have made a crude oil market, and have therefore helped to lower its price. But their position is inherently difficult because their crude costs them much more than it does an integrated producer-refiner. Those independents have prospered who have bought at the best price the market afforded. They have been able to undersell a product price level set by the integrated producers' formulae of posted prices (or posted less "reasonable" discount) plus AFRA freight (less "reasonable" discount) plus customary refining margin, etc. But life has not been prosperous for all, and it has been easy for none.[118] The integrated producer-refiner, unwilling though he is to start any price cutting, can stand a great deal of it.

The stereotype of large world-diversified oil companies drawing on their "vast financial reserves" in a "ruthless price war" to eliminate their rivals is as ludicrously wrong as it is passionately believed. The truth is much more simple. For the integrated producer, a cut in product prices is in effect a cut in the price of crude oil netted back to the producing area. Since the crude oil profit has always been more than sufficient, he has always been able to meet any lower product price.

A government that wants locally owned refineries therefore finds plenty to worry about. Two courses of action are open to it: lower the price of crude, or prevent "excessive competition" and raise the price of products. The first is difficult, because there is no obvious way to do it, and because the complexities of oil price overwhelm a civil servant with limited knowledge and plenty of other things to do. Hence the second course almost imposes itself. Yet consuming-country governments are not eager simply to fix and hold prices. Not only do consumers have some small voice, but the effects of energy costs on domestic cost levels and international competitiveness cannot be ignored.[119]

Income Tax. Government difficulties in dealing with oil prices are shown in the taxation of refining profits earned by local subsidiaries of the integrated companies. The higher the price at which crude is transferred to the local subsidiary, the less its profits. The transfer price set by the company itself has no public validity. The principle in the United States is that a valid transfer price is that which would be set by arm's-length bargaining between independent buyer and independent seller, and it has recently been affirmed and much elaborated.[120] Some part at least of the impetus to this action was the problem of finding a price for crude imported by American companies from their production abroad. The tax law of Australia seems broadly the same, and is modelled on the British.[121] It is hard to see what other principle a democratic government could follow consistently.

The refining-marketing subsidiaries of the integrated companies in Western Europe have declared little or no taxable profit; there have been frequent declared losses. Edward Symonds, of the First National City Bank, has estimated that the "downstream profits" of the seven large companies, calculated on the basis of posted prices for integrated transfers, first turned into losses in 1958. Mr. Symonds was not slow to point out:

[118] See Albert, Brondel, and Ascari, "Situation Financière . . . ," EEC, No. 1948/II/68-F. They estimate that during 1958-67, CFP and Gelsenberg lost refining market share; Petrofina was stable; ENI, Wintershall, and Antar went from 6 to 12 percent. Deutsche Erdöl Aktiengesellschaft was later bought out by Texaco.

[119] *U.K. Fuel Policy, 1965*, p. 9; EEC, *L'Influence Economique du Prix de l'Energie*, Serie Economie et Finances No. 4, 1966.

[120] Section 482 of the Internal Revenue Code. See guidelines published in *Federal Register*, Apr. 16, 1968.

[121] Australia, Taxation Board of Review Decisions (New Series), Case No. N69, Mar. 20, 1963. Pp. 270-71, 275-84 set out the theory.

> . . . The tax gatherer . . . is . . . struck by the fact that the value added in processing and distributing oil within the national boundaries gives rise to little or no income on which he can levy tax. . . . [And even in the United Kingdom, France, and the Netherlands, the only countries where] dividends and other advantages [are] gained from the participation of national companies in the international oil business . . . the tax credits allowed to the companies on their payments to foreign governments remove much of the tax revenue that would otherwise result.[122]

It is not for us to say that the "downstream losses" have or have not any validity in law. But if derived from posted prices and AFRA rates (less unexplained "reasonable" discounts) they are not market facts. Moreover, "downstream losses" are utterly incompatible with the billions of dollars invested in refining since 1958. No sane management would do so. Any set of prices and tanker rates compatible with "downstream losses" during a period of heavy investment are discredited by that fact alone.

Chapter VI can be used to show the difference in landed cost computed from posted prices and AFRA, or from current prices and current long-term tanker rates.[123] These figures are only approximate and anyone using them must consider some theoretical and statistical problems. First, the use of a term charter rate implies that an independent importer (or seller) would find it necessary and prudent to sign up tonnage in advance and use spot freights only as a supplement. Second, the price estimates are only approximate: some oil was moving at less, and some at more. Were the oil market a freely competitive one, we would expect prices to fluctuate in response to random supply-demand changes, just as spot tanker rates fluctuate. Hence it would be as unreasonable to make the lowest or the highest price granted anywhere within any time period *the* arm's-length price for anyone, as to say that the lowest or highest spot rate granted during the year should be considered as *the* arm's-length spot tanker rate. But since the oil market is not a freely competitive one, it is incorrect to try to strike a balance between ups and downs by taking the average. If others can get a persistently low price, there is no reason why the taxpayer cannot.

Every country has its own methods of tax administration, rules of legal procedure, and standards of proof. The tax collector will have wide fact-finding powers in one country, few in another. Hence, the data used here must not be considered as the f.o.b. and landed cost that a treasury would calculate. But the method will be the same.

Income tax in the consuming countries will probably become more acute an issue. Even if actual market prices do not decline, increasing knowledge of them is coming into the public domain, and therefore a government's estimated market price will decline. The issue has been raised at least once over the past decade in the United States, Great Britain, France, Germany, Italy, Australia, New Zealand, and Japan. In the absence of any verifiable knowledge, one may surmise that usually additional taxes due have been settled by negotiation up to at least a recent year, but that the parties are not committed for the future.

Governments are not likely to retreat from their position that all taxpayers must be treated alike, and that the only acceptable price is an arm's-length price or equivalent. But they are also very much aware that the companies have already paid taxes to the producing countries on the basis of artificially high prices for oil. To the extent that a consuming-country government presses for what it considers its due, it challenges the

[122] Edward Symonds, *Oil Prospects and Profits in the Eastern Hemisphere* (1961); *Oil Advances in the Eastern Hemisphere* (1962); both New York: First National City Bank.

[123] Even in 1967, imports into the United States and Britain seemed largely entered at posted prices f.o.b., (*Platt's Oilgram Price Service*, Aug. 26, 1968.)

producing-country governments to take even as much as they do now, let alone the larger amounts they will claim in the future.

For the years through 1964, the United States government in effect settled the issue in favor of the producing countries and against itself. In principle, at least, it no longer does so. "Any such international generosity carried into the future would simply be a complete concession by the United States that other countries may unilaterally assert any jurisdictional rules they desire. . . . No sovereign country can give this blank check to the rest of the world. . . ."[124] Not in principle, anyway; we know nothing of the practice. But the difficulties are revealed by a later action. In mid-1970, the Treasury announced a procedure whereby an American taxpayer threatened with double taxation under Section 482 could obtain consultation between the "competent authorities" of the two treaty countries, and agreement on allocations of income.[125]

The extremely slow pace of these investigations, of which practically nothing is public knowledge, shows how difficult the consuming-country governments find the problem. Their reluctance to tackle it tends to deflect them to the option of higher oil product prices.

Insistence by a large consuming country on collecting full income tax could depress prices. An integrated producing-refining company might be unable to sell profitably in *A* the oil produced in *B* because the total taxes due to both countries was more than the pretax profit. The consuming-country government could in effect tell refiners to bring in oil from countries where the tax-plus-cost was low enough to be profitable. This would provide a notice and incentive to producing-country governments: if they raised taxes above a certain point, already reached in one place, they would lose outlet and revenue; while if they shaded taxes to permit price shading, there would be plenty of incremental output and revenues to be gained, at the expense of other governments.

Producing companies will attempt to stay in the producing countries as concessionaires as long as possible. Were they to withdraw and buy the oil from the government that had expelled them for refusing demands they could no longer meet, the price they paid would be a fairly clear-cut item and the basis for a higher tax assessment on their refining affiliates.

One minor aspect of American tax policy may be mentioned at this point: the depletion allowance, which permits a producing company to deduct indefinitely from taxable income a fixed percentage of the sales value of the oil, or alternatively half of the net profit, thus reducing taxes by about a fourth. Since our subject is the world market, we are interested only in the provision which permits an American company to apply percentage depletion to production outside the United States. Income earned outside the United States is subject to American taxation, but the tax payments to the producing countries are credited against liability to the United States. Hence, if the foreign tax payments were less than American income taxes due, the company could use its percentage depletion allowance on foreign output against the net tax liability due to the United States. But foreign taxes (including royalties) are so much greater than what U.S. taxes would be, that the depletion allowance against foreign production is not used. "The major oil companies have foreign tax credits coming out of their ears."[126]

124 Remarks by Stanley S. Surrey, Assistant Secretary of the Treasury, Nov. 1, 1967 (multi.), p. 30.

125 Revenue Proceeding 70–18, dated July 20, 1970. Technical Information Release 1037, June 26, 1970.

126 *Platt's Oilgram News Service*, May 5, 1969.

Locally Owned Crude Production. A consuming-country government, aware of the competitive disadvantage of the nonintegrated refiner, and desiring local refinery ownership, will naturally consider helping them to enter the production business. Buying a participation in known exploited reserves may not be possible, but has not been discussed. The only method contemplated is to have local companies explore and develop crude oil. The process can be traced step by painful step in the EEC, which for over a decade has been struggling toward an energy policy. The 1962 memorandum drawn up by the Council of Ministers[127] was rejected because it was too permissive to oil imports; only Italy and the Netherlands finally supported it. Reconcilement with France, Germany, and Belgium-Luxembourg was only possible by getting them into the circle of oil producers. Accordingly, the 1964 agreement took a significant step forward[128] in naming as its objective Community-owned production. In February 1966 the EEC Commission filed the "Marjolin memorandum" on how the accord was to be carried out. (It is not clear whether this memorandum was ever published, but it has been widely paraphrased.)[129] For the sake of both security and competition, it proposed an important place for Community oil companies, and proposed subsidies and other financial aids to help them explore and develop crude oil reserves. The Community would be there as an outlet for oil not fully price-competitive.

This immediately set off a debate on the proper definition of a Community company: Where was its main office? Who owned it? In time, this will doubtless be resolved. The only point of interest here is the steady progression in the EEC toward the French concept of state-supported integrated companies.[130] A glance at this system will be worthwhile, not because the EEC is expected to go all the way—it may or may not—but because by showing the extreme one can see the direction. Because "independence" has been such a basic French policy for so long, we may seek in French experience an indication of what other governments will do as they, too, become devoted to it. Of help in this is the good old French tradition of clear expression, where other nations mumble and mutter.

The Case of France. Since 1926, French oil importing and refining have been by a special grant or privilege allocated through licensing. The Compagnie Française des Pétroles (CFP) and Compagnie Française de Raffinage (CFR) have had a certain reserved share of the market. Competition has been sluggish at best, and the government has kept prices high. When head of the Direction des Carburants, M. Giraud explained that "[French] ex-refinery prices are higher than in other countries . . . [in order] to maintain the self-financing of the petroleum industry. Otherwise, the French petroleum companies . . . would no longer be able to combat the foreign companies and after a very short time this would be harmful to our economic security. . . ."[131] Deferring the issue of "security," one need only point out that high profit margins are precisely what draw the

[127] *Bulletin de la CECA*, unnumbered, Aug. 1962.

[128] "Protocole d'accord du 21 avril 1964," *Journal Officiel*, No. 69, Apr. 30, 1964, p. 1099.

[129] I have found the most complete account to be in *Bulletin de l'Industrie Pétrolière*, Feb. 28, 1966. See also Albert, Brondel, and Ascari, "Situation Financière . . . " EEC No. 1948/II/68–F; also published in *Direction*, Apr. 1968.

[130] For an able analysis of French oil policy generally, see Farid W. Saad, "France and Oil" (Ph.D. thesis, M.I.T., 1969).

[131] André Giraud, "La Politique Pétrolière Française," speech before the Conseil Economique et Social, Aug. 8, 1968 (duplicated).

detested foreigners into the French market like bees to the jam pot. It then becomes necessary to limit their expansion, and the government must always be contending with the companies. But the tension is not seen as the result of keeping price far above cost, but rather as a struggle for independence.

Even before World War II there was some search for oil in the Sahara, and after the war it was accelerated. The cost has never been publicly reckoned, but it was heavy; Hartshorn's estimate of $1.2 billion[132] may be somewhat low. We will not try to indicate whether the return was positive. But the 1963 Evian agreements providing for Algerian independence gave the new government enough from oil production that only the maintenance of high internal French prices could leave enough margin for a profit to the French and the few foreign operators in Algeria.

The comment of *Le Monde* is relevant, because the sober honesty of this newspaper is well known, and because it was strongly anti-Gaullist. The agreement was expensive, but it was worth something to give the international oil cartel a shove: "*pour bousculer le cartel.*"[133] In 1965 the treaty was revised, more favorably to Algeria,[134] but this supplied new bones of contention. In mid-1970, nearly all non-French companies having left, Algeria made a "definitive and irrevocable" increase in the per-barrel tax payable by French companies from 79 cents to $1.21. *Le Monde* considered this as a clear violation of the accord, and regretted the threatened breakdown of what had seemed a promising new method of association between producing and consuming countries, including French support to the Algerian economy, and also with the operating companies getting a "less leonine share" of the profits. (Cf. above, p. 209.) But they were puzzled over what to do in response. Boycotting Algerian oil would largely hurt the French companies.[135] Perhaps there might be a cut in imports of Algerian wine. They did not ask whether future investment in Algerian oil was a net benefit or a net burden to the French economy, i.e., to make or at least suggest a comparison of costs with market prices of oil available elsewhere.

By early 1971, Algeria had raised its demands in the ultimately successful attempt to build a transitory tanker shortage into a permanently higher tax rate for crude oil. André Fontaine, the foreign editor of *Le Monde*, wrote an excellent analysis of French policy, which placed it in its world political setting. In 1962, General de Gaulle reckoned he would keep Algeria inside the "mouvance française," and set the world an example of a new type of relation between former metropolis and colony. This would be an implied rebuke to the United States. "At a moment when the swift deterioration of relations with Cuba presented a serious threat to American security at the Florida border, Paris was eager to give an example, thanks to the wisdom of 'the old guide,' of the happy resolution

[132] J. E. Hartshorn, *Oil Companies and Governments* (London: Faber & Faber, rev. 1967), p. 263.

[133] *Le Monde*, Aug. 29, 1962.

[134] *Revue Pétrolière*, Oct. 1965, pp. 33–36; *Le Monde*, July 30, 1965.

[135] In fact, the non-French companies (American, British, and German) had already been almost completely expelled. This was the last step in the process which began with requirements that first some then all profits be retained in Algeria. Then the government asked that they become 51–49 partnerships, with an Algerian majority holding. Getty accepted the demands, Shell and Phillips did not and were taken over (*Petroleum Intelligence Weekly*, June 22, 1970). At that time, Mobil was still negotiating. Earlier, the Sinclair properties had been seized because the American parent, Sinclair Oil Co., had been bought by another American company (Arco) in the United States, without prior Algerian permission. Arco made no attempt to buy off the expulsion, which in any case involved only a tiny part of Sinclair assets.

of the French-Algerian conflict. From one end of the third-world to the other, they applauded the boldness and generosity of him whom they gladly honored, in the Arab countries, with the title of 'sheikh'. Formerly the butt of UN resolutions, France now had a starring role." One after the other, the principal elements of the accord were disregarded or repudiated. "Harassment and plunder of all kinds, disregard of many agreements, make the cost look heavy of the self-imposed French burden of buying Algerian crude oil at prices above market, importing unneeded wine, and giving Algeria considerable economic aid of various kinds." But M. Fontaine, and *Le Monde*, are not willing to give up the Gaullist foreign policy, whose costs they recognize.[136]

In 1965 several French government oil companies—as distinguished from the part-owned CFP and CRP—were merged to form ERAP, a name now widely known. It has become a concessionaire or contractor in Iran, Iraq, Saudi Arabia, Libya, Canada, probably soon in Indonesia and Venezuela.[137] That ERAP will find oil is certain. How much is worth finding, and whether there has been a good use of the capital taken out of the French economy depends, of course, on what it would have cost to procure the oil from other sources. We are back again at the reference price. It would be an unjust exaggeration to say that oil prices in France will be simply marked up to whatever is needed to make ERAP look profitable. But certainly the government will stretch a point (and then another) to prevent the appearance of too large a deficit.[138]

ERAP (and to a lesser extent CFP) will have great staying power in the producing countries. I have pointed out earlier that host governments would appropriate anything over and above the minimum return on investment, enough to keep the companies from leaving. But ERAP will for political reasons stay in some though not all places, even when the rate of return is zero or negative; whether its losses are absorbed by a simple government grant or by raising prices higher, or cancelling indebtedness, etc., hardly matters.

In 1967 the president of ERAP remarked that its Canadian subsidiary was an instrument of French foreign policy. "That created some consternation in Canada, but was regarded by students of these matters as an obvious truth."[139] That same year, when ERAP was obtaining the Iraq concession on acreage formerly held by IPC, *Le Monde* was thrilled by the contest of wills. To negotiate for this acreage, someone was needed "who would not flinch" when mighty IPC—including the four biggest Anglo-Saxons—"showed their teeth," i.e., someone "capable of braving the anger of the members of IPC." The Iraqis, it noted, found that someone in France. Thus "the Anglo-Americans [lose] any chance of expansion into the hitherto unexplored parts of the country." They have been outmaneuvered; they cannot "block France, as ERAP, . . . from a place in the untouched zones of Iraq without provoking a grave political crisis." This is their just reward because "on the morrow of the last war they would not let France into the game in this region. The Anglo-Americans also would run risks in breaking up the negotiations which France, as CFP, is conducting in Baghdad" on North Rumaila. "Not the least result of this

[136] Fontaine, "La Tentation de la Rupture", *Le Monde*, Jan. 14, 1971.

[137] *Le Monde*, July 23, 1970.

[138] A Frenchman in partial dissent: "The French love deficits. The Left is pleased because they think they are making the rich pay. The Right is pleased because it proves that government enterprise is a shady mess." Another Frenchman replies: "Do not exaggerate the Right-Left division over nationalized enterprise. I would regard myself as on the Left, but with no particular love for them. The French State has been in business for three hundred years, and the more they mismanage the more clever they think they are." A third Frenchman sums up: "French business is divided into a private sector, rigidly controlled, and a public sector, completely free."

[139] *New York Times*, Mar. 21, 1970.

agreement would be the start of discord among the Anglo-American oil companies," since a deal on North Rumaila would be less repugnant to the non-American than to the American companies.[140] So "Gaullist" a statement appeared in the same issue as a signed editorial denouncing the general. To consider French policy as having been a vagary or personal eccentricity of General de Gaulle is mistaken.

Three unusually important statements on French oil policy were published that same month,[141] by the Minister of Industry, his immediate subordinate who heads up energy matters, and the president of ERAP. They are summarized here very briefly.

Their object is to have French companies produce an amount equal at least to French consumption. This does not imply that they will displace everyone else. They do intend to have about 50-60 percent of consumption covered by public or private French companies; and a barrel imported by a non-French company must be balanced by a barrel produced by a French company sold somewhere else. This implies investments of the order of a billion dollars per year. (The whole Middle East plus all Africa have never absorbed that much; but of course this is only a proposal; it is for the government to approve the spending.) Another object is "to cultivate more just relations with the producing countries" and also diversification. Furthermore, they desire a Community policy of this type which will be based on European companies, i.e., with decision centers in Europe.

This seems admirably clear as a statement of tactics. But there is no hint of any economic strategy these tactics serve—of the economic benefit to France. This does not make the policy wrong.[142] It means only that the benefit is not economic. France and the rest of the Community are to undertake oil production and sale for its own sake and in spite of possible cost. Logically, they should try to minimize that cost by trying to get as high a price for their crude as possible. It also follows that they will stay on as contractors or concessionaires when the private profit seekers have fled.

The former Minister of Industry, and professor of economics, later Minister of Social Affairs, Jean-Marcel Jeanneney,[143] in discussing security of supply, lists four types of risk: (1) overcharging by sellers—*chantage de prix*; (2) accidents, such as earthquakes; (3) strategic security, i.e., on whom is one willing to depend over the long term; (4) military security.

Now, there is no *economic* distinction between (1) and (3). For if there is no control of supply by sellers acting in concert, there is no economic dependence, no way of holding buyers to ransom. Given enough competition among sellers—for that is the only economic question—one does not depend on any one or any group of them over the short or long term.

It is unlikely that M. Jeanneney would waste time on superfluous definitions. He must have a noneconomic distinction in mind. The fact of buying from others, or the need to buy from others, regardless of the terms at which one buys, is itself a dependence. And this fits perfectly with the policies stated by MM. Giraud, Guichard, and Guillaumat. The

[140] *Le Monde*, Nov. 23-29, 1967.

[141] A. Giraud, P. Guillaumat, and O. Guichard, "Trois Documents Definissent la Politique Pétrolière Française," *Revue Pétrolière*, Nov. 1967, pp. 21-27.

[142] It neither proves nor disproves the verdict of M. Servan-Schreiber, in *Le Défi Americain* (Gallimard, 1967), p. 119: "The expensive contrariness of a so-called patriotic policy." For a careful and unfavorable economic appraisal, see Farid W. Saad, "France and Oil" (Ph.D. thesis, M.I.T., 1969).

[143] *Actes du Premier Colloque Franco-Italien d'Economie de l'Energie* (Paris: Dalloz, 1963), p. 399.

need to be self-sufficient in oil, or to have imports no more than exports, is an end in itself.

Another expert, looking from a very different angle, also confirms this thesis. M. Huré of BP-France notes that in 1959 newly discovered Sahara and Gabon crude was taken up by French refiners, who received a declaration of "commercial non-aggression . . . in return for the service being rendered."[144] Thus the interests of old-timers and newcomers were reconciled by suppressing competition. The sales of the "orphan production" were assured without any need for them to fight for outlet–i.e., add to the supply of open-market crude. In 1962 larger amounts were arranged, though at lower prices. But the private companies were resentful in 1963 when the State companies received a larger share of the French market despite the private firms' "heavy obligations of taking finished products and crude" from the State companies. However, further negotiations improved matters, though it is not specified how.

M. Huré does not object to fully integrated State or State-controlled companies, and expects the exploration program, begun at $300 million annually in 1965, to escalate to $380 and $560 million in 1970 and 1975. (Possibly these have been superseded by the much larger amounts mentioned above. Actual subsidies have been in the range of $50-$60 million per year.)[145] Nobody, he notes, can say what return there will be on this investment. But "it is perfectly defensible to aim at compensating the drain, however modest, that the international groups necessarily impose on the French economy to supply it, by a profit obtained under the same conditions in supplying foreign countries." The "necessary drain," to be compensated by exports, makes sense only if loss of autarky is bad in itself. For if the price of the imported oil makes it as good a choice, or better, than any other energy source, there is no drain but a benefit. Contrariwise, the French economy is drained in real terms if high-cost oil is found at a poor return on investment. But if world prices can be kept at current levels, the problem disappears; for even high-cost oil will cost no more than is needed to buy oil from abroad today.

Thus the various glimpses we get of the aims of French oil policy are consistent: self-sufficiency is an end in itself, to be secured by owning concessions abroad. This will be made easier by maintaining high oil prices. To the extent that French policy becomes a model for EEC policy, therefore, the influence of the whole Community will be thrown on the side of higher prices.[146]

Other EEC Countries. Much attention has been paid in *Germany* to the plight of domestic companies, and considerable sums have been loaned to help them to explore for crude.[147] In 1964-69 $200 million was budgeted, but presumably not all was borrowed.

[144] Joseph Huré, speech to Institut des Hautes Etudes de la Défence Nationale, Jan. 24, 1966.

[145] *Petroleum Press Service*, Feb. 1970, p. 57.

[146] In October 1969, a long-term sales contract for about $500 million worth of oil was signed between CFP and Hispanoil. Previously, "French approaches . . . had always been turned down as if, in the eyes of the authorities of Madrid, French capacity had no 'credibility,' oil being by nature the affair of the Anglo-Saxons." . . . "Hispanoil, controlled directly or indirectly by the state to the extent of 55.8 percent, will become one of the principal instruments of Spanish oil policy, which has a certain family resemblance to French policy started in 1928. [Apparently Hispanoil is to] provide at least 50 percent of the oil needs of the Spanish economy. One can therefore expect that Spain will take her place among the states impatient to be involved directly with the international sources of oil without passing by the usual route of the international oil groups." (*Le Monde*, Oct. 9-15, 1969, p. 11.) Hispanoil has also secured an oil concession off Kuwait in return for a guaranteed market for its partner KNPC, once oil is found. So far, none has been.

[147] See *Petroleum Press Service*, Mar. 1968, p. 95; Sept. 1968, p. 324.

This gives the government a vested interest in their success, and therefore, in a higher crude price.

In *Italy*, the national company Ente Nazionale Idrocarburi (ENI) has been in exploration and production for years,[148] endowed by the Italian government with half a billion dollars, of which $327 million were paid in by 1967. In 1966-67, its capital expenditures on exploration and production were $219 million, or just over 10 percent of total expenditures by all companies in Europe, Africa, and the Middle East. Production has not been commensurate. Leaving Italy aside—for both economic and political reasons—ENI production capital expenditures abroad in 1965-66 (the 1967 total is not divided as between Italy and other) amounted to $100 million. Output was 121 TBD in 1965; 103 TBD in 1967, to which we may add 90 TBD to allow for lost production in the Egyptian and Nigerian fields after May 1967.[149] The net growth in capacity was thus 72 TBD (= 103 + 90 - 121). If 1965-66 output is assumed to be respectively 121 and 136 TBD, and to decline at 10 percent per annum, capacity lost would have amounted to 26 TBD over two years, and the gross capacity increase would be 98 TBD. Thus the investment would be about $1,020 per initial daily barrel. But since much of the capital expenditure was for exploration, the true development figure must have been lower, say in the neighborhood of $700 per daily barrel.

Of course, the two years could have been unusually good or poor. An alternative is to take producing expenditures outside Italy during 1956-68, $410 million. Compared with capacity at the end of 1967 of 153 TBD, the investment per initial daily barrel is $2,680. This figure is biased both up and down. Since the expenditures were made over a decade, their 1968 value should include the cumulative earnings sacrificed by not being invited elsewhere. The amount would be much larger than $410 million. Offsetting this is the value of interim production. Moreover, since much of the outlay was for exploration, the figure would overstate development expense. Unfortunately, the data are too sparse to permit any better calculation. Obviously ENI costs are many times the Persian Gulf supply price, and capital costs alone—exclusive of operating costs and taxes—seem to be above market price.[150]

ENI has been "mending fences with the big private companies,"[151] but has criticized their excessive and "violent competition" which should be stopped by a world commodity agreement by producing and consuming countries. Higher prices are predicted.[152]

[148]The following paragraph is based on ENI, *Consolidated Balance Sheet and Statement of Income: 1966*, and *Annual Report, 1967: Abridgement of Board of Directors' Report*; CMB annual *Capital Investment* report; *Oil and Gas Journal*, annual "World Wide Oil" issue; and *ENI News*, Feb. 5 and Apr. 2, 1969.

[149]Lost Egyptian production is credited as the full 90 TBD produced in the first five months of 1967 before the June war, in the Belayim offshore and onshore fields where ENI had only a half interest. In Nigeria, ENI has a half interest (with Phillips) in the Ebocha field. By June 1970, production had not yet begun. Hence no allowance is made for 1967 capacity.

[150]At a 20 percent rate of return and 5 percent decline rate, the present-barrel-equivalent factor would be 4.46, capital cost would be $1.65. Adding an optimistic 10 cents for operating cost and 65 cents tax, total cost-plus-tax would be $2.40—far above any market price. But again the reader should be warned of the scanty available data. If investment per initial daily barrel were only $700, then on the same assumptions the crude oil cost ENI would be only $1.18, which would have been a good price for medium to low quality crude oil in years past, though not in 1967 or later years.

[151]*Platt's Oilgram News Service*, May 12, 1965, p. 3; *Wall Street Journal*, Jan. 23, 1968, p. 32. See also "ENI's New Image," *Petroleum Press Service*, Dec. 1965.

[152]See, e.g., ENI, *Energy and Petroleum in 1965*, Milan, pp. 14-17.

In April 1967 ENI, ERAP, and eight German producers presented a joint memorandum to the EEC Commission, proposing that they be given financial aid and protection from the competition of the international companies.[153] There have been discussions about merging ENI with ERAP and perhaps the German companies to form one grand "truly European" company. Less competition would be favorable to the profits of the company or companies, to whose success the EEC would be committed. Mr. Yamani, in the speeches already quoted (above, pp. 211ff.), said that European national oil companies "have a good future." Conflict of interest may arise if they "try to obtain crude supplies at the lowest possible cost. However, this problem is likely to be mitigated by the fact that their national oil companies will be in partnership with ours at the producing end."

As for the smaller EEC members, the *Netherlands* is a net seller of oil because of Shell. Moreover, the government is itself a royalty-owner on gas from the great Gröningen field. Gas prices are set to meet oil prices plus a premium for being clean, convenient, and controllable. Large buyers pay about 40 cents per million Btu.[154] Other Community countries have objected that Dutch gas prices are too high, but to no avail.[155]

Belgium has for years been burdened by a high-cost coal industry in the Walloon areas, exacerbating the language dispute which has been so bitter in recent years. It may finally have decided to close out its coal industry.[156] One compelling reason was to keep its steel industry competitive. The conclusion is not minced: "From now on, none of our coal production has any further interest for our economy on which it is a very heavy charge; only social/regional considerations indicate some slowing-down in the cutback of production." (Pp. 7-8.) Each Community country has its own coal problem and must work out its own rate of cutback, but it is a Community problem to see that the cutback does not last too long.

Coal should be quickly replaced by nuclear energy, but not by Dutch natural gas unless it is treated as really a Community resource. (Unless, we may read, the Dutch substantially reduce their prices.) But every obstacle to the cutback in coal should be removed.

The emphasis on nuclear energy is price-supporting, but the others are price-depressing.[157] Belgium has one large refiner, Petrofina, which produces only 14 percent of the crude it consumes, but is able to buy (presumably at prices comparable with the integrated firms), from BP on long-term contract, and predictably dislikes "price wars brought on in part by American independent oil companies. . . . This has had dramatic repercussions on thinking on oil policy" in the EEC.[158]

Great Britain. We have seen that commitment to coal and to nuclear energy tends to make Britain support high prices. We must also add North Sea gas, which the Gas Council buys at 28.7 cents per million Btu, with higher prices for later production. But the selling price to be fixed by the Gas Council has been little discussed. The Council will be able to do what the oil companies could not do with respect to coal: price a little below the

[153] *Petroleum Intelligence Weekly*, Apr. 5, 1967; *Petroleum Press Service*, Nov. 1967, p. 408.

[154] *Oil and Gas Journal*, July 22, 1968, p. 33, reckoning 935 Btu/cf.

[155] *Platt's Oilgram News Service*, May 10, 1965, p. 1; May 14, 1965, p. 1; July 14, 1965, p. 3.

[156] The rest of the paragraph is a paraphrase of: Royaume de Belgique, Ministère des Affaires Economiques, Direction Générale des Etudes et de la Documentation, *La Politique Energétique en 1967-1976* (Brussels, Dec. 1967).

[157] For the notable comment on security of supply, see below, p. 247.

[158] *New York Times*, July 20, 1967, p. 47.

nearest alternative, which will be oil, including town gas from light products. In 1968 the Council asked the nationalized British Steel Corporation to pay 50 cents per million Btu, or $20.70 fuel-oil equivalent,[159] just double the current European market price of fuel oil. A year later, it sold to two petrochemical manufacturers at something in the 40–45 cent range, $16.60–$18.60 fuel-oil equivalent.[160] Either the sights were lowered, or private companies were less accommodating than nationalized companies.

The interest of the Gas Council is therefore in a high oil price. It would afford high profits, to be reinvested in gas distribution, at no apparent cost to the Treasury or to consumers. It seeks monopoly profits not for private owners but to do good, and earn the plaudits of their fellow citizens. So little does price-output policy depend on the "true motives" of decision makers. And the unbuttoned romantic outburst that North Sea gas could be the most important thing in Britain since the Industrial Revolution reveals a state of mind more important for policy than any assessment of costs and benefits. High reference prices are also used by the Prices and Incomes Board, which has warned the Gas Council "not to entice new customers with promises of cheap North Sea gas now, when steep price increases may be necessary in a few years."[161]

Professor Odell points out that Dutch gas (above, p. 240) was being priced as though by an intelligent monopolist, whereat "France, Germany and . . . Italy were relatively content, for each had energy interests to protect." The British Gas Council buying price was "sufficiently below the Dutch price to compete in the Netherlands . . . , Belgium, France, and North Germany."[162] But the Gas Council selling price is even considerably above the Netherlands selling price; it is acting like a monopolist of gas enjoying an additional favor: the British oil market is less competitive than is the Continental market.

In 1967 the National Executive Committee of the Labor Party recommended a National Hydrocarbons Corporation to find and produce gas and oil offshore in the North Sea, and also to refine and possibly go international like ENI and ERAP. This was to be part of a comprehensive fuel policy, "at [whose] core will remain the need to maintain a robust and efficient coal industry." There is no claim that a nationalized company will produce anything better or cheaper. It is rather that "the public sector should advance where it was most needed—at the growing points of the British economy and in the new industries based on science."[163] State enterprise in oil and gas is considered a good in itself (although it has been pushed hard by conservative regimes in Continental Europe). As it turned out, the National Hydrocarbon Corporation was never formed, but the Gas Council and the National Coal Board have been awarded drilling licenses. If the National Coal Board's working partners are successful, an indirect subsidy may supplement the more visible direct one of high fixed coal prices. Finally, Britain, like the Netherlands, is (through domestic companies) a net seller of crude oil as well as natural gas.

Much and perhaps most of the "truly European" oil and gas production will be done by State-owned or State-supported companies. *The companies must not lose money lest the governments lose face.* The products they sell must be valued at a high enough price

[159] *Platt's Oilgram News Service*, Sept. 18, 1968.

[160] *Oil and Gas Journal*, Aug. 25, 1969, p. 60.

[161] Sir Henry Jones, quoted in *Wall Street Journal*, June 26, 1966, p. 15; Prices and Incomes Board, paraphrased in *Platt's Oilgram News Service*, Feb. 14, 1969.

[162] Peter R. Odell, "Natural Gas in Western Europe," Inaugural Lecture, University of Rotterdam (pub. Haarlem, 1969), pp. 10–12, 15, 16.

[163] Labor Party Fuel Study Group, *A National Hydrocarbons Corporation* (London, Aug. 1968), pp. 1, 9, 14.

to make them look successful. The competition of substitute products must be prevented, or their price advantage ignored. That is why heavy fuel oil prices in Europe have been ignored in the British and Continental coal industry accounts. If the coal had been valued at anywhere near the thermal equivalent of heavy fuel oil, the true losses would have been revealed and this was unbearable.

Let us look now at the two largest consuming countries outside of Europe.

The United States has an obvious interest in high crude oil prices. It supports a high-cost domestic oil industry, and the lower the world price becomes the more difficult it is to resist a flood of imports, or to justify the protection against it.[164] When the "Kennedy Round" negotiations for lower tariffs were begun in 1962, President Kennedy made a prior agreement with Senator Kerr that oil imports would be excluded.[165] Moreover, the United States is a very large net seller, because most of the crude sold to the world market is produced by American companies. Its real gain and foreign exchange gain is therefore large.

Japan[166] has become the largest importing nation in the world. As an export-oriented economy, it has a strong interest in low energy prices; but it also has a coal industry. The reluctance to allow direct investment by foreign companies led to most of the Japanese oil industry being conducted by joint ventures, half domestic and half foreign. The usual pattern was for the foreign producer to put up half of the total capital needed, and receive total crude supply rights indefinitely. But a considerable fraction of Japanese refining is done by wholly domestic firms (see Chapter III), and the government has been concerned with keeping them viable, promoting mergers of those that are too small. Price control of refined products has occasionally been resorted to, but only for a limited time. The Ministry of International Trade and Industry (MITI) seems well aware of the conflicting considerations—though awareness is not the same as solution.

Until 1962 crude oil prices to Japanese buyers were probably the lowest in the world. Allocations of foreign exchange were granted inversely proportional to the price (in foreign exchange) charged to the Japanese buyer; hence, an international company wishing to expand outlet had a good reason to reduce the price to its affiliates or to arm's-length customers. But with the abolition of this control in 1962, the price advantage was speedily lost, since at this very time prices to Europe were declining. By 1965 it was generally accepted that prices to East-of-Suez buyers were somewhat higher. The government set up a Comprehensive Energy Policy Commission, with members from industry, government, and universities. In 1966-67, the Commission suggested two lines of policy, both of which were speedily accepted. One was to lower the price of imported crude oil; subsidiary to this was a greater degree of independence in buying by the Japanese partners in joint ventures, possibly giving the foreign partner supply rights only in proportion

[164] This is not to deny that the United States has a security problem, or to assume that it should import freely. We need only the proposition that the U.S. government considers restricted imports and high world oil prices to be in its own national interest. Recall the reception and shelving of *The Oil Import Question: A Report on the Relationship of Oil Imports by the Cabinet Task Force on Oil Import Control*, Feb. 1970. See also the discussion on import controls, chap. V, pp. 150ff.

[165] *Wall Street Journal*, Oct. 30, 1962, p. 2.

[166] The following paragraphs are based largely on: Ching C. Chen, "Crude Oil Prices and the Postwar Japanese Oil Industry" (Ph.D. thesis, M.I.T., 1966); Petroleum Association of Japan, *The Petroleum Industry in Japan*; annual; Japan Energy Research Center, *Quarterly Reports on Energy*, and English translations of some of the reports and recommendations of the Comprehensive Energy Policy Committee. The Japanese industry would well repay a special study, and I regret the need to slight it here.

to his ownership. The other policy was to have a considerable fraction of Japanese consumption covered by Japanese (private) companies. Loans and subsidies were to be provided for exploration.[167]

This policy really goes back to 1958, and to its encouragement we probably owe the very successful Arabian Oil Company (AOC), which found an extension of the great Safaniya field off Saudi Arabia, and then four other large but so far undefined and largely unexploited fields.

In the long run these two Japanese policies are contradictory; whatever reasons support one condemn the other. If crude prices are expected to rise, Japanese buyers should not seek independence but sign long-term contracts at current prices or even higher. If crude prices are expected to decline, it is best to keep commitments to a minimum and be a buyer, not a seller. But in the short run these opposing policies may be viewed as a way of keeping open two options, and even as complementary. " 'We cannot negotiate with the majors to secure cheap crude now,' argues an official of MITI. 'We haven't any bargaining power.' " And according to one of those searching, oil produced abroad by Japanese companies "must be cheaper than Mideast oil"[168]—cheaper, that is, when account is taken of taxes as well as costs. Depending on developments over the next few years, the Japanese government may decide to push vigorously in one direction and withdraw from the other. In either event, what happens in Japan may be more important for the world oil trade than what happens anywhere else.[169]

India seemed in 1960 to be choosing a low price policy. But this was subordinated to independence per se. Refining growth was largely reserved to the public sector by limiting private refineries' expansion. Publicly owned refineries have been slow to go on stream. Their crude contracts had been expensive.[170] Because the crude oil supplier provides much or most of the refinery investment, the refinery is not free to bargain, and product prices will be set to cover a multitude of overpayments. A 1963 contract, giving a U.S. company supply rights, was wildly extravagant and was renegotiated to $1.48 for Iran light,[171] some 10-15 cents above the current market price, and 20-22 cents above the late 1968 price. Subsequently the transfer price of crude for private refineries has been marked down; but the price has at all times been well above the market. In 1968 it was reduced to $1.38 f.o.b.[172] but market prices were 5-15 cents less. In early 1970 a further reduction to $1.28 brought them close to market levels.[173]

[167] *Platt's Oilgram News Service*, Aug. 27, 1968.

[168] *Wall Street Journal*, Aug. 20, 1969, p. 28.

[169] One anomaly may be mentioned, all the more striking because of the generally high level of industry and government discussion: the building of very small and uneconomic nuclear power plants. The first, by Kansai Electric, is only 350 mw, less than half the size of Dresden II.

[170] Except as noted, this paragraph is based on *Petroleum Press Service*, *Oil and Gas Journal*, and *Platt's Oilgram News Service* for the 1960 price negotiations between India (particularly Mr. Malaviya's use of Russian oil offer); Government of India, Ministry of Steel, Mines and Fuel, *Report of the Oil Price Enquiry Committee* (Delhi, 1963) ("Damle Report"); Ministry of Petroleum and Chemicals, *Report of the Working Group on Oil Prices* (New Delhi, 1965) ("Talukdar Report"); and B. Das Gupta, "The Supply and Price of Imported Crude Oil to India," *Journal of Development Studies*, vol. 3 (1967), pp. 249-66.

[171] *Platt's Oilgram News Service*, Aug. 17, 1965. A high processing margin was also guaranteed (*Petroleum Press Service*, Jan. 1969, p. 4).

[172] *Petroleum Intelligence Weekly*, July 1, 1968.

[173] *Petroleum Intelligence Weekly*, Feb. 9, 1970. Later in the year, Arabian light was accepted at $1.29 (*Petroleum Intelligence Weekly*, July 13, 1970). But others were buying at $1.25 (*Oil and Gas Journal*, July 13, 1970, p. 36).

There is a small, very high-cost domestic producing industry (see below); offshore exploration has been discussed but not begun. A joint company (Iminoco), with ENI and others in offshore Iran, did find the commercial Rostam field, but no market could be found quickly. At the same time, therefore, as the transfer price to foreign-owned refineries was being negotiated down, the oil ministry let it be known that it had spurned some offers as too low, and had to "resist the temptation" to sell at low prices which would make it difficult to get the higher prices expected later. Hence India prefers not to lift its Rostam crude, leaving its investment unproductive.[174] The very small and expensive nuclear reactors have already been mentioned. To minimize foreign exchange costs, expensive coal is used, especially on the railways. India seems committed to self-sufficiency and high prices.

A Note on Indian Crude Oil Production Costs. Under Indian law, oil production and related statistics are classified as defense secrets. However, it is official that during 1957-66 inclusive, the Oil and Natural Gas Commission (ONGC) spent \$306.6 million on oil exploration and production. In 1960 the Commission discovered the Ankleshwar and Cambay fields. At the end of 1966, producing capacity was stated as 52 TBD (thousand barrels daily).[175] Cumulative production was approximately 34.5 million barrels,[176] or 94.5 TBD; hence at a 10 percent decline rate, lost capacity was 9.45 TBD, and gross capacity created in 1961-66 was 61 TBD (=52 + 9), at a cost of \$5,030 per daily barrel. Assuming 10 percent return (no political risk),[177] and a 10 percent decline rate, the PBE factor for a 25-year development period is 4.94 (continuous compounding) and capital cost per barrel is \$2.79.[178]

Assume the currency to be overvalued by 50 percent through 1965 and by 20 percent thereafter, and about one-sixth of total expenditure to be made in 1966. Average overvaluation must therefore be 45 percent. If domestic content of oil production investment made in India was 60 percent, overvaluation of the total investment expenditure must be 27 percent. Applying it, the capital cost in domestic resources was \$2.18. Operating expenses may be guessed at 30 cents, or around \$2.48 total.

During this period the official import cost of crude oil was probably around \$1.75–though this figure is a crude guess. Hence oil exploration-cum-development was a large net loss. But the adverse verdict must be tempered by considering that exploration is a probe of the unknown. Had large prolific fields been found, the return might have been handsome.

A separate question has to do with investment in the development of the existing ONGC fields. Of the 1957-66 expenditures, about one-third may be guessed as being for exploration and two-thirds for development, since these were the approximate proportions for the whole area east of the Persian Gulf during 1957-66.[179] Indian drilling

[174] *Petroleum Intelligence Weekly*, July 20, 1970.

[175] ONGC special report, summarized in *Oil and Gas Journal*, Nov. 6, 1967.

[176] The mid-point between cumulative production in the first six months of 1966 and of 1967. *Oil and Gas Journal*, Dec. 26, 1966 and Dec. 25, 1967.

[177] By the formula in chap. II: $c = I/365q_0 \int_0^T e^{-(a+r)t}\, dt$, where I = investment, q_0 = initial daily output, r = rate of return, a = decline rate, t = time in years, $T = 25$. The integral is the PBE factor.

[178] Edward S. Mason, "Transport and Energy in India's Development," in Edwin T. Haefele, ed., *Transport and National Goals* (Brookings, 1969), p. 17, referring to some 1963-64 studies.

[179] Chase Manhattan Bank, *Capital Investments of the World Petroleum Industry*, annual.

activity was a substantial and growing proportion of the total during this time—over one-third in 1962-66.[180] Development expenditures were therefore roughly 0.67 × \$306.6 = \$204 million, or \$3,340 per initial daily barrel. Using the same cost of money and decline rate, development capital cost is \$1.855 per barrel. Applying the 27 percent currency correction factor yields \$1.46; operating costs would bring the total to about \$1.76.[181] Hence, further development investment would not seem worth while, though some or even most existing wells are worth operating until operating costs rise too high. As partial corroboration, one may note the opinion of the ONGC that 1964 prices did not cover their costs. It is true that the Talukdar committee appeared skeptical of this claim, or complaint.[182]

One writer has called the ONGC program "a major economic success."[183] He reaches this conclusion by calculating the discounted cash flow that would take place from producing and selling the discoveries made through 1961, valuing the oil by an import parity of \$1.75, less 25 cents per barrel "real social cost to the economy of producing the discovered oil."[184] The meaning of this "cost" is not stated, but since no discounting is mentioned, and no overvalued currency, Mr. Tanzer must be referring only to current operating costs. Let us assume them to be correct. But he assumes that the 1966 production rate builds up by a factor of about 2.5 *net* by 1972, *with zero development investment*. It is a massive error. As was indicated in Chapter I, the error is allied to that of considering oil production as an industry of increasing costs.

Mr. Tanzer is fortified in his belief by the hosannas to ONGC by a journal favorable to private enterprise. This kind of reasoning is dangerous at best, and some familiarity with the oil industry and public policy in other countries would have imposed more caution. Conservatives, like radicals, treat oil as a good in itself, particularly when their implicit oil reference price is far above its true value.

Security of Supply

The literature on European energy policy has been heavily infused with "security," and "assurance of supply." The obvious meaning would seem to be that there is a risk of interruption of fuel deliveries which would impose a heavy economic cost. It is precisely analogous to buying insurance—or becoming a self-insurer—against fire, hurricane, theft, etc. In any year the likelihood of the event is small, but when multiplied by the expected loss if it does occur, the danger is worth buying off through a small current outlay, fixed and known in advance.[185]

One expects, therefore, to find elaborate analyses of the types of risk of interruption; their chances of occurrence; measures that would prevent or offset interruption—and their relative cost. This is precisely what one does not find. Even *Perspectives* is no exception,[186] and in a special report in 1968 the EEC Commission notes that the Com-

[180] *World Oil*, annual "International Outlook" issue.

[181] Even the official f.o.b. price to India was only \$1.28 for 34° Iranian light in early 1970. A likely long-term tanker rate (Worldscale 40) makes the Persian Gulf–Bombay cost about 9 cents per barrel; to Calcutta, nearly 18 cents. Import parity would be \$1.37 to \$1.46.

[182] *Talukdar Report*, p. 31.

[183] Michael Tanzer, *The Political Economy of International Oil and the Underdeveloped Countries* (Little, Brown, 1969), p. 228.

[184] Ibid., p. 405.

[185] See also Appendix to Part Three.

[186] European Coal and Steel Community, *Nouvelles Réflexions . . .* , p. 29.

munity only produces 5 percent of the oil it uses, unlike any other "*grande espace économique*" except Japan. Hence it must pay attention to security "while taking account of the economic aspects of the problem."[187] Security is not explained.

The Belgian memorandum of December 1967, discussed above, is a notable first in viewing security as a problem in probabilities of interrupted supply. Only when it is understood that coal is no longer a useful product, but a burden to the economy to be shed as rapidly as possible, then and not sooner does it become possible to look directly at security as a problem in insurance against delivery stoppage.

"Security" usually means fear and distrust of the large international oil companies. Only once, to my knowledge, has this aspect of "security" found its way into print: a 1965 speech by then EEC Commissioner von der Groeben, who stated that if the oil companies run European coal out of business, they will have a monopoly of fuel and possibly exploit it.[188] Mr. von der Groeben did good service by airing this viewpoint, but tact overbore truth, and nobody carried on the discussion. It is a durable legend the world over that large businesses are usually on the prowl for smaller rivals whom they can ruin through local price cutting. Mixed with this is the common complaint that since American oil companies enjoy the advantage of a light tax burden on a sheltered, profitable American market, they have a competitive advantage in European markets, which European governments must somehow offset. Or else the governments must prevent or at least limit competition.[189] This is the worldwide popular wisdom about large diversified companies. Little is gained by pointing out that capital is never spilled but always drawn to where a good return beckons. The more competition is limited, and the more profits are guaranteed, the more incentive for entry by foreign companies. Moreover, the less economic power the oil companies have, the more they are forced to compete and drive prices down, the more they are execrated for their "ruthless economic power."

There is also a fear that the oil companies can be used by the American and British governments to help or hurt or exert pressure against other governments. This belief will, but should not, survive the events of June 1967, when the United States and Britain were precisely the objects of the brief Arab boycott, which ended when the local governments decided it should end.

The interested reader would do well to note how often a press item will mention a nation's "need" for oil, or desire for "assured supplies," as *obviously* explaining some economic or political policy. Political influence is sought for the sake of "access" to oil. The economic cost of tying in with a producing nation is justified by the hope of political influence. Thus "independence," "security," and "influence" form a closed system, excluding economic costs and benefits. Something very important is assumed to be in

[187]European Economic Community, *La Situation Actuelle du Marché de l'Energie dans la Communauté*, July 31, 1968, p. 53.

[188]Summary in *Platt's Oilgram News Service*, Sept. 17, 1965, p. 3; the complete text (mimeo.) is well worth reading.

[189]"In a market free from government intervention, competition is subject to the risk of *abusing competition*, which is as bad in the long run as no competition. *Independent companies are threatened or disappear* in the face of companies infinitely more powerful because of the excessive profit leverage which can only be resisted by companies enjoying a particularly favorable treatment in their home countries, as well as access, in the same country, to an interior market with wide profit margins." Pierre Desprairies, "L'Europe et son Pétrole Quotidien," *Revue Française de l'Energie*, vol. 202 (1968), pp. 222–27. (Emphasis in original.)

that circle, but it remains a mystery to the writer.[190] In the following passage, observe that "security" is first mentioned without explanation and then referred to as though it had just been explained:

> The six countries of the Common Market cannot fail to attach a major interest to the security and the price of their supply. . . . [Discovery and diversification] solve their problems only in part and in the long run. New methods must be sought, different from those which leave only to the oil companies the care and responsibility of this security. . . . Relations . . . will not cease to progress between oil producing and consuming countries, and most particularly between the Arab world and Europe. . . . We are approaching, no question about it, a growing cooperation between two parties whose interests, often contradictory in appearance, are in fact tightly linked.[191]

For the present analysis, it does not much matter whether this kind of "security" has any meaning at all. It is believed; therefore it is. If one fears to buy nearly all fuel from the large oil companies, if this is "dependence," if one feels "insecure," then one *is* insecure, and something must be done to ease the feeling. Some large part of fuel supply "must" be provided from "one's own" companies. Therefore heavy investments will be made, and commitments to higher prices, and these facts must be incorporated into the forecast.

In the long run, if their producing ventures in oil, nuclear energy, etc., turn out badly enough, some nations may give up the attempt and try to minimize the real cost of energy. Indeed, to some extent they will have a foot in both camps: try to decrease the cost of oil to the national economy while charging high prices at home and transferring large amounts to the government producing interest, as does the British Gas Council. But this kind of two-faced role is limited, for what keeps the price up is the maintaining of barriers to competition.

The United Nations and the Prospects of a World Commodity Agreement

The underdeveloped countries are extremely concerned with price maintenance for primary commodities: it was incorporated in the UN resolution noted above (p. 211) and has been repeatedly demanded since then at all official and nonofficial levels. The former chairman of the Council of the Food and Agriculture Organization has called for higher prices as part of the worldwide struggle against hunger: "We must de-mystify the problem. . . . We must have done with neo-colonialism and must reestablish trade equilibrium, constantly degraded by the fall in the prices of primary materials."[192] The United Nations Conference on Trade and Development (UNCTAD) has demanded greater economic equality by price supports. Although the United States and Britain opposed the creation of UNCTAD, they agree with this aim: President Kennedy originally called for support of primary product prices as a basic part of U.S. foreign economic policy; and the

[190] There has also been some academic discussion, associated chiefly with the late Professor Byé, of security of supply in the form of policies to reconcile the interests of both groups of countries by industrializing the producing countries, who are to have "a true mastery of oil exploitation within their borders." See J. M. Martin and Edith Roth, "La Motion de Sécurité des Approvisionnements . . . dans le Cadre de l'Association avec les Pays Producteurs," Convegn o Internazionale sull'energie, Rome, Mar. 11–13, 1968 (multi.).

[191] Essences et Lubrifiants de France/Entreprise de Recherches et d'Activités Pétrolières (ELF/ERAP), *Bulletin Mensuel d'Informations*, May 20, 1969.

[192] *Le Monde*, Mar. 20, 1963, p. 17, interview with Josué de Castro.

promotion of the world wheat, sugar, and coffee agreements fitted the action to the word. The Council of Ministers of the Organisation for Economic Co-operation and Development (OECD) has urged "concerted policies" to "increase the receipts [of] the less developed nations [from] their exports, as much from basic raw products as manufactured goods."[193] The Soviet Union had for some time been in favor of a world economic conference for this purpose. As a petroleum exporter, and courting the underdeveloped nations, the Soviet political and economic interest is in supporting prices. At an August 1962 conference of Marxist economists, the principal paper estimated that the developed nations were getting $14-16 billion annually from the underdeveloped by means of "unequal exchanges."[194] Since the total exports of the developed capitalist countries to the underdeveloped amounted to around $20 billion in 1961,[195] this implies either that their "equitable" value was only $4-6 billion, a 400 percent overcharge, or else that the "equitable" value of the underdeveloped nations' exports was really around $34-36 billion, a markdown of 40 percent. It is a good round sum either way, but one might like a little more detail on which way.[196]

If price support for primary commodities is sought, oil will not be last on the list. The underdeveloped countries are mostly oil consumers, but solidarity will make them favor price maintenance for oil, at least in the short run. Furthermore, some of them have, or hope for, their own oil industries, and the higher the price, the better these look.

France has long favored price-supporting world commodity agreements as an economic policy.[197] At the 1968 New Delhi UNCTAD conference, the disagreement between France and the United States was confined to orders of priority.[198] At the request of 14 African governments and France, the World Bank made a study of commodity stabilization plans, and in mid-1969 announced that it would be more active "in helping countries . . . to participate more effectively in international commodity arrangements aimed at stabilizing world prices."[199]

Although it is always difficult and often impossible to translate the policy of price support into an actual agreement, the forces supporting it are strong, though diverse. In

[193] See *Le Monde*, Nov. 30, 1962, p. 22.

[194] *Current Digest of the Soviet Press*, Sept. 19, 1962, pp. 13-14.

[195] Organisation for Economic Co-operation and Development, *Foreign Trade*, Series A, Aug. 1962. The "developed" nations are taken as the United States, Canada, Japan, OECD Europe, Australasia, South Africa, and Israel; the Soviet and the Chinese blocs are excluded; the rest of the world is taken as "underdeveloped."

[196] An interesting dialogue between Emmanuel Arghiri and Charles Bettelheim appeared in *Le Monde*, Nov. 27-Dec. 3, 1969. Arghiri argues that the higher wages in the developed countries are reflected in higher prices which form an "unequal exchange," oppressing the less developed countries. Bettelheim argues that prices determine wages, not vice versa. Hence the workers in the developed countries are only passive beneficiaries of the "unequal exchange." But he does not question the fact of "unequal exchange." This disagreement among Marxists would not have been published in *Le Monde* had there not been widespread interest by non-Marxists. Neither tries to explain what is meant by "unequal exchange," though both are deeply and sincerely troubled by its implication. Should either have recalled their master's phrase: "the fetishism of commodities"?

[197] France, Ministère d'Etat, *La Politique de Cooperation avec les Pays en Voie de Developpement. Rapport de la Commission d'Etude . . . remis au Gouvernement, le 18 juillet 1963*. See especially vol. I, pp. 94-100, and vol. II, appendices 4, 19, and 23.

[198] See *New York Times*, Feb. 6, 1968, p. 55: "In France's view, the accent must be on stabilization of commodity prices, the expansion of technical cooperation, the development of human resources, and the 'judicious application of preferences,' more or less in that order. The U.S. priorities begin with tariff preferences, commodity agreements, and diversification of products, and move on to areas the French never mention. . . ."

[199] International Bank for Reconstruction and Development, press release, July 9, 1969.

addition to the desire of both producing and consuming countries for higher oil prices, many in the West believe, like the Russians, that price ratios have been "unfair" and are getting worse. They feel that the West has been grinding the faces of the poor nations by paying too little and charging too much. Government officials find increasing difficulty in obtaining appropriations for aid to the underdeveloped world; they look to commodity agreements as a way of getting aid labelled as something else, which bypasses the budget and apparently costs nobody anything.

A UN report on exploitation of underseas mineral resources outside territorial boundaries is revealing.[200] One might expect at least a perfunctory acknowledgment that prices may be (in any sense) too high or too low. But the only danger mentioned is disorderly marketing, whereby small production increments break the price level. This is, of course, one of the surest symptoms of a price above competitive levels, which is what the nations favor.

The mere existence of much favorable opinion does not mean that there will be any agreement for any specific commodity. Experience with wheat, sugar, tin, coffee and olive oil agreements, as well as the attempts to set up others, shows that the participation of the consuming countries is absolutely necessary. Mr. Yamani, in the speech quoted above (p. 212), recognized this. Since most of the consuming countries, accounting for the overwhelming bulk of world consumption, do want high oil prices for one reason or another, the prospects of an oil commodity agreement being made do seem at least fairly good: whether good enough one cannot say.

[200]UN Economic and Social Council, *Resources of the Sea*, Part I, Report of the Secretary General, E/4449 Add. 1, Feb. 19, 1968.

CHAPTER VIII

Conclusions: The Tehran-Tripoli Agreements of 1971

The events leading up to the Tehran and Tripoli agreements of early 1971 mark a new stage in the evolution of the market. The crude oil price decline has been reversed, in real terms, back to 1963.[1] The producing nations now have a cartel tolerated by the consuming countries and actively supported by the United States. But there is as yet no effective limit on output, nor division of the market, and hence no change in the long-term outlook.

The oil companies have been further demoted, and are now the cartel's "tax collecting agency," as BP's chairman has well said. The dramatic impact of the price increase is important. Government and public opinion are much impressed by statements too vague for analysis about how the tide has turned and we now face a long-term shortage. These are the latest but surely not the last version of "growing demand will dry out the surplus"—no more relevant than it was during 1957-70. Through the middle of 1971, at least, there is nothing emanating from industry or government that has not been dissected above. The higher prices have come as a relief, justifying the costly attachment to coal or uneconomic nuclear power, the search for "their own" oil, and other price-supporting reflexes. The restraint on competition will help the cartel in the future as in the past.

The Events: A Brief Summary

During 1969-70 demand for oil products increased somewhat more rapidly than had been expected in Western Europe.[2] As Table VIII-1 and Appendix Table VI-C-7 show, the increase was mild and well within the normal range of year-to-year fluctuation. The percent of refining capacity utilized was slightly higher in both Europe and Japan, thus raising incremental cost.[3] While there was no shortage in refining capacity overall, there

[1] The price of a barrel of 31° Iranian was about $1.16 at the end of 1969 or early 1970. Assume a tax increase of 50 cents transferred to price, or a new f.o.b. of $1.70. The general price level in the United States in the last quarter of 1970 was about 27 percent higher than the 1963-64 average. (See Tables VI-3 and VI-6.) Therefore, a price of $1.70 equates to about $1.34 at 1963-64 prices. The derived Persian Gulf netbacks for 1963 and 1964 were $1.59 and $1.29, respectively. In a 1963 paper, I estimated the arm's-length price of a 34° crude at the Persian Gulf at about $1.40 per barrel, and considered a long-run price of $1.00 as more likely than not. ("Les Prix Pétroliers à Long Terme," *Revue de l'Institut Francais du Pétrole*, Dec. 1963.)

[2] Good accounts of the 1970-71 events may be found in the trade press, especially *Petroleum Press Service* and *Petroleum Intelligence Weekly*. The daily newspapers' coverage, notably *Platt's*, was of course much more detailed, hence harder to follow. Tanker rates are from Platou.

[3] The percent of refinery capacity used in Japan averaged 95.0 and 96.5 in the respective fiscal years 1968 and 1969 (ending on March 31 of the next calendar year). Petroleum Association of Japan, *The Petroleum Industry in Japan* (1970), part 4.

TABLE VIII-1. 1970 Consumption Increase, Principal Areas

	Consumption			
	Average annual change (*percent*)			1970
Area	1960-70	1965-70	1969-70	(*million BD*)
Western Hemisphere	4.5	5.3	4.3	18.6
Western Europe	11.8	10.0	11.2	12.7
Japan	21.3	18.5	18.0	4.0

Source: *BP Statistical Review of the World Oil Industry, 1970.*

was one in heavy fuel oil, to feed a growing demand for low-sulfur fuel on the U.S. East Coast. Perhaps more important, this aggravated the chronic tanker shortage; during January-May 1970 the spot tanker rate was at Worldscale 120, the highest since late 1967 (1969 average 85), and the Rotterdam composite price was $2.24 (1969 average $2.04). (See Tables IV-2 and VI-B-1.)

The higher prices and profits for refined products went largely to the integrated companies who were also producing the oil, above all in North Africa, where the producing country governments demanded they pay higher taxes. Then in May 1970 the Trans-Arabian Pipeline (Tapline) was blocked by Syria,[4] and the Libyan government began production cutbacks for most of the companies operating there in order to force them to agree to higher taxes. The direct effect of the cutback was small, for relatively little crude was involved. But the effect on tanker rates surpassed all former experience: the *average* single-voyage rate for July-December 1970 was about Worldscale 230, and Rotterdam product prices rose to $3.33 per barrel for the second half of the year. The Libyan cutbacks, helped by the Tapline closure, multiplied the very shortages which had created the cutbacks and the soaring prices and profits.

The companies producing in Libya speedily agreed to a tax increase that was only a fraction of the price increase. The Persian Gulf producing countries then demanded and received the same increase, whereupon Libya, which had not rescinded the cutbacks, demanded a further increase, and the Persian Gulf countries followed suit. The final results of the two successive negotiations, in November 1970 and in February 1971, were agreements increasing tax-royalty payments at the Persian Gulf as of June 1971-December 1972 by about 47 cents per barrel, rising to about 66 cents by 1975. (The specific example is Iranian heavy 31°, used here as standard.) North African increases were larger; over the same period Libya gained nearly 80 cents, though perhaps 15 cents is only temporary.[5]

The additional payments are best reckoned from 1969, since 1970 was affected by the temporary agreements of November, and by disturbances of production growth patterns. The per-barrel increase is approximately 50 percent over 1969 as of the middle of 1971, and escalates to about 75 percent as of 1975. It is very conservative to assume that the agreements will last until then, with only a modest further increase to 100 percent by 1980. Assuming a continued 11 percent growth per annum in production, the factors of increase are, on the average, about as follows:

$$\begin{aligned} &\text{for 1971:} \quad (1.11)^2 \times 1.50 = 1.85 \\ &\text{for 1975:} \quad (1.11)^6 \times 1.75 = 3.28 \\ &\text{for 1980:} \quad (1.11)^{11} \times 2.00 = 6.12 \end{aligned}$$

[4]The shortage would have been alleviated to a minor degree by the use of the Trans-Israel Pipeline to full capacity. For political reasons, this was ruled out.

[5]See *Petroleum Press Service*, Mar. 1971, p. 162; *Petroleum Intelligence Weekly*, Supplement, Apr. 5, 1971.

Payments to governments in the Persian Gulf, Libya, and Venezuela in 1969 were \$6,723 million.[6] Hence the payments now envisaged or programmed for those three years are \$12.4 billion, \$22.0 billion, and \$41.2 billion. Aside from the omission of Algeria, Indonesia, and minor countries, these are *minimum* estimates of the payments if current market control is maintained.

The cost to consumers will be greater, however. The companies intended to raise crude and product prices not only to cover the excise tax increases but "actually to increase their margins and return on investment";[7] to "leave some over," as was said in Great Britain, where the February tax increase, of 28 cents per barrel, was matched by a price rise of 42 cents.[8] There has been great diversity from country to country, but the best summary of the results was doubtless that given by Kenneth E. Hill: the agreement "truly is an unexpected boon for the world-wide industry . . . As a result earnings in the downstream function in the Eastern Hemisphere are for the first time in recent memory quite attractive . . ."[9] Mr. Hill rightly emphasizes product prices, since the great bulk of world oil goes through integrated channels, but in arms'-length crude oil transactions the price increase has apparently more than covered the tax increases.[10] The tax increase, like the American price increases of 1953 and 1957, gave the clear signal to which the group could respond without collusion. If they show the same prudence, the retreat from the new peak will again be slow.

We can therefore expect the following pattern during the early 1970s. From time to time, either in pursuance or in violation of the Tehran-Tripoli agreements, the tax is increased, whereupon prices increase as much or more, but then tend to erode as the companies compete very slowly at the crude level and less slowly at the products level. Thus over the near term prices increase, in steps, yet at any given moment there is a buyer's market–i.e., more is available than is demanded at that price.

The companies' margin will therefore wax and wane, but as Mr. Hill and others have pointed out, they benefit by the new order. They cannot, even if they would, mediate between producing and consuming nations. As individual competitors, they are vulnerable to producing-nation threats to hit them one at a time. As a group, they can avoid paying for a settlement and even profit by it through raising prices in concert, for the settlement is that clear signal to which they respond without collusion. The Secretary-General of OPEC has said truly that there is no basic conflict between companies and producing nations. As the head of Shell put it, there is a "marriage" of companies and producing governments. Most precise of all was the chairman of BP, who called the companies a "tax collecting agency" for both producing- and consuming-country governments.[11] There is, however, a difference in kind between serving a government in its own country to collect revenue from its own citizens, and serving a government to collect revenue from other countries. As agents of foreign powers, the multinational companies are vulnerable in most consuming countries.

6 *Petroleum Press Service*, Sept. 1970, p. 324.

7 *Middle East Economic Survey*, Feb. 19, 1971.

8 *Platt's Oilgram Price Service*, Feb. 22, 1971.

9 *Oil and Gas Journal*, May 10, 1971, p. 46.

10 *Petroleum Intelligence Weekly*, June 28, 1971.

11 Dr. Pachachi of OPEC, quoted in *New York Times*, Feb. 11, 1971, p. 14; Sir David Barran of Shell, speech to Fuel Luncheon Club, London, Feb. 16, 1971, p. 2; and Sir Eric Drake of BP, statement to stockholders, in BP *Annual Report 1970*.

Irrelevance of Supply and Demand. The higher crude oil and product prices have no connection with world supply and demand for crude oil. They reflect no scarcity of crude oil present or foreseen. The 1970 shortages in refining capacity and tankers were not only temporary; if permanent, they would tend to depress the demand for crude oil. (On a longer-term view, if pollution proves a resistant demon and requires heavy taxation or restriction on oil uses, that will also lessen demand.) When prices are raised by increasing scarcity, agreements are superfluous. None were needed to raise tanker rates or refining margins.

As prices declined in the 1960s, there were no reports of long-term contracts at higher prices than current. Indeed, the longer-term offers or deals tended to be at slightly lower prices than the shorter. (See Appendix VI-I-2.)

If there had been an increasing long-run crude oil scarcity, with discovery falling behind consumption, there would have been a gradual rise in the investment per unit of new oil producing capacity needed to offset decline and increase output. In fact, over the ten-year period from 1958-59 to 1968-69 the investment needed per barrel of new crude oil capacity *fell* substantially at the Persian Gulf and even in Venezuela, despite the rise in the general price level. (See pp. 62-63.) Of the 74 fields now operating in the Persian Gulf, 31 were found during the 1960s. (See pp. 69ff.) Whether the new fields are bigger or smaller than the old will not be known until there have been years of development history. (See p. 37.)

Even if crude oil investment and labor requirements were to increase in real terms several fold, they would be unimportant. The cost of providing capacity and operating it in the big Persian Gulf concessions does not exceed 10 cents per barrel, assuming a 20 percent rate of return after tax. By using some drastic (unfounded) assumptions of zero new discoveries, a faster growth rate, etc. the 1980s cost could be projected up to 20 cents. (See pp. 69-73.) But such figures are so small as to be lost in the error of estimate of the tax increases.

Therefore, one cannot account for the crude tax and price increases by a "surge in demand" against supply constraints, or by any kind of competitive hypothesis. It is not the first case of water running uphill. During 1950-1957, the price increased despite its being many times costs. (See pp. 156-58.) The "real price" increase (adjusted for general price level changes) was not much different from the current boost, but it was spread over nearly a decade, while there was as much talk with as little basis of an Eastern Hemisphere "energy gap" and growing long-run scarcity.

Supply and demand are as irrelevant to the future price as to the past. A price set by supply equalling demand means a price in the neighborhood of incremental cost, which in crude oil is chiefly the return on the new investment. In prudence, the worst possible supply situation has been imagined—that of zero discoveries; even so, the price equating demand and supply is still only a negligible fraction of the actual price. Hence the reader can safely put out of mind any forecasting of supply and demand conditions over the next 10 to 15 years. They are the tail, not the dog. The only thing that matters is whether the current market control, which explains the enormous margin, will flourish or fade. As will be seen later, theory and experience both suggest that if and when the United States becomes a larger importer, the effect will be to lower prices. (Of course in a competitive market the effect would be to raise prices.)

The Producing Nations Extend Their Market Control. It has been shown above that the newcomers have been a minor though not negligible factor in the declining price level after 1957. (See pp. 199–202.) The old internationals were perhaps reluctant rivals; but competition kept breaking or creeping in, and long before the end of the decade, they had lost control of the market. The price-cost gap lay overwhelmingly in taxes and royalties. The producing nations exerted most of the market control—passively. By refusing to consider lower taxes, they put a floor of tax-plus cost under the price.

The important new fact is the active cooperation of the producing nations. Nobody foresaw the extent to which they would act together and threaten to withhold supply if their terms were not met.[12] They could do this because the industry was integrated. The profits of higher product prices went, not to independent transporters and refiners-marketers, but to the producing companies whose operations and properties were within their reach.

The multinational oil companies are the agents not only of oil-exporting governments, but also of the United States, without whose active support OPEC might never have achieved much. When the first Libyan cutbacks were decreed, the United States could have easily convened the oil companies to work out an insurance scheme whereby any single company forced to shut down would have crude oil supplied by the others at tax-plus-cost from another source. Had that been done, some or most companies might have shut down, but the Libyan government would have faced a loss of production income. Bank deposits abroad could have been frozen in retaliation for the seizures, but such a step, however helpful, would not have been strictly necessary. OPEC unity would have been severely tested at a time when they were unprepared for conflict. The revenue losses of Libya would have been gains to all others, and all would have realized the danger of trying to pressure the consuming countries. Any Libyan commander of a division, brigade, or perhaps even a regiment could consider how he might gain a billion or several billions of dollars a year by issuing the right marching orders. In the interval before the coup and the new government, the shortage of tankers might have inflicted upon consumers some tiny percentage of the costs they were to suffer for failing to resist.

Mere failure to act does not necessarily imply that the United States favored the result. But there was unambiguous action shortly thereafter. A month after the November agreements with Libya, a special OPEC meeting in Caracas first resolved on "concrete and simultaneous action," but this had not been explained or translated into a threat of cutoff even as late as January 13.[13]

The turning point came on January 16, when the companies submitted their proposals for higher and escalating taxes.[14] The United States convened a meeting in Paris of the OECD nations (who account for most oil consumption) on January 20. The record of the meeting has never been made public and probably never will, but (as will become clear

[12] I testified in March 1969: "In company-government disputes . . . the obvious tactic of host governments would be to stop all production in order to avoid being bargained down one at a time. Host governments may also pressure the companies' home governments by the threat to cut off oil supplies. . . ." *Government Intervention in the Price Mechanism, Hearings before the Subcommittee on Antitrust and Monopoly of the Committee on the Judiciary* (U.S. Senate, 91st Cong., 1st sess.), p. 17. This is not far from what actually happened, but it was not a forecast, only a warning (made also in 1967) that a concerted shutdown was likely enough to warrant some insurance against it. During early 1971 there were many public statements to the effect that the OPEC action had often been predicted by American oilmen. An examination of examples offered in the trade press shows them all to be vague talk of "pressure," "demands," etc. If demands were dollars, we could all be rich. But certainly I was much mistaken in thinking the chance of a threatened cutoff was small.

[13] *New York Times*, Jan. 14, 1971, p. 2, interview with Iranian Finance Minister, Dr. Amouzegar. *Wall Street Journal*, same date, has no reference to any retaliation.

[14] *New York Times*, Jan. 17, 1971, p. 1; Jan. 19, 1971, p. 2.

below) there is no doubt that the American representatives, and the oil companies who were present as observers, assured the other governments that if they offered no resistance to higher oil prices they could at least count on five years' supply at stable or only slightly rising prices.

The OECD meeting could have kept silent, thereby keeping the OPEC nations guessing, and perhaps moderating their demands for fear of counteraction. Or the spokesman might have said that he was sure the OPEC nations were too mature and statesmanlike to do anything drastic because after all the OECD nations had some drastic options open to them too—but why inflame opinion by talking about those things? Instead, the OECD spokesman praised the companies' offer, and declined to estimate its cost to the countries whose interests he claimed to represent. He went even further in stating publicly: "Contingency arrangements for coping with an oil shortage . . . were not discussed."[15] Thereby the consuming countries revealed that there were no plans for rationing, for cooperation in managing stockpiles which were in any case inadequate, or for any other common defensive action against the threat of a concerted shutdown. Before January 20, an open threat by the OPEC nations would have carried little credibility in view of the previous failure of even mild attempts at production regulation. After the capitulation, threats were credible and were made often,[16] culminating in a resolution passed on February 7 by nine OPEC members, including Venezuela, but not Indonesia, providing for an embargo after two weeks.[17] This was a world away from Caracas. Dr. Amouzegar, the Iranian Finance Minister, who in effect was chief of the producing nations' team, said: "There is no question of negotiations or resuming negotiations. It's just the acceptance of our terms."[18] The companies were resigned to this, but wanted assurances that whatever they agreed to would not be changed for five years.[19] This was the substance of the "agreement" on February 14. Soon thereafter, the North African nations and Nigeria reached similar agreement, while Venezuela superseded its earlier agreements, legislating new taxes without the formality of negotiation.

The United States government was active throughout this period. Under Secretary of State Irwin arrived in Tehran the day after the Paris meeting, publicly stating his government's interest in "stable and predictable" prices,[20] which in view of the meeting meant higher prices. A personal letter from President Nixon to the Shah was not delivered because it was feared the Shah might not like it. Mr. Irwin explained the plight of Europe and Japan if oil supplies were cut off.[21] Perhaps this explains why the Shah soon thereafter threatened a cutoff,[22] since there is no more effective incitement to drastic action than the assurance that it will damage your opponents severely. The State Department thought "Mr. Irwin's mission persuaded the oil producing nations of the need to enter real negotiations."[23] They did not try to reconcile this with the public record, nor with Dr. Amuzegar's statement that there was nothing to negotiate.

[15] *Wall Street Journal*, Jan. 21, 1971, p. 26; *New York Times*, Jan. 21, 1971, p. 11; *Oil and Gas Journal*, Jan. 25, 1971, p. 82; *Platt's Oilgram News Service*, Jan. 21, 1971.

[16] *New York Times*, Jan. 24, p. 18; Jan. 25, p. 1; Jan. 26, p. 5; Feb. 3, p. 1; Feb. 5, 1971, p. 3; *Wall Street Journal*, Feb. 2, 1971, p. 11; Feb. 4, 1971, p. 2.

[17] *New York Times*, Feb. 8, 1971, p. 9; *Wall Street Journal*, Feb. 8, 1971, p. 18.

[18] *Wall Street Journal*, Feb. 9, 1971, p. 4; see also *Wall Street Journal*, Feb. 12, 1971, p. 11.

[19] Ibid., and see *New York Times*, Feb. 13, 1971, p. 7.

[20] *New York Times*, Jan. 18, 1971, p. 5.

[21] *Platt's Oilgram News Service*, Mar. 1, 1971; the meeting had been six weeks earlier.

[22] *New York Times*, Jan. 25, 1971, p. 1.

[23] *New York Times*, Jan. 26, 1971, p. 5.

On the morrow of Tehran, while other capitals maintained a glum silence President Nixon said he was pleased.[24] The State Department had previously stated the objective as "stability, orderliness, and durability . . . We don't want anybody . . . badly hurt."[25] At a special press conference, and later, it called the Tehran agreement acceptable, referring many times to "stability" and "durability." "*They expected the previously turbulent international oil situation to calm down following the new agreement.*"[26] They must really have believed this. Otherwise they would not have tried to claim credit for Mr. Irwin[27] or for Secretary of State Rogers.[28] They must have said the same thing at the Paris meetings in January, and probably in May 1971; the oil companies were still saying so. We now live with the consequences.

The Instability of Unexploited Market Power

Nobody knows, even approximately, how high the OPEC nations can raise taxes, and thereby prices, before sales drop off so rapidly as to reduce their total take. But clearly taxes can be increased several times without reducing sales, because oil consumers have few alternatives. Synthetic crude oil from coal or shale or tar sands appears very expensive. In automotive transportation, there is no substitute for gasoline, whose price is only a minor fraction of the total cost of buying and operating a vehicle. (In the longer run—and increasing air pollution may make it shorter—higher prices would mean smaller, cleaner-burning cars.) Nuclear power sets a fairly high ceiling for heavy fuel oil for electric power production; for general industrial use, European or Japanese or even American coal set a far higher ceiling. In industry, fuel costs are usually a rather small part of the total cost and, even when they are substantial, a higher cost imposed on all the consuming industries in all the industrial nations has no effect on competition among them.

The average price in Europe of a barrel of oil products in 1969 was between $9.00 and $13.00 per barrel. (Some respectable authorities disagree with each other.)[29] If the new tax rates were doubled to, for example, $2.80 per barrel, a straight pass-through into prices would be an increase of only 10-14 percent. It is doubtful that such an increase would have an appreciable effect on oil consumption. The price increase in 1969-1970, far from slowing down the growth in consumption did not even prevent an unusual increase.

Moreover, about half of the European oil price consists of taxes levied by the various consuming-country governments. The producing nations have long insisted that in justice they *ought* to receive some or most of this amount. One can argue that forever. But given sufficient control of supply, most or all of this tax take *can* be transferred from consuming to producing countries. It is a much bigger prize than what the producing nations have yet achieved.

[24] *Wall Street Journal*, Feb. 16, 1971, p. 4.

[25] *Platt's Oilgram News Service*, Jan. 19, 1971; also Jan. 28, 1971.

[26] *New York Times*, Feb. 17, 1971, p. 3; *Platt's Oilgram News Service*, Feb. 17, 1971; Feb. 18, 1971.

[27] *New York Times*, Feb. 17, 1971, p. 3.

[28] *Wall Street Journal*, Mar. 5, 1971, p. 1.

[29] Sir David Barran, chairman of Shell, gives $9.00 as "the ultimate selling price" of a barrel. (Speech to Fuel Luncheon Club, London, Feb. 16, 1971, p. 2.) But the Direction des Carburants gives the 1969 European average as Fr. 235 or $7.60 per ton ex-tax, with the tax slightly more. (Quoted in *Le Monde*, Jan. 29, 1971.) Then the price including tax would be slightly over $95.20 per ton, or $12.70 per barrel.

This is an odd, unstable condition. The OPEC nations still have great unexploited power because price can be greatly increased without provoking a loss in sales and in total revenue. Therefore they are likely to try again soon. Yet, with no prorationing system in sight they cannot control the level of output and allocate markets. Their market power can only be exercised through brinkmanship.

The Tehran and Tripoli events have wrought an irreversible change. Oil supply is now much less secure than ever before. In order to keep and extend their gains, the producing nations must keep the consuming countries insecure. The genie is out of the bottle, the producing countries have been extremely successful in using the weapon of a threatened concerted stoppage, and they cannot be expected to put it away. What we must expect from the nations, singly and then jointly, are such actions as new tax increases, insistence on retroactive payments, and penalties for violating real or fancied regulations. The producing nations may refrain from raising posted prices, thus observing the letter of the agreement; but even so, there is a conflict between the Tehran-Tripoli agreements and the OPEC 1968 resolution (see p. 211), which make any fiscal arrangement subject to change because of "changing circumstances." Taxes and prices will be raised again, and again.

One need not waste time with assurances that the Tehran and Tripoli "agreements" are stable and durable or that they will last for five years. If such agreements were worth anything the present crisis wouldn't exist.[30] But, as Johnson told Boswell, it is all right "to say to a man, 'Sir, I am your most humble servant.' . . . You may *talk* in this manner; it is a mode of talking in Society; but don't *think* foolishly." One does not suspect the oil companies of thinking foolishly.

Were the producing nations content to remain the owners of the oil concessions, using the companies as their tax collectors, they could raise taxes much more and hold the line for years before chiselling on tax rates to gain additional volume began to make inroads into prices. But they appear more firmly committed than ever to participation, and the next campaign may be for a share in the established concessions. Their better income and cash position will speed up the participations and takeovers. Possibly first on the list is the Iranian Consortium concession, which ends in 1979. After Tehran, the Consortium's option to renew is worthless. But the Consortium cannot simply be allowed to cease drilling years before 1979 and run down its facilities. Investment decisions must be made years in advance. For the sake of an orderly transition, which is in the interests of both parties, the takeover must begin around 1975, and negotiations will probably start before this study is published. Where Iran leads, Saudi Arabia may not be far behind. One need not speculate how many other independent producing Persian Gulf nations will still exist at that time—certainly, fewer than today. There may be no concessions left by 1975; none by 1980.

Takeover by a government company does not necessarily mean departure by the local international oil company, who will most likely stay on as contractor for production and sale, although some or all of the concession might be reoffered. It is only a difference in degree from the international company's function today. "Nationalization" or "expropriation" is not a cataclysmic change, but a continuum. There are many profitable ways in which a private company serves a government as agent. The Tehran and other agreements mark a movement along the spectrum, "takeovers" still another; neither need be painful to the companies involved. But we must expect considerable turnover. Success has

[30] *Petroleum Intelligence Weekly*, Feb. 8, 1971.

hitherto been with companies owning large low-cost deposits. But the 1970s will see the entry of the can-do oil company, which can outbid the incumbent for the use of those deposits because of its lower costs or better market connections.

The only real break is when a government simply expels a company, or insists on conditions so unprofitable that the company is better off leaving, and there is no replacement. The producing countries know that the departure of oil companies would speed up the competition and hence do not wish to expel them. A big refiner-marketer which is no longer a producer can be a formidable buyer. If Algeria, Libya, and Iraq were direct sellers, offering some 6 million or more barrels daily in the Mediterranean, world prices would crumble.

Whether it operates directly or through a foreign contractor paid in money or oil, a national company does not have a price floor at cost-plus-tax, but at cost alone. At current prices even a mediocre field (by Persian Gulf standards), producing for sale at cut prices, returns several hundred percent per year to its government-owner; a good one, 600–700 percent.[31] It is hard to resist such an incremental investment and return, especially when a rival is ready to enjoy it if you do not.

Unless the producing nations can set production quotas and, what is more important, obey them, they will inevitably chisel and bring prices down by selling incremental amounts at discount prices. This becomes uncontrollable if national companies go downstream. For any one of them, downstream operations seem irresistible, both to show the world what they can do and also to sell incremental output. Iran seems determined on this course, and it may be very profitable. But what is good for the individual is not good for the group. There are too many ways to chisel. One may transfer oil to downstream subsidiaries or partners at high f.o.b. prices, at lower tanker rates, or with generous delivery credits. There are refinery deals whereby the producing nation puts up most of the money but takes a minority participation, or lends at less than market interest rates. One can arrange buy-back deals, barter deals, and exchanges of crude in one part of the world for availability elsewhere. The world oil cartel of the 1930s was eroded by this kind of competition, and so will be the new one in the 1970s.

The Passive Consumers. The consuming countries are impotent because they are divided among themselves, and within each country.

The United States has a large producing interest in the world market. Estimates for 1971 show a production by American companies of about 6.4 billion barrels outside the United States.[32] As noted earlier, the price increases exceed the tax hike, and this is normal in a noncolluding oligopoly, which uses the tax hike as a signal. For every cent of increase in the price above that in tax, there is an additional $64 million in profit. For example, if the British experience (see p. 252) were universal, a gain of 14 cents per barrel would mean $900 million more. Since the United States is an importer, there is an offset roughly estimated at $250 million. One need not guess how great will be the net gain; there may not even be any after a short time. But the United States will probably be unscathed and even benefited by price increases. Moreover, higher energy costs will now

[31] As shown in table II–8, p. 76, current operating costs at the Persian Gulf are about 5 cents per barrel, and another 5 cents provide a 20 percent rate of return. Hence a price of (say) $1.60 yields $1.55 over operating cost, or 620 percent per year.

[32] See Chase Manhattan Bank, *Balance of Payments of the Petroleum Industry* (1966), appendix III. Since American companies were said to produce 55.3 percent of non-U.S. non-Communist output, their 1970 production may be approximately reckoned at 0.553 x 28.7 million barrels daily x 365 = 5.8 billion barrels; and 1971, if 10 percent higher, would be 6.4 billion barrels.

be imposed on its competitors in world markets—in petrochemicals, higher raw material costs as well. The United States has a large domestic oil-producing industry. The less the difference between domestic prices and world prices, the less is the political tension between producing and consuming interests.

The United States desired the goodwill of the producing nations. Buying popularity with someone else's money is always appealing, but there must also have been a desire to mitigate the tension caused by the strife between Israel and the Arabs. Yet if the Arab-Israeli dispute were settled tomorrow, the oil market would be no less turbulent. The producing nations would have no reason to slow down for one minute their drive for ever-higher taxes, and the rewards and dangers of a concerted shutdown would be no more and no less. The acknowledged leader of the Persian Gulf nations in February 1971 was Iran, which has collaborated with Israel in supporting TIP; in contrast, the United States early and late opposed the pipeline (p. 336).

The Arab nations in 1970 accounted for less than 60 percent of OPEC oil. As in 1967, partial boycott is a ritual gesture, a sign of frustration, not a policy. When he was Secretary General of OPEC, Francisco Parra stated that he could not conceive of any situation arising from Middle East politics that would lead to oil company expulsion (which is or should be much more worrisome to the United States than any boycott). Nor did he believe oil could be used as a political weapon by withholding supplies from market: " . . . there just can't be an effective selective embargo."[33]

One cannot say how important these various reasons were, because there has been no public discussion or exposure of them. Giving them all full credit, American support of OPEC seems badly mistaken for one outstanding reason: it has severely damaged security of supply to friendly countries, and before very long, to ourselves as well. (This aspect is considered further below.)

Not only do the consuming countries have divergent interests, but for reasons given earlier each nation is divided within itself. Most of them are producers and sellers of energy and are much more concerned with apparent success as producers than with economic benefits. Higher oil prices make expensive coal and nuclear power projects look better, or at least less bad. Some countries have or seek producing interests in the Persian Gulf or other oil producing areas, avowedly for the sake of "control" or "security" of supply. In view of the recent events, this sounds more strange than ever, but the consuming countries will not soon overcome their obsessive fear and dislike of oil companies as big and foreign, which they are, and as a cartel, which they are not. No large consuming country appears ready to accept that: (1) the *only* threat to security of supply is a concerted stoppage of the oil flow by the producing governments; (2) the international companies can do nothing good or bad for security because they have no control of supply or power to cut off anybody; (3) "owning reserves" is meaningless because nobody owns reserves any more except the governments who have the physical force above ground. But producing interests have an attraction going far beyond any calculation of economic costs and benefits. When President Pompidou explained the complexity of the French interest in the world oil market, he did not mention as a possible objective the minimizing or lowering of energy costs to the French economy.[34] Abandonment of Algerian oil production as costing more than it is worth would acknowledge the failure of a 10-year-old policy. (See pp. 235-36.)

[33] *Wall Street Journal*, Sept. 8, 1968, p. 9.
[34] *Le Monde*, Jan. 22, 1971.

The important consuming countries have export interests which will benefit by the higher oil prices because of the oil producing countries' greater purchasing power. Those segments of industry profiting are relatively small, but the gains are large for them, the losses diffused over very many, and hence their political influence will be much greater.

The balance-of-payments impact of higher oil prices will be small or favorable in nations that have large money markets—chiefly the United States, United Kingdom, and Switzerland. For the less developed nations there are no such offsets, but the most likely reaction is a renewed demand for price supports for their own raw materials.

All things considered, it is not surprising that the consuming countries have done nothing in response to the threats of a concerted shutdown. They will remain passive for at least a time, content to waste more money on high-cost oil substitutes and make a transfer payment of over $12 billion in 1971, an amount far larger than what they transfer as aid to less developed countries[35] and rising to over $40 billion in 1980. But the producing nations cannot help making them increasingly insecure, because only the threat of cutoffs can maintain or further increase oil prices. The need for security will require large stockpiles, emergency plans for severe rationing and taxation, and, most important and difficult, concerted action among at least some countries, since it is impossible for all to act together. Consuming countries may learn in time that lower prices and secure supply are one and the same. Power to withhold is power over price. Once the producing nations cannot cooperate through their tax collection agents, and must compete, prices will be less and security more.

Moreover, some though by no means all of the OPEC nations will use their funds in ways the consuming countries will find uncomfortable—buying jet fighter-bombers, for instance, or supporting "liberation" movements in and outside the Middle East. Ceasing to support the price of oil might have political as well as economic dividends.

It is even possible that this schooling in real security problems will drive out the obsession with "access" to petroleum. Probably no word is more familiar in newspaper accounts; and it should be, for the concern over "access" is great. Rarely has a word been so compact of error and confusion. Nobody has ever been denied access to oil: anyone willing to pay the current price could have more than he wanted. One may assume what he likes about future demand, supply, and market control, and conclude that the future price will be high or low, but that price will clear the market in the future as in the past. The worry about "access" assumes something very queer indeed: that *all* of the producing countries will join in to refusing to sell to some particular buyer—for what strange motive is never discussed. One catches hints that such a boycott will favor one consuming nation over another. Why a producing country should engage in this expensive pastime is never discussed, although it takes only one other country, with a desire for gain, to cure this irrationality. "Access" is damaging to rational policy because it is incessantly repeated, never explained. Explicit discussion may work a cure.

Finally, the position of the United States will probably change in time as it becomes a more substantial importer. Were the price of oil set today by supply and demand, this change would be a force for higher prices through a causal chain: greater output leads to

[35] Angus Maddison, *Economic Progress and Policy in Developing Countries* (W. W. Norton, 1970), pp. 233–34, estimates government aid to all less developed countries in 1967 at $7.7 billion. But about one-fourth must be deducted as loans at normal interest rates; hence the net transfer was $5.8 billion. During 1950–67 aid increased 9.3 percent per year. At that rate, the 1971 transfer to less developed countries would be $8.2 billion.

more intensive development of known fields, requiring greater investment and labor inputs per barrel of new capacity. (It will also induce more exploratory effort, but this should not be added to the higher development cost, for it would be double counting: more intensive development and more exploration are substitutes, not complements.) Higher capital and labor requirements per barrel of new capacity are another way of saying that higher incremental cost is necessary for greater output.

But this causal chain is not relevant because the price is already so many times higher than even the higher incremental cost. The rules applicable to a competitive market will not work in so awesomely noncompetitive a place. Greater American imports mean a lower price for two rather different reasons. First, the prospect of greater sales and profits is a potent lure for sellers to shade the price and give away part of a very wide margin over cost for higher total profit. There are American refiners with substantial crude deficits, who will be ready to buy large amounts if the price is right—and large buyers are a great help to competition. Experience confirms this. When quotas on heavy (residual) fuel oil were lifted in the United States in 1966, imports increased considerably and price declined. In Europe, the rapid expansion of the fuel oil market had had the same effect some years earlier.

The other reason is that a rapid expansion of American imports would mean an additional drain on the balance of payments. The United States would no longer be the gainer on rising oil prices because of the favorable effect on oil company profits. On the contrary: oil profits will probably be held to a limit representing an acceptable rate of return on the oil company development. Lower oil prices would not mean lower oil company profits, since those profits would already be close to the minimum. The American oil deficit would overwhelmingly be payments to the producing nations. These payments, as monopoly revenue, are perfectly dispensable.

Thus in the end the United States and the other consuming nations can have the price of oil they want. To lower it, all they need do is take Sheik Yamani's advice (see p. 215) and remove the multinational oil companies as the OPEC nations' tax collection agents. Their producing affiliates might well remain in place as contractors paid in money or oil. But if the companies became buyers of crude oil the indispensible cartel machinery for monitoring compliance with the excise tax agreement would disappear. There would no longer be a floor of tax-plus-cost, but only one of cost. The producing nations would be forced into competition with each other and, as Sheik Yamani correctly warned, there would be a steep decline in the world price.

The consuming countries are not about to give up their search for "their own" oil and their attempts to have their own companies be sellers of crude oil. But some consuming governments may take a few modest steps to buy a little cheaper than others. The basic principle is to provoke competition among suppliers. The more often they face the market test, the greater the likelihood that their coordination will slip and that they will be forced to give some ground on prices. Some governments may allocate foreign exchange inversely to prices, as Japan did until 1962. In issuing refining permits, they may favor independent refiners and unite them along with electric power companies in buying consortia. They may limit or forbid long-term contracts, disallow "excessive" transfer prices, and find ways of taxing international companies a bit more heavily than domestic. Heavier excise taxes on oil products would probably be absorbed by sellers when margins are very wide. The most important result might be a learning effect. If the countries consuming most of the oil act together, they can use an excise tax to force the price *down* quite as well as the producing nations have used an excise to force prices *up*. Instead of cost-plus-tax to the producing nation becoming a floor to price, price-less-cost-less-tax to the consuming country would be a ceiling to tax payable to the producing nation. Moreover, producing governments reducing their own taxes most quickly will be able to gain large increases in sales and profits as compensation. Private companies will have hard going in the middle. The most important move is direct dealing with producing governments to obtain special low prices by offering incremental sales. Downstream investment by producing nations is a particularly good way to hide price discrimination. The international companies may be involved, but without the power of decision, since one must settle price with principals not agents.

The companies are needed and will survive, as a class, as contractors for the producing governments if nothing else. It would be a grievous waste of resources to displace them as a group, and not because of the "huge amounts of capital" they furnish. This is an industry slogan which will serve the industry badly. Capital requirements in the large producing areas are very modest, and the producing nations will have many times more than they can possibly use. But the real strength of the companies, and their value to society, lies in their expertise with the process and logistics of producing, transporting, refining, and marketing oil, and in the personnel resulting from decades of a search for excellence. Human capital accumulation is the most important in oil as in every other activity.

A Summing Up

The high reference prices and continued heavy investment in oil and gas exploration and development, and in nuclear power, mean an ever greater potential supply pressing against limited outlets. The only means of lasting oil price stability is by international agreement, whereby the inevitable long decline in prices might be spread out over a generation. Of course, even the agreement must gradually buckle and fail as companies and governments evade it, find loopholes, chisel and cheat.

Host government revenues per barrel will at first be stable but then will decline almost cent for cent with prices, as these governments in fact, and later in form, emerge as the owners. Some concessions will be expropriated for a time or for good, and some companies will find it cheaper to buy from other companies or from government companies than to produce from their own wells. Profit margins will for some companies be dangerously squeezed, but in the long run they will all or nearly all survive and prosper, for their services are essential and the margin allowed them cannot be narrowed past a certain return on investment.

By 1985 the world energy outlook will have changed in ways we cannot foresee, and the current problems may be moot—or more urgent than ever. In the meantime, world oil looks to be more and more a politically dangerous struggle among companies and governments over markets, and between companies and governments over profits. The more they invest and make oil cheap, the more they will hold it dear. There will be much to-and-fro by eminent gentlemen to meetings wherein will be discourse of justice and oppression, threats and promises, dramatic exits and quiet reentries. Patched-up deals will last awhile; painfully drafted "final" agreements will be forgotten before the ink is dry. Somehow the oil will continue to be produced at ever-expanding rates. It will be an interesting story. No mere economist can hope to do it justice, but some will try.

* * * * * *

FROM THE TIME MACHINE

[The following document was retrieved from a trip in a Time Machine. Unfortunately, the letterhead and date are missing, and there is no identification of the writer or addressee, but the text is almost completely legible. –The Author]

As you are aware, the United States Congress recently changed the law to permit a maximum of [illegible] of the country's crude oil consumption to be supplied by imports, starting two years hence. This means an increase of 3 million barrels daily over and above the current rate. Our country and others have therefore been approached by a number of American refiners, who are actively seeking the best terms on which they could buy very large amounts of crude oil under long-term contracts. It is obviously an attractive prospect, but has raised a problem which calls for immediate decision.

The Alpha Petroleum Corporation, of New York and Houston, has put in a firm bid for half a million barrels daily, delivery to begin on a smaller scale but escalating up to that figure over 30 months. Our own contractor company has estimated that they would need to drill 50 new wells, half of them in sites already chosen, and the other half either by closer drilling or in some pools discovered earlier but not altogether delimited. No exploratory drilling is needed.

Since the comprehensive contract revision of [illegible–probably a date] whereby the concession was formally changed to a service contract, our contractor has been paid his operating expenses plus enough to return him 20 percent per year on a discounted-cash-flow basis. We have had his estimates audited, and can confirm that the payment which would be due him on this new project would be about 17 cents per barrel. It might be less, depending on some drilling estimates which cannot be altogether precise. At the current price for the stated grade of oil, the net return to our national treasury would be $1.25 per barrel, $230 million per year.

The problem is that Alpha is offering 10 cents less than the current price, alleging that they have received an offer at 8 cents less, for oil of admittedly slightly higher quality, from [illegible]. The ambassador of that country has denied any such offer, and so has their minister of petroleum. Our problem is whom to believe. In our opinion, Alpha is telling the truth. Their reputation is excellent, while the minister in question has not been candid with us in the past–you are familiar with the matter–and we consider him hostile to our government and nation. They would like us to let the good bids pass by. This agrees with the account which we have obtained from our usual and hitherto reliable source, whom you know.

The Alpha bid would bring us $15 million a year less than what we would have at the current price. But in seeking this small addition we seem to be endangering the other $215 million, since it is our judgment that Alpha will get the offer from others. The small price reduction does not endanger the revenues we receive under existing contracts. Furthermore, if we close the deal quickly, we may be able to persuade Alpha to increase the amount to as much as a million barrels daily, since we have made some rather careful estimates of the requirements of their European and Japanese subsidiaries, and of the pattern of their production, much of it from more expensive sources.

We therefore recommend: that Alpha be asked to pay 5 cents per barrel under the market, and if they persist in demanding 10 cents off, to propose that the contract be for 750,000 barrels daily, perhaps over a longer buildup period than 30 months. But if necessary, we should take the original terms of 500,000 daily at 10 cents off.

Appendix to Part Three

Security of Supply For The Eastern Hemisphere

[This appendix was first circulated as a working paper of the Department of Economics, Massachusetts Institute of Technology. Completed in October 1967, it has not been published in English, but has appeared in Direction *(Paris),* Economia Internazionale delle Fonti di Energia *(Milan), and* Enerugi *(Tokyo). It is followed by the author's "Afterword"* *(p. 274)**, written in 1971.]*

Security of fuel supply for Europe, Japan, and the rest of the non-Communist Eastern Hemisphere can be had, not only without additional cost, but at a large net saving. Indeed, the 1967 crisis will be a blessing if it forces a hard look at some facts that governments have not been able to recognize or act upon because of domestic political pressures and fixed positions.

What is the Security Problem? There have been two sudden reductions of supply ten years apart, and there can be more at any time on short notice. In one way, these crises are like fires or accidents–we want assurance against being struck without warning. But assurance of fuel supply actually reduces the chances of "fire." For if the Area is secure, the threat to deprive it of fuel is empty, and the attempt is unlikely.

Supply was reduced by the Suez Canal closure, and the cost of being unprepared will not finally be paid for some years. But by 1969-70, the Canal will be only a minor instrument and its closure a minor nuisance, since large tankers will in any case be carrying much of the load, and west-of-Suez productive capacity will be much increased. It was largely a happy accident. Contrary to the fable that will be accepted as future history, no statesman's foresight provided those tankers. On the contrary, they were built by Japanese (and Swedes) and bought by shipowners and oil companies for the sake of profit and in the teeth of accusations of unfair competition and unpatriotic conduct. The shipbuilding capacity is in place, however, and the 1967 shutdown will keep it going harder and longer.

Our concern now must be with production. The worst possible is a shutdown by the concerted action of all or nearly all producing countries. No single producing country matters. Even in 1951-54, Iraq, Kuwait, and Saudi Arabia moved quickly to fill the 35 percent gap left by the Iranian shutdown. In the winter of 1966-67, when the Syrians stopped the flow from Kirkuk in Iraq to the Mediterranean, there was not a ripple in the slightly changed flow pattern. Yet the loss of Iranian output was 660 TBD, that of northern Iraq nearly 900,000.[1] It is a measure of the growth and maturation of the industry in the 15 years' interval.

[1] In 1950 Iran output was 663 TBD, of a total Middle East output of 1,760 TBD, or 37.6 percent. See annual reviews, recently entitled "World Wide Oil," in the *Oil and Gas Journal (OGJ)* and recently *Oil and Gas International (OGI)*.

Loss of two of the Big Five of the Eastern Hemisphere (Iran, Iraq, Saudi Arabia, Kuwait, Libya) would be at least a minor nuisance; three might be serious, at least for a year to 18 months. Within that time, the lost capacity could, if necessary, be replaced by the rest of the oil-producing nations. But as it became apparent that the producing country was risking a permanent loss of its chief or only source of revenue, either the government would allow production to resume, or else it would be overthrown by those who wanted the revenue.

Hence the extreme of the security problem is clear enough: *be prepared for a total cessation for a limited period.*

Traditional Ways of Attaining Security. Diversified supply of oil, as well as domestic coal production, have been the favored means to security, but diversification has been the accidental result of oil companies seeking oil. Policy had nothing to do with it. The widespread impression that the 1957 Suez crisis led to exploration in North and West Africa is not true; Nigeria had been explored for years, and its first major field had already been brought in.[2] Libya acreage had been taken up in 1955, and in Algeria two major fields (Hassi Messaoud and Edjele) had been found after decades of search and some minor finds dating back to 1918 in what was then French North Africa.[3] Diversification has been rather a disappointment, though not altogether useless. It has helped to establish substantial new oil-producing countries since 1957, and the more such countries there are the harder it is to plan and enforce a total shutdown. But anything which makes cooperation likely, or nonconformity difficult, makes diversification less effective. The opening of great new areas in North Africa threatened to be no help and has been of limited help, since these areas joined in the temporary embargo, such as it was. The quarrel of one nation in the Middle East-North African area has a good chance of being the quarrel of all. The Libyan oil workers' union headquarters in Cairo delayed resumption elsewhere. Only the Nigerian shutdown could be called an unlucky accident, unrelated to the Middle East crisis. Venezuela, Iran, and Indonesia, unaffected by the embargo, are all old-timers in oil. Any search for diversification due to the 1956-57 crisis, which would not otherwise have come about, has been a waste.

But while a new petroleum area may add to security once it has large-scale production, there is no advance assurance that any particular exploration effort in a new area, or even several taken together, will be anything but a dead loss. The odds are always against finding anything, and they are very long against finding anything worth finding. Hence the feverish discussion of crash exploration programs in new areas "to diversify supply sources" is foolish. If money is spent in new untried areas, it will probably be lost. On the other hand, if the exploration is in areas now producing, there may possibly be a commercial profit, but there can be no gain in security. And even the improbable combination of large new discoveries in new areas will not give any security until many years are past and the need also may have passed. Oil exploration for security is like the situation of a man trying to provide for his old age by going to the race track to wager his hard-earned pay.

Another kind of diversification is altogether worthless: that is when a consuming country decides to import from more than one of the existing producing countries. Any sacrifice or higher cost incurred in this way is a dead-weight loss because it does not in the least diminish the threat of a concerted shutdown.

Whether wholly or partly nationally owned, a company like BP in Great Britain, or CFP or ERAP in France, or ENI in Italy, cannot provide security of supply any more

[2]See *Oil and Gas Journal*, Dec. 26, 1966, p. 122, giving discovery dates of Nigerian fields, each one of which represented an effort of at least a few years, perhaps of many.

[3]*Oil and Gas Journal*, Dec. 31, 1956, p. 154, on the beginnings of Libyan exploration, and on Morocco, Algeria, and Gabon. There is also mention of an encouraging oil show in Nigeria, other than the discovery recorded in the footnote immediately above.

than a privately owned company. Indeed, it is a more tempting target. But public or private ownership is simply irrelevant to the chances of a concerted shutdown. Nor does a private company owned by nationals of a given country make that country any more secure than if the company were owned by foreigners.

Unfortunately, in every security crisis a cry goes up to "diversify" within or without the established areas by subsidizing local companies to explore. Suppliers and contractors will be kept busy, and some private concerns will take long risks with public money. Some may become rich; nobody else will gain.

But this is not quite the story—indeed it may be the lesser half. For there is a perfectly sincere belief, particularly in Continental Europe, that oil is somehow special. Oil is not a vulgar commodity like the others, but must be the stuff of high strategy and national policy. One must not be "at the mercy of the companies"—whatever that means—especially since they are huge companies, huge international companies, huge "Anglo-Saxon" companies. To many Continental Europeans this is the security problem. Their politeness in rarely saying this publicly has not served them well, nor anybody else. For if the fear were voiced and freely discussed, it would be seen to be groundless. "The companies" can only cut off or threaten or exploit a given country if they can act together as a unit. But even the loose cartel of the 1930s has been dead nearly 30 years; like John Brown's body, it is moldering in the grave, but the myth goes marching on. Like other delusions, it harms those who believe it.

Perhaps 40 years ago or more the handful of Anglo-American companies who were then the international industry could have been used as the tool of British and American policy. Hence there may have been sense in laws like the French act of 1928. But to imagine the companies as tools of Anglo-American policy today is far-fetched. The producing countries would not permit it, and they have the physical force on the spot. Indeed, the Anglo-Americans were singled out during the brief 1967 boycott. Fear of "dependence on the oil companies" is just another distressing example of prejudice against big business, and against foreigners. Xenophobia is not only "wrong" but, like most prejudices, expensive.

Turning to coal, domestic coal production obviously gives *permanent* assurance of a *part* of the fuel supply, which is badly out of joint with the need of *temporary* assurance of *all* or nearly all its fuel supply. What is the price of this limited security? One can make some approximate calculations for Western Europe.

The cheapest grade of coal in the European Coal and Steel Community is priced at about $16 per metric ton at the mine.[4] Subsidies come to over $5, so the total cost to the economy can be no less than $21.[5] A metric ton of fuel oil has nearly 1.5 times the heat value of a metric ton of coal, and hence would be no more or less expensive than if it cost $31.[6]

Since 1958, heavy fuel oil has been freely available at the Channel ports at about $12 per metric ton, varying perhaps 10 percent up or down. Past mid-1966, up to the outbreak of the Suez war, it was steady around $10.50.[7] The loss to the EEC countries'

[4]ECSC, *1967 Annual Report*, Statistical Appendix, table 13. Taxes are excluded; they range from 1 percent in Belgium to 11 percent in France. An average of prices weighted by total output in the various producing basins is $16.10.

[5]ECSC, *Nouvelles Réflexions sur les Perspectives Energétiques de la Communauté Européene* (1966), p. 21. These are largely supplementary labor costs. A weighted average (*1966 Annual Report*, Statistical Appendix, table 2) is $5.11. Hence total average cost per ton is $16.10 + $5.11 = $21.21.

[6]Conversion factors can only be approximate. Those used here are from *Petroleum Press Service*, giving fuel oil 18.3 thousand Btu/lb., and bituminous coal from 10.2 to 14.6 thousand Btu/lb., we calculate with a middle value of 14.35. Then 18.30/12.35 = 1.48, and $21.21 × 1.48 = $31.40.

[7]There is a variety of sources, the periodic reports being in *Platt's Oilgram Price Service* and *Europa Öl-Telegramm.* During the first half of 1967 it has been around $10.70 for barge lots; cargoes are about 25 cents cheaper. Hence $31.40 less $10.50 is $20.90. Moreover, the operating cost is somewhat less in burning oil than coal.

economy is $20–$21 per metric ton of oil equivalent. British costs *seem* much lower, in the neighborhood of $14.75 per metric ton. Taken together, the weighted average cost is $18.30 per ton, $27/ton oil-equivalent, and average loss to OECD Europe is about $16.35 per ton oil-equivalent.[8]

So bruising is this simple fact to so many commercial, political, and intellectual egotisms that many ingenious explanations are offered and eagerly accepted why oil prices are "abnormally" and "temporarily" low. By confusion over "marginal cost," the "depressed" prices are seriously ascribed to a surplus of refining capacity, which in fact was so chronically short of demand that it doubled in the six years 1960–66. But truth like cheerfulness will break through. The experts of the European Communities, who in 1962 had projected a long-term value for heavy fuel oil of $18.00 per metric ton, revised it in early 1966—showing a commendable independence of spirit—to $12.50, thus wiping out most of the wishful reckoning.[9]

To be sure, the coal cost is an aggregate or average. Some mines cost much less than others to operate, and indeed the refusal of governments to let the whole range of cost be calculated and published, their insistence on average costing and on prices to cover average costs, deserves more attention than it has received. But retrenchment is too little too late. Lord Robens of the British National Coal Board is sufficient authority. His cheapest pits, he avers, can produce at 3 pence per therm (35 cents per million Btu) or $14.50 per metric ton oil-equivalent. If he is only given more time and much more money, he can some day produce much more coal at this rock-bottom figure.[10] Unfortunately, it is many years since oil was this high at the Channel ports.

In 1966 coal production in the OECD countries of Europe was about 212 million tons oil-equivalent, excluding coking coal.[11] Their total replacement by oil would have saved $3.5 billion per year. That was the deadweight loss to Western Europe.

[8]According to the U.K. National Board for Prices and Incomes, *Report No. 12: Coal Prices* (Cmd. 2919, 1966), total estimated colliery expenditure 1966–67 was £810 million plus £10 million for fixed asset replacement. Total debt was £960 million (p. 2); an interest charge of 7.5 percent is applied, rather than the official one of 4.8 percent, which means £73 million. Although interest is a fixed cost, it must be used as a proxy for the capital cost of maintaining a given rate of output. (The Prices and Income Board notes that electricity earns 6.75 percent, gas higher than 6 percent, and "industry generally" 12 to 14 percent. Surely the last figure is a closer approximation of the true drain on the British economy. However, I use the 7.5 percent of the U.K. Atomic Energy Authority as a conservative estimate of capital cost.) Total cost is then £893 million, which comes to £5.27 or $14.75/ton for 170 million tons. EEC plus U.K. production totals 380 million tons, averaging $17.30 per ton, or $27.05 per ton. Subtracting $10.70 gives $16.35. No account is taken of the very small production in other countries.

This estimate seems consistent with that of Turvey and Nobay, in the *Economic Journal*, vol. 75 (Dec. 1965), p. 792, of coal sold "to industry" in 1964 at £5.8 ($16.25) per ton, since a delivery charge is presumably included. (Sources and methods are not explained.) Brechling and Surrey, in "An International Comparison of Production Techniques: The Coal-Fired Electricity Generating Industry," *National Institute Economic Review*, May 1966, p. 33, gives the 1963 average price of coal delivered to generating plants as 42 pence (49 cents) per million Btu, a much better measure of price. Assuming 29 million Btu per long ton (12,400 Btu/pound and 2,240 pound tons), the price per ton would be £5.02 ($14.05). This again seems consistent, since coal delivered to electric generating plants would be expected to be cheaper than the average for all coal at the mine.

[9]Compare *Nouvelles Réflexions* (1966), p. 27, with ECSC, *Etude sur les Perspectives Energétiques à Long Terme de la Communauté Européene* (1962), chap. 9, sec. 4. The *Etude* was reviewed in the *Economic Journal*, vol. 74 (1964), by E. F. Schumacher, identified only as living in London. He criticized projection 13 years ahead, to 1975: "These figures are not worth the paper they are written on. They are a case of spurious verisimilitude bordering on mendacity." An article in the London *Times* (Apr. 11, 1963), had predicted 37 years ahead to A.D. 2000: a steeply rising real cost or even physical shortage of fuel. The writer of the article was economic adviser to the National Coal Board, E. F. Schumacher.

[10]Address to Coal Industry Society, Mar. 6, 1967.

[11]U.S. Bureau of Mines, *International Coal Trade*, May 1967, p. 17, gives 398 million tons, or 269 million tons oil-equivalent (m.t.o.e.). Coking coal consumption in 1964 was estimated by OECD,

The figure may seem too bad to be true. As a near-term projection, it is an underestimate. First, coal costs are increasing every year. Second, heavy oil costs less in Southern Europe ($1 per ton less in Italy). Third, if coal were phased out to be replaced by oil, the price of oil would almost certainly decrease. This is because the Middle East reserves are so vast that additional capacity can be created to produce several times the current output and at a cost so low that it would be vastly profitable to do so. Freedom by European buyers to buy in the cheapest market would send oil company salesmen rushing to every electricity company as the first step in expanding sales, and the resulting competition would send prices down. The lessons of recent experience are plain: in those countries where trade in fuel is freest and sales most buoyant, prices are lowest.

Fourth, and perhaps in the long run most important, the price of coal is being used as a reference price or standard by which to judge new energy sources, such as nuclear power. In Britain, the Dungeness B power station will produce electric power, according to the original estimates, at $15.60 per metric ton oil-equivalent.[12] It is painful to see the near-euphoria this produces among British observers who simply pay no attention to oil because it is excluded by hypothesis–it is some kind of odd stuff which, as everyone knows, sells at a temporarily abnormally low price. Late in 1966 the estimated Dungeness B cost was further increased.[13] Even looking beyond to the next generation of reactors, and assuming the best, Sir William Penney estimates that *if* the later advanced gas-cooled reactor stations perform as hoped for, generating costs will by the mid-1970s equate to fossil fuel at 2.25 pence/therm (25.3 cents/thousand btu), heavy fuel oil at $10.90, which is not even as good as what is available now, and takes no account of advancing technology in fossil fuel use (which Sir William, like all observers, considers as very impressive in the recent past).[14] But with fictitious coal prices as a standard, huge amounts of scarce capital may be wasted on uneconomic nuclear power stations to match the near quarter of a billion dollars (£89 million) which the National Coal Board pours annually down holes in the ground.[15] The EEC is even more wasteful because their coal is even more expensive.

Of course one cannot tolerate the abrupt dismissal of close to a million mine employees. Once this is understood, the whole problem of fuel cheapness and security is bathed in light: *European coal production is no longer an industry, it is only a means of social insurance*. Awkward and wasteful, it can be abolished to the immense gain of the miners themselves before anyone else. To see why, we should first reckon the costs of an adequate security program by stockpiling crude oil.

Stockpiling Crude Oil. The cheapest and best place to store crude oil is at the ocean terminals where it arrives. These are imposing enough today, but not compared with the terminals for supertankers of 300,000 tons, the first of which is going up at Bantry Bay in Ireland, others in Japan. Storage and oil to fill it should be provided at government

---

Energy Policy (1966), p. 32, at 65 m.t.o.e. or 96 million tons coal-equivalent (m.t.c.e.), and it has not changed appreciably since. Imports in 1966 were just under 25 million metric tons, and if one assumes that half was for coking coal, then European production of coking coal (8 percent or less) was about (96 less 12) 84 m.t.c.e. or 57 m.t.o.e. Hence European coal produced for other than coke was in 1966 about 314 m.t.c.e. or 212 m.t.o.e. Reckoning at $16.35 per m.t.o.e., this comes to $3.47 billion.

[12] Tentative estimates are presented in detail in my letter to *The Economist*, July 17, 1965, p. 272. Revised estimates based on C.E.G.B. data were presented to the Tokyo meeting of the World Power Congress in October 1966.

[13] Testimony presented to the Select Committee on Science and Technology in March 1967 by Mr. Brown; the calculations are as of September 1966.

[14] Sir William Penney, *Nuclear Power* (the Citrine Lecture, 1967), pp. 8–9.

[15] During 1960–65, NCB expenditures were £532 ($1,490) million, £89 ($242) million annually, specific colliery expenditure was £462 ($1,295) million, £77 ($216) million annually. But even noncolliery expenditures are for coal products. Hence the total is coal investment, and totally wasteful.

expense, but for the sake of economy private enterprise should manage the facilities and commingle oil freely with theirs. For additional capacity is a valuable right to an oil company. The reason in brief is that larger tankers are much cheaper than small ones, but require much more storage capacity ashore. The interval between tanker arrivals increases in strict proportion to the increased size of the ships. But the amount of inventory needed increases somewhat more than proportionately. In effect, many small tankers are a spreading of risk, and fewer large tankers a concentration. Therefore, if an oil company managing a given amount of oil in storage were permitted to draw upon the government stock within a range of, say, 10 percent, provided only that replacement was made within a short period, it might be worth their while to bid for the right to manage the inventory.

In any case, the operating storage cost would be very low, but the capital outlay on the facilities and of the oil to fill them would be heavy, and the annual expense would essentially be the interest on the capital employed. The writer's calculations of storage cost, made some years ago, seem to have been taken seriously by other observers,[16] but the new conditions have made them obsolete. Today, storage facilities can be provided at a big ocean terminal for about $1.25 per barrel.[17] Oil can be purchased f.o.b. the Persian Gulf today by big credit-worthy buyers for less than $1.25 (the coincidence of the two figures is purely accidental, of course) and shipped, emphasizing the cheaper summer seasons, at 43 cents per barrel to Rotterdam and 37 cents to Marseille or northern Italy. (Under the usual method of rate quotation, this would be Intascale less 65 percent through Suez, less 67.5 percent around the Cape.) An average delivered cost to Europe would then be about $1.65, North and South taken together. There should be no undignified hassle over this price. The oil companies are selling for less to some buyers and realizing less from crude devoted to their refining-marketing operations. The value of a barrel of products sold in Europe, less marketing and refining costs (which must include a market rate of return on the capital employed) does not return them as much as $1.25 today.

Thus the capital outlay needed to store a barrel is about $1.25 + $1.65 = $2.90. The notional interest rate should not be mere interest cost to the government, but rather the return that the funds would fetch in private industry. Or, what comes to the same thing, the burden should be reckoned as the amount that would be needed to pay the holders of debt securities and equity securities to advance the money to a private low-risk enterprise–including also that part of the profit enjoyed by government as tax receiver. By this standard, the 4.5 percent used to reckon atomic power projects in the United States, or coal in the United Kingdom, is nonsensical, and even the 7.5 percent used in the United Kingdom is too low. If one uses 10 percent discount, an allowance for the limited life of the facilities (25 years) would raise the effective annual capital charge to nearly 11 percent. Then the annual capital cost of storing a barrel of oil is 30 cents, and adding 5 cents for operating costs, the total is 35 cents. In other words, if a whole year's supply had to be kept in stock, the cost would be 35 cents per barrel; six months' would cost half.[18]

[16]"The World Oil Outlook," in *Natural Resources and Economic Development*, ed., Marion Clawson (Johns Hopkins University Press for Resources for the Future, 1964), pp. 121–23. Cited in *Political and Economic Planning, A Fuel Policy for Britain* (undated but assumed to be 1965), p. 183.

[17]Compare the $1.27/barrel at Kharg Island in Iran. Construction costs at Bantry would be lower, land costs higher. The first Japanese central terminal system will include storage facilities for 3 million tons (22.2 million barrels). The cost of the entire project, including sea berths, docks, pumps, etc., is estimated at $32.2 (£11.5) million, or $1.45 (£0.518) per barrel. *Zosen*, Aug. 1967, p. 18. Clearly a doubling or more of storage alone would cost only a fraction of the average cost of the whole operation. Hence my investment figure may be much too high.

[18]That is, only half as many barrels would be stored.

How many months' supply do we need? The French Minister of Industry in November 1966 estimated six months' because the economies of the producing nations could not support a longer shutdown.[19] If M. Marcellin meant that none of the supplying countries could hold out any longer, he was surely wrong; but if he meant that not all of them could hold out even that long and that the chain was as strong as the weakest link, he was right. The producing nations involved in the 1967 crisis never were able to close ranks even at the start, and then the embargo began crumbling almost as soon as it began. Hence, six months seems much longer than necessary, but it will serve as the upper end of the range. Six months' special storage costing 18 cents per barrel plus the normal commercial stock of about 45 days, plus at least one month by stretching the stock through rationing, gives Europe nearly nine months'.

The government of South Africa has had to make similar calculations, but its danger is of course much greater since it could conceivably find both producing countries and consuming countries lined up against it, and as a relatively small market, it could not count on the producing nations being subject to unbearable pressure because of lost revenues. There are no official estimates, but the *Rand Daily Mail* has reported that the government was providing 18 months' supply,[20] and I believe the report has not been denied.

Although six months' inventory atop the normal two seems adequate, it can be backstopped very cheaply with two years' supply for the electric power industry. Dual-firing is cheap to install when going from coal to oil, but not the reverse. Henceforth all new power stations should be oil, but as a security measure they should either be made double-firing from the outset (as are coastal stations in the United States and Scandinavia) or at least be required to provide the stoker space needed in case of a later conversion to dual-firing. Coal production can then make a last contribution to the welfare of Europe. The coal itself is costless, for it will in any case be produced as the industry is phased out. The problem is only the cheapest and least unsightly place to store it, taking due account of where it will be eventually used. The electric power industry of OECD Europe used 153 million metric tons oil-equivalent in 1964, or 240 million short tons of coal.[21] In the United States, ground storage is provided, and profitably, for private companies at 15 cents per short ton during the peak December-July period, with a movement in and out; so 20 cents per short ton per year of dead storage seems more than adequate.[22] Two years' supply under 1964 conditions would mean a full year's supply in 1974 since the industry has been approximately doubling every decade, and would cost $96 million.

Thus Europe could be assured of from one to two years' electricity supply and well over nine months' supply of oil (for some, though not all, of the heavy fuel oil could be diverted to non-electricity consumption, and to a significant extent the slack would be transmitted to the lighter fuels) at an annual cost under 1965 conditions of about $872

[19] République Française, Assemblée Nationale, 2e séance du novembre 1966, p. 4321.

[20] *Rand Daily Mail* (Johannesburg), Aug. 24, 1966, p. 1.

[21] In 1965 (later figures not available) OECD Europe consumed 918.4 m.t.c.e. non-coking coal plus oil. (U.S. Bureau of Mines, *International Coal Trade*, Feb. 1967, p. 15.) Conversion of the total to oil is on the basis of equivalence of coal to crude oil, not fuel oil; the proportion is a barrel of oil of 34° gravity equal to 0.207 metric ton of coal, hence 4.44 billion barrels.

[22] Interstate Commerce Commission Tariff 1355-A, Bessemer and Lake Erie R.R. That the operation is profitable is shown by the later expansion of the original facilities and by another railroad setting up a similar installation. Another terminal was announced early in 1967, with initial capacity of 1.2 million and ultimate capacity of 4.5 million short tons, costing $5.75 million. *NCPC Newsletter*, Feb. 2, 1967, p. 4. Hence the capacity cost per ton lies between $2.85 (£1.02) and $1.28, but much nearer the smaller figure. At 11 percent capital charge, the cost would be about 14 cents (1 shilling) per short ton per year, which is consistent with the other estimates. Capital and operating costs would be lower for dead storage, but these are at best first approximations.

million. (This is 17.5 cents per barrel multiplied by 4.4 billion barrels of oil-equivalent of total oil and coal energy used excluding coking coal and adding the $96 million for coal storage.) Since the annual cost of supporting a superfluous coal industry was in 1964 $3.5 billion, the substitution of adequate security for inadequate security actually saves Europe $2.6 billion per year.

If, as now seems more likely, only three months' special supply need be stored, the cost would be $484 (£176) million and annual savings over $3 billion.

The Path to Follow. Once we turn from where we should be going to how to get there, the academic researcher's knowledge runs into sharply diminishing returns. But three problems are worth a quick glance: the time period, the coal miners, and the balance of payments.

Even if storage for security were accepted tomorrow as a policy of objective, it would take at least a year to perfect actual programs, to find likely sites, etc. Indeed, the size of the stockpile would have to be carefully reckoned. The estimates made in this paper have taken no account, for example, of European natural gas as part of the energy supply. Yet within a short time it will not be negligible. It might also be worth a last effort to inquire into American coal, which would be only slightly more expensive to buy (and much cheaper to store) if the high U.S. railroad freight rate, discriminatory against export sales, were lowered. Chances of success do not look too good. The writer had occasion to warn, on the basis of data ending in 1963, that if the discrimination did not cease, steam coal exports to Western Europe would dwindle.[23] They have in fact dropped by over 30 percent in three years, at a time when total EEC fuel consumption is up 11 percent.[24] But the official optimism about a big market for U.S. steam coal at only slightly less than current prices remains unshaken and based on the same comfortable illusion that oil prices are temporarily abnormally low.

If it takes two to three years before plans are drawn up and storage built, no time is lost because it will take that long for prices to come back to mid-1967 levels. At the time of writing (beginning October 1967) there is no sign of an early reopening of the Suez Canal. Mr. George Brown was apparently unsuccessful in trying to get Norwegian support for his proposal to have clearing work done at the expense of the maritime nations, but with all receipts going to the United Arab Republic and taking no notice of the problem of the navigation of Israeli ships in the Canal.[25]

Probably the Canal will some day be reopened under circumstances not foreseeable today. But its importance will be much less, and even the absolute volume of shipments may never regain the mid-1967 level. There has already been much silting, and a decrease in maximum permissible draft from 38 to 34 feet means that the largest ship acceptable drops from about 60,000 to about 34,000 deadweight tons, which in view of the distribution of ship sizes is drastic indeed. Hence a new equilibrium must wait on the addition of enough large tankers (175,000 tons and upward) to round the Cape at total costs somewhat lower than the old Canal transit by smaller ships.

The time needed to perfect plans and build facilities could also be used for the redeployment of mine labor. By the end of 1967, Western European underground and surface workers taken together will number about 900,000,[26] and their average wage is

[23] M. A. Adelman, "American Coal in Western Europe," *Journal of Industrial Economics*, vol. 14 (1966).

[24] For exports of steam coal, see *International Coal Trade*, July 1967, p. 6. For total Community consumption see ECSC Annual Reports: 1964, p. 60, 1970, p. 65.

[25] *Journal de la Marine Marchande*, Aug. 31, 1967, p. 1931.

[26] According to the *Colliery Guardian* (London), Jan. 20, 1967, U.K. coal manpower fell from 510,556 end-1963 to 446,788 end-1965 and 413,667 end-1966. The annual decline rate over the three years was thus 7 percent, and in 1966, 7.5 percent. The *Report* of the National Board for Prices and

around $2,500 per year. Hence, even to pay them all their current wages for their lifetime would cost about $2.2 billion a year, leaving a clear economic gain, which would increase rapidly over time. In practice, coal employees age fifty-five and over would probably be retired forthwith on full salary, while younger men could be released with either current wages guaranteed for a time ahead, or a lump-sum payment, so that either way they were sure of not losing out. Generosity should be the order of the day. Society benefits from changing these men from pensioners to productive workers and should therefore stand the costs of changing them. Of course I assume here a certain value judgment: that we owe certain duties to our fellow citizens as individuals or as families, but that we owe nothing to a corporate personality known as "the coal industry," and nothing more than thanks to those who, like Lord Robens, have tried their considerable and commendable best to do the impossible. Others will feel insulted at the proposal to put away "their" coal industry, but there is no arguing about tastes.

This brings us to the balance of payments. Getting rid of coal means a large addition to the import content of fuel. Furthermore, four months' supply, say, when Area fuel needs are about 14 million barrels daily, means 1.7 billion barrels of storage, costing, if my estimates are correct, nearly $5 billion. Import content of both oil supply and storage varies widely among nations. Only one general remark is in order.

The balance of payments can be considered as a short-term liquidity *constraint*, like the cash management of a private firm. Expenditures profitable to a business enterprise must either be postponed or else covered by special financing arrangements if the necessary funds are not otherwise available. But to refrain from profitable expenditures permanently because cash is not available immediately is the kind of ultraconservatism which assures the death of the enterprise.

For a nation, the balance of payments may be regarded as not a temporary liquidity constraint but as a permanent policy objective: autarchy. It is an expensive luxury for rich countries, but not ruinous. However, given fixed exchange rates, it means permanent, incurable foreign-exchange deficits. For if an economy is to accept expensive food, cement, energy, or what not, for the sake of saving foreign exchange, the level of domestic costs is so high that exports cannot find markets. The plight of underdeveloped countries, undone by their passion for import substitution, ought to serve as a warning to those more fortunate.

* * *

There is no dilemma of cheap *versus* secure fuel for Europe, nor for Japan, Australia, and other Asiatic nations. The only way to cheap and secure fuel is to stockpile oil and get rid of coal. The measurable economic gain is huge, but the noneconomic gain is not to be despised: the end of a filthy scar on the landscape.

Income (p. 6), expects an increasing loss rate and the end-1966 employment noted above is lower than its estimate. Hence to subtract 7 additional percent for an end-1967 estimate of 385,000 seems conservative.

The attentive reader will have noticed that British labor requirements per ton are much higher than ECSC, yet prices are much lower. This anomaly might repay further study.

According to the Annual Reports of the European Coal and Steel Community, the 1963–66 decline (as of end-September) averaged 5.6 percent per year, but was down 11 percent in 1965–1966 alone (1967, table 42). From the first half of 1966 to that of 1967, the decline in underground workers was 13.5 percent. (*International Coal Trade*, Aug. 1967, p. 9.) Projecting the September 1966 figure forward by 15 months at that rate comes to 525,000. Added to the British total, this is 910,000 for all Western Europe.

The highest paid workers in the ECSC coal mines receive respectively $2,640, $2,900, $2,800, in Germany, Belgium, and France, respectively. ECSC, *Annual Report*, 1967, table 53.

AFTERWORD

The foregoing was written in October 1967, and is included in this book because it bears directly on an important aspect of the world oil market and has been quoted in several publications. Today, in 1971, it is useful to examine the argument critically in the light of hindsight.

1. A serious error of omission was to ignore the need for strict rationing and for a stiff excise tax to reduce demand and transfer windfall gains to the government. Such measures are costless if never used, and they stretch considerably the time the stockpile will last.

In 1967, a 90-day stockpile, stretched for another 30 days by rationing and use of commercial stocks, would probably have sufficed to break a boycott. But Tehran has given the producing nations much greater confidence, and a taste of the rewards of a threatened boycott. Hence the needed necessary inventory is greater; I do not try to estimate by how much. It is doubtful that the consuming countries today are willing to accumulate a larger stockpile than 90 days', however; which makes it more likely that they will again be powerless during the next confrontation.

2. Lord Robens's hope of coal at threepence per therm is not 35 cents, but 30 cents per million Btu when we take account of the devaluation of sterling. That hope has not been given public burial, but has been privately interred. (An overvalued currency is often given as a reason why investment, uneconomic at nominal exchange rates, really pays as a use of domestic resources.)

American steam coal exports declined further: in 1969 they were only 32 percent of the 1963 figure.[27] As for European coal, an excellent study has been written by Richard L. Gordon.[28]

3. The price of heavy fuel oil in Europe decreased much more quickly than was expected here. By October 1968 it was below its prewar level, and by October 1969 it was about \$9.30. (By the second half of 1968, cargo lots were being delivered in Denmark in tanks ashore for \$9.75-\$10.50.)[29] Late in 1969 the effects began to be registered of both an upward movement in oil tanker rates and a creeping reduction in European fuel oil supply. This was because (1) more of the Libyan crudes were being dewaxed into middle distillates, rather than dumped into the residuals, to meet increasing imports by the United States, and (2) there was an accelerated decline in American and European coal supply. These effects were brutally reinforced by the developments described in Chapter VI (pp. 190-91) and Chapter VIII. By late 1970 hopes were rekindled for the many in Europe to whom twelve years of stable prices in the \$9-\$13 range were *conjoncturel* while \$26 was *structurel*, and who saw high-cost coal and nuclear power as vindicated. A more sober guess at a future price level would be to begin with the \$9.30 level. It may have been somewhat below equilibrium because of the large amounts of Libyan crude oil dumped into residual for lack of dewaxing facilities. It is even more difficult to measure the effect of the tanker rate, which in 1969 was between a spot level of Worldscale 85 (\$7.50/ton) and a long-term level of Worldscale 38 (\$3.35/ton). A convergence toward Worldscale 70 (\$6.20) may be expected toward the mid-decade, but I believe it will decline further. The safest procedure is probably to add the increased tax of 50 cents per barrel, or \$3.30 per ton of heavy fuel oil, alternatively to the early 1967 level of \$10.50 and the later 1969 level of \$9.30, for a range between \$12.60 and \$13.80 per ton. This is probably on the high side, for the tax will tend to be shifted least on to

[27] *International Coal Trade*, June 1970, p. 5.

[28] Gordon, *The Economics of Decline: The Western European Coal Industry*, 1946-1968 (Praeger, 1970).

[29] Isefjord Power Company, *Annual Report, 1968*.

heavy fuel oil, where competition is keenest and demand most elastic, and predominantly to naphtha products, including gasoline, and to middle distillates.

If, as in Chapter VI, one uses the U.S. price level as a crude indicator of the world price level, the 1967 fuel oil price of $10.50 would in 1971 dollars be about $12.50. In other words, if the price of coal and of nuclear power follows broad price trends, the relative attractiveness of heavy fuel oil at $12.50 would be about the same as it was in 1967 at $10.50. The relative economics of heavy fuel oil seem little changed.

4. The Suez Canal is no longer of any relevance to price forecasts. (See pp. 128ff.)

5. No account was taken of the possibility of pipelines across Egypt and Israel. (See, Appendix IV-C.) However, since the cost to Europe is not much different from the round-Cape voyage, there is no effect on our conclusions.

6. The estimation of crude oil storage costs is not of major substantive importance, but is interesting as a case study in how estimates are made in oil economics. In the 1967 paper a global construction cost of $1.25 per barrel was used. In *The Oil Import Question*,[30] estimates are given of $2.66 per barrel in steel tanks, $1.01 per barrel in salt domes. The ultimate sources were two: the U.S. Department of the Interior estimated $4.95 in steel tanks; an oil company, $2.75. Unfortunately, such estimates are incompatible with data in the public domain:

a) Cited above was the estimate of $1.27 per barrel for steel tank storage at Kharg Island; of $1.45 for a Japanese terminal, including more than storage facilities.
b) In the *Petroleum Processing Handbook*[31] the cost of three types of steel tanks is given, for a maximum size of 280,000 barrels, and it varies from $1.00 to $1.18. Larger tanks would be cheaper, while "noncode" tanks up to one million barrels in size save 25-30 cents "over generally accepted present costs." This is, of course, exclusive of land and of facilities needed to operate the tanks.
c) In 1968 storage costs associated with a Newfoundland ocean terminal are estimated at $1.00 per barrel.[32] At the ENI refinery at Ingolstadt, storage for 720,000 barrels cost $1.32 per barrel. Because land was the scarce factor a double wall design was necessary, at higher cost.[33] Hence $1.32 contains an indirect allowance for land cost.
d) In April 1968 salt dome storage near Marseille was estimated at 75 cents per barrel.[34] Salt dome storage in Germany, near Bremen and near Wilhelmshafen, was estimated variously at 79 cents or 83 cents.[35]
e) In Venezuela in 1970 an open storage pit for heavy fuel oil cost 55 cents per barrel, "including all pumping, piping and auxiliary equipment." Steel-tank storage for intermediate and final products cost $1.17 per barrel. The tank sizes ranged from below 170,000 to 530,000 barrels.[36] Hence, if only crude had been stored in the largest tanks, the cost of steel-tank storage would be less than $1.17.

Hence the estimate made above, of $1.25, is probably higher than the average investment per barrel that will actually need to be made.

7. Construction costs and interest rates have much increased since 1967; i.e., the program should have been started earlier.

[30] Cabinet Task Force on Oil Import Control, *The Oil Import Question* (Washington, 1970), p. 54.

[31] *Petroleum Processing Handbook* (McGraw-Hill, 1962), pp. 8-24 to 8-28.

[32] *Petroleum Intelligence Weekly*, Sept. 9, 1968.

[33] *Oil and Gas International*, Dec. 1968, p. 78.

[34] *Europe and Oil*, Apr. 1968, p. 11.

[35] *Oil and Gas International*, Nov. 1968, p. 139; *Petroleum Intelligence Weekly*, Sept. 16, 1968.

[36] *Oil and Gas Journal*, Aug. 24, 1970, pp. 80-83.

TECHNICAL APPENDIXES TO THE CHAPTERS

Appendix II-A

United States: Operating Costs

Criticism, Comparison, and Evaluation of Appendix Tables II–A–1 to II–A–5

We shall consider first how and to what extent the figures presented in the five Appendix II-A tables (pp. 283-87) may be biased. Criticism naturally falls into two parts according to our fraction: the numerator, expenses per well per day; and the denominator, barrels of daily capacity per non-stripper well.

Daily expense per well. We have excluded stripper wells from our capacity figures, but have not been able to segregate their costs. In effect, stripper wells have been treated as though they were originally non-stripper, and had average costs of drilling and completion. But, as I have estimated,[1] something like half of the new wells drilled in Texas in 1954-61 were around or not much over 10 barrels per day in *initial* capacity. A well of this type would on the average be cheaper to drill, equip, and operate, than one producing several times as much. If so, some kind of adjustment is needed whereby the stripper wells are reckoned at a lower daily operating cost per well. If we had data on the original capacity of all wells and on their operating costs, we could deduct the strippers' operating costs from total operating costs of all wells. Since this deduction would be less than proportional to the number of wells, this would tend to increase the total operating costs of non-marginal wells. Unfortunately, data are not available to make any such correction. If we assume that strippers were on the average 25 percent cheaper, the cost per non-stripper well per day is not $7.82 but $9.28; our cost per barrel, not 16.7 but 19.8 cents. (See the last column in Appendix Table II-A-3.)

We now make a comparison of our estimates of per-well costs with those from other sources. Arps[2] has estimated that the great bulk of all wells cost between $200 and $400 a month, or $6.58 to $13.18 daily, with much clustering around $300 monthly, or $9.88 daily. Our figure of $7.82 is within the range, but is decidedly near the low end, being one-fourth lower than Arps's middle value.

Another check is a sample compiled by Eggleston[3] which is published here with his permission as Table II-A-4. The general average, computed by multiplying the mid-point value for each well class (except the pathological top class) by the number of wells, adding up the products and dividing by 2,460, is $10.50 daily per well. Average production per well, again excluding the open-ended class, is 44.8 barrels daily which is quite close to the average capacity calculated here. Average cost would be 23.5 cents per barrel;

[1] M. A. Adelman, "Efficiency of Resource Use in Crude Petroleum," *Southern Economic Journal*, vol. 31, 1964, pp. 101, 120.

[2] J. J. Arps, "Valuation of Oil and Gas Reserves," in Thomas C. Frick, ed., *Petroleum Production Handbook* (New York: McGraw-Hill, 1962).

[3] W. S. Eggleston, unpublished communication to the author.

very slightly less if the largest (open-end) class is included. If Eggleston's figures of production refer to capacity, or are based only on capacity operations, this would serve as a strong confirmation of the critique, and would confirm that in Table II-A-3 the last column is a better estimate of average daily operating cost for non-stripper wells, as I believe.

Stekoll[4] notes eight wells on a West Texas lease, ranging from $3.19 to $8.03; the median $6.23, somewhat below our average. For Pennsylvanian sandstones in Oklahoma between 2,000 and 8,000 feet, oil wells producing no water would incur $5.00 daily expenses. Elsewhere, pumping wells from 5,000 feet producing 33.8 barrels daily are assumed to incur $5.00 of operating expenses per well daily. Where depth does not exceed 8,000 feet, a permissible rule of thumb is given for a flowing well as $2.46 per month and for a pumping well as $4.94 for the first well on the lease and $3.29 for subsequent wells. However, there is no allowance for water disposal.[5] Terry and Hill show a typical project containing wells of 36 barrels initial capacity, and per-barrel expenses are estimated at 18.9 cents.[6]

Operating costs of non-marginal East Texas wells are "relatively low," and are reckoned at about $50 monthly, or only $1.67 daily.[7] A plausible assumption is said to be $150 monthly per flowing well, $200 per pumping well; with closer spacing and slim-hole techniques, only $100 and $150, respectively.[8] In a North Texas waterflood, it seems to be around $9.00 daily per well.[9] In the Poth field, expenses per well-day can be calculated as $3.66 in the A sand but $8.50 in the B sand.[10] In North Texas the economic limit was 3.29 barrels daily[11] when the median North Texas crude price was $2.75,[12] which would correspond to $9.01 per well-day.

Capacity. Reliance on capacity data from the Independent Petroleum Association of America (IPAA) or the National Petroleum Council (NPC) may also involve error. The NPC does not give statewide breakdowns, but their capacity estimate for the U.S. total at the beginning of 1964 is about 10 percent higher than that of IPAA, and for District III, it is 8.4 million barrels daily as against 7.4–about 14 percent higher. I may mention in passing my opinion that the discrepancy between NPC and IPAA cannot be explained by allowing for differences in concept; indeed, such an adjustment makes it even larger.[13]

[4] Marion H. Stekoll, "Cutting Costs of Well Completion and Operations," in *Economics of Petroleum Exploration, Development, and Property Evaluation* (Prentice-Hall for Southwestern Legal Foundation, 1961), p. 145. Eight wells constituting a single lease, cost in dollars per month: 94, 112, 153, 170, 209, 235, 239, 244. Median is $189.50, mean is $183.40 per month, $6.24 per day; high $8.03, low $3.19.

[5] John M. Campbell, *Oil Property Evaluation* (New York, Prentice-Hall, 1959), pp. 402, 484.

[6] Lyon F. Terry and Kenneth E. Hill, "Valuation of Producing Properties for Loan Purposes," *Journal of Petroleum Technology*, July 1953, reprinted in J. J. Arps, ed., Petroleum Transaction Reprint Series No. 3, *Oil and Gas Property Evaluation and Reserve Estimates*, SPE, n.d. (apparently 1960).

[7] *OGJ*, Aug. 22, Sept. 26, Dec. 5, 1960: a good account of the East Texas Proceedings. See also: *Dallas Morning News*, June 17, 1965, p. A-22.

[8] F. A. Garb and J. J. Gruy, "Practical Application of Digital Computers to Economic Analysis of Producing Properties," *JPT*, Feb. 1965, pp. 143-50.

[9] Joseph D. Moyer, "Multiple Zone Water Flooding Pays Off for Cities Service," *World Oil*, Nov. 1964, p. 83.

[10] Robert W. Hopf, "Shallow Path Attracts Drill," *OGJ*, Dec. 27, 1965, p. 202.

[11] Moyer, "Multiple Zone Water Flooding. . . ."

[12] *POPS, Crude Oil Supplement* (monthly); and NPC, *Report of the National Petroleum Council*, Committee on Proved Petroleum and Natural Gas Reserves and Availability (Washington, 1965), p. 3.

[13] Cf. Wallace F. Lovejoy and Paul T. Homan, with Charles O. Gavin, *Cost Analysis in the Petroleum Industry* (Dallas: Southern Methodist University, 1963); Executive Office of the President, Bureau of the Budget, *Petroleum Statistics Report*, Mar. 22, 1965, p. 18; and API, *Technical Report No. 2: Organization and Definitions for the Estimation of Reserve and Productive Capacity of Crude Oil* (1970), pp. 20-23.

Low-productivity wells (7 to 11 barrels daily) are estimated to cost about $8.20 daily.[14] In the Neches field in East Texas in 1966, the average wells costing $13.33 daily produced 22.6 barrels. But the number of wells could have been reduced by more than half. With less salt water to dispose of, and saving of the fuel needed for artificial lifting, average cost per well would have fallen and output per well doubled. With cost under 30 cents and wells producing more, per-barrel cost would be considerably less.[15] In Kansas and Oklahoma production costs were estimated at 25 to 35 cents per barrel.[16] But we do not know whether production was restrained by allowables.

Garb and Gruy[17] consider as plausible assumptions: operating costs excluding taxes just below $5.00 per flowing well and $6.60 per pumping well. With better spacing, slim-hole techniques, and lower fracture costs, operating costs can be brought down to about $3.30 per flowing well and $5.00 daily per pumping well.

But even if we had no doubts of our aggregate capacity figures, attainable capacity per well would probably be underestimated. The sum of the individual capacities of all wells must substantially exceed the capacity of the total population of wells. The extreme case is in such overdrilled areas as East Texas, where many wells could make several thousand barrels per day if all their neighbors were shut in, but the whole group of such wells could not do this at the same time. In 1960 it was proposed to shut in permanently, with no reduction in capacity, about seven-eighths of the East Texas non-marginal wells in order to save operating costs. Here the sum of individual capacities must have been at least seven times the aggregate capacity for the non-marginal wells as a whole. Indeed, about nine-tenths of the wells could have been closed in,[18] but those proposing the change were restrained by prudence, watering down their proposal in the vain hope of getting it accepted.[19] Five years and millions of wasted dollars later, a yet more modest proposal was rejected.[20] It advocated shutting in only 9,500 out of the 17,200 non-marginal wells, those with leaky casings, costing an average of $3,500 per well to repair. Again the local governments and chambers of commerce and labor unions opposed the measure, and again they succeeded.[21] East Texas is, of course, so extreme an example that one should not consider using it in order to adjust capacity for the understatement of average well capacity. But allowables in the big producing states of Texas, Louisiana, Oklahoma, and Kansas contain a very substantial well factor. So long as this is the case, there is an incentive to drill additional wells in order to get higher allowables, even when these yield no corresponding capacity increases, or even no additional capacity whatever.

In order to make the appropriate correction, we would need (1) the sum of the individual capacities of all wells in an area, for comparison with (2) the aggregate capacity for each producing area. However, the data of (1) do not now exist, at least not in the public domain.

The writer has made some very rough estimates of the distribution of *new* District III oil wells during 1954-61 inclusive.[22] If we divide total new non-marginal[23] capacity of

[14]SPE 1432 (1966), p. 101.

[15]*OGJ*, Jan. 30, 1967, p. 168; Mar. 13, 1967, p. 76.

[16]Ibid., Sept. 4, 1967.

[17]Garb and Gruy, "Practical Application of Digital Computers. . . ."

[18]Thomas C. Frick (JPT, vol. 12, 1958, p. 102) estimates that no more than 1,500 wells are needed to drain East Texas. The drilling of additional wells has undoubtedly increased daily capacity, but just as surely the increase has not been anything like proportional to the number.

[19]*OGJ*, Aug. 22, Sept. 26, and Dec. 5, 1960.

[20]Ibid.

[21]*OGJ*, Nov. 27, 1965, p. 42.

[22]Adelman, "Efficiency of Resource Use in Crude Petroleum," p. 117.

[23]There are some unavoidable inaccuracies in equating stripper with marginal wells, but here too the statistics leave us no choice.

3.4 million barrels daily by 46,000 wells, the average is 74.1 barrels per day. This is 1.72 times the observed capacity per non-marginal well derived from IPAA data. In other words, the average capacity of individual new wells in District III was 72 percent higher than the average capacity existing there. However, this figure of 1.72 is too rough to use as an adjustment factor.

Because we cannot tell the qualitative relationship between full capacity and total well capacity, we cannot appraise what seems like an interesting development: the percentage of non-stripper capacity utilized seems to be gradually improving. Between 1960 and 1963, the national percentage went from 66.4 to 69.2. The improvement in Louisiana was from 63.2 to 66.7, in Texas only from 54.3 to 54.7 percent, respectively. It is also encouraging that in Texas the output of stripper wells seems to be on the decline, having gone from 465,200 to 438,200 barrels in the three years, while output of non-stripper wells increased from 2,000,069 to 2,228,000 barrels.

Because these two errors tend to offset each other, we must be content with tables of the form of Table II-A-3, which for 1962 shows per barrel costs varying from 7 cents in Louisiana to 32 cents in Arkansas. This confirms the general impression about the relative attractiveness of petroleum operations in these two states.

Test of square-root hypothesis. In Table II-A-5 the lowest line must be eliminated as a pathological case, since costs are higher while average output is actually less than in the next lower cost class. This indicates wells with special problems, not wells where increasing output is worth getting at higher cost per well. The results of the test indicate that above the median value the actual cost rises further than the predicted. A better fit than with the square-root factor would be given by the function $Y = \$1.068X^{0.639}$, where Y is daily cost per well and X is average daily output. In effect, 0.639 fits better than 0.5. Ought we to adjust accordingly? I think not, because unusual operating expenditures may push up costs a little faster than the square root of output. The more productive a well, the more does it pay to spend extra money to overcome unusual operating difficulties. But this factor is one which we want to eliminate or hold constant at this stage, since we need to treat it as a random variable, approximately equal on the average from one region of the world to another. It is not random but systematic in this table. Hence, while the table would by no means *prove* a square-root relationship, it is at least no cause to *reject* it.

Let us try another check with our estimate of $9.28 per day for a 46.5 barrel well. If operating cost per well increases as the square root of output, then a well making 1,000 barrels daily should incur expenses 4.64 times as high as the average, or $43 a day, $1,300 monthly. Since Arps gave $1,000 a month as the absolute maximum for wells in the United States it would appear that for the biggest wells there, working under the most difficult conditions, our cost estimate might be high.

Miller and Dyes[24] calculate operating costs on a flowing well producing 1,000 barrels daily as $300 per month for each half of a dual completion, i.e., $600 per month per well. The same well on the pump costs $400 per completion, $800 per well. (For smaller wells, going on pump may triple operating costs.)[25] In terms of our formula, since these wells are about 25 times as productive, they should cost about five times as much: not quite $40 a day, or $1,190 per month. Thus our formula greatly overstates the cost, despite the fact that the Miller and Dyes calculation is for "a foreign operation, [hence] the costs assumed for the calculations are higher than those encountered domestically." This article is directed to setting up an optimum spacing and production plan "which

[24] C. C. Miller and A. D. Dyes, "Maximum Reservoir Worth–Proper Well Spacing," *Petroleum Transactions*, AIME, vol. 216, 1959, pp. 334-40.

[25] J. K. Jordan, W. M. McCardell, and C. R. Hocott (Humble Oil), "Effect of Rate on Oil Recovery . . . ," Petroleum Publishing Co., May 13, 1957. Pages not numbered.

gives the maximum return. In addition, we do not want to lose recovery. These considerations of maximum return and maximum recovery present no serious conflict."

For larger wells, it would seem that costs increase more slowly than our square-root relationship, and hence our estimate of overseas well costs may be on the high side. Fortunately the absolute cost per barrel becomes much less in this range, and therefore even very large relative errors are of little significance.

Steele[26] has recently combined the data in Table II-A-4 with new data from the Bureau of Mines[27] to estimate operating costs by five production rates and as many depth classes. His is a notable "first" in looking at the whole spectrum of reservoirs to derive a short-run supply curve. As of 1965, the median non-stripper cost per barrel is about 25½ cents.[28]

TABLE II-A-1. United States: Operating Costs, 1959-1963

(dollars in millions, except lines 16-17)

	1959	1960	1961	1962	1963
1. Development costs, excluding overhead	2,313	2,082	2,070	2,266	2,039
2. Operating costs, excluding overhead and taxes	1,450	1,390	1,455	1,535	1,581
3. Total operating and development costs (1 + 2)	3,763	3,472	3,525	3,801	3,620
4. Operating costs as percent of total (2/3)	38.5	40.0	41.2	40.3	42.4
5. Overhead costs, development plus operating	442	424	457	478	470
6. Operating overhead costs (5 × 4)	170	170	188	193	199
7. Operating costs including overhead, all wells, oil plus gas (2 + 6)	1,620	1,560	1,643	1,728	1,780
8. Operating oil wells, average during year (thousands)	579	587	593	596	595
9. Operating gas wells, average during year (thousands)	82	87	94	99	101
10. Appalachian oil wells, average during year (thousands)	117	116	112	107	100
11. Annual operating costs of Appalachian oil wells (Line 10 × 365 × $1.25)	53	53	51	49	46
12. Operating oil wells, non-Appalachian (8 − 10)	462	471	481	489	495
13. Annual operating costs, non-Appalachian wells (7 − 11)	1,567	1,507	1,592	1,679	1,734
14. Operating non-Appalachian oil wells as percent of all operating non-Appalachian wells, oil plus gas (12/12) + (9)	84.9	84.4	83.7	83.2	83.0
15. Annual operating costs, non-Appalachian oil wells (14 × 13)	1,330	1,272	1,333	1,397	1,440
16. Cost per non-Appalachian oil well per year (15/12) in dollars	2,879	2,701	2,771	2,856	2,910
17. Cost per non-Appalachian oil well per day (Line (16)÷ 365) in dollars	7.89	7.38	7.59	7.82	7.98

Sources and Notes (by line):

1, 2, 5—API et al., *JAS, Part II: Estimated Expenditures and Receipts of the United States Oil and Gas Producing Industry*, various years.

8, 9, 10—*World Oil*, mid-Feb. issue, various years.

11—Rough estimate made as follows: in 1962 production in the four Appalachian states (New York, Ohio, Pennsylvania, West Virginia) was 16.2 million barrels. The total value of this production, at the well, was $63.5 million, for an average value at the well of $3.92 per barrel. Average production per well was 157 barrels per year or 0.4 barrels per day; and average sales value per operating day was $1.57. Average cost per day must have been somewhat lower, since many wells must have earned something over base operating cost.

[26]Henry B. Steele, testimony in U.S. Senate, Committee on the Judiciary, Subcommittee on Antitrust and Monopoly, *Governmental Intervention in the Market Mechanism* (1969), pp. 208-33, 439-44.

[27]U.S. Bureau of Mines, *Information Circular 8362: Depth and Producing Rate Classification of Oil Reservoirs in the Fourteen Principal Oil Producing States* (1967).

[28]Ibid., table V, assuming stripper production is 20 percent of total.

TABLE II–A–2. Capacity per Well of Non-Stripper Wells by States, 1962

State	Production (*1,000 BD*)			Average number of producing wells during 1962 (*1,000*)						Average output per well (*BD*)			
	Total	Stripper	Non-stripper (1) – (2)	Total	Stripper	Non-stripper (4) – (5)	Productive capacity (*1,000 BD*)	Non-stripper productive capacity (7) – (2)	Non-stripper production as a percent of capacity (3) ÷ (8)	Total (1) ÷ (4)	Stripper (2) ÷ (5)	Non-stripper (3) ÷ (6)	Non-stripper capacity per well (8) ÷ (6)
	(1)	(2)	(3)	(4)	(5)	(6)	(7)	(8)	(9)	(10)	(11)	(12)	(13)
Arkansas	76	24	52	6.1	4.2	1.9	78	54	96.3	12.5	5.7	27.4	28.4
California[a]	841	166	675	38.8	21.4	17.4	983	817	82.6	21.7	7.8	38.8	47.0
Colorado	116	6	110	2.1	0.6	1.5	119	113	97.3	55.2	10.0	73.3	75.3
Illinois	216	206	10	31.0	29.3	1.7	208	2	(b)	7.0	7.0	5.9	(c)
Indiana	33	31	2	5.6	5.2	0.4	30	(b)	(b)	5.9	6.0	5.0	(c)
Kansas	307	192	115	43.8	40.4	3.4	338	146	78.8	7.0	4.8	33.8	42.9
Kentucky	49	37	12	19.7	14.3	5.4	48	11	109.1	2.5	2.6	2.2	(c)
Louisiana	1,307	25	1,282	25.6	9.2	16.4	1,994	1,969	65.1	51.1	2.7	78.2	120.1
Michigan	47	11	36	4.3	3.9	0.4	48	37	97.3	10.9	2.8	90.0	92.5
Mississippi	153	2	151	2.7	0.2	2.5	152	150	100.7	56.7	10.0	60.4	60.0
Montana	87	10	77	3.5	2.5	1.0	84	74	104.1	24.9	4.0	77.0	74.0
Nebraska	68	2	66	1.8	0.4	1.4	68	66	100.0	37.8	5.0	47.1	47.1
New Mexico	300	29	271	15.7	6.7	9.0	359	330	82.1	19.1	4.3	30.1	36.7
North Dakota	69	1	68	1.7	0.1	1.6	90	89	76.4	40.6	10.0	42.5	55.6
Oklahoma	555	314	241	80.1	68.7	11.4	661	347	69.5	6.9	4.6	21.1	30.4
Texas	2,585	473	2,112	197.0	94.3	102.7	4,390	3,917	53.9	13.1	5.0	20.6	38.1
Utah	85	0	85	0.8	0.0	0.8	88	88	96.6	106.3	(b)	106.3	110.0
Wyoming	372	17	355	7.7	3.4	4.3	395	378	93.9	48.3	5.0	82.6	87.9
Total U.S. Excl. Appalachian	7,266	1,546	5,720	488.0	304.8	183.2	10,133	8,588	66.6	14.9	5.1	31.2	46.9

See next page for sources.

TABLE II–A–2. Continued.

Sources (by columns):

(1) and (4)–U.S. Bureau of Mines, *Minerals Yearbook, 1963*, vol. II, *Mineral Fuels*, pp. 401, 419.

(2) and (5)–Interstate Oil Compact Commission, *National Stripper Well Survey, Statistical Summary, 1941–1963* (1965), pp. 2, 4.

(7)–IPPA, *Report of the Productive Capacity Committee*, 1965.

Notes: We aim at a producing cost that would be incurred with wells operating at or near capacity. Unfortunately, as Stekoll says (see p. 280, fn. 4), although the true unit of cost is the well, costs are almost invariably calculated by leases.

By definition, stripper wells produce at capacity, so that if we subtract from total capacity (col. 7) the output of stripper wells (col. 2), we have (in col. 8) the capacity of non-strippers. For the United States as a whole, non-stripper excess capacity is about one-third of the total, but it varies widely. In New Mexico, it is only 18%, and California also does well at less than 20%. In the big market-demand–prorationing states, Texas is worst, with excess capacity at about 46 percent, while Louisiana has only 35%. Average U.S. output per non-stripper well was 31 barrels daily and capacity was 46 barrels. This too varies widely, from a low of 31 barrels in Oklahoma to a high of 120 barrels in Louisiana, much better than 38 in neighboring Texas.

[a] Includes Alaska.

[b] Not meaningful.

[c] All wells are strippers.

TABLE II–A–3. Operating Costs per Barrel for Non-Stripper Wells, by States, 1962

State	Daily capacity of Non-stripper wells	Operating costs per barrel at a cost of $7.82 per well/day	Operating costs per barrel at a cost of $9.28 per well/day
	(1)	(2)	(3)
Arkansas	28.4	27.5	32.7
California[a]	47.0	16.6	19.7
Colorado	75.3	10.4	12.3
Kansas	42.9	18.2	21.6
Louisiana	120.1	6.5	7.7
Michigan	92.5	8.5	10.0
Mississippi	60.0	13.0	15.5
Montana	74.0	10.6	12.5
Nebraska	47.1	16.6	19.7
New Mexico	36.7	21.3	25.3
North Dakota	55.6	14.1	16.7
Oklahoma	30.4	25.7	30.5
Texas	38.1	20.5	24.4
Utah	110.0	7.1	8.4
Wyoming	87.9	8.9	10.6
Total U.S. Excl. Appalachian	46.9	16.7	19.8

Sources and Notes: Col. (1) is taken directly from col. (13) in table II–A–2. Operating costs of $7.82 were calculated in table II–A–1. The $9.28 operating costs per day are computed as follows: In 1962 there were 305,000 stripper wells and 185,000 non-Appalachian, non-stripper wells. If we assume a stripper to have daily operating costs equal to 75% of those for a non-stripper, 34 have 412,000 equivalent non-stripper wells in 1962. Annual operating costs for non-Appalachian oil wells were $1,397 million (line 15, table II–A–1), and this gives an annual operating cost of $3,386 per non-stripper well ($1,397 million divided by 412,000). Dividing $3,386 by 365 gives a per-day operating cost of $9.28 per non-stripper well.

[a]Including Alaska.

TABLE II–A–4. Frequency Distribution of Well Operating Costs, United States

Range of costs per well-day (*dollars*)	Mid-point	Number of wells	Estimated production per well (*BD*)	Estimated total production (*BD*)	Average
(1)	(2)	(3)	(4)	(5)	(6)
1.50 to 3.00	2.25	15	2 to 5	50	3.3
3.01 to 6.00	4.50	296	5 to 15	3,000	10.1
6.01 to 9.00	7.50	727	15 to 30	12,000	16.6
9.01 to 12.00	10.50	635	30 to 60	25,000	39.3
12.01 to 18.00	15.00	559	60 to 100	40,000	71.5
18.01 to 25,00	21.50	163	100 to 300	20,000	123.0
25.01 to 35.00	30.00	65	100 to 1000	10,000	152.0
Over 35.00	....	45	150 to 1000	6,000	133.5
		2,505		116,050	

Source: W. S. Eggleston, unpublished communication to the author. Cols. (2) and (3) were computed by the author.

TABLE II–A–5. Test of Square-Root Hypothesis on Operating Costs

Average output (*BD*)	Ratio to median class	Square root of ratio	Predicted daily cost (col. 3 × $10.50)	Actual daily cost
(1)	(2)	(3)	(4)	(5)
3.3	.084	.290	$ 3.04	$ 2.25
10.1	.256	.505	5.31	4.50
16.6	.423	.650	6.82	7.50
39.3	1.000	1.000	...	10.50
71.5	1.840	1.355	14.22	15.00
123.0	3.130	1.775	18.65	21.50
152.0	3.871	1.960	20.29	30.00

Source: Table II–A–4.

Appendix II-B

Kuwait, Libya, and Iran: Calculation of Operating Costs

Kuwait

Operating costs, 1966. The government of Kuwait gives 4.1 cents per barrel as cost of production and gathering and 0.66 cents as transport to loading point.[1] We are not told the time period to which the figures relate. Since average output per employee rose by 68 percent in 1960-66, and continued to decline thereafter, any averaging of years has an upward bias. Kuwait operating costs for the year 1966 would therefore be lower than the published figure. (The reckoning of 1.5 cents "as cost of amortizing the capital employed" is unexplained. It is probably the result of applying an interest rate to a depreciated original investment cost; if so, it is irrelevant to cost for the purpose of an investment decision.)

Libya

TABLE II-B-1. Operating Costs, 1964-1965

	1964	1965
1. Total oil company expenditures (million $)	349	370
2. Oil company capital expenditures (million $)	319	336
3. Difference (million $)	30	34
4. Oil production (million barrels)	315	445
5. Operating costs (line 3)/(line 4) (cents per barrel)	9.5	7.6

Sources (by line):
1–Bank of Libya, *Monthly Bulletin*.
2,4–*AAPG Bulletin*, vols. 49-50 (1965-66), Aug. issues.

[1] MEES, May 20, 1966, p. 3.

IRAN CONSORTIUM

TABLE II-B-2. Operating Costs, 1962-1969

Year	Non-capital expenditures (*million $*)	Proportion of employees in producing	Annual production (*MB*)	Cost per barrel (*cents*)
	(1)	(2)	(3)	(4)
1962	96.6	.398	475	8.09
1963	87.4	.400	527	6.63
1964	91.6	.397	606	6.00
1965	94.6	.398	660	5.70
1966	113.1	.403	736	6.19
1967	112.8	.402	900	5.04
1968	114.8	.402	990	4.66
1969	135.1	.404	1134	4.81

Sources and Notes: Iranian Oil Operating Companies, *Annual Report*, various years.

Expenditures converted at $2.80 to the pound, 1962-66; $2.68, 1967; and $2.40 subsequently. Proportion of production operating expenditures to total operating expenditures assumed to be the same proportion as the number of employees, thus col. (4) = (1) × (2) / (3).

In both Libya and Iran operating costs include pipelines.

Appendix II-C

United States: Development Expenditures, 1959–1970

Sources of Error or Bias in Estimated U.S. Development Investment

In the calculation of development expenditures set down in Table II-C-2 at the end of this Appendix, there is a source of what I would consider error, but others might not. Operating costs of marginal wells were estimated separately in Appendix II-A, because such wells are a political curiosity or liability, not an economic asset, and because they are so far different from normal non-marginal wells that they could not be used in estimating from one area of the world to another. Unfortunately there is no basis for our separating out development expenditures for marginal wells. Yet these may be large. Up to 55 percent of oil wells completed in Texas in 1961 could have been stripper wells from the outset, averaging no more than 10 barrels daily.[1] Matters have improved since then because of wider spacing rules, but while we may be sure that the strippers' percentage of development *expenditures* was considerably less than 55, there is no way of saying how much less.

Since we are trying to estimate production costs from non-marginal wells, disregarding the effects of cost-raising market-demand prorationing, we assume an 8 percent decline rate in the future, even if the 4 percent or other lower decline rates have characterized the past. In other words, we need to assume a *past* decline rate as close as possible to actual in order to make the best possible estimate of loss of capacity made good by development expenditure; this renewed capacity must be added to the net increment of capacity in order to get the total (gross) increments. But in order to estimate what development costs would be if wells were not restricted, we need to use the natural decline rate as it existed in the early 1960s.

It is a revealing fact that the latest (and last) attempt to enumerate the whole range of crude oil production costs used dates from the period just before market-demand prorationing became effective.

[1] M. A. Adelman, "Efficiency of Resource Use in Crude Petroleum," *Southern Economics Journal*, vol. 31 (1964), p. 124.

TABLE II-C-1. Supply Curve of Crude Petroleum, 1931-1934

(cents per barrel)

Range of cost	Midpoint	Percent of total output	(2) × (3)	Most expensive output eliminated: Highest class	Most expensive output eliminated: Two highest classes
(1)	(2)	(3)	(4)	(5)	(6)
Under 40	20	10	200	200	200
40-79	60	50	3,000	3,000	3,000
80-119	100	25	2,500	2,500	...
120-440[a]	280	15	4,200	...	...
		100	9,900	5,700	3,200
Average cost per barrel (cents)			80	53	43
Savings per barrel eliminated (cents)				280	136

Source: U.S. Department of the Interior, Petroleum Administrative Board, *Report on the Cost of Producing Crude Petroleum* (1935).

[a]Some higher.

TABLE II-C-2. Development Expenditures, United States, 1959-1970 (Part 1)

(wells in thousands, expenditures in millions of dollars)

Year	Productive wells drilled: Oil	Productive wells drilled: Gas	Productive wells drilled: Total	Drilling expenditures productive wells: Oil	Drilling expenditures productive wells: Gas	Drilling expenditures productive wells: Total	Equipping leases for productive wells: Oil	Equipping leases for productive wells: Gas	Equipping leases for productive wells: Total	Dry holes drilled: Exploration	Dry holes drilled: Development	Dry holes drilled: Total
	(1)	(2)	(3)	(4)	(5)	(6)	(7)	(8)	(9)	(10)	(11)	(12)
1959 ...	25.4	5.0	30.4	1,321	509	1,830	404	79	483	10.6	8.5	19.1
1960 ...	21.3	5.3	26.6	1,111	540	1,651	345	86	431	9.5	8.1	17.6
1961 ...	21.2	5.7	26.9	1,087	537	1,624	355	91	446	9.0	8.1	17.1
1962 ...	21.4	5.9	27.3	1,161	569	1,730	421	116	537	8.8	7.9	16.7
1963 ...	20.7	4.8	25.5	1,071	442	1,513	428	99	527	8.7	7.6	16.3
1964 ...	21.0	4.9	25.9	1,063	510	1,573	500	119	619	9.0	8.5	17.5
1965 ...	18.9	4.8	23.7	1,067	486	1,553	461	119	580	8.0	8.0	16.0
1966 ...	15.9	4.1	20.0							8.1	6.5	14.6
1967 ...	14.9	3.6	18.5							7.2	5.8	13.0
1968 ...	13.8	3.3	17.1	New data, allocation not needed						7.3	5.2	12.5
1969 ...	12.9	3.9	16.8							7.4	5.2	12.6
1970 ...	12.5	3.8	16.3							6.15	4.65	10.8

TABLE II–C–2. (Part 2)

Year	Development dry hole drilling expense			Development overhead			Total development expenditures		
	Oil	Gas	Total	Oil	Gas	Total	Oil	Gas	Total
	(13)	(14)	(15)	(16)	(17)	(18)	(19)	(20)	(21)
1959 . . .	285	82	367	214	72	286	2,224	742	2,966
1960 . . .	264	92	356	190	80	270	1,910	898	2,708
1961 . . .	268	98	366	201	85	286	1,911	811	2,722
1962 . . .	294	107	401	213	71	304	2,089	883	2,972
1963 . . .	282	89	368	210	74	284	1,991	701	2,692
1964 . . .	311	106	417	204	81	285	2,078	716	2,894
1965 . . .	316	109	425	220	74	294	2,064	788	2,852
1966 . . .						168	1,800	760	2,560
1967 . . .						192	1,850	747	2,597
1968 . . .	New data, allocation not needed					199	1,820	818	2,638
1969 . . .						207	1,920	846	2,766
1970 . . .						220	1,940	911	2,851
							23,597	9,721	33,318

Sources and Notes:

Cols. (1), (2), (4), (5), and (9) from *JAS* sections I and II, respective years.

Until 1965

Col. (3) = (1) + (2); col. (6) = (4) + (5).

Cols. (7) and (8) allocate oil and gas expenditure on equipment leases [Col. (9)] by number of wells drilled, i.e. (7) = (1)/(3) × (9); (8) = (2)/(3) × (9).

Cols. (10)–(12)–(12) from *OGJ* Forecast and Review, Jan. 30 annually; (10) from AAPG *Bulletin*, annual article on "Exploratory Drilling"; (11) is difference.

Cols. (13)–(15). Col. (15)–Total dry hole expenditures are found in *JAS* section II annually. The proportion attributable to development is determined by number of wells drilled, i.e. (15) = total dry hole costs × (10)/(12). This is allocated between oil and gas in the proportion of footage drilled (obtained from *JAS*).

Cols. (16)–(18)–"Development and Production Overhead" from *JAS* is allocated in (18) to development by proportion of non-overhead expenditure, i.e. in the proportion

$$\frac{(6) + (15)}{(6) + (15) + \text{Producing Costs}} = (18).$$ This is in turn divided between oil and gas by non-overhead expenditures,

$$\text{e.g. } (16) = \frac{(4) + (7) + (13)}{(6) + (9) + (15)} \times 18$$

Col. (19) = (4) + (7) + (13) + (16).
Col. (20) = (5) + (8) + (14) + (17).
Col. (21) = (19) + (20).

From 1966 On

Data reported in *JAS* as total development expenditures, including allocated overhead. Col. (12) is *JAS* total. In (10) and (11) this is allocated between exploratory and development dry holes in the ratio of the *QRDS* data. Dry hole development expenditures are allocated according to numbers of gas or oil wells.

Appendix II-D

Venezuela: Development Cost

A check of the Chase Manhattan Bank (CMB) expenditure data against the Ministerio de Minas e Hidrocarburos data shows a very close agreement. Therefore, since CMB data are more widely available, they are used here.

For 1966-68, total expenditures were $450 million and gross new capacity supplied was 1,080 TBD; hence the investment per initial daily barrel was $417. These numbers are the basis for Table II-7.

TABLE II-D-1. Venezuela: Production Capital Outlays, 1947-1968

(*dollars in millions*)

Year	Production outlays	Lease acquisitions	Pipeline outlays	Production capital outlays adjusted
	(1)	(2)	(3)	(4)
1947	250	1	20	269
1948	315	16	45	344
1949	345	2	10	353
1950	145	...	5	150
1951	210	...	17	227
1952	255	...	25	280
1953	245	...	10	255
1954	245	...	10	255
1955	285	...	15	300
1956	680	338	35	377
1957	900	370	70	600
1958	500	...	110	610
1959	375	...	30	405
1960	225	...	20	245
1961	165	...	5	170
1962	160	...	5	165
1963	170	...	5	175
1964	155	...	5	160
1965	175	...	5	180
1966	120	...	10	130
1967	125	...	5	130
1968	185	...	5	190
Total				5,970

Sources: Cols. (1) and (3)–CMB, *Capital Investments of the World Petroleum Industry*, various years. Col. (2)–Venezuela, Ministerio de Minas e Hidrocarburos, *Memoria y Cuenta*, 1964, p. I-189. Converted at 1 bolivar = 32.36 cents. Col. (4) = (1) + (3) –(2).

TABLE II-D-2. Venezuela: Gross Capacity Change and Initial Investment Cost, 1947-1968

Year	Oil well completions	Average initial production per newly completed well (*BD*)	Added capacity (*BD*)	Production capital outlay (incl. transport) (*million dollars*)	Outlay per initial daily barrel (*dollars*)
	(1)	(2)	(3)	(4)	(5)
1947 ...	661	817	540	269	498
1948 ...	752	722	543	344	634
1949 ...	548	935	512	353	689
1950 ...	536	853	457	150	328
1951 ...	1,054	408	430	227	528
1952 ...	1,181	459	542	280	517
1953 ...	773	512	396	255	644
1954 ...	683	486	332	255	768
1955 ...	1,016	711	722	300	416
1956 ...	1,262	554	699	377	539
1957 ...	1,574	607	955	600	628
1958 ...	969	874	847	610	720
1959 ...	548	1,392	763	405	531
1960 ...	342	1,137	389	245	630
1961 ...	372	1,143	425	170	400
1962 ...	455	914	416	165	397
1963 ...	425	824	350	175	500
1964 ...	560	741	415	160	386
1965 ...	592	693	410	180	439
1966 ...	349	710	248	130	524
1967 ...	277	1,269	352	130	369
1968 ...	398	1,207	480	190	396
Total			11,223		

Sources: (1) Venezuela, Ministeria de Minas e Hidrocarburos, *Memoria y Cuenta 1968*, p. I-A-63 (and earlier years).

(2) Unpublished data kindly supplied by Ministerio de Minas e Hidrocarburos.

(3) = (1) × (2).

(4) : Col. (4) of table II-D-1.

(5) = (4) ÷ (3).

TABLE II-D-3. Venezuela: Distribution of Daily Output per Well, 1964

Output	Fields	Wells	Percent of total Venezuelan production
Over 3,000 *BD* per well	2	152	14.6
2,000-2,999	...	...	...
1,000-1,999	3	60	2.3
600-999	4	136	3.2
400-599	5	434	6.3
300-399	3	4,150	47.5
Less than 300	n.a.	5,929	26.1
Total		10,861	100.0

Source: *OGJ*, Jan. 4, 1965, pp. 149-50.

n.a. = Not available.

Appendix II-E

Middle East and Africa: Development Expenditures, Capacity, and per-Barrel Investment

Allocation of Chase Manhattan Bank Expenditures Among Countries

The method used here is also applicable to separate areas and fields within a country. The procedure consists of the segregation and then the addition of three costs: drilling, for all wells; completion, for all wells other than dry holes; and equipping leases, for all non-dry wells. Our basic assumption is that the three types of cost of drilling and equipping wells outside the United States are governed by the same physical factors as in the United States, and hence are systematically related in the same way. In the basic formula given here, the subscript *i* represents a given country; the sum of the individual country expenditures must add up to the total capital expenditures for the region, and each country is allocated the fraction x_i/X of the total regional expenditures. In symbols:

$$x_i = AD_i(DH_i a_i) + AD_i(PW_i b_i) + \frac{PW_i}{PW}(.25\,\Sigma PW_i b_i)$$

where: x_i = cost units, $\Sigma x_i = X$

AD_i = average well depth

DH_i = dry holes, $\Sigma DH_i = DH$

PW_i = non-dry wells, $\Sigma PW_i = PW$

a_i = cost index for dry holes of depth AD_i

b_i = cost index for non-dry wells of depth AD_i

The first step is to determine dry hole cost, AD_i $(DH_i a_i)$. The average depth and the number of dry holes drilled are taken from *World Oil*. The index a_i is taken from Table II-E-2, which shows the relation of dry hole cost per foot to depth of well, as given in the *Joint Association Survey (JAS)*. Cost per well for any given depth may arbitrarily be denoted as unity. Then the cost per foot for every other depth is expressed as a ratio. For example, taking the 3,751-5,000-foot range as the standard, cost per foot was \$6.80, while in the 10,001-12,500-foot range cost per foot of a dry hole was \$18.03. Then a well in the deeper size range is assumed to cost 2.65 (=18.03/6.80) times as much as a well in the shallower range.

The second element AD_i (PW_ib_i) is the cost of drilling and completing all non-dry wells through the control manifold known as the Christmas tree. We are forced to lump together oil and gas completions with service wells. The factor b_i is taken from Table II-E-3, also based on the *JAS*, while number and average depth of non-dry wells is again from *World Oil*. If a productive well was completed at a depth of 10,001 to 12,500 feet, each foot cost 3.15 (21.46/6.80) times the arbitrary standard unit.

The third element, that of equipping leases, must be approximated more loosely. In the United States, about 25 percent of the combined tangible and intangible costs of the productive well must be added for equipping leases. Hence we add together for all countries in the region the cost units on productive wells, PW_ib_i, and multiply it by 0.25. This is allocated to each country in proportion to its share of total productive completions.

To simplify the explanation of this procedure, a sample calculation for Saudi Arabia for the year 1963 is presented in Table II-E-4. Lines 1-14 of the table relate to the whole Middle East.[1] Production capital expenditures were $150 million (line 1), and as we have no published record of lease acquisition expenditures (line 2) we consider the total sum a cost (line 3). But we need to separate out exploration from development. Dry hole cost units and non-dry well cost units are computed as explained earlier, and the total is given in line 5. Drilling and completing account for 82.7 percent of cost units, and are therefore $124 million out of $150 million, and equipping leases account for the remaining $26 million—$136,000 average equipment per productive completion.

Lines 15 through 25 then apply a share of the exploration and development expenditures to Saudi Arabia. The country's average well depth determines a dry hole cost index of 1.17 and a non-dry index of 1.79, and when these are applied to footage drilled, we arrive at the respective cost units (lines 20 and 21). These are added together and multiplied by average Middle East cost per drilling-completing unit to yield a drilling-completing outlay of $8.5 million. To this is added $2.4 million estimated equipment expense for the non-dry wells (line 24). Thus total production capital expenditures for Saudi Arabia are reckoned at $10.9 million or 6.7 percent total Middle East—not surprising, since their 20 wells drilled were some 8 percent of 254 in the whole region. Of the 20, twelve were producers and five were service wells; two were dry holes and one was gas. We consider the service wells as productive completions, but need to subtract the gas well and the dry holes as not being oil development expense; this is done in lines 26-31. Oil development expenditures might, of course, be quickly estimated as the ratio of Saudi Arabian productive and service completions to all Middle East wells, multiplied by Middle East production capital expenditures, or (17/254) × $150 million, or $10 million, close enough to the more precisely calculated $9.8 million. Indeed, the chief virtue of our elaborate calculation may be to demonstrate that we can do without it.

We can get some notion of the biases in our figures by considering some of the individual types of well. Two of the productive completions were the discovery wells of the offshore Abu Sa'fah field, which was really an exploratory expenditure, and must have cost more than the average productive completion. Two producers were in a borderline area, since they found a new zone (Fadhili) in a known field (Qatif), and they were completed offshore. One at least of the dry holes was a dry wildcat, but one other *might* have been a dry development well. Hence there is one possible well which should perhaps be included and is not; two which should be excluded but are included; and two doubtful, both included. Moreover, the included productive wildcats and doubtful wells must cost considerably more per well than the one well which possibly should be included.

[1] Persian Gulf plus Israel, Syria, and Turkey.

Of course depth is not the only cost-determining factor. Probably the authors of the *JAS* were just as desirous as we of introducing additional variables, but they could not. I must, however, take sole responsibility for not making a distinction between onshore and offshore wells. For South Louisiana, the difference is very great, of the order of 3 to 1. But I think it would be wrong to apply such a factor outside the United States. The total cost of drilling a well is so much higher in these other areas that it swallows up the difference between onshore and offshore. The average well drilled in the United States cost in round numbers $50,000, and in the Middle East it was ten times that much. A well of the same depth drilled offshore the United States might well cost $150,000 or more. There is no reason to suppose that the percentage increase would even be significant in the Persian Gulf, though the absolute difference might be as great. The only data on this are of two types. One, if I may anticipate, is in the project announcements for some offshore fields, notably Safaniya, Khafji, and Darius. These estimates are higher than for the Persian Gulf as a whole, but the order of magnitude is not clearly outside the limits of the margin for error.

Cameron[2] reckons the cost of a well drilled in the Persian Gulf, where one rig drills 12 per year, as $730,000; equipping the well for production is expected to cost about another $140,000 (19 percent), or $870,000 per productive well. It can be seen in Table II-E-5 that we reckon $150.1 million for 191 productive completions, or $785,000 per productive completion, so that the ratio of offshore to total wells is 1.11. It may be that some such adjustment ought to be made. If so, we would have higher estimates for Abu Dhabi, Saudi Arabia, and the Neutral Zone, and lower estimates for Iraq and Kuwait. Iran might go either way. But while this experiment might be well worth making, it appears obvious that it could have no important effect on the estimates here.

The oil industry is being drawn, not pushed, into the sea by the well-founded expectation of lower costs. "We have heard many frightening words over the years about excessive costs of doing business offshore,"[3] says Weeks, but he notes that the *unit* finding cost is lower offshore because of better success ratios, larger fields, and easier drilling in softer sediment, where "fifty or more development wells may be drilled from a single three-to-five million dollar platform."

So much for the expenditures set forth in Table II-E-5 for the specimen year 1963, and the results in Table II-E-6.

New capacity developed. For the United States we estimated that 8 percent of capacity would be lost in any year. Hence the gain in capacity secured by development expenditures could be reckoned by adding the cumulative low to the net gain over any given period. For Venezuela, we had data from which to measure the gain in gross capacity. For the newer producing areas in the Middle East and Africa, we have no basis on which to estimate any decline rate. Bradley[4] shows widely varying decline rates in certain Persian Gulf fields, but the largest ones seem to show zero decline, and they account for most of the oil. Hence he assumes alternative rates of 2 and 5 percent per year. Although I agree with his method, the basis (the Abqaiq field) seems too narrow for generalization. We simply have no way of analyzing the experience, on the basis of *published* data, in a large enough sample of Persian Gulf and African fields. When these data become available, Bradley's method will come into its own.

The problem of capacity loss is further complicated because there have been a few large-scale injection projects in the Persian Gulf, notably in the Kirkuk field of Iraq, and

[2]A. Bryce Cameron, "The Petroleum Prospects Under the Marine Areas of the World," in Peter Hepple, ed., *Petroleum Supply and Demand* (London: Institute of Petroleum, 1966).

[3]Lewis G. Weeks, "Offshore Development and Resources," *Journal of Petroleum Technology*, Apr. 1969, pp. 377-85.

[4]Paul G. Bradley, *The Economics of Crude Petroleum Production* (Amsterdam: North Holland Publishing Co., 1967), tables 4.2, 6.2.

in some of the Saudi Arabian fields. In Abqaiq, a decline was halted and production stabilized. In Kirkuk, the 1968 injection rate was 1,050,000 barrels daily and the production rate 1,070,000 barrels daily.[5] One cannot say how much of the stability in Persian Gulf output per well is due to the fact that only a small part of the reservoir is being currently drained, and how much is due to these injection projects.

A complete water drive could insure practically a zero decline rate in any given well. But it would be consistent with a decline of average output per well for the field as a whole, because of water encroachment and the increasing water cut of the edge wells. But this can be offset by drilling more wells closer to the fault line or other boundary of the reservoir toward which the water drive is moving the oil.

One indicator would be the decline in output per well. In Kuwait, between 1951 and 1955, output per well averaged 5,900 barrels daily, then declined to 4,490 in 1959, an average of 7.1 percent per year. There was an increase in 1960, but this was the first development year for the big Raudhatain field,[6] making total Kuwait production data noncomparable.

In 1964 Raudhatain was producing 328,000 barrels daily from 29 wells, or 11,320 barrels daily per well.[7] Hence the rest of Kuwait (nearly all Burgan) produced 1,791,000 barrels daily from 428 wells, or 4,190 barrels daily per well.[8] Hence the Burgan decline must have been only about 1.3 percent annually, and appears to have ceased. We cannot be sure, since the basis of well classification, as between those operating and producing, and service wells and abandoned wells, has not been consistent. This invalidates the comparison of successive years as given in the *Oil and Gas Journal*. Hence we are forced to the working conclusion, or assumption, of 1 percent per year, corresponding to the production-reserve ratio, as more realistic than zero. The decline must have been partly deliberate, reducing water cut by plugging back the lower perforations, also thereby gaining more information on the oil-water contact. Indeed, "a maximum allowable production rate has been calculated for each producing well in order to prevent premature production of water."[9]

Capacity has also been maintained by workovers. During 1962-68 the Iran Consortium drilled 138 new development wells and worked over 134 old ones: 12.2 percent of the number in any one year. Of course a workover costs on the average only a fraction of the time and money needed for a new development well. In Saudi Arabia, development well drilling totalled 16 rig-years in 1962-66, of which workover rig-years totalled 6, which gives perhaps a fair idea of the relative importance of the two types. Annual reports of the Kuwait Oil Co. indicate about 18/76 of all rig-months (1967), or 21/57 (1968) are workovers. About 15 percent (1964) or 10 percent (1966) of all existing wells are worked over in the course of a year.[10]

"Overhang" and upward bias. As noted in the text, to treat all productive completions as development wells overstates the development component of the total well costs. "At some remote location the cost of an access road may be greater than the cost of a deep well in the U.S.A."[11] But the access road was an exploratory cost, unnecessary for later

[5] *PPS*, Aug. 1969, p. 285.

[6] KOC, *Annual Report, 1960*, pp. 9-11.

[7] D. I. Milton and C. C. S. Davies, "Exploration and Development of the Raudhatain Field," *Journal of the Institute of Petroleum*, vol. 51 (Jan. 1965), pp. 17-28.

[8] *OGJ*, "Worldwide Oil," Dec. 28, 1964.

[9] Bassam F. El-Ghussein, "Reservoir Control in the Burgan Field," Serial No. 43 (B-1), Fifth Arab Oil Petroleum Congress, Cairo, Mar. 1965; A. F. Fox, "The Development of the South Kuwait Oilfields," *Institute of Petroleum Review*, vol. 15 (1961), pp. 373, 378; on water encroachment, which has forced plugging back of the lower perforations, NPC, 1964.

[10] *AAPG Bulletin*, Aug. respective years; Annual Reports.

[11] John C. Dunlop, "Petroleum Exploration in Foreign Countries," in *Economics of Petroleum Exploration, Development, and Property Evaluation* (Prentice-Hall for Southwestern Legal Foundation, 1961).

wells. In the great Safaniya field, the first wells were drilled from a fixed platform designed to withstand maximum wind and water pressures, and took nearly 80 days to complete. Once commercial production was established, drilling tenders could be used, and much less elaborate installations; completions took only 14 days.[12] In 1966 they took 15 days,[13] indicating that the learning process for that field had largely exhausted itself. Safaniya wells need no access roads, to be sure, but one can hardly doubt that the first well cost a good deal more than 5.7 (=80/14) times the cost of a development well. In the Raudhatain field No. 1 well took 310 days to drill and set casing, Nos. 2 and 3 took 176 and 66 days. "It was then considered that the main drilling problem in the new area had been mastered," and an independent contractor took over. No. 23 was drilled and completed in 31.4 days, just one-tenth of what the first required.[14] Obviously, the expense of the first few producers vastly overstated development expense per well.

One estimate[15] is that offshore finding and development may cost "as much as $500 to $1,000 per barrel daily of capacity." With a poor discovery, the cost is often much higher. But the data cited earlier in the same article point to prevailing or average investment costs much lower. Contracting of big mobile offshore rigs costs from $5,000 to $10,000 daily, and supporting aircraft another two or three thousand, which gives a range between $7,000 and $10,000 daily, or from $2.55 million to $4.75 million annually. Positioning fees and mobilization and demobilization outlays are perhaps overgenerously provided for in reckoning the "venture cost of a one-string exploration operation" at between $5 million and $7 million annually, or between $1.5 million and $3 million per wildcat well (assuming three or two wildcats per year respectively). Taking a middle value of $2.25 million per wildcat, and 5,000 initial daily output, this would give $450 per initial daily barrel on a wildcat well. Development wells would be considerably cheaper. If the pattern of cost reduction were anything like Safaniya, development cost would be at or below $200 per initial daily barrel. A different source estimates that it costs $12,000 to $15,000 daily to drill an offshore well, or $3.7 to $5.5 million annually—which is considerably less.[16] Nor can there be much doubt that future offshore costs of making a foot of hole will be considerably below present ones.[17]

Leicester gives an example of an exploratory program which costs a total of $40.3 million, including $10.4 million for geological-geophysical surveys and core drilling, $12.0 million for overhead, and $17.9 million for drilling the first well.[18] Since we exclude survey work and core drilling, we would include $29.1 million development cost if this well was a producer. Alternatively, let us suppose that this was a dry hole or badly placed and that a second well was drilled at a cost of only one-third of the first, or $5.8 million, and it was successful. Then total expenditures would be $34.9 million, of which we would reckon only half as development, since the dry hole is not thus counted. Hence, we would reckon either $29.1 million or $17.5 million as development cost, when in fact the costs of subsequent development probably would be soon down to $3 million or less per well. Indeed, an average Middle East well, taking all types together, costs somewhere around $500,000 to $700,000. As the field was developed, this original large overestimate, which could hardly be less than a factor of 10, and would probably be nearer a

[12] Quincy J. Lowman, "Safaniya: World's Largest Offshore Field," *OGI*, vol. 5, Nov. 1965, pp. 47–51.

[13] *AAPG Bulletin*, 1967, p. 1634.

[14] Milton and Davies, "Exploration and Development . . . Field," pp. 17–28.

[15] *PONS*, Oct. 15, 1964, pp. 58ff., reporting interview with Mohammed Geroushi, Libyan member of OPEC Board of Directors.

[16] *OGI*, Dec. 1965, p. 58.

[17] Ibid.

[18] P. Leicester, "The Risk and the Reward," in United Nations, Department of Economic and Social Affairs, *Techniques of Petroleum Development* (1964), 64.II.B.2, pp. 134–37.

factor of 60, would of course be successively diluted, until, with wells drilled in the number of several dozen, the overestimate would gradually disappear for that field.

But in any area where a substantial number of successful wildcats are completed, relative to development wells, our method of reckoning development cost does result in a fairly substantial overstatement. It does not seem possible, given the present basic data, to say anything more precise.

Table II-E-9 presents a compendium of project announcements. The variation is enormous, and the details often less than clear. We can only hope for a confirmation of orders of magnitude.

Land transport investment. We wish to compare costs of oil as laid aboard the ship at a Persian Gulf port; hence we exclude pipelines to the Eastern Mediterranean. Cost of port jetties and facilities are counted as production capital expenditures.[19] The only Persian Gulf country where pipeline costs present a real problem is Iran, for in the other countries the pipeline costs to Persian Gulf loading points are very low (e.g., the Kuwait figure cited earlier) and lost in the margin for error. In Iran, however, pipeline costs are substantial. Fortunately, there are excellent current data because the whole export capacity of the Iran Consortium was recently replaced. Cookenboo[20] estimated cost at $1,258 million per 100 miles for a 32-inch line in 1952. If we suppose the same amount necessary today for a 42-inch line, this would amount to 0.35 cents per barrel; or 1.05 cents in total. (See Appendix II-I, where the operating cost of the pipeline is taken into account more fully.)

Decline rates. The Persian Gulf decline rate is taken as 1 percent. This is not inconsistent with the observed lack of decline rate in the Persian Gulf. For, as just seen, fluid injection and workovers were needed to offset a decline, and their cost is included in development of total expenditures. Moreover, some part of the stable production rate must be ascribed to the occasional "refreshing" of Persian Gulf production by new producing fields as with Raudhatain in Kuwait. If we rather simple-mindedly considered the decline rate as being essentially the reciprocal of the reserve production ratio, as in the United States, then we might compare the current (1969) 4.4 billion annual Persian Gulf output with approximately 332 billion barrels of published proved reserves, and put the decline rate at 1.3 percent. To avoid the appearance of precision, we round it to 1 percent.

As for Libya, it is prudent to assume that the reservoirs are not of Persian Gulf dimensions or driving mechanisms. During 1967 production in the Oasis fields topped out and about equalled that of 1966. However, nine oil development wells had been drilled in 1966, compared with a previous total of 324, or 2.7 percent. A steady-state replacement rate would be higher. It is only an average: in the Gialo field, the rate was about 4 percent, in others zero. In the Esso Zelten field, 1966 production was about equal to that of 1965, despite the addition of two new wells to the previous 48–some 2 percent. However, output increased considerably the next year, and later statistics are garbled.

Also indicative is the ratio of reserves to annual production, which at the end of 1969 was 31. This would suggest a figure of around 3 percent.

[19] Continental Oil Co., *Annual Report, 1965*, p. 18.

[20] Leslie Cookenboo, Jr., "Costs of Operating Crude Oil Pipe Lines," *Rice Institute Pamphlet*, vol. 41 (Apr. 1954), pp. 87–88.

TABLE II-E-1. Published Concession Payments and Bonuses, Middle East, 1948-1968

Year	Area	Amount and payer	Source and remarks
1948 . . .	Neutral Zone (Kuwait)	$7.5 million American Independent Oil Co.	Longrigg,[a] p. 215.
1949 . . .	Neutral Zone (Saudi Arabia)	$9.5 million Pacific Western Oil Co. (later Getty Oil)	Longrigg,[a] p. 215.
1952 . . .	Qatar	£260,000 = $728,000 Shell Overseas Exploration Co.	Longrigg,[a] p. 231.
1958 . . .	Iran	$25 million Pan American Petroleum Corp.	*OGI*, Feb. 1961, p. 43. Offshore concession, payment recoverable against "producing taxes."
1958 . . .	Neutral Zone (Kuwait)	$5 million Arabian Oil Co.	*OGI*, Feb. 1961, p. 43. Payment due when production is 50,000 BD.
1961 . . .	Kuwait	£30 million = $84 million Kuwait Shell Petroleum Development Co.	*OGI*, Feb. 1961, p. 43.
1965 . . .	Iran, et al.	$195 million–5 groups	*PPS*, Feb. 1965, pp. 65-66.
n.a. . . .	Saudi Arabia	$500,000 Petromin/ERAP/Tenneco	*PPS*,[b] Mar. 1969, p. 100.
n.a. . . .	Saudi Arabia	$2 million Petromin/ENI/Phillips	*PPS*,[b] Mar. 1969, p. 100.
n.a. . . .	Abu Dhabi	$1 million Maruzen/Daikyo/Nippon Mining	*PPS*,[b] Mar. 1969, p. 100.
n.a. . . .	Sharjah	$1.5 million Shell	*PPS*,[b] Mar. 1969, p. 100.

[a] S. H. Longrigg, *Oil in the Middle East* (London: Oxford University Press, 1961, 2nd ed.)

[b] The *PPS* summary of Middle East agreements in March 1969 does not give dates. The purpose of this table is to adjust the Chase Manhattan Bank capital expenditure figures (e.g., table II-F-3). The amounts involved in the last four agreements are negligible compared with total Middle East expenditure, so no adjustment was needed and it was not necessary to find the dates.

TABLE II-E-2. Dry Hole Cost Index

Depth range (*feet*)	Average depth (*feet*)	Average cost per foot	Index	Average cost per well (2) × (3)	Index
(1)	(2)	(3)	(4)	(5)	(6)
0-1,250	812	$ 6.46	0.95	$ 5,246	0.18
1,251-2,500	1,805	5.87	0.86	10,595	0.36
2,501-3,750	3,081	6.24	0.92	19,225	0.65
3,751-5,000	4,357	6.80	1.00	29,628	1.00
5,001-7,500	6,141	7.93	1.17	48,698	1.64
7,501-10,000	8,687	12.05	1.77	104,678	3.53
10,001-12,500	11,204	18.03	2.65	202,008	6.82
12,501-15,000	13,604	25.50	3.75	346,902	11.71
15,001 and over	16,261	36.98	5.44	601,332	20.30

TABLE II–E–3. Non-Dry, Non-Gas Wells Cost Index

Depth range (*feet*)	Average depth (*feet*)	Average cost per foot	Index	Average cost per well (2) × (3)	Index
(1)	(2)	(3)	(4)	(5)	(6)
0–1,250	825	$ 8.66	1.27	$ 7,144	0.24
1,251–2,500	1,808	8.45	1.24	15,278	0.52
2,501–3,750	3,102	9.80	1.44	30,400	1.03
3,751–5,000	4,361	10.34	1.52	45,093	1.52
5,001–7,500	5,171	12.13	1.78	62,724	2.12
7,501–10,000	8,679	15.54	2.28	134,872	4.55
10,001–12,500	11,152	21.46	3.15	239,322	8.08
12,501–15,000	13,451	27.80	4.09	373,938	12.62
15,001 and over	16,035	39.20	5.77	628,572	21.21

Source and Note, tables II–E–2 and II–E–3: API et al., *JAS, Part I, 1955–56, 1959–62*. The average is a simple average of the two series. For severe and merited criticism of the cost factors, see Franklin M. Fisher, *Supply and Costs in the U.S. Industry: Two Econometric Studies* (Resources for the Future, 1964).

TABLE II–E–4. Saudi Arabia, 1963: A Sample Calculation

1. Production capital expenditures, Middle East	$150 million
2. Reported lease acquisitions, Middle East	0
3. Difference (lines 1–2)	$150 million
4. Dry hole units, Middle East	554,601
5. Non-dry hole units, Middle East	2,806,989
6. Total drilling and completing units (lines 4 + 5)	3,361,590
7. Equipping units (.25 times line 5)	701,748
8. Total units (line 6 + 7)	4,063,338
9. Share of expenditures to drilling and completing of dry and non-dry wells (line 6 ÷ line 8)	82.7 percent
10. Drilling and completing expenses (line 3 × line 9)	$124 million
11. Drilling and completing cost per unit (line 10 / line 8)	$36.9
12. Non-dry wells drilled, Middle East	191
13. Equipping expenses (line 3 minus line 10)	$26 million
14. Equipping costs per well (line 13 / line 12)	$136 thousand
15. Dry holes drilled, Saudi Arabia	2 feet
16. Non-dry wells drilled, Saudi Arabia (12 oil, 5 service, one gas)	18
17. Average depth of wells, Saudi Arabia	6,691
18. Dry hole index	1.17
19. Non-dry hole index	1.79
20. Dry hole units (line 15 × line 17 × line 18)	15,657
21. Non-dry well units (line 16 × line 17 × line 19)	215,584
22. Total drilling and completing units (line 20 + line 21)	231,241
23. Drilling and completing costs, Saudi Arabia, (line 22 × line 11)	$8.5 million
24. Equipping costs, Saudi Arabia (line 16 × line 14)	$2.4 million
25. Production capital expenditures, Saudi Arabia (line 23 / 24)	$10.9 million
26. Oil well drilling units (line 16 × line 17 × line 19)	203,607
27. Total oil well drilling units, Middle East	2,795,012
28. Oil units as a percentage of total units	6.06 percent
29. Oil well drilling costs (line 28 × line 10)	$7.50 million
30. Oil equipment expense (line 14 × line 20)	$2.30 million
31. Oil development expenditure (line 30 + line 31)	$9.80 million

Sources (by line):

1. Chase Manhattan Bank.
2. No published reports found of lease acquisition payments in 1963.
4. See line 20 below; total for all countries in Middle East is line 4.
5. See line 21 below; total for all countries in Middle East is line 5.
6. See line 22 below; total for all countries in Middle East is line 6.

12, 15, 16, 17: *World Oil*, Aug. 15, 1965.
18. Table II–E–2.
19. Table II–E–3.

TABLE II–E–5. Derivation of Oil Development Capital Expenditures, Middle East, 1963

(*millions of dollars*)

Country	Average depth of wells	Dry holes drilled	Depth factor	Dry hole units (1) × (2) × (3)	Non-dry well completions	Cost factor	Non-dry well units (1) × (5) × (6)	Total units (4) + (7)	Share of total units	Country drilling expense
	(1)	(2)	(3)	(4)	(5)	(6)	(7)	(8)	(9)	(10)
Abu Dhabi	10,032	3	2.65	79,754.4	21	3.15	663,616.8	743,371.2	22.11	27.4
Bahrain	3,140	...	...	...	9	1.44	40,694.4	40,694.4	1.21	1.5
Iran	9,433	5	1.77	83,482.0	37	2.28	795,767.9	879,249.9	26.16	32.4
Iraq	1,621	...	...	...	7	1.24	14,070.3	14,070.3	0.42	0.5
Israel	7,074	5	1.17	41,382.9	10	1.79	126,624.6	168,007.5	5.00	6.2
Kuwait	6,296	18	1.17	132,593.8	36[a]	1.79	405,714.2	538,308.0	16.01	19.9
Neutral Zone	5,982	2	1.17	13,997.9	14	1.79	149,908.9	163,906.8	4.88	6.1
Qatar	6,007	2	1.17	14,056.4	7	1.79	75,267.7	89,324.1	2.66	3.3
Saudi Arabia	6,691	2	1.17	15,656.9	18[b]	1.79	215,584.0	231,240.9	6.88	8.5
Turkey	5,495	22	1.17	141,441.3	30[a]	1.79	295,081.5	436,522.8	12.99	16.1
Misc.	6,888	4	1.17	32,235.8	2	1.79	24,659.0	56,894.8	1.69	2.1
Totals		63	...	554,601.4	191	...	2,806,989.3	3,361,590.7	100.01	124.0

TABLE II–E–5. Continued.

Country	Equipping costs/well	Country equipping expenses (11) × (5)	Total expenses (10) + (12)	Oil well completions	Oil well units (1) × (6) × (14)	Oil units as percent of total units	Oil well drilling expenses	Oil well equipping expenses (11) × (14)	Oil well total expenses (17) + (18)
	(11)	(12)	(13)	(14)	(15)	(16)	(17)	(18)	(19)
Abu Dhabi	0.136	2.9	30.3	21	663,616.8	19.74	24.5	2.9	27.4
Bahrain	0.136	1.2	2.7	9	40,694.4	1.21	1.5	1.2	2.7
Iran	0.136	5.0	37.4	37	795,767.9	23.67	29.4	5.0	34.4
Iraq	0.136	1.0	1.5	7	14,070.3	0.42	0.5	1.0	1.5
Israel	0.136	1.4	7.6	10	126,624.6	3.77	4.7	1.4	6.1
Kuwait	0.136	4.9	24.8	35	394,444.4	11.73	14.5	4.8	19.3
Neutral Zone	0.136	1.9	8.0	14	149,908.9	4.46	5.5	1.9	7.4
Qatar	0.136	1.0	4.3	7	75,267.7	2.24	2.8	1.0	3.8
Saudi Arabia	0.136	2.4	10.9	12	143,722.7	6.06	7.5	2.3	9.8
Turkey	0.136	4.1	20.2	29	285,245.4	8.49	10.5	3.9	14.4
Misc.	0.136	0.3	2.4	2	24,659.0	0.73	0.9	0.3	1.2
Totals	0.136	26.1	150.1	183	2,714,022.1	80.74	100.1	25.0	125.1

Sources (by column):
Col. (1), (2), (5) and (14) from *World Oil*, Aug. 15, 1964.
Col. (3) and (6) from API et al., *JAS, Parts I and II, 1963*. (See text for method of calculation.)
Col. (10) calculated on the basis of drilling units allocated to a country times the cost per unit.

[a] 1 service.
[b] 1 gas, 5 service.

TABLE II-E-6. Oil Development Capital Expenditures, Middle East, by Countries, 1947-1967

(millions of dollars)

Country	1947	1948	1949	1950	1951	1952	1953	1954	1955	1956
Abu Dhabi	...	...	...	...	...	...	...	...	...	...
Bahrain	5.5	...	1.5	1.4	1.9	3.2	2.5	8.6	12.7	9.0
Iran	12.8	15.5	17.2	13.0	11.3	...	...	...	5.3	2.6
Iraq	3.7	5.0	9.1	15.2	14.6	17.9	22.4	23.3	33.5	21.9
Israel	...	...	...	...	...	...	...	...	...	11.7
Kuwait	8.8	46.1	77.7	8.1	8.8	17.7	9.6	10.7	20.8	22.7
Neutral Zone	...	...	...	...	...	...	1.9	9.3	11.9	13.3
Qatar	...	7.2	...	6.7	3.7	10.4	5.2	7.9	8.9	1.5
Saudi Arabia	52.8	67.0	37.6	17.6	21.2	48.3	26.5	11.6	5.0	17.0
Turkey	...	...	...	1.5	0.5	4.5	...	5.9	7.8	9.9
Misc.	...	...	...	...	...	...	1.6	...	...	29.9
Total	83.6	140.8	143.1	63.5	62.0	102.0	69.7	77.3	105.9	139.5

Country	1957	1958	1959	1960	1961	1962	1963	1964	1965	1966	1967
Abu Dhabi	...	...	4.7	2.9	15.1	20.5	27.4	17.4	23.0	33.2	17.8
Bahrain	6.7	1.0	...	0.6	3.5	7.3	2.7	3.5	1.9	2.5	2.7
Iran	7.4	26.2	58.6	18.9	26.6	58.5	34.4	30.3	29.2	23.7	45.6
Iraq	9.0	6.0	21.5	38.2	22.4	3.3	1.5	0.9	2.8	0.8	1.4
Israel	31.7	2.0	3.0	2.9	1.1	3.6	6.1	3.3	3.8	0.9	...
Kuwait	34.8	81.8	48.1	17.8	27.0	30.1	19.9	30.6	10.3	29.4	52.7
Neutral Zone	16.9	20.1	28.3	47.5	56.2	48.8	7.4	3.5	2.6	8.9	6.6
Qatar	5.5	4.1	...	3.4	4.7	4.3	3.8	4.8	4.6	1.5	2.1
Saudi Arabia	23.8	31.6	16.6	15.6	27.7	23.2	9.8	21.0	27.5	33.2	34.5
Turkey	18.0	17.1	18.1	13.5	16.9	17.7	15.0	21.5	26.8	26.5	15.2
Misc.	6.4	7.0	6.7	5.9	1.5	15.8	1.2	15.3	6.6	1.6	2.2
Oman	...	...	...	...	...	...	...	...	...	4.7	6.6
Syria	...	...	...	...	...	...	...	...	...	6.4	12.2
Total	160.2	196.9	205.6	167.2	202.7	233.1	129.2	152.0	139.1	173.3	215.4

Sources and Calculations: As in table II-E-5.

TABLE II-E-7. Oil Development Capital Expenditures, Africa, by Countries, 1955-1967

(millions of dollars)

Country	1955	1956	1957	1958	1959	1960	1961	1962	1963	1964	1965	1966	1967
Algeria	3.1	8.1	14.7	60.3	96.9	125.5	105.1	63.0	65.5	40.2	29.3	25.0	30.7
Angola	...	...	6.5	4.3	9.5	1.7	4.4	1.3	...	...	...	1.2	4.0
Egypt	2.2	12.7	12.9	24.7	21.9	26.8	22.9	28.1	10.6	11.3	10.6	5.7	12.9
Gabon	...	...	...	...	10.6	12.9	5.5	6.2	6.1	4.8	12.4	9.2	5.6
Libya	...	...	...	3.5	16.1	36.3	83.4	70.4	66.4	107.0	87.6	62.3	79.3
Morocco	1.6	4.5	5.5	4.0	3.5	2.0	2.4	2.1	...	...	...	...	...
Tunisia	...	...	...	...	...	...	...	...	...	...	5.4	6.3	9.9
Nigeria	...	...	9.1	25.4	47.0	29.2	42.6	33.8	37.6	87.5	115.3	147.7	94.9
Misc.	5.9	14.7	33.1	28.3	3.6	0.6	...	...	...	3.4	...	4.1	5.0
Total	12.8	40.0	81.8	150.5	209.1	235.0	266.3	204.9	186.2	255.0	260.6	261.5	242.3

Sources and Calculations: As in table II-E-5.

TABLE II-E-8. Summary of Oil Field Development Capital Expenditures, Well Method, 1947-1966, Five-year Periods, Selected Countries

(millions of dollars)

Years	Iran	Iraq	Kuwait	Saudi Arabia	Algeria	Libya	Nigeria
1947-51	59[a]	48	150	196	...	...	...
1952-56	...	119	82	108	...	...	...
1957-61	138	97	291	115	403	139	153
1962-66	176	9	120	115	223	394	422

Sources: Tables II-E-6, II-E-7.

[a]1947-1950.

TABLE II–E–9. Published Investment per Barrel in Persian Gulf Development Projects

Development expenditures (000 $)	Incremental capacity (000 BD)	Investment per daily barrel of initial capacity ($)	Source
(1)	(2)	(3)	(4)
300	30	10	*OGI*, July 1961, p. 85
550	35	16	Ibid.
2,350	50	47	Ibid.
10,500	150	70	*OGI*, Jan. 1966
11,800	150	79	Ibid.
2,355	30	78	*OGJ*, Sept. 24, 1962, p. 112
12,000	150	80	*AAR*, 1964, p. 6
11,500	125	92	*OGI*, Jan. 1965, pp. 30–33
11,500	120	96	*OGI*, Jan. 1964, p. 39
11,500	120	96	*OGJ*, Mar. 25, 1963, p. 84
97,000	1,000[a]	97	*WSJ*, Apr. 5, 1968, p. 8
6,625	65	102	*AAR*, 1963, p. 7
12,800	125	102	*OGJ*, Aug. 6, 1962, p. 102
2,750	25	110	*OGI*, May 1961, p. 70
10,000	50	143	*OGJ*, Apr. 22, 1957, p. 98
16,800	105	164	*N.Y. Times*, 11 Apr. 1968
100,000	220	173[b]	*Offshore*, Nov. 1968, p. 54
25,000	138	181	*PIW*, Mar. 6, 1967
9,400	46	204	*MEES*, Feb. 1965
5,000	22	230	*OGJ*, Dec. 27, 1965; *WWO* Supplement, p. 95
23,957	104	230	*OGI*, Jan. 1962, pp. 34–35
15,000	112	234	*OGI*, Apr. 1964, pp. 54–56
89,600	350	256[c]	*PPS*, Nov. 1968, p. 407
23,900	88	272	*PPS*, Jan. 1965, p. 12
11,000	40	275	*OGJ*, May 4, 1964, p. 37
11,000	40	275	*OGJ*, May 4, 1964, p. 67
28,000	100	280	*OGJ*, July 12, 1965, pp. 71–72
42,000	120	350	*OGJ*, July 30, 1962, p. 113
20,000	50	400	*OGJ*, June 30, 1969, p. 3
280,000	700	400	*OGI*, Feb. 1962, p. 32
11,000	27	407	*OGI*, June 1964, p. 118
180,000	300	600	*PIW*, July 20, 1964
37,500	40	937	*OGJ*, Sept. 25, 1963, p. 111
163,000	170	958	*OGI*, Jan. 1962, p. 30
25,000	17	1,429	*OGJ*, June 22, 1959, p. 85

Median investment per initial daily barrel of capacity = $181

[a]Estimated.

[b]Expenditures $100 million for 21 wells, of which 16 were under development. AAPG *Bulletin*, 1967, "Developments in Foreign Fields," p. 1632; and ibid., 1968, p. 1554. Drilling time per well originally ten weeks. Compare ibid., Nov. 1966, p. 1747, with *Offshore*, Nov. 1968, p. 54. Current completions take only 24–39 days; *Offshore* explains one-time difficulties, delays, and expense. Hence a conservative estimate is: half of (16/21) × $455, or $173 BD.

[c]"The fields are not bonanza ones, and the oil is not low cost."

Appendix II-F

Rig-Time Method of Estimating Total Production Investment and Development Investment Expenditures

In Appendix II-E expenditures for the Middle East and Africa were allocated by developing cost factors based chiefly on average well depth. Total production investment was then divided between exploration and development by the proportion of new oil, gas, and service wells on the one hand, and dry holes on the other.

As has been indicated, depth is not the only cost-determining factor. The hardness of the formation, and drilling problems like losing mud circulation, or pulling the whole string to replace the drill bit, may have much or more to do with relative costs. "Lost circulation is the most troublesome and costly problem in drilling for oil . . . [because of wasted] materials, lost rig time, and lost holes."[1] Of course, lost holes are registered as rig time.

An alternative approach is therefore to allocate drilling and related expenditures according to rig time spent in each area. Rig time registers not only depth but all other factors which increase the days necessary to complete a well. Furthermore, rig time is sometimes recorded separately for exploration and development, and we can therefore get a much more accurate division.

A test of possible bias in the rig-time method is made in Table II-F-1. For the heavy-duty rig, most nearly relevant to Persian Gulf conditions, the maximum error is 2 percent at 8,000 feet, less at shallower horizons.

In principle, therefore, the rig-time method must be considerably more accurate than the methods used earlier. Unfortunately, the data are often unavailable, particularly for small areas. But in Table II-F-2, the exploration and development rig time in five large producing countries is stated; and in II-F-3, the estimated field development expenditures are given.

The estimate for the Iran Consortium is 25 percent below what can be calculated from its annual reports, but the true error is very much less and may be one of overstatement. There have been large-scale pipeline expenditures since the inauguration of the Kharg terminal in March 1966, as well as much investment in natural gas and gas processing facilities. A rough guess is that these account for about $47 million (of a total difference of $66 million). Were we able to estimate and subtract these two types of expenditure, the figure for 1967 might well be less than that for 1964.

The estimate for Iraq is a slight overshoot (6 percent), that for Abu Dhabi is twice the actual, but the sampling error is very great. The estimate for all Libya is too low, and may be only partly explained by some non-oil projects; most of the apparent error is real. But the estimate for Oasis (Libya) is higher than the actual. Thus of six tests, three rig-time estimates are too high, two are too low, and one may be. There seems no reason to expect a bias in the rig-time method.

[1] J. V. Messinger, "How to Combat Lost Circulation," *OGJ*, May 13, 1968.

Except for Iran (see Appendix II-I), the expenditure is matched with our estimate of gross capacity developed, which is altogether different from that in Appendix II-E. Instead of assuming that each new well has average capacity, and multiplying the average by the number of new wells, expenditures in any given year are related with the increase in production from the first half of the current year to the next year. Similar estimates were made for areas where no independent expenditure data were available.

Where output expands strongly, its growth is a good measure of capacity growth. But as indicated in the text, capacity lost in any time period as percent of net growth is a/n, and as n becomes small in relation to a the error of neglecting a becomes intolerable. Hence at low growth rates estimates may be far off.

In Iraq, the highest rate of output was reached in February and November 1966. It has since stayed within a very narrow range, except for deliberate shutdowns. One could therefore view the expenditure during 1966-68, $9.2 million, as the steady-state capital cost of producing about 1.5 million barrels daily, or 1.65 billion barrels over the three-year period, at a capital cost of 0.53 cents per barrel. This does not appear satisfactory.

In Kuwait, the January 1967 peak was only modestly surpassed in December 1968, and this was barely surpassed in December 1969. In Saudi Arabia, the March 1967 peak was only surpassed in February 1969. If the Kuwait and Saudi Arabian decline rates exceed 1 percent, gross capacity additions are seriously understated, and costs overstated. (See Table II-E-4.)

It becomes a very minor matter that the use of an exponential function tends to overstate costs because the decline curve will probably be closer to the hyperbolic, especially for water drive reservoirs, the most important type, and hence production will not lessen as rapidly.[2] Moreover, we are unable to make any estimate of the "economic limit," the rate of output where production is no longer worthwhile. Hence the probable life of the project is unknown. However, the error is small for 25-year projects and large fields can be expected to last well beyond them. Of course, the higher the discount rate, the quicker the convergence.

Tables II-F-5 to II-F-9 summarize calculations of estimated development expenditures for the Iran Consortium, Iraq Petroleum Co., Abu Dhabi Petroleum Co. (onshore), total Libya, and Oasis in Libya. In each of these cases we are able to check the estimates against actual expenditures. Tables II-F-10 and II-F-11 give estimates for Nigeria, Kuwait, and Saudi Arabia, but with no check available.

TABLE II-F-1. Relation of Offshore Rig-Time to Drilling Cost

	Rig Types							
	A		B		C		D	
Depth (*000 feet*)	Drilling time	Drilling cost	Drilling time	Drilling cost	Drilling time	Drilling cost	Drilling time	Drilling cost
4	100	100	100	100	100	100	100	100
5	105	107	111	107	108	107	109	107
6	116	114	121	115	118	116	121	117
7	126	123	132	124	136	125	144	129
8	139	133	147	134	154	136	169	143

Source: E. F. Klementich (Shell Oil Co.), study reported in *OGJ*, Mar. 10, 1969, p. 44.

Types: A, heavy-duty rig; B, high performance shallow-depth rig; C, smaller shallow-depth rig; D, converted workover rig.

Rig types B-D were designed especially for the experiment; only A is in current use, and accounts for about 75% of the offshore rigs.

[2] Robert Mannon, "Oil Production Forecasting by Decline Curve Analysis," SPE 1254 (1965), preprint (subject to revision), pp. 6-7, calls the exponential "a pessimistic forecast." Also, J. J. Arps, "Analysis of Decline Curves," *Transactions*, AIME (1945), 160, 228. Reprinted in J. J. Arps, ed., Petroleum Transaction Reprint Series No. 3, *Oil and Gas Property Evaluation and Reserve Estimates*, SPE, n.d. (apparently 1960), and Morris Muskat, *Physical Principles of Oil Production* (McGraw-Hill, 1949).

TABLE II-F-2. Rig Months, Five Principal Producing Countries, 1959-1970

Area, country	1959 E	1959 D	1959 T	1960 E	1960 D	1960 T	1961 E	1961 D	1961 T	1962 E	1962 D	1962 T	1963 E	1963 D	1963 T	1964 E	1964 D	1964 T
Africa	n.a.	n.a.	1,098	n.a.	n.a.	1,236	n.a.	n.a.	1,224	n.a.	n.a.	1,194	n.a.	n.a.	1,224	n.a.	n.a.	1,272
Libya	...	n.a.	192	...	n.a.	344	372	42	414	365	72	432	421	108	529	389	146	535
Middle East	n.a.	n.a.	600	n.a.	n.a.	660	n.a.	n.a.	648	n.a.	n.a.	624	n.a.	n.a.	678	n.a.	n.a.	732
Iran (consortium)	n.a.	n.a.	n.a.	24	66	90	36	90	126	48	102	150	48	90	138	30	66	96
Iraq (IPC)	75	27	102	54	54	108	12	42	54	0	15	15	0	12	12	0	12	12
Kuwait (KOC)	21	33	54	20	28	48	30	42	72	35	49	84	48	48	96	40	50	90
Saudi Arabia (Aramco)	58	48	106	27	49	76	17	48	65	17	48	65	16	48	64	9	72	81

Area, country	1965 E	1965 D	1965 T	1966 E	1966 D	1966 T	1967 E	1967 D	1967 T	1968 E	1968 D	1968 T	1969 E	1969 D	1969 T	1970 E	1970 D	1970 T
Africa	n.a.	n.a.	1,248	n.a.	n.a.	1,062	n.a.	n.a.	942	n.a.	n.a.	1,050	n.a.	n.a.	1,230	n.a.	n.a.	1,200
Libya	274	192	466	153	142	295	84	174	258	125	208	333	156	297	453	105	145	249
Algeria	...	...	...	n.a.	n.a.	168	n.a.	n.a.	102	n.a.	n.a.	150	148	108	256	b	b	b
Middle East	n.a.	n.a.	798	n.a.	n.a.	756	n.a.	n.a.	732	n.a.	n.a.	648	n.a.	n.a.	594	n.a.	n.a.	666
Iran (consortium)	12	42	54	18	48	66	24	54	78	30	60[a]	90	30	72	102	30	78	108
Iraq (IPC)	0	12	12	0	12	12	0	6	6	0	6	6	0	9	9	1	12	13
Kuwait (KOC)	12	12	24	12	39	51	0	76	76	6	51	57	2	10	12	4	8	12
Saudi Arabia (Aramco)	12	46	58	9	45	54	18	38	56	13	45	58	9	55	64	0	38	38

See next page for notes on sources and methods.

TABLE II–F–2. Continued.

Notes on Sources and Methods:

1) E, exploratory; D, development; T, total.

2) Workover rig-time included with development. "Appraisal" wells are assumed to be development, except when stated to be the contrary. Usage in Kuwait differs from elsewhere, since many "appraisal" wells are called exploratory, and this is confirmed in the text (see note 5 below).

3) Libya development rig-months are overstated, because they include all exploratory wells other than new field wildcats. This is not true of the other countries, but nomenclature is not fully consistent.

4) Estimates for the total Middle East (Persian Gulf plus Turkey, Israel, and Syria) and total Africa are made by adding the number of wells drilling at the beginning and end of the year (more precisely, December 1 of the preceding year and December 1 of the given year), as given in the annual supplement "World Wide Oil" to the *OGJ*, and multiplying by six.

This assumes an even increase or decrease over the year. For example, four wells drilling at the start of the year and two at the end are reckoned as an overage of three, hence 36 rig-months. Yet two rigs may have been stopped and removed at the start of the year, leaving only two operating for 24 rig-months; or they may have been stopped and removed at the end, with the true rig-month total 48. Moreover, rigs may have been moved into and then out of the company's area during the year, or vice versa. In general, the smaller the number of rigs, the greater the chance of error from these irregularities.

5) For individual countries, the *OGJ* data were used only as a last resort. The basic sources were the AAPG *Bulletin*, and the annual reports of the various companies, which often differ widely from the *OGJ*. Libya AAPG totals were greater than *OGJ* four times, lower three times, identical once. Iranian Consortium is always substantially lower than total Iran, of course, since there has been much drilling by other companies. Original Saudi Arabian (Aramco) totals are above *OGJ* six times, lower three times, identical once. Since the data for these three companies are very good throughout, this suggests that the *OGJ* tends somewhat to understate or at least not overstate total rig time.

In Iraq, our estimates are higher than *OGJ* in 1959 and 1960, but much lower in 1961. This is as it should be, since exploratory drilling was stopped there on April 7, 1961. For the remaining seven years, the estimates are well below those of the *OGJ*-based measure, and it must be admitted that the room for possible error is wide.

Kuwait is very difficult to estimate during 1959–65. During 1959–63, our estimates are higher than the *OGJ* average four times, lower once. But for 1964, there is no way of estimating the total. Hence the *OGJ* average is used, allocating as follows. Kuwait Oil Co. 1964 *Annual Report* mentions 33 development wells, 70 workovers, 15 exploratory holes. We assume that rig-months per well, exploratory : development : workover is about 9 : 3 : 1, thereby assigning 50 months to development, 40 to exploration. For 1965, the *OGJ*-based average is 84 total rig-months, while we estimate only 24. We may well be off the mark, but it seems impossible to reconcile our sources with anything much higher than 24. Starting with 1966, fortunately, the KOC annual reports give a direct count of total rig-months, which are twice larger, once smaller than the *OGJ*-based ones. Exploratory rig-months are estimated from the spudding and completion dates, and subtracted from the total.

[a]Includes natural gas.

[b]Sources not consistent.

n.a. = not available.

TABLE II-F-3. Estimated Field Development Expenditures, Five Principal Producing Countries, 1962-1968

(*millions of dollars*)

Year	Total production capital expenditures	Development expenditures	Total production capital expenditures	Development expenditures			
	Africa	Libya	Middle East	Iran consortium	Iraq	Kuwait	Saudi Arabia
1962	335	20.0	275	45.0	6.6	21.6	21.2
1963	275	24.0	150	19.9	2.7	10.6	10.6
1964	325	37.3	175	15.8	2.9	11.9	17.2
1965	340	52.3	210[a]	11.1	3.2	3.2	12.1
1966	355	47.6	255[b]	16.2	4.0	13.2	15.2
1967	355	65.4	290[b]	21.4	2.4	30.1	15.1
1968	575	117.9	300[b]	27.8	2.8	23.6	20.8

Sources and Notes: Expenditures from Chase Manhattan Bank, *Capital Investment of World Petroleum Industry*, respective years; percentage factors from table II-F-1.

The ratio of development rig-months for the country to all rig-months for the area is multiplied by total expenditures for the area. Thus in Libya (see table II-F-2) in 1962, 72 development rig-months are 6% of all Africa rig-months; 6% of $335 million equals $20 million.

[a]Excluding bonuses $195, see table II-E-1.
[b]Including natural gasoline plants.

TABLE II-F-4. New Capacity Installed, Five-year Periods, 1947-1969

(*thousand barrels daily*)

	Iran	Iran (consortium only)	Iraq	Kuwait	Saudi Arabia	Algeria	Libya	Nigeria
1. 1947-51								
Net increase	240	...	288	704	471	...	...	...
Capacity decline....	22	...	6	13	56	...	...	...
Total	262	...	294	717	527	...	...	...
2. 1952-56								
Net increase	...	...	343	391	165	...	...	...
Capacity decline....	...	...	29	48	46	...	...	...
Total	...	...	372	439	211	...	...	...
3. 1957-61								
Net increase	590	...	317	690	530	432	91	66
Capacity decline....	48	...	40	72	57	27	6	3
Total	638	...	357	762	587	459	97	69
4. 1962-66								
Net increase	1,386	1,152	729	717	1,254	340	1,598	513
Capacity decline....	85	85	61	103	93	149	126	10
Total	1,471	1,237	790	820	1,347	489	1,724	523
5a. 1966-68								
Net increase	1,347	973	155	277	414	329	1,576	...
Capacity decline....	75	72	41	70	78	125	175	...
Total	1,422	1,045	196	347	492	454	1,751	...
5b. 1966-68								
Net increase	1,481	1,318	nega.	515	576	n.a.	1,582	n.a.
Capacity decline....	75	72	...	70	78	...	175	...
Total	1,556	1,390	...	585	654	...	1,757	...

See next page for sources and notes.

TABLE II-F-4. Continued.

Sources: World Oil, "International Outlook," 1958; *British Petroleum*, "Statistical Review of the World Oil Industry," various years. *OGJ*, "World Wide Oil," various years; *PIW* monthly output figures around the middle of the following month.

Notes:

NET INCREASE

Lines 1, 2, 3–1947–61: Increase in annual output first year of period to the year following end of period (e.g., 1952 less 1947).

Line 4–1962–66: First five months 1967 less first six months 1962.

Line 5a–1966–68: First six months 1969 less first six months 1966.

Line 5b–1966–68: Peak month January–June 1969 less peak month January–June 1966.

Exceptions: Iran first period 1950 less 1947.

Iraq second period, mean of 1956 and 1958 outputs less 1952. Sixth period, November 1966 less first six months 1962.

Algeria and Nigeria fourth period, first six months 1967 less first six months 1962.

CAPACITY DECLINE For all Middle East countries, 1% of total output for period; Libya and Nigeria, 3%; Algeria, 5%.

TABLE II-F-5. Iran Consortium: Tests of Rig-Time Method and Calculations of Development Investment per Initial Daily Barrel, 1963–1968

1. Iran Consortium, rig time as percent of Middle East 522/4344	12.0%
2. Production and "exploration" expenditures, Middle East, CMB	$1,625 million
3. Iran Consortium, estimate (line 1 × line 2)	$ 195 million
4. Iran Consortium, producing company capital expenditures, actual	$ 306 million
5. Less pipeline outlays, Kharg II project (1963–65 inclusive)	$ 45 million
6. Equals producing-exploration expenditures, plus unknown amount of pipeline and non-oil production expenditures	$ 261 million
7. Maximum error of estimate [(line 6–line 3) / line 3]	34%

Sources (by line):

1. Table II-F-1.
2. Chase Manhattan Bank, *Capital Expenditures of World Petroleum Industry*, adjusted for bonus expenditures II-E-1.
4. Table II-I-1.
5. *PPS*, May 1966, p. 176.

TABLE II-F-6. Iraq Petroleum Company: Tests of Rig-Time Method and Calculations of Development Investment per Initial Daily Barrel, 1963–1966

1. Iraq rig time as percent of Middle East 48/2964		1.62%
2. Production expenditures Middle East, CMB		$ 790 million
3. Iraq, estimate (line 1 × line 2)		$12.8 million
4. Actual expenditures: IPC	$12.4 million	$13.6 million
Mosul	$ 0.1 million	
Basrah	$ 1.1 million	
5. Error of estimate [(lines 3–4) / line 4]		5.9%
6. Increase in output November 1966 over 1963 first half		629 TBD
7. Decline in capacity (1% of total output 1963–66)		51 TBD
8. Total capacity increase (line 6 + line 7)		690 TBD
9. Investment per daily barrel of capacity (line 4 / line 8)		$19.7 million

Sources (by line):

1. Table II-F-1.
2. Chase Manhattan Bank. Exploration expense excluded because there was none in Iraq during these years.
4. Compiled by Thomas R. Stauffer from file in Company's Registration Office (London) and reproduced with permission.
6. *OGJ*, World Wide Oil, 1963; *PIW*, Dec. 19, 1966.
7. *World Oil*, "International Outlook," respective years.

TABLE II-F-7. Abu Dhabi, Onshore: Tests of Rig-Time Method and Calculations of Development Investment per Initial Daily Barrel, 1965-1966

1. ADPC rig time as percent of Middle East 48/1554	3.08%
2. Production and exploration expenditures, Middle East	$ 550 million
3. Abu Dhabi Petroleum Co., estimate (line 1 × line 2)	$16.9 million
4. Actual expenditures, ADPC	$ 8.2 million
5. Error of estimate [(line 3–4) / line 4]	+106%
6. Development rig-time as proportion of total rig-time (19/48)	39.6%
7. Development expenditures (line 4 × line 6)	$3.25 million
8. Increase in output January–May 1967 over first half 1965	113 TBD
9. Decline in capacity (1% of total output 1965-66)	6 TBD
10. Total capacity increase (line 8 + line 9)	119 TBD
11. Investment per daily barrel of capacity (line 7 / line 10)	$27.3 million

Sources (by line):

1. Table II-F-1 and AAPG *Bulletin.*

2. Chase Manhattan Bank, *Capital Investment of World Petroleum Industry*, respective years, adjusted for bonus expenditures (II-F-1).

4, 6. AAPG *Bulletin.*

8. *OGJ*, "World Wide Oil," respective years; *PIW*, June 19, 1967.

9. *World Oil*, "International Outlook," 1967, p. 156.

Note: Since only two years are covered, the possible sampling error is very great, both as to expenditures and as to the growth in capacity. Development began in August 1962. [AAPG *Bulletin*, 1964, p. 1368.] If we add 5/12 of 1962 expenditures to 1963-66, the total is $53.1 million. Unfortunately, we cannot divide this into exploration and development for years before 1965. But we can at least set an upper bound by assuming it was *all* development, in which case ($\$53.1 \times 10^6)/(264 \times 10^3)$ = $205 per initial daily barrel. The true development cost lies somewhere between these extremes.

TABLE II-F-8. Libya: Tests of Rig-Time Method and Calculations of Development Investment per Initial Daily Barrel, 1963-1965

1. Libya total rig time as percent of Africa (1530/3744)	40.9%
2. Production capital and exploration expenditures (940 + 330)	$1,270 million
3. Libya estimate (line 1 × line 2)	$ 520 million
4. Pipeline expenditures	$ 210 million
5. Total Libyan estimated capital expenditures (line 3 + line 4)	$ 730 million
6. Actual total Libyan capital expenditures	$ 955 million
7. Error of estimate [(lines 6-5) / line 6]	23.5%
8. Production capital expenditures (line 6-line 4)	$ 745 million
9. Error of estimate, production-exploration expenditure [(lines 8-3) / line 8]	30.2%
10. Production-capital expenditure/production capital and exploration all Africa (940/1270)	74.0%
11. Development rig time Libya as percent of total Libya rig time	29.1%
12. Field development expenditures [lines (8 × 10) × line 11]	$ 159 million
13. Production increase first half 1966 over first half 1963 (1,429 — 350)	1,079 TBD
14. Decline in capacity (3% of total output 1963-65)	76 TBD
15. Total gross capacity installed (line 13 + 14)	1,155 TBD
16. Field development investment per daily barrel (line 12 ÷ line 15)	$ 138 million

Sources (by line):

1, 11, table II-F-1.

2, 4, 10, Chase Manhattan Bank, *Capital Investment Expenditures of World Petroleum Industry*, respective years. *N.B.* All pipeline expenditures for Africa assigned to Libya because no such construction noted elsewhere by AAPG.

6. *AAPG Bulletin*, respective years.

13. *OGJ*, "World Wide Oil," respective years.

14. *World Oil*, "International Outlook."

Note: Total capital expenditures by BP-Bunker Hunt up to the start of Sarir-L operations in early 1967 were $168 million (PONS, Feb. 13, 1967), of which $73 million was for the pipeline [AAPG *Bulletin*, 1967, p. 1575] and $95 for field exploration and development. Approximately 100-158 rig-months were for development (ibid.) so we estimate $60 million. The producing capacity, as shown by 1967 output, was 320 TBD, or $187 per BD, which at 3% decline, 25-year life, and 20% return (PBE 1,750) implies 10.7 cents per barrel. The pipeline capacity is 600 TBD, or $122/DB, which at 20% return (PBE ratio 1,980) implies 6.2 cents per barrel. Adding average Libya operating cost, the total is 10.7 + 6.2 + 8.4 = 25.3 cents total cost at tidewater.

TABLE II-F-9. Oasis (Libya): Tests of Rig-Time Method and Calculations of Development Investment per Initial Daily Barrel, 1963-1964

	1963	1964
1. Total Libyan production capital expenditures (million $)	308	319
2. Total Libyan rig-months	529	535
3. Oasis rig-months	159	146
4. Oasis as percent of total	30.5	27.3
5. Estimated Oasis production capital expenditures (line 1 × line 4) (million $)	92.6	87.4
6. Actual Oasis production capital expenditures (million $)	60.9	75.1
7. Error of estimate (line 5 less line 6, as percent of line 6)	+52.0	+16.5

Sources (by line):
2, 3: *AAPG Bulletin*, Aug. 1964 and 1965.
6: Continental Oil Co., *Annual Report, 1964*.

TABLE II-F-10. Nigeria: Estimated Development Investment per Daily Barrel, 1965-1966

1. Nigeria rig time as percent of Africa (432/2310)	18.70%
2. Nigeria development rig time as percent of Africa (174/2310)	7.54%
3. Production capital expenditures, Africa	$ 695 million
4. Estimated Nigeria production capital expenditures (line 1 × line 3)	$ 130 million
5. Estimated Nigeria development capital expenditures (line 2 × line 3)	$52.4 million
6. Production increase first five months 1967 over first half 1965 (566 TBD less 270 TBD)	296 TBD
7. Decline in capacity (3% of total output 1965-66)	21 TBD
8. Total capacity increase (line 6 + line 7)	317 TBD
9. Field development investment per daily barrel (line 5 / line 8)	$ 165 million

Sources (by line):
1, 2: *AAPG Bulletin*, table II-F-1.
3: Table II-F-3.
6: *OGJ*, "Worldwide Oil," respective years; *OGI*.

TABLE II-F-11. Kuwait and Saudi Arabia: Estimated Development Expenditure per Daily Barrel

	Kuwait[a]	Saudi Arabia[b]
1. Development rig months, 1962-68[c,d]	325	342
2. Total rig months, Middle East	4,968	4,968
3. Country percent of total (lines 1 / 2 × 100)	6.55	6.87
4. Production capital expenditure, Middle East, 1962-68 (million $)	$1,655	$1,655
5. Country development capital expenditure (lines 4 × 3)	$108.4	$113.7
6. Increase in average daily production Jan.-June 1969 over Jan.-June 1962 (TBD)	716	1,333
(6a. Increase in peak production Jan.-June 1969 over Jan.-June 1962) (TBD)	(837)	(1,581)
7. Total of annual average production 1962-68 (TBD)	15,027	14,621
8. Estimated decline rate (percent)	1	1
9. Capacity lost and replaced (TBD) (line 7 × line 8)	150	146
10. Total gross capacity installed (TBD) (line 6 + line 9)	866	1,479
(10a. Line 6a + line 9)	(987)	(1,727)
11. Development investment per daily barrel ($) (line 5 / line 10)	$125	$77
(11a. Line 5 / line 10a)	($110)	($66)

Sources (by line):
1 and 2–Table II-F-1.
4–Table II-F-3.
6 and 7–*OGJ*, World Wide Oil, respective years.
6a–*PPS*.

[a] Kuwait Oil Co. only.
[b] Aramco only.
[c] Excludes stratigraphic drilling.
[d] Includes workovers, appraisal wells.

Appendix II-G

Kuwait: Computed and Actual Investment, 1950–1955

As a check, the Gulf totals of productive and other completions for the six-year period are compared in Table II-G-1 with those published by *World Oil* and by Kuwait Oil Co.; and the agreement is obviously close. Then, in Table II-G-2 our estimates of expenditures and of new capacity are compared with those calculated from the Gulf statements.

As compared with our estimates, Gulf expenditures fluctuate rather violently from year to year, probably because they are reckoned as committed or as funds advanced. For the three years 1950–52, our estimates are more than three times actual, but adding in the next two consecutive years puts them close together. For the period as a whole, calculated expenditures are nearly 20 percent higher than actual.

TABLE II-G-1. Wells Completed in Kuwait, 1950–1955

	Productive completions[a]			All completions	
Year	*World Oil*	GOC	KOC	*World Oil*	GOC
	(1)	(2)	(3)	(4)	(5)
1950	13	14	17	13	14
1951	20	20	31	21	20
1952	21	22	17	23	22
1953	17	18	23	17	18
1954	12	14	14	13	16
1955	21	18	22	22	18
Total	104	106	124	109	108

Sources:

Cols. (1), (4)–*World Oil*, "International Outlook" issue, respective years.

Cols. (2), (5)–Gulf Oil Corp., *Registration Statement, Form S-1*, filed with S.E.C., Jan. 10, 1956 [doubled].

Col. (3)–Kuwait Oil Co., *Annual Review of Operations*, respective years.

[a]Including successful wildcats.

TABLE II-G-2. Kuwait Production Investment and Capacity Increase, 1950-1955

Year	Production expenditures (GOC) (*million $*)	Development expenditures (Table II-E-6) (*million $*)	Annual increase in output (GOC) (*TBD*)	Investment in dollars per initial daily barrel (Table II-E-5)	(GOC)
	(1)	(2)	(3)	(4)	(5)
1950	0.456	8.1	73	111	6
1951	0.636	8.8	117	75	5
1952	11.540	17.7	186	95	62
1953	18.606	9.6	115	85	162
1954	27.680	10.7	90	119	307
1955	5.152	20.8	140	147	37
Total	64.070	75.7	821	92	78

Sources:

Cols. (1), (3) from Gulf Oil Corp., *Registration Statement, Form S-1,* filed with S.E.C., Jan. 10, 1956 [doubled].

Col. 4 = (2)/(3).

Col. 5 = (1)/(3).

Appendix II-H

Saudi Arabia: Accounting Costs, 1962–1968

An estimate of total production costs for Saudi Arabia, which approximate the accounting costs, is shown in this appendix. Since there is a notional 50-50 profit split between the operator and government, the published payment to the government is doubled in order to obtain the accounting profit. This is deducted from the notional sales revenues at posted prices in order to derive the total production costs as they must have been entered on the accounting records of Arabian American Oil Company in order to explain and justify the particular payments to the government. Since cost is a small residual from a much larger total, it is not surprising that it fluctuates rather strongly, and has in recent years been close to zero. Moreover, in 1964 there was a change in accounting procedures whereby capital expenditures were depreciated rather than currently expensed. In addition, both receipts and costs are affected by the large Ras Tanura refinery. Finally, the use of straight posted prices overstates receipts to 1966, when Aramco used realized prices on sales to nonaffiliates. The table should be regarded as only a specimen, or even a warning, of how not to estimate producing costs.

TABLE II-H-1. Accounting Cost of Saudi Arabian Oil, 1962–1968

	1962	1963	1964	1965	1966	1967	1968
1. Value of exported crude oil at posted prices (million $)	784	879	958	1,119	1,350	1,460	1,623
2. Average posted crude price, (dollars per barrel)	1.70	1.78	1.82	1.79	1.78	1.78	1.82
3. Refined products exported (million barrels)	78	86	92	104	105	112	138
4. Posted crude value of refined products (line 2 × line 3) (million $)	132	153	168	186	187	199	252
5. Total crude value, at posted prices (line 1 + line 4) (million $)	916	1,032	1,126	1,305	1,537	1,659	1,875
6. Total oil company payments to Saudi Arab government (million $)	410	455[a]	524	617[a]	760[a]	880[a]	927
7. Total accounting profit (line 6 doubled) (million $)	820	910	1,047	1,234	1,520	1,760	1,854
8. Cost of production, defined as posted sales value less profits (line 5 less line 7) (million $)	96	122	79	71	17	−101	23
9. Total crude production (million barrels)	600	652	694	805	950	1,024	1,114
10. Production costs per barrel, in cents (line 8 ÷ line 9)	16	19	11	9	1.8	...	2.1

Source: Saudi Arabian Monetary Agency, *Annual Report*, various years.

[a]Excluding special nonrecurring payments.

Appendix II-I

Iran Consortium: Development Costs

Iran production capital expenditures are shown in columns (1)-(3) in Table II-I-1. Total oil development costs are calculated as follows:

1) Development rig-months are from two-thirds to three-fourths of total rig-months and development drilling expenditures are estimated from them in column (5). For the whole period 1963-69, \$85.7 million was spent drilling for 2,201 thousand barrels daily (TBD) of gross new capacity, i.e., the difference between 1970 and 1963 output plus 1 percent of cumulated 1963-70 output. This amounted to \$38.9 per initial daily barrel. At a PBE factor of 5.2 (see Table II-2) and 1,865 PBEs, drilling costs come to 2.10 cents per barrel.

2) The second development cost is that of producing equipment in the neighborhood of the well: settling tanks, gas-oil separators, gathering systems, and the like. Here we have almost no data anywhere. In the United States in 1966, lease equipment (exclusive of fluid injection projects) was 28.8 percent of development drilling cost.[1] This is subject to wide errors in both directions. Since drilling cost per barrel of capacity is so low in Iran, the ratio of non-drilling costs should be much higher than in the United States. On the other hand, production is so atomized in the United States, with 700,000 oil wells operating as against the Consortium's 190, that the equipment expenditures per barrel of production must be much higher than in Iran, since so much equipment must be furnished to take care of minuscule crude streams.[2] Therefore, the U.S. factor has been rounded to 30 percent, multiplied by the drilling cost of \$85.7 million, yielding a figure of \$25.7 million for lease equipment, or \$11.7 per initial daily barrel and 0.63 cents per barrel. Total field development outlay is \$111.4 million. This is \$50.5 per initial daily barrel and 2.7 cents per barrel produced.

3) Development costs include pipelining the crude to tidewater, and storage, jetties, and pumping equipment. There are no annual data on this very large element. But we are fortunate in that during 1962-65, in switching to a new export terminal on Kharg Island, the whole operation was rebuilt at a cost of \$92 million. If we assume conservatively that the capacity provided was that of early 1967, not quite 2.4 million barrels daily (MBD), then the cost is \$39 per initial daily barrel. At a PBE factor of 5.3 (since there is no decline rate) and 1,935 PBEs, the cost per barrel is 2.0 cents. It is assumed that the costs of looping the pipelines and expanding the Kharg terminal, which has been considerably enlarged since its opening in 1966, are no less per barrel than the original installation—an obvious overestimate. We add this 2.0 cents to the field development cost to obtain a total of 4.7 cents per barrel for the seven years 1962-69 inclusive.

[1] *JAS*, 1966, Part II.

[2] A single production facility, serving 10 Iranian wells, accounts for 480 TBD; "a large but ordinary Christmas tree" on the biggest well accounts for 95 TBD. See *World Oil*, Apr. 1970, p. 53.

4) The remaining "other production capital expenditures" are excluded. Additional pipeline expenditures, looping the systems from the producing areas to Kharg, have been excluded because additional pipeline capacity is equated to zero by assuring a limit of 2.4 MBD (see par. 3). Other expenditures are not related to crude oil production. The IOCC notes that sulfur and gas liquids are recovered and sold from the natural gas stream that is sold to the National Iranian Oil Co. In 1968 the Consortium started "a mammoth Natural Gas Liquids project," which was "a major investment."[3] It was expected that by 1970 it would supply dry gas to the Soviet Union and liquid feedstocks for export, and export began in 1970.[4] Three extraction plants in the Agha Jari field and the world's largest such plant at Marun were completed in 1969.[5]

In 1968 about 19 rig-months (of a total of 60) were for gas, and in 1969 five gas wells were drilled.[6]

One can only guess at the amounts involved.

In Table II-I-2, columns (1)-(3) set out the cost per additional daily barrel developed, the PBE factors, and the per barrel costs. For the 1985 estimate, total expenditures on all drilling are used, plus exploration. Over the seven years 1963-69, the amounts spent were $86 million and $125 million respectively; hence we adjust by a factor of 125/86, or 1.46. It is assumed (see text, pp. 72-73) that wells are only 68 percent as productive, hence a correction of 1.47.

Accordingly, investment per additional daily barrel increases by a factor of 1.46 × 1.47 = 2.15, and is entered in column 4. Since a 6 percent decline rate is assumed, the PBE factor increases as shown in column 5, and the final result is that field development capital cost goes from 2.7 to 6.9 cents. There is no effect on pipeline and terminal capital costs.

As for operating cost, it was just under 5 cents per barrel in 1967–69. Adjustment is needed by the square root formula (1.21 is the square root of 1.47) and for the higher present value of the equated outlays.

TABLE II-I-1. Iran: Production Capital Expenditures, 1962–1970

(*millions of dollars*)

Year	Exploration and drilling	Other	Total	Development rig-months as percent of total rig-months	Development drilling expenditures (1) × (4)
	(1)	(2)	(3)	(4)	(5)
1963	30.2	13.4	43.6	65.2%	19.7
1964	11.8	25.5	37.3	68.8	8.1
1965	12.9	77.9	90.8	77.8	10.0
1966	9.5	26.6	36.1	72.7	6.9
1967	16.6	24.4	41.0	69.2	11.5
1968	22.6	34.8	57.4	66.7	15.1
1969	21.6	57.1	78.7	66.7	14.4
1970	17.3	25.9	43.2	66.7	11.5
Total	142.5	285.6	428.1	...	...

Source: IOOC, *Annual Report*, various years. Sterling converted at $2.68 in 1967, $2.80 previously, $2.40 thereafter.

[3]*Annual Report, 1966*, pp. 4, 7; *1967*, p. 10; *1968*, pp. 4, 20; *1969*, p. 4.

[4]*OGJ*, Dec. 28, 1970, p. 94.

[5]*Annual Report*, 1969, p. 7.

[6]Ibid., 1968, p. 18; 1969, pp. 6-7.

TABLE II-I-2. Iran Consortium: Capital Cost and Operating Cost 1963-1969, and Increase Under Assumed Zero Discoveries, 1969-1985

A. CAPITAL COST

Element of cost	1963-1969 Investment per initial daily barrel	1963-1969 PBE factor	1963-1969 Cost per barrel (*cents*)	1985 Investment per initial daily barrel	1985 PBE factor	1985 Cost per barrel (*cents*)
	(1)	(2)	(3)	(4)	(5)	(6)
Exploration and drilling	$38.9	5.2	2.1	$ 83.6	4.3	5.3
Lease equipment	11.7	5.2	0.6	25.2	4.3	1.6
Pipeline, terminal	39.0	5.3	2.0	39.0	5.3	2.0
Total	89.6	...	4.7	141.0	...	8.9

B. OPERATING COST

(*cents per barrel*)

Element of cost associated with:	1967-1969 Percent of total	1967-1969 Amount	1985 Well productivity factor	1985 Present-worth factor $(a+r)/r$	1985 cost
Exploration and drilling	43	2.15	1.21	26/20	3.38
Lease equipment	13	0.65	1.21	26/20	1.02
Pipeline, terminal	44	2.20	1.0	1.0	2.00
Total	100	5.00	...	...	6.40

Source: Table II-I-1.

Note: PBE factor 5.2 implies 1% decline, 20% return. PBE factor 5.3 implies zero decline, 20% return. PBE factor 4.3 implies 6% decline, 20% return.

Appendix III-A

Aramco Offtake Prices, 1949–1955

On May 3, 1949, the Economic Cooperative Administration (ECA) later known as the Mutual Security Administration (MSA) promulgated section 201.22(E)(2) providing that the price for which it would finance must be no higher than the price charged anywhere else. In late 1950, "ECA undertook to obtain company documents covering shipments of Middle East crude oil to the Western Hemisphere. The agency then discovered that all shipments to the Western Hemisphere since the date of its regulation had been billed not at \$1.75 but at a uniform f.o.b. price of \$1.43 per barrel."[1]

One's first impression is that Esso, Mobil, and Caltex began transferring at \$1.43 on the date of the regulation. But the statement is equally compatible with the transfer at \$1.43 not only after but before the date of the regulation; it is only that ECA did not have documents for shipments at an earlier period.

In form, the price was the posting of \$1.75 less an 18.5 percent marketing allowance.[2] But apparently sales by Aramco offtakers were made at \$1.75. In 1955, Saudi Arabia demanded and obtained taxes based on the posted rather than the discounted price. Apparently the discount had no market significance.

[1] *Statement of Mutual Security Administration Concerning Its Relations with International Oil Companies with Respect to Petroleum Prices*, Aug. 15, 1952 (mimeo.).

[2] *Arbitration between the Government of Saudi Arabia and Arabian American Oil Company–Memorial of Arabian American Oil Co.*, vol. IV, pp. 2282–98. I am indebted for this reference to my former student, Dr. Farid W. Saad.

Appendix III-B

TABLE III-B-1. Crude Oil Production, Availability, and Self-Sufficiency Outside North America: Eight Largest Companies, 1957-1966

(*thousand barrels daily*)

	1957	1958	1959	1960	1961	1962	1963	1964	1965	1966
ESSO										
1. Production	1,782	1,769	1,854	1,921	2,067	2,328	2,610	2,845	3,062	3,352[a]
2. Long-term contracts	105	144	156	164	196	210	215	217	231	255[b]
3. = 1 + 2	1,887	1,913	2,010	2,085	2,263	2,538	2,825	3,062	3,293	3,607
4. Crude processed	1,389	1,446	1,625	1,759	1,912	2,181	2,341	2,526	2,799	3,003
5. = 3 − 4	+498	+467	+385	+326	+351	+357	+484	+536	+494	+604
6. = 1 − 4	+393	+323	+229	+162	+156	+147	+269	+319	+263	+347
MOBIL										
1. Production	392	459	485	535	559	585	640	729	838	953
2. Long-term contracts	45[c]	68	88	95	106	104	124	107	66	63
3. = 1 + 2	437	527	573	630	665	689	764	836	904	1,016
4. Crude processed	294	337	391	433	487	525	582	676	757	844
5. = 3 − 4	+143	+190	+182	+197	+178	+164	+182	+160	+147	+172
6. = 1 − 4	+98	+122	+94	+102	+72	+60	+58	+53	+81	+109
SOCAL										
1. Production	529	530	534	631	704	745	776	838	1,007	1,166
2. Refining throughput	277	283	295	340	373	405	433	482	543	604
2 − 1	+252	+247	+239	+291	+331	+340	+343	+356	+464	+562
TEXACO										
1. Production	617	646	708	760	837	887	985	1,084	1,258	1,440
2. Refining throughput	401	446	586	650	703	737	729	817	950	1,058
2 − 1	+216	+200	+122	+110	+134	+150	+256	+267	+308	+382

Note: Late in 1965 an Esso official stated that Aramco overlifting rules had been somewhat eased. This may be taken to mean that Esso was intending to overlift to a noticeable degree, and one may therefore assume that in 1966 contract purchases increased by 10.35%, the 1957-65 average. If so, they were 255 TBD in 1966, while 202 TBD represented overlifting, which is added to gross production.

[a]Includes estimated 203-TBD overlifting from Aramco.

[b]Excludes estimated 203-TBD overlifting from Aramco.

[c]Rough estimate based on assumption long-term contract geared to refinery throughput.

TABLE III–B–1. Continued

	1957	1958	1959	1960	1961	1962	1963	1964	1965	1966
SHELL										
1. Production	1,497	1,397	1,608	1,603	1,618	1,783	1,813	2,029	2,180	2,380
2. Long-term contracts	356	434	452	548	539	640	702[d]	770[d]	869[d]	861[d]
3. = 1 + 2	1,853	1,831	2,060	2,151	2,157	2,423	2,515	2,799	3,049	3,241
4. Crude processed[e]	1,207	1,281	1,486	1,624	1,753	1,989	1,983	2,245	2,441	2,593
5. = 3 – 4	+646	+550	+574	+527	+404	+525	+532	+554	+608	+648
6. = 1 – 4	+290	+116	+122	–21	–135	–115	–170	–216	–261	–213
GULF										
1. Production	774	930	952	1,101	1,115	1,257	1,341	1,439	1,554	1,633
2. Long-term contracts[f]	–412	–458	–542	–540	–640	–640	–707	–700	–735	–747
3. = 1 + 2	362	472	410	561	475	617	634	739	819	886
4. Crude processed	146[g]	163	177	217	231	305	331	379	427	451
5. = 3 – 4	+216	+309	+233	+334	+244	+312	+303	+360	+392	+435
6. = 1 – 4	+628	+767	+775	+884	+884	+952	+1,010	+1,060	+1,127	+1,182

[d]Includes purchases from Amerada Libya. Amerada *Annual Report*, 1964 and 1966, not fully consistent on amounts sold to Shell.

[e]Strictly speaking, the above numbers are the estimated volume of crude oil processed by Shell for its own account, i.e. excluding that processed by Shell for others and including that processed by others for Shell. Since precise information is not available, the volume of refined products *sold* by Shell is used as an approximation. These volumes are roughly 5% below the actual amount of crude oil charged because of refining losses.

[f]Does not include crude oil processed by others for Gulf's account; information not available.

[g]Gulf-Shell contract renewal set minimum figures: 400 TBD for 1958, 600 for 1962. See *Prospectus*, May 6, 1963.

TABLE III–B–1. Continued

	1957	1958	1959	1960	1961	1962	1963	1964	1965	1966
BP										
1. Production	977	1,232	1,296	1,480	1,570	1,773	1,834	2,106	2,216	2,423
2. Less long-term contracts[h]	–150[j]	–209	–347	–298	–288	–330	–336	–307	–296	–293
3. = 1 + 2	827	1,023	949	1,182	1,282	1,443	1,498	1,799	1,920	2,130
4. Refining throughputs . .	642	730	780	940	1,009	1,134	1,230	1,369	1,505	1,585
5. = 3 – 4	+185	+293	+169	+242	+273	+309	+268	+430	+415	+545
6. = 1 – 4	+335	+502	+516	+540	+561	+639	+604	+737	+711	+838
CFP										
1. Production	183	255	301	384	428	484	579	629	699	829
2. Long-term contract . . .	+10	+9	+8	+9	+28	+44	+61	+61	+80	+37
3. = 1 + 2	193	264	309	393	456	528	640	690	779	866
4. Refining throughputs . .	182	221	237	278	316	335	389	438	479	538
5. = 3 – 4	+11	+43	+72	+15	+140	+193	+251	+252	+300	+328
6. = 1 – 4	+1	+34	+64	+6	+112	+149	+190	+191	+220	+291

Notes:

Line 1. Gross production, includes royalty oil. In annual report sometimes called "net production" to indicate that only the company's share of joint production is included. This is confusing, since "net" usually means excluding royalty. To make matters worse, "net production" sometimes means excluding condensate reinjected, irrespective of ownership.

Line 2. Long-term contracts: Does not include over- and under-lifting from joint ventures, except as noted.

[h]Long-term contracts with Mobil and Esso only.
[j]Includes rough estimate for Mobil.

Sources: Annual Reports of the eight companies, including supplements of financial and operating data.

Appendix III-C

TABLE III-C-1. Refining Capacity by Company by Country, December 1957

(thousand barrels per stream-day)

Country	Total	Largest companies' total	Esso	Mobil	Shell	SoCal	Texaco	Gulf	BP	CFP
MIDDLE EAST										
Aden	120.0	120.0	...	...	...	...	...	...	120.0	...
Bahrain	211.0	211.0	...	...	...	105.5	105.5	...	...	...
Iran	538.3	505.0	37.0	37.0	74.0	37.0	37.0	37.0	214.0	32.0
Iraq	55.8	9.1	1.1	1.1	2.3	...	...	...	2.3	2.3
Israel	85.0	...	...	...	...	...	...	...	...	...
Jordan	...	...	...	...	...	...	...	...	...	...
Kuwait	30.0	30.0	...	...	...	...	...	15.0	15.0	...
Lebanon	24.0	22.9	1.3	7.6	2.6	3.1	3.1	...	2.6	2.6
Neutral Zone	...	...	...	...	...	...	...	...	...	...
Qatar	0.6	...	...	...	...	...	...	...	...	...
Saudi Arabia	189.0	189.0	56.7	18.9	56.7	56.7	...	...	...	...
Syria	...	...	...	...	...	...	...	...	...	...
Turkey	6.9	...	...	...	...	...	...	...	...	...
M.E. Total	1,260.6	1,087.0	96.1	64.6	135.6	202.3	145.6	52.0	353.9	36.9
Percent	100.0	86.2	7.6	5.1	10.8	16.0	11.6	6.1	28.1	2.9
AFRICA										
Algeria	...	...	...	...	...	...	...	...	...	...
Angola	...	...	...	...	...	...	...	...	...	...
Egypt	75.0	13.6	...	...	6.8	...	...	...	6.8	...
Ethiopia	...	...	...	...	...	...	...	...	...	...
Ghana	...	...	...	...	...	...	...	...	...	...
Ivory Coast	...	...	...	...	...	...	...	...	...	...
Kenya	...	...	...	...	...	...	...	...	...	...
Libya	...	...	...	...	...	...	...	...	...	...
Malagasy Rep.	...	...	...	...	...	...	...	...	...	...
Morocco	2.3	...	...	...	...	...	...	...	...	...
S. Africa	21.1	21.1	10.6	10.5	...	...	...	...	...	...
Tunisia	...	...	...	...	...	...	...	...	...	...
Africa Total	98.4	34.7	10.6	10.5	6.8	...	...	...	6.8	...
Percent	100.0	35.3	10.8	10.7	6.9	...	...	...	6.9	...

TABLE III-C-1. Continued

(thousand barrels per stream-day)

Country	Total	Largest companies' total	Esso	Mobil	Shell	SoCal	Texaco	Gulf	BP	CFP
ASIA-PACIFIC										
Australia	207.4	192.2	14.6	14.6	56.0	18.5	18.5	...	70.0	...
Burma	6.5	...	...	...	...	...	...	...	...	...
British Borneo	50.0	50.0	...	...	50.0	...	...	...	...	...
Formosa	28.0	...	...	...	...	...	...	...	...	...
India	98.7	67.8	13.9	13.9	25.0	7.5	7.5	...	...	...
Indonesia	243.8	243.8	35.5	35.5	172.8	...	...	...	...	...
Japan	430.1	101.2	13.5	13.5	20.5	26.8	26.9	...	...	...
Pakistan	6.0	...	...	...	...	...	...	...	...	...
Philippines	18.0	18.0	...	...	...	9.0	9.0	...	...	...
Total	1,088.5	673.0	77.5	77.5	324.8	61.9	61.9	...	70.0	...
Percent	100.0	61.8	7.1	7.1	29.8	5.7	5.7	...	6.4	...
EUROPE										
Austria	46.9	8.0	...	3.5	4.5	...	...	...	...	...
Belgium	162.9	88.0	36.0	...	3.0	...	...	...	49.0	...
Denmark	0.6	...	...	...	...	...	...	...	...	...
Finland	18.0	...	...	...	...	...	...	...	...	...
France	633.6	530.0	85.0	43.0	85.0	12.5	12.5	...	102.0	190.0
Greece	30.0	...	...	...	...	...	...	...	...	...
Italy (incl. Sicily)	580.5	336.0	121.6	65.0	80.0	4.0	4.0	...	17.0	44.4
Netherlands	349.0	349.0	...	...	300.0	24.5	24.5	...	...	...
Norway	1.8	1.8	1.8	...	...	...	...	...	...	...
Portugal	22.9	...	...	...	...	...	...	...	...	...
Spain	132.7	...	...	...	...	...	...	...	...	...
Sweden	47.5	14.5	...	...	14.5	...	...	...	...	...
United Kingdom	519.3	504.2	160.0	30.0	214.2	...	...	...	100.0	...
W. Germany	357.2	130.2	53.8	16.0	15.0	...	...	...	45.4	...
Total	2,902.9	1,961.7	458.2	157.5	716.2	41.0	41.0	...	313.4	234.4
Percent	100.0	67.6	15.8	5.4	24.7	1.4	1.4	...	10.8	8.1

TABLE III-C-1. Continued

(thousand barrels per stream-day)

Country	Total	Largest companies' total	Esso	Mobil	Shell	SoCal	Texaco	Gulf	BP	CFP
LATIN AMERICA										
Argentina	197.3	55.3	32.3	...	23.0	...	...	...	...	...
Bolivia	11.7	...	...	...	...	...	...	...	...	...
Brazil	113.8	...	...	...	...	...	...	...	...	...
Chile	20.0	...	...	...	...	...	...	...	...	...
Colombia	70.7	29.7	26.5	1.0	...	...	2.2	...	...	...
Cuba	84.2	82.0	35.0	...	27.0	...	20.0	...	...	...
Ecuador	5.6	9.0	9.0	...	...	...	...	...	...	...
Mexico	444.0	...	...	...	...	...	...	...	...	...
N.W.I.	650.0	650.0	440.0	...	210.0	...	...	...	...	...
Peru	49.2	47.0	47.0	...	...	...	...	...	...	...
Puerto Rico	72.3	6.5	...	...	...	...	...	6.5	...	...
Trinidad	141.0	133.0	...	...	33.0	...	100.0	...	...	...
Uruguay	28.0	...	...	...	...	...	...	...	...	...
Venezuela	755.8	696.0	412.0	...	215.0	...	10.0	59.0	...	...
Total	2,643.6	1,708.5	1,001.8	1.0	508.0	...	132.2	65.5	...	...
Percent	100.0	64.6	37.9	0.0	19.2	...	5.0	2.5	...	...
World Total	7,993.6	5,464.9	1,644.2	311.1	1,690.9	305.2	380.7	117.5	744.1	271.3
Percent	100.0	68.4	20.6	3.9	21.2	3.8	4.8	1.5	9.3	3.4

Source: *Oil and Gas Journal*, "World Wide Oil," 1957.

TABLE III–C–2. Refining Capacity by Company by Country, December 1966

(*thousand barrels per stream-day*)

Country	Total	Largest companies' total	Esso	Mobil	Shell	SoCal	Texaco	Gulf	BP	CFP
MIDDLE EAST										
Aden	175.0	175.0	...	...	...	...	...	...	175.0	...
Bahrain	200.0	200.0	...	...	...	100.0	100.0	...	...	...
Iran	514.0	475.0	35.0	35.0	70.0	35.0	35.0	35.0	200.0	30.0
Iraq	77.9	8.0	1.0	1.0	2.0	...	...	...	2.0	2.0
Israel	100.0	...	...	...	...	...	...	...	...	...
Jordan	7.5	...	...	...	...	...	...	...	...	...
Kuwait	360.0	250.0	...	...	...	...	...	125.0	125.0	...
Lebanon	36.2	34.9	2.3	10.6	4.6	4.1	4.1	...	4.6	4.6
Neutral Zone	50.0	...	...	...	...	...	...	...	...	...
Qatar	0.6	...	...	...	...	...	...	...	...	...
Saudi Arabia	285.0	255.0	76.5	25.5	...	76.5	76.5	...	...	...
Syria	25.0	...	...	...	...	...	...	...	...	...
Turkey	124.0	81.7	...	36.4	17.6	8.3	8.3	11.1	...	...
M.E. Total	1,955.2	1,479.6	114.8	108.5	94.2	223.9	223.9	171.1	506.6	36.6
Percent	100.0	75.7	5.9	5.5	4.8	11.5	11.5	8.7	25.9	1.9

TABLE III-C-2. Continued

(thousand barrels per stream-day)

Country	Total	Largest companies' total	Esso	Mobil	Shell	SoCal	Texaco	Gulf	BP	CFP
AFRICA										
Algeria	46.0	37.8	7.4	2.5	10.1	...	...	...	4.4	13.4
Angola	14.0	...	...	...	...	...	...	...	...	...
Egypt	189.0	...	...	...	...	...	...	...	...	...
Ethiopia	10.0	...	...	...	...	...	...	...	...	...
Ghana	29.0	...	...	...	...	...	...	...	...	...
Ivory Coast	19.0	12.6	0.2	3.4	2.8	...	1.6	...	2.2	2.4
Kenya	44.0	44.0	11.2	...	11.2	5.2	5.2	...	11.2	...
Libya	11.4	9.0	9.0	...	...	...	...	...	...	...
Malagasy Rep.	12.0	...	...	...	...	...	...	...	...	...
Morocco	34.0	...	...	...	...	...	...	...	...	...
Mozambique	12.0	3.6	...	...	...	...	...	...	...	3.6
Nigeria	42.0	21.0	...	...	10.5	...	...	...	10.5	...
Rhodesia	20.0	16.2	...	3.6	4.2	1.6	1.6	...	4.2	1.0
Senegal	12.6	7.6	0.1	1.5	1.5	...	1.5	...	1.5	1.5
South Africa	153.2	149.3	...	39.0	40.0	15.1	15.2	...	40.0	...
Sudan	20.0	20.0	...	...	10.0	...	...	...	10.0	...
Tanzania	13.8	...	...	...	...	...	...	...	...	...
Tunisia	22.5	...	...	...	...	...	...	...	...	...
Africa Total	704.5	321.1	27.9	50.0	90.3	21.9	25.1	...	84.0	21.9
Percent	100.0	45.6	4.0	7.1	12.8	3.1	3.6	...	11.9	3.1

TABLE III–C–2. Continued

(*thousand barrels per stream-day*)

Country	Total	Largest companies' total	Esso	Mobil	Shell	SoCal	Texaco	Gulf	BP	CFP
ASIA-PACIFIC										
Australia	503.3	426.3	26.1	74.2	83.0	51.5	51.5	...	140.0	...
Burma	26.4	...	...	...	...	...	...	...	...	...
Formosa	52.0	...	...	...	...	...	...	...	...	...
India	346.5	114.5	50.5	...	42.0	11.0	11.0	...	...	...
Indonesia	268.2	80.0[a]	40.0	40.0	...	...	...	...	...	...
Japan	2,211.2	482.4	116.1	57.6	118.5	95.1	95.1	...	...	...
Malaya/Singapore	183.0	165.6	22.1	21.0	106.5	...	...	...	16.0	...
New Zealand	67.0	40.3	...	12.9	11.5	2.9	2.9	...	10.1	...
Pakistan	76.0	23.4	8.5	...	8.5	3.2	3.2	...	...	...
Philippines	121.0	109.9	31.7	14.3	18.8	17.5	17.5	10.1	...	...
South Korea	50.0	12.5	...	...	...	...	...	12.5	...	...
Thailand	46.2	5.0	5.0	...	...	...	...	...	...	...
Asia-Pacific Total	3,955.8	1,459.9	300.0	220.0	388.8	181.2	181.2	22.6	166.1	...
Percent	100.0	37.0	7.6	5.6	9.8	4.6	4.6	0.6	4.2	...

TABLE III-C-2. Continued

(thousand barrels per stream-day)

Country	Total	Largest companies' total	Esso	Mobil	Shell	SoCal	Texaco	Gulf	BP	CFP
EUROPE										
Austria	97.0	9.0	...	4.5	4.5	...	...	...	...	...
Belgium	346.6	198.6	90.0	...	8.6	...	...	...	100.0	...
Denmark	116.8	116.8	46.4	...	40.4	...	...	30.0	...	...
Finland	111.0	...	...	...	...	...	...	...	...	...
France	1,704.5	1,157.5	121.9	115.0	270.8	15.9	15.9	35.8	192.2	390.0
Greece	94.0	31.4	27.5	3.9	...	...	...	...	...	...
Ireland	48.0	38.4	19.2	...	...	4.8	4.8	...	9.6	...
Italy	2,038.0	817.2	319.7	115.0	158.0	25.0	25.0	...	78.5	96.0
Netherlands	667.4	661.0	160.0	...	340.0	50.5	50.5	60.0	...	...
Norway	63.0	63.0	63.0	...	...	...	...	...	...	...
Portugal	37.0	5.2	...	2.0	...	...	...	...	...	3.2
Spain	429.9	55.4	25.0	...	...	15.2	15.2	...	...	...
Sweden	80.0	40.0	...	...	40.0	...	...	...	...	...
Switzerland	95.0	72.5	13.5	...	50.0	...	...	...	9.0	...
United Kingdom	1,736.5	1,598.8	462.0	65.0	447.5	50.0	50.0	1.8	522.5	...
West Germany	1,794.1	1,023.2	378.8	32.0	261.0	29.5	113.9	...	208.0	...
Yugoslavia	68.0	...	...	...	...	...	...	...	...	...
Europe Total	9,526.9	5,888.0	1,727.0	337.4	1,620.8	190.9	275.3	127.6	1,119.8	489.2
Percent	100.0	61.8	18.1	3.5	17.0	2.0	2.9	1.3	11.8	5.1

TABLE III-C-2. Continued

(*thousand barrels per stream-day*)

Country	Total	Largest companies' total	Esso	Mobil	Shell	SoCal	Texaco	Gulf	BP	CFP
LATIN AMERICA										
Argentina	435.5	159.0	69.0	...	90.0	...	...	...	...	...
Barbados	3.0	3.0	...	3.0	...	...	...	...	...	...
Br. West Indies	11.0	...	...	...	...	...	...	...	...	...
Bolivia	14.5	...	...	...	...	...	...	...	...	...
Brazil	370.4	...	...	...	...	...	...	...	...	...
Chile	91.0	...	...	...	...	...	...	...	...	...
Colombia	129.3	53.3	47.2	2.1	...	...	4.0	...	...	...
Costa Rica	8.0	...	...	...	...	...	...	...	...	...
Ecuador	11.5	1.5	1.5	...	...	...	...	...	...	...
El Salvador	13.5	12.8	8.1	...	4.7	...	...	...	...	...
Guatemala	20.5	20.5	...	...	4.8	7.2	8.5	...	...	...
Jamaica	28.0	28.0	28.0	...	...	...	...	...	...	...
Mexico	394.0	...	...	...	...	...	...	...	...	...
N.W.I.	795.0	795.0	460.0	...	335.0	...	...	...	...	...
Nicaragua	5.6	5.6	5.6	...	...	...	...	...	...	...
Panama	55.0	...	...	...	...	...	...	...	...	...
Paraguay	5.0	...	...	...	...	...	...	...	...	...
Peru	63.6	59.5	57.0	...	...	2.5	...	...	...	...
Puerto Rico	155.0	40.0	...	...	...	...	...	40.0	...	...
Trinidad	390.0	390.0	...	...	65.0	...	325.0	...	...	...
Uruguay	40.0	...	...	...	...	...	...	...	...	...
Venezuela	1,194.2	1,150.0	465.0	80.0	388.0	48.0	63.0	106.0	...	...
Virgin Islands	50.0	...	...	...	...	...	...	...	...	...
Latin America Total	4,283.6	2,718.2	1,141.4	85.1	887.5	57.7	400.5	146.0	...	...
Percent	100.0	63.5	26.6	2.0	20.7	1.3	9.3	3.4	...	...
World Total	20,426.0	11,866.8	3,311.1	801.0	3,081.6	675.6	1,106.0	467.3	1,876.5	547.7
Percent	100.0	58.1	16.2	3.9	15.1	3.3	5.4	2.3	9.2	2.7

Source: Oil and Gas Journal, World Wide Oil issue, 1966.

[a]Shell Indonesia was sold to government in 1966.

Appendix IV-A

Calculations of Spot Equivalent, Persian Gulf–Rotterdam, Round Trip

Cost per ton = (fuel cost + ship hire + port charges)/(cargo)
M = round-trip distance in miles

(1) Through Suez Canal in 1966

52,000 tons, \$1.735/ton/month[1]

Assume: 15.7 knots; 60 tons heavy fuel (HVF) daily underway, 10 tons daily in port; 1.5 tons diesel every day. HVF \$11.90/ton, diesel \$16/ton. 3-1/2 days in port including 1/2 day delay. One-day contingency fuel reserve. Water and stores, 300 tons.

Cargo (tons):
$52{,}000 - [60M/(24 \times 15.7)] - [(5 \times 10) + 60 + 300] = 51{,}590 - 0.159M.$

Fuel cost (\$):
$11.90\,[60M/(24 \times 15.7) + (5.0 \times 10)] + 16[1.5M/(24 \times 15.7) + 5(1.5)]$
$= 1.958M + 715.$

Ship hire:
(\$1.735 × 52,000)/30.4 = \$2,968 per day
Days at sea = M/(15.7 × 24). Days in port = 5.0
Hire per voyage (\$): 2,968 (M/377 + 5) = 7.87M + 14,840.

Cost per voyage per ton:

$$\frac{\$\,1.958M + 715 + 7.87M + 14{,}840 + 10{,}000}{51{,}590 - 0.159M}$$

$$= \frac{\$\,9.83M + 25{,}555}{51{,}590 - 0.159M}$$

= \$3.07, M = 13,000 miles

[1] Platou, *Annual Report, 1966*, p. 23.

= Intascale − 55 (Intascale flat = \$6.80).

(2) Around Cape in 1968[2]

Ship: 253,000 tons, \$1.07/ton/month.

Assume speed: 15.75 knots; fuel consumption, 135 tons daily Bunker C underway, 225 tons total in port. Fuel cost, \$10/ton. Port charges, \$31,500. Days in port, 3. Contingency fuel reserve, 1 day. Water and stores, 300 tons.

Cargo (tons): $253{,}000 - [135M/(24 \times 15.75)] - (225 + 135 + 300)$
$= 252{,}350 - 0.357M$.

Fuel cost (\$): $10[135M/(24 \times 15.75)] + 10(225) = 3.57M + 2{,}250$.

Ship hire: $(\$1.07 \times 253{,}000)/30.4$ per day $= \$8{,}905$ per day.
Days at sea $= M/378$; days in port $= 3.0$.
Hire per voyage in (\$), $8{,}905\left(\frac{M}{378} + 3.0\right) = 23.558M + 26{,}715$.

Cost per voyage per ton:

$$\frac{\$\,3.57M + 2{,}250 + 23.558M + 26{,}715 + 31{,}500}{252{,}340 - 0.357M}$$

= \$2.75, M = 22,500 miles (Kharg-Rotterdam round trip)

= Intascale − 75.2 percent predevaluation (Intascale flat = \$11.08)

= Intascale − 71.8 percent postdevaluation (Intascale flat = \$9.76)

= Worldscale 31 (Worldscale flat: \$8.87).

[2] Clarkson, Nov. 8, 1968, using Platou factors Mar. 1968, except as shown.

Appendix IV-B

Flat Rates on Major Voyages

Table IV-B-1. Intascale and Worldscale Flat Rates on Major Voyages

Dollars per Long Ton

Voyage	*(Before Nov. 1967)* *Intascale*	*(1969)* *Worldscale*
Caribbean-Northwest Europe	4.62	3.74
Caribbean-U.S. East Coast	2.11	1.85
Libya-Northwest Europe	3.36	2.68
Libya-U.S. East Coast	5.13	3.94
Persian Gulf-Northwest Europe (around Cape)	10.60	8.93
do., through Suez Canal	6.80[a]	5.18[a]
Persian Gulf-Japan	6.60	5.47
Persian Gulf-U.S. East Coast (around Cape)	11.04	9.29
do., through Suez Canal	8.38[a]	6.43[a]
Nigeria-Northwest Europe	5.09	3.99
Nigeria-U.S. East Coast	5.60	4.42

[a]Add 89 cents for Suez tolls.

Appendix IV-C

Cost and Value of Pipelining Crude Across the Suez Isthmus (Israel or Egypt)

As in Chapter II, the general cost formula is: $c = I/q_0 \int_0^T e^{-(a+r)t}\,dt$. Since there is no pipeline decline rate $a = 0$. For the Trans-Israel Pipeline (TIP), there were three successive installments of investment and capacity: the first 400 TBD cost $60 million, the next two cost $33 and $20 million respectively.[1] It is not necessary to allow for the carrying cost of work in progress, i.e. consider it as an inventory, and an interest charge. If we assume that the first installment is paid out evenly over 18 months, and the next two each over a year, then assuming a 10 percent borrowing rate, the end-of-1969 equivalent investment is practically the same as the simple unadjusted figure.

For TIP, we assume that the first two years' average use is only 200 TBD, the next two years, 800 TBD; thereafter, full utilization at 1.2 MBD. Assuming 15 percent discount and a 25-year life, we must compute three present-barrel equivalent (BPD) factors, as follows:

$$\int_0^2 e^{-0.3}\,dt, \quad \int_2^4 e^{-0.3}\,dt, \quad \int_4^{25} e^{-2.4}\,dt, \text{ or } 1.74, 1.32, \text{ and } 3.84.$$

Then we can add three sub-totals of PBEs:

$$365[(1.74 \times 2 \times 10^5) + (1.32 \times 8 \times 10^5) + (3.84 \times 1.2 \times 10^6)]$$

$$= 2.20 \times 10^9 \text{ PBEs.}$$

Hence $113 million corresponds to 2.20 billion PBEs, or 5.2 cents per barrel capital cost. A rule of thumb of 2.5 to 3 cents per barrel for current operating pipeline costs makes total cost about 8 cents per barrel.

The most likely source of error is political risk. Companies operating in Arab countries dare not use TIP. This is often belittled by the well-informed. As *Le Monde* put it in commenting specifically on this pipeline: "Who can breach the boycott of the Arab League? In theory, no Arab country nor any international company with interest in any such country. . . . In fact, experience shows that so long as appearances are kept and some ingenuity is used one can go through the net . . ."[2] However, the attitude of the United States government has been consistently unfavorable for a long period of time,[3] and this

[1] *The Economist*, June 20, 1968, and *New York Times*, June 24, 1968. There are some slight discrepancies. The estimate of $113 million may be too high; see *PIW*, Oct. 27, 1969, for an estimate of $100 million.

[2] *Le Monde*, Oct. 14–18, 1968.

[3] *WSJ*, Feb. 20, 1957; *PONS*, Apr. 22, 1968.

may have an effect. During the desperate tanker shortage of 1970, when single-voyage rates touched Worldscale 300, utilization of TIP was only 265 TBD despite initial capacity of 440. However, more traffic was expected and capacity was being expanded to 600 TBD.[4]

For the Suez-Mediterranean (SUMED) pipeline across Egypt, a much higher construction cost is assumed, \$200 million, in part because of strong inflation during 1968–72. Assuming immediate full use of capacity, PBEs provided are $365 \times 6.94 \times 10^6 = 2.53$ billion, or nearly 8 cents per barrel capital cost; 6 cents is a guess at operating cost, which must be higher than for TIP, in part because a number of oil companies will need to provide individual storage facilities.

The value of the pipeline (or the Canal) lies in saving time. The round-Cape round trip Persian Gulf–Rotterdam (22,000 miles at 15.5 knots) takes 59 days' sailing and 3 days in port. The broken voyage using the pipeline (12,000 miles) takes 32.2 days' sailing and 6 days in port. Hence the saving in time is about 24 days, or 38.5 percent.

In 1966, as noted in the text, the break-even rate was reckoned at about Worldscale 35, or 42 cents Persian Gulf–Rotterdam. In other words, the charterer saved about 15 cents by the shorter passage, and it was reasonable to pay 12 cents for the two-way passage. The Canal and pipeline are little different because waiting and forming up convoys and slow steaming through the Canal resulted in a net loss of about 48 hours.[5]

The writer estimated in early 1968 that, political risk aside, TIP or SUMED were commercially sound and would pay 12 percent at a tariff of 8.75 cents.[6] Later, the U.S. Department of the Interior stated that neither pipeline would be economic, but did not reveal their cost calculations. However, they did state that they assumed a round-Cape rate of 39 cents, or Worldscale 33.[7]

[4] *PPS*, May 1971, p. 176.

[5] M. E. Hubbard, "The Comparative Costs of Oil Transport to and Within Europe," *Journal of the Institute of Petroleum*, vol. 53, Jan. 1967, p. 17.

[6] *POPS*, Apr. 2, 1968.

[7] *PIW*, Jan. 19, 1970.

Appendix V-A

Crude Oil Price Information

The changes in posted prices of Middle East crudes and one Venezuelan, from 1950 to 1960, are outlined for reference in Table V-A-1. The rest of this appendix summarizes the available information on actual market prices. Unfortunately, most of the data refer to two rather special markets, Argentina and Japan.

Argentine offers. Table V-A-2 shows the details of seven offers of Middle East crude made to Yacimientos Petrolíferos Fiscales (YPF) from April 1954 to October 1956. Not all bids resulted in contracts and in some cases the outcome is not known, but it has been assumed that the bids were all genuine. Where details of the contract are not available, lowest bids have been selected. While not large by the standards of international companies, quantities were more than short-term distress sales. Six of the seven cases ranged from half a million to two million cubic meters. (A cubic meter = one kiloliter = 6.3 barrels.) In only two of the seven offers were there no substantial discounts. The first was in March 1954. This was close to the time that crude oil was said to be in tight supply, in sharp contrast to the situation six months earlier (*POPS*, Feb. 19, 1954). The other was the September 1956 bids for four cargoes at f.o.b. prices (usually bids were made on a c.i.f. basis).

The other five bids showed discounts ranging from about 15 cents to 60 cents. However, calculation of a "true" price depends entirely on what a competitive freight rate would have been. This can be estimated by looking at shipping contracts actually made around the date of the offer. Sometimes offers were submitted for only the transport of YPF's requirements. This ought to be an ideal indicator of the appropriate freight rate, but there were sometimes quite wide variations in the offers which cannot be explained with published data only. Table V-A-3 lists those shipping contracts that were made at about the time of the offers and whose terms correspond closely to the delivery terms of the bid; it also lists any direct transport offers associated with the bids. Where there seem to be wide variations, alternative estimates are given in Table V-A-2. Although these variations make quite a difference to the particular offers, in general it can be deduced that effective discounts varied from about 15 cents to 60 cents.

Japan. Prices of Middle East crudes to Japan have been discussed in considerable detail by Suzuki.[1] He distinguishes between actual discounts (possibly secret at the time) and disguised discounts in the form of low freight rates, etc.

Japanese buyers were given freedom of choice in their purchases of crude in 1952. The first discount mentioned, in June 1953, was by CFP in Iraq crude (this discount was increased when the posted price was raised the following month). By 1955, discounting had become universal except for Arabian crude. Early discounters after CFP were Gulf,

[1]M. Suzuki, in "Competition and Monopoly in the World Oil Markets," supplement to *Energy in Japan*, Quarterly Report No. 4, Institute of Energy Economics, Tokyo, Feb. 1968.

BP, and Shell. Amounts of discount were: Iraq 2-9 cents, Kuwait 6 cents, Iran, which returned to the market in October 1954 and began to be discounted a year later, 10-15 cents. In the case of Iran the discounters were Tidewater and Tampinex, a nonproducer.

In 1955 a new foreign exchange allocation was instituted by the government which gave preference to low f.o.b. prices. This may have encouraged open discounts instead of disguised or secret ones in the form of low freights or favorable loans. In any case, it introduced a new factor not applicable to other markets. According to Suzuki, in 1955-57 discounts continued and the range began to expand. Kuwait crude was discounted 8-14 cents in 1956 and 10-30 cents in 1957. BP was the largest discounter followed by Shell and Gulf. Iraq crudes were discounted from 0-16 cents during both years. Iranian crude suffered larger discounts than the others. Discounting was done mainly by independents. In 1956 the range was 10-20 cents and 13-33 cents in the next year.

Other information. In March 1953 a contract for a Persian Gulf crude was made with a Brazilian refinery for 20,000 barrels daily at 12-15 cents below posted (*POPS*, Mar. 20, 1953). In July 1953 CFP was said to be dumping crude from time to time because it could not dispose of production. In August 1953 "French interests" (presumably CFP) were selling crude at 25 cents per barrel off the market.

By February 1954, however, there was a shortage probably until Iranian crude came back in October (*POPS*, Feb. 19, 1954).

In October 1954 two companies made charterback deals and "French interests have offered outright discounts of 5-10 cents in the Far East," presumably CFP to Japan (*POPS*, Oct. 27, 1954). The next month *POPS* commented on the "new crude oil marketing trend" that may have begun with the recent offers to Uruguay for their 1955 needs (the details of the offers do not seem to have been published). Also in recent contracts there had been loans of technical personnel and offers to purchase products as well as the usual phantom freights (*POPS*, Nov. 19, 1954).

In April 1955 it was said that only one Middle East producer had a reputation for explicit price cutting, but most others had given concessions in the form of free technical advice, monies placed in escrow for refinery rehabilitation, charterback deals, etc. (*POPS*, Apr. 20, 1955).

Shell admitted commencing discounts in December 1955 to Japan; it had been "forced to follow other suppliers who had been granting secret discounts." This was revealed in an Australian federal tax case on crude prices (Taxation Board of Review Decisions (New Series) 1962-1963, Case N-69, pp. 270-327).

We have only two pieces of information on Europe, both for mid-1955. An independent refiner is quoted as saying, "I never paid full posted price yet for one barrel of Middle East oil" (*POPS*, Apr. 20, 1955). A rebate of 36 cents per barrel was granted to an Italian refiner (*PW*, July 22, 1955, pp. 10-11).

An article on Brazil's need for 90,000 barrels a day on a five to ten year basis gives some suggestion that the South American market was unusually competitive. "Expectation is that principal offerings will originate from Venezuela. But some Caribbean sellers feel that South American sales in the past few years have been 'extremely competitive' and not all suppliers will make offerings. . . . Several contracts call for spiked crudes which make an easy out for surplus motor fuel, but run up crude cost to seller by as much as 40 cents a barrel." (*POPS*, Jan. 3, 1957.) It may be that South America was unusually competitive compared to the other open markets that existed or it may be that the Caribbean suppliers were simply not used to supplying anyone but captive affiliates.

Venezuelan prices. Table V-A-4 shows realizations in Venezuela for crude and products from 1950. This can only be an approximate indicator of prices because of variation in the type of crude and of products sold. It seems that the realization of crude did vary with posted prices, rising steeply in 1953 and again in 1957. From 1960, however, there has been a gradual slipping of prices, suggesting that more and more crude was sold at less than posted prices.

The realization on products might be a better indicator of market prices before 1957. About two-thirds in volume, perhaps one-half in value, the product exported was heavy fuel oil. The course of product realizations, stable from 1951 to 1954 and rising in each of the following two years, follows closely the trend of heavy fuel oil prices. (See Appendix VI-A.) Although taxes after 1966 are based on fixed "reference prices," companies have a reason to quote actual price, in order to show it as below the reference price. For if the realized price were higher, it would serve as the basis for taxation.

Summary. Notwithstanding the posted crude price increases in June 1953, there seems to have been considerable competition in Japan and, measured in size of discount, even more in South America (except for a short period in 1954). There was also discounting to independents in Europe in 1955, but it is not clear how, if at all, prices to Europe changed.

TABLE V-A-1. Posted Prices for Selected Crudes, 1950-1960

(*dollars per barrel*)

	Tia Juana 31° Amuay Bay (Venezuela)	Arabian 34° Ras Tanura (Persian Gulf)	Iraq 36° Fāo	Kuwait 31° Mena al Ahmadi	Iran 34° Bandar Mah Shahr
Dec. 1950	...	1.71	1.75	1.65	...
July 1952	2.30	...	...	...	...
Apr. 1953	...	...	...	1.50	...
June-July 1953	2.55	1.93	1.92	1.72	...
Oct. 1954	...	...	...	...	1.91
Feb. 1956	...	...	1.87	...	...
Jan. 1957	2.80	...	...	...	...
May-June 1957	...	2.08	2.00	1.85	2.04
Feb.-Mar. 1959	2.65	1.90	1.82	1.67	1.86
Aug.-Sept. 1960	2.55	1.80	1.74	1.59	1.78

Source: Petroleum Press Service. Adjustments made for gravity differentials, where necessary for continuity, at 2 cents per degree API.

Note on 1960 cuts. Esso cut Arabian 34° by 14 cents, Iraq 36° by 12 cents, and Kuwait 31° by 9 cents in early August. Shell followed these cuts, but BP on Aug. 16 announced smaller reductions which were finally adopted by all companies as in the last line of this table.

TABLE V-A-2. Argentine Offers, Middle East Crudes, March 1954-October 1956

Date of offer	Lowest offer: company	Type	Amount	Delivery period	C.I.F.	Assumed ft. rate (and range or alternative)	Rate offered	Discount	Equiv. Ras Tanura 31°	*POPS* ref.
			(000 cu. m.)		(...................*$ per barrel; percent*			...................)		
Mar. 1954	Cal. Trans.	Arab 34°	544	Jun-Dec. '54	2.93	USMC-45% 0.99	USMC-45% 0.99	0	1.88	5.3.54
Oct. 1954	Shell	K 31°	850	? 9 months	2.60	USMC-40 1.13	USMC-53 0.88	0.25	1.40	10.26.54 10.27.54
Oct. 1954	BP	K 31°	...	...	2.64	USMC-35 0.45 1.03-1.22	USMC-51 0.92	(0.15-0.34)	...	...
Oct. 1954	Gulf (offered to renegotiate if lower price offered)	K 31°	...	...	2.98	...	USMC-38 1.26	...	...	...
May 1955	BP	K 31°	1,100	Jun-Dec. '55	2.49	USMC-52 0.90 (USMC-20) 1.51	USMC-59 0.77	0.13 (0.74)	1.62	5.23.55 5.24.55
Oct. 1955	BP	K 31°	1,300	Dec. 55-Oct. 56	2.54	USMC-25 1.41 (USMC-35)	USMC-56 0.82	0.59 (0.40)	1.17	10.26.55 10.28.55 11.23.55
April 1956	?	Ar 34° reduced	125	June-Dec. '56	3.27	(USMC Flat?) 1.80	USMC-26 1.34	0.56	1.42	4.18.56
Sept. 1956	BP	K 31°	4 cargoes	Oct.-Dec. '56	None	...	...	0	...	9.28.56 10.1.56
Oct. 1956	BP	K 31°	2,000	Dec. '56 Jan. '58	3.53	(Scale + 100) 2.40	(Scale + 50) 1.81	0.59	1.21	10.12.56 10.15.56

Note: See Table V-A-3 for basis of freight estimates.

TABLE V-A-3. Argentine Offers, 1954-1956: Basis of Estimates of Freight Rates

Offer No.:	Consecutive voyage fixtures	Rate used
1. Apr. 19-May 18, 1954	Several consecutive voyage offers for 11-15,000 tankers for May/June to end Dec. at USMC-45 to -50%. Most at USMC-45%.	USMC-45%
2. Sept. 22, 1954	9 months from Oct. USMC-44%, 11,500 dwt.	
Oct. 8	Oct.-May USMC-35%, 19,000 dwt.	
Nov. 3	Jan.-end May USMC-36%, 15,000 dwt.	
Oct. 8	12 months USMC-40%, 12,000 dwt.	USMC-40%
Oct. 26	1 or 2 years USMC-42%, 14-15,000 dwt.	(USMC-35-45%)
3. May 1955	Lowest offer for transport alone was USMC-52%.	USMC-52%
	Other offers were from USMC-40 to -20%.	(USMC-20%)
4. One transport offer which was at USMC-25%. Also:		
Oct. 13, 1955	12 months at USMC-25%, 11,000 dwt.	
Oct. 19	12 months at USMC-26%, 9-12,000 dwt.	USMC-25%
Oct. 19	12 months USMC-35%, 14,600 dwt.	(USMC-35%)
5. Apr. 6, 1956	1 yr. USMC-17½%, 17,500 dwt.	
Apr. 6	1 yr. USMC-10%, 17,500 dwt.	
Apr. 20	1 yr. USMC-5%, 26,000 dwt.	
Apr. 20	early May-Sept. USMC + 7½%	(USMC flat)
6. Oct. 5, Nov. 1956-Mar. 1960	USMC + 42½% NWI-USNH = Scale + 83%, 15,500 dwt.	
Oct. 5, 1956	Scale + 100%, 1 yr., 16,000 dwt.	
Oct. 10, 1956-Aug. 1957 for 1 yr.	Scale + 100%, 16,000 dwt.	
Oct. 12, 1956	18 months from July 1958. Scale + 100%, 15,000 dwt.	(Scale + 100%)

Note: For an explanation of rates used see chap. IV, pp. 103-4.

TABLE V-A-4. Posted and Realized Prices, Venezuela, 1950-1970

(dollars per barrel)

Year	Average crude posting	Average realized price		
		Crude	Products	Crude and products
1950	2.27	2.12	2.09	2.11
1951	2.27	2.00	2.32	2.05
1952	2.17	2.11	2.29	2.14
1953	2.33	2.30	2.34	2.30
1954	2.40	2.31	2.33	2.31
1955	2.41	2.29	2.49	2.34
1956	2.45	2.30	2.56	2.36
1957	2.91	2.59	2.92	2.65
1958	2.83	2.48	2.54	2.50
1959	2.69	2.19	2.37	2.23
1960	2.70	2.08	2.22	2.12
1961	2.72	2.10	2.21	2.13
1962	2.71	2.06	2.13	2.08
1963	2.70	2.02	2.08	2.04
1964	2.73	1.94	1.96	1.95
1965	2.63	1.89	1.90	1.89
1966	1.67	1.86	1.89	1.87
1967	2.67	1.84	1.90	1.85
1968	2.62	1.83	1.95	1.87
1969	2.69	1.79	1.84	1.81
1969, 1H	...	1.82	1.85	1.83
1970, 1H	...	1.74	1.87	1.78

Source: República de Venezuela Ministerio de Minas e Hidrocarburos. *Petróleo y Otros Datos Estadísticos* 1970, p. 185, and press release, Dec. 1, 1970. (Original data from companies' accounts submitted to Department of Finance.)

1H = first half of year.

Appendix VI-A

Heavy Fuel Oil Prices, 1951–1967

In this appendix heavy fuel oil (HFO) prices are examined at three locations: the Caribbean (1951-67), the Persian Gulf (1957-67) of small relative importance, and Northwest Europe (1953-67), with some data on Italy when available. Our information is derived from a few regular series, such as those recently published in *Platt's Oilgram Price Service* (*POPS*), but no single series covers the entire period. In addition, therefore, a large number of individual observations have been culled from the trade press. Finally, Tables VI-A-7 and VI-A-8 form the background for the discussion in Chapter VI on the rise of HFO prices in the period 1950-56.

Caribbean Prices

U.S. imports–U.S. Bureau of the Census *Reports Nos. FT110* (this series is now numbered FT125), *U.S. Imports of Merchandise for Consumption*. Monthly and annual figures are given; quantities are in barrels and value is f.o.b. port of origin. There are two headings, "Residual Fuel Oil" and "Residual Fuel Oil for Ships and Aircraft." Taking exports from both Venezuela and the Netherlands West Indies, we have four series. These prices are given in Table VI-A-1, together with the prevailing posted price for HFO at Curaçao.

The bunker prices are generally 10-20 cents lower than the nonbunker imports. This could be related to the difference between spot and contract prices. "Ordinarily contract accounts would be billed at ten shillings per ton less than spot." (*POPS*, Oct. 26, 1956; 10s. per long ton equals 21 cents per barrel). Normally the posted price of fuel for bunkering is 10 cents per barrel *higher* than Bunker "C" quoted in cargo lots because of the smaller quantities involved. We should note that there was a usual 5 percent trade discount even when supplies were apparently tight. (*POPS*, Dec. 1, 1961, Dec. 18, 1961). The nonbunker prices are very close to posted. This does not make them suspect, as it would for a current series. Rises and falls in the monthly figures often preceded changes in posted prices, so it seems that the posted prices were at that time responsive to market changes. We have two outside observations. In April 1954 Navy Special was offered at Aruba at \$2.23. Navy Special is a mixture of 70 percent Bunker "C" and 30 percent No. 2 fuel oil (*PW*, Dec. 4, 1959). If we take the marine diesel quotation at the time as an approximation to the price of No. 2 oil, 8.1 cents per gallon, the equivalent price for Bunker "C" is \$1.72. This is very close to the prevailing import prices. However, if the actual marine diesel price were somewhat lower than the posted price, the Bunker "C" would of course be a little higher. In November 1956 *Petroleum Week* gives Bunker "C" f.o.b. Caribbean at 25 cents per barrel over posted.

Taking the lowest series of import prices one can outline briefly the general course of prices up to 1957. In 1951 it was steady around \$1.55, 20 cents under posted price. The only real departure from the trend of posted prices occurred in the first half of

1952, when bunker price rose from \$1.55 to \$1.80 while the posting stayed at \$1.75. This rise was followed by an equally sharp fall to \$1.46 by the beginning of 1953, while the posting fell to \$1.60. Except for a brief period of stability in 1954, at around \$1.70 (posting \$1.80), all series rose until 1956. The prices immediately before the Suez closure were: nonbunker imports \$2.30, bunker imports and posted price about \$2.10.

The timing of the weakness of the end of 1951 is of interest because the Korea tanker boom collapsed by November 1951, but f.o.b. prices in the Caribbean tended if anything to strengthen a little. The decrease to mid-1953 is reflected both in postings and in the other observations. The sharp jump in 1953 reflects the price increase in the United States. But since that increase was produced by regulation not scarcity, and since heavy fuel was not regulated in any case, the apparently "natural" increase could be explained by a follow-the-leader tactic which the few Caribbean sellers were able to enforce—although prices may not have increased quite as much as posted. But this explanation does not hold for the 1951-52 rise, nor the fairly steady increase which is visible from mid-1953 to about mid-1956. It is true that U.S. import figures are not particularly reliable since they are only importers' declarations, and that the posted price can be viewed with suspicion. Yet the movement is persistent and it lasted for three years. Moreover, there would have been no motive to overstate receipts on sales of heavy fuel oil, as filed with the Venezuelan authorities—quite the contrary, since taxes were based upon realized prices. Indeed, since there was a clear rising trend during 1951-56, it would be arbitrary to make a special explanation for that limited period. At this point the fact is merely noted; a hypothesis will be suggested later.

There is little point in following the prices through the rest of 1956-57 except to note, as above, that the failure of bunker imports to rise casts some doubt on the previous figures.

From 1957 there are some checks. *PW*'s "Market Trends" specifies prices on a number of occasions. A criticism of this and all other prices from journals (except regular series) is that prices tend to be mentioned only when something unusual is happening and so may give a distorted picture of the normal course of prices. For instance, a number of references are to shipments out of the Persian Gulf, just over \$1.00; while it is made clear that these are extraordinary, there is no way of knowing what an ordinary price would be.

Canadian import statistics—Canada, Bureau of Statistics, *Trade of Canada, Imports by Commodities*. Volume figures are given in Imperial gallons and value in Canadian dollars, which have been translated here into dollars per barrel. The heading up to 1963 is: "Heavy Fuel Oil, Nos. 4, 5 + 6"; from 1964: "Fuel Oil NES/Heavy Oil." This means that higher quality fuel oils are included, so the price may be overestimated at times. Fuel for bunkering is not included in the import figures. Quantities were small (10,000-50,000 tons) in 1958, but from 1959 a typical monthly volume was 50,000 tons.

Table VI-A-2 shows annual and monthly prices. Particularly in the early years, there is a wider range of prices than in the U.S. counterpart. This suggests that some nonmarket prices are being included. The proportion of these diminishes with time, so that prices would appear to fall even if they did not actually do so. From 1961 to 1964 the difference between the average monthly prices and the minimum for N.W.I. imports for each year fell from 14 cents to 4 cents.

Further information on U.S. and Canadian imports. From all series it is clear that prices fell from 1957, but the extent of this fall varies. In 1958 the U.S. and Canadian figures do not fall below \$2.00. Posted price was brought down to \$2.00 in October 1958, where it has remained ever since. In March 1959 import controls were strengthened in the U.S., but heavy fuel exports continued to increase under a flexible formula distinct from the more rigid crude oil import regulation. From 1960, U.S. annual import prices diverge sharply from Canadian imports and other indications, and it is clear that the American series is no longer reliable. In 1958 and 1959 the two series are close together, but for 1958 at least they both look high because they are close to two New York prices reported in *PW*. At the end of 1958 some lots were sold in the Persian Gulf at $1.15 and $1.09. Before the winter started surpluses were evident at the U.S. West Coast, Caribbean, and Persian Gulf. (*PW*, Jan. 30, 1959). "Ships bunkering was a no-discount market, but today shading is possible in many world ports." (*PW*, Nov. 14, 1958). (This last supports the position that the import prices are too high.)

In 1959, after some recovery in the cold 1958–59 winter, heavy fuel oil still seems to have been in worldwide surplus. Again, *PW* indications are below U.S. and Canadian *averages*, but they are not far below the Canadian *minimum*. The two June references were for small quantities. In August, "Whaling fleets can get bunker oil at about . . . $1.60 in the Caribbean." Here again quantities would not be large. There are few further quotations for 1959. The price stayed at $1.60 from April to September 1960, and did not reach $1.70 before February 1961.

Some observations can be made from the sale of Navy Special to the U.S. Navy, (*PW*, Dec. 4, 1959). But there is no way of finding the Bunker "C" equivalent without making assumptions about the market price for No. 2 fuel oil. In December 1959, 5 million barrels were offered at $1.80. Assuming equal percentage discounts on the two fuels, this would be equivalent to an HFO price of $1.58. Similarly the May 1960 price of $1.73 (after discount) for 2½ million barrels Navy Special would be equivalent to $1.52 for heavy fuel oil.

At the end of 1959 there was talk in *PW* of discounts "as high as five percent" for Bunker "C," i.e., a price of $1.90. There is a sharp divergence between short and long term, or it may be that this bunker market is not subject to arm's-length bargaining, or maybe it is misleading reporting. Moreover, Bunker "C" does not necessarily mean oil sold to ships; in *POPS* it simply means heavy fuel oil. The same situation occurred in 1966, when prices reported for bunker fuel seemed considerably higher than the residual fuel exported to New York.

The European Coal and Steel Community (ECSC) *Tenth Annual Report* (1962) shows spot prices up to January 1962 for the Caribbean (see Table VI–3). The February 1958 price of $1.67 suggests that spot prices fell earlier than is implied by import figures, but the 1959 and 1960 prices are close to our other sources.

In mid-1961 *PW* ceased publication. Between then and 1966, I have found only three references in the other journals, perhaps because prices changed very little in that period.

At the beginning of 1966, liberalization of U.S. residual fuel oil imports and the Venezuelan government's attempt to raise the price, at least for tax purposes, brought much press discussion of current and past prices.

"Heavy Fuel Oil has for the past several years been posted at $2.00/barrel at Venezuelan ports and also in Netherlands Antilles. In actual practice, however, the oil frequently has been subject to discounts of as much as 25 percent or to about $1.50." (*POPS*, Jan. 3, 1966, p. 1). This suggests that $1.50 was the low end of the range, hence a representative or average figure would be higher. Yet the Canadian figure was hardly over $1.50. Moreover: "For years the posted price of Venezuelan has stood at $2.00 a barrel. And for as many years the realized price has hovered at $1.50–$1.55 per barrel." (*OGJ*, Jan. 10, 1966, p. 49). Taken literally, this would mean a price of $1.50 going back to 1958, but it is probable that it refers to the period since 1960.

"Realized prices, which have been falling simultaneously with crude, are about 50 cents less."[1] Deducting 50 cents from the crude prices given, we have: 1960, $1.58; 1961, $1.60; 1962, $1.56; 1963, $1.52; 1964, $1.47. This is too approximate to compare in detail with our other figures, but the level is about right.

[1] Venezuela, Ministry of Mines and Hydrocarbons, *Carta Semanal*, Jan. 8, 1966, pp. 4–6.

For 1965 *Petroleum Intelligence Weekly* (*PIW*) suggests $1.48 as a representative price. The developments from 1960 to the beginning of 1966 amount therefore to a fall from about $1.60 to about $1.50. We cannot trace this fall in any detail—there must have been rising periods also. It seems from the Canadian statistics that much of the fall took place in 1965.

Liberalization of U.S. heavy oil imports in 1966 and efforts by the Venezuelan government to legislate a price increase might have been expected to push prices up during 1966. By November and December, Canadian imports were down to $1.49. It seems that if prices did rise, the increase could not be maintained.

At the beginning of 1967, *POPS* set up a regular fortnightly report of Caribbean prices. The year started at $1.60, and the price was $1.55 from January 30 to mid-May, when it fell to $1.50. These prices refer to HFO with no sulfur limits stated; 2 percent maximum was 5 cents higher; 1 percent was $1.90 for the brief period it was quoted.

In April and October 1966, two more Navy Special tenders were reported. The lowest offer in April was $1.705 f.o.b. Curaçao. Heating oil was 5.6 cents a gallon at that time, so the HFO equivalent was $1.43. Similarly in October, the lowest bid was $1.68 f.o.b. Aruba. Heating oil was 5.8 cents a gallon, and the HFO equivalent was $1.35.

Summary. The picture that emerges from Caribbean data is of prices generally rising over a period of around five years, broadly from 1953 to 1957, then a sharp decline through 1959 and a long recessional to levels in mid-1967, which were back to about those of 1951.

Persian Gulf Prices

Very little information can be obtained and what is there is difficult to interpret.

Prices were first posted in Abadan in 1957. Before that, prices at the Persian Gulf were usually taken as Platt's low[2] so Caribbean posted prices were entered. Ras Tanura quotations are usually 10 cents higher than those for Abadan (because the former is closer to markets, and tanker voyages are less), so 10 cents has been added to get a Ras Tanura equivalent.

From U.S. and Canadian import figures (which give f.o.b. values) one can get some prices for Kuwait and Saudi Arabia. They are much less reliable than the Caribbean counterparts, because there are only a few isolated shipments. Some look like c.i.f. prices, or quantities are too small. The complete information is given in Table VI-A-4.

The other information consists mainly of reports from *PW* during 1958-59. Some sales were made at very low prices between $1.00 and $1.15, while the posted price was $1.70-$2.00. A long-term contract between Texaco and Cawood was said to be at $1.14. By September 1960 the posted price was $1.65-$1.75, and *PW* said that discounts were negligible. A similar statement was made in *PIW* in November 1961. The next month, *POPS* was puzzled by the cut in Bunker "C" prices when there was no apparent surplus. The Persian Gulf was "the only full price bunkering market in the world" (*POPS*, Dec. 1, 1961). These statements were confirmed by some Japanese transactions. In October 1961, two shipments were made from Abadan (posted price $1.65) at $1.58 and $1.60. At Ras Tanura (posted $1.75) there were eight shipments at $1.69 and eight from Bahrein. In December 1962 heavy fuel oil was 6 cents off posted. In July 1963 again it was reported that no rebates were available.

However, in January 1962 Aminoil was reported to be selling heavy fuel oil to Japan at $1.35. This low price perhaps indicated an attempt to break into the Japanese market. In February the next year companies were receiving $1.50 for fuel oil to Europe and Japan.

[2] "F.o.b. prices at Caribbean or Middle East refineries are normally equal to the lower end of the range [low of Platt's] for comparable products." UN-ECE, "The Price of Oil in Western Europe" (E/ECE/205, Geneva, 1955), p. 25, n. 3.

The ECSC *Tenth Annual Report* (1961-62) gives actual HFO prices for the Persian Gulf and Naples at four dates between February 1958 to January 1962 (see Table VI-A-3), which we cannot compare exactly with a press report. The Persian Gulf prices look roughly halfway between the posted price and the $1.00 prices reported after Suez. The Italian prices seem to be in line with those from other sources.

From 1960, Japanese power companies were discussing burning Khafji crude, since it could be bought 20 cents more cheaply (*PPS*, July 1963, p. 255). Khafji was being quoted at $1.46 in 1963, and heavy fuel oil at $1.65 at Ras Tanura, so the 20 cents may refer to posted prices. However, Khafji was available to Japanese refiners at about $1.25, and this would make fuel oil about $1.45 at Ras Tanura if market prices were being discussed.

In early 1964 spot prices in the Eastern Mediterranean were reported to be at $1.25. This could reflect competition from Italy rather than the Persian Gulf.

In January 1965 there was a report of a Caltex-U.S. military contract under which Navy Special was supplied at $1.20 in the Persian Gulf (*PIW*, Jan. 25, 1965). Assuming (as for the Caribbean) that Navy Special is 70 percent Bunker "C," 30 percent No. 2 fuel oil, and the percentage discount on both components is the same, this is equivalent to 99 cents for heavy fuel oil alone. It is impossible to say how far back this contract may date.

Singapore import figures for 1965 and 1967 (*POPS*, May 12, 1967, May 8, 1969) show 900,000 tons of heavy fuel imported in each year from Kuwait at c.i.f. prices of $1.43 and $1.45. Assuming Intascale less 50 percent for freight, the Ras Tanura equivalent is $1.13-$1.15. At Intascale less 70 percent, the Ras Tanura price would be $1.25-$1.27; this may be closer to the actual price, since Singapore can take tankers of up to 42-feet draft.

There are two Canadian import figures for $1.25 in August 1966 from Kuwait and Iran. In November *POPS* (Nov. 6, 1966) reported that independents and national companies would sell for ship's bunkers at the equivalent of $1.57 a barrel. This would be the same as $1.47 for a cargo load, since the usual cargo-bunker differential is about 10 cents per barrel.

In 1967 *POPS* began to report Persian Gulf prices regularly once a fortnight. In the first report at the end of January the price was quoted at $1.40-$1.50; in April it was reduced to $1.35-$1.50.

Japanese power companies have been importing substantial quantities of Khafji crude for direct burning. For the year beginning April 1967, it was planned to buy 2 million kilolitres. The Khafji price was still under negotiation, but it would not be more than the current $1.24 and not likely to be much less. The companies would not buy the crude unless it were the same price or lower than the prevailing HFO price. Assuming crude and HFO are equivalent ton for ton, heavy fuel would be available at $1.32 or more. (A barrel of heavy fuel oil is 7.02/6.59 the weight of a barrel of Khafji 27° crude.)

Summary. The post-Suez reaction took Persian Gulf spot prices down from almost $2.40 to $1.00. By September 1960, there was a recovery to $1.65-$1.75. For more recent years the situation is confused. We have several sources saying the pre-Suez price was around $1.35-$1.40 or more (*POPS* regular reports, *POPS* bunker price, Japanese buying of crude for burning); others suggest it was around $1.25 or less (Navy Special contract, Singapore imports, Canadian imports). The Singapore imports are too large to be dismissed as some temporary dislocation. Also there were imports from other countries, mainly Far Eastern, at just over $1.40, so this would seem to be the competitive c.i.f. price there.

Prices for Northwest Europe and Italy

The information is very mixed, but one can see in broad outline the same picture as in the Caribbean, i.e., a sharp fall after Suez I which was virtually complete by 1959. Fluctuations since then have been wider than in the Caribbean. Varying freight rates are partly responsible. An important factor also is local distress, e.g., the Ruhr in 1959, Italy in 1961, and Strasbourg and Munich in 1964.

The discussion will take the form of an approximate chronological study of the different sources after a study of Danish power station costs, which is the only series that spans most of the period.

Danish power station fuel costs. The figures shown in Table VI-A-5 are extremely valuable but must be used with caution, especially as one goes farther back in time. There is a considerable lag because prices are based on average value of the inventory from which the fuel is drawn. Probably around 1955, the low series can be taken to represent fuel oil since the purchasers were coastal power stations who would not only pay the lowest prices, but also be in the best position to choose oil or coal as the better bargain was offered. Therefore, if much or even most of the fuel they burned was coal, the price they paid could not help but be close to that paid for oil alone. In 1963 oil prices were listed separately, and it is interesting that they averaged almost the same as for all fuel, while the low oil price was somewhat higher than the low price generally, but later came to within a few cents of it.

Table VI-A-5 shows the low Danish price converted back to an f.o.b. Rotterdam equivalent. Compared with the other series from *POPS* and *Europa Öl-Telegram* (*EOT*) (see below), it is apparent that the agreement is close, but that the f.o.b. Rotterdam equivalent of the Danish price is generally lower. It may be that the Rotterdam price tends to be a little higher than what would be paid by a power station able to offer an attractive long-term contract.

An approach to continuous data may be found in an article from *Oil and Gas International*, Dec. 1961 (Figure VI-A-1). This presumably was written in a government or international agency. It is interesting that the series for Belgium and the Netherlands show a gradual increase—abstracting from the Korea boom—from the beginning of 1950 through 1956 (the boom and collapse of Suez), and then a fall to levels below those of 1950. Thus they tend to confirm the Caribbean series, but they depict a somewhat sharper decrease, and a fairly stable point after that. The German data are only available from the middle of 1955 and show the post-Suez price declining rapidly and continuously through about the beginning of 1960. Market prices broke away from list prices after Suez; if there were any rebates earlier, they must have been very small compared with those that followed. The breaking away began in Germany in early 1958; later that year in Holland and Belgium. Similarly, German prices reached their lowest point at the beginning of 1960 and the Dutch towards the end of that year. In August 1961 the Ruhr price was $1.75, while at Rotterdam it was $1.90.

European Coal and Steel Community (ECSC-EEC). Prices in various Common Market centers are published in the annual reports of the Community and in some of its other publications (see Tables VI-A-6, 7, and 8). Prices for Antwerp are quoted ex-refinery, while those for Hamburg and Rotterdam included, through 1965, delivery to consumers' premises. Although the footnote explaining this first appeared in the 1964 report, the fairly constant difference between Antwerp and Rotterdam suggests that it applied earlier. The 1966 report gave ex-refinery prices for Rotterdam for that and the previous year. *Nouvelles Réflexions sur les Perspectives Energétiques à la Long Terme de la Communauté Européenne* (1966, p. 10) warns that the prices are not comparable from one year to another; one should be cautious in using them.

Further confirmation that prices before Suez did not depart much from import parity is the fact that for 1956 only posted price and AFRA rates are given. (See Table VI-A-3.) In 1960 the Hamburg price was $11.00 a ton delivered, while the Antwerp price ex-refinery was $14.00, and the Rotterdam price $12.65; however, from 1961 the Antwerp prices are always the lowest.

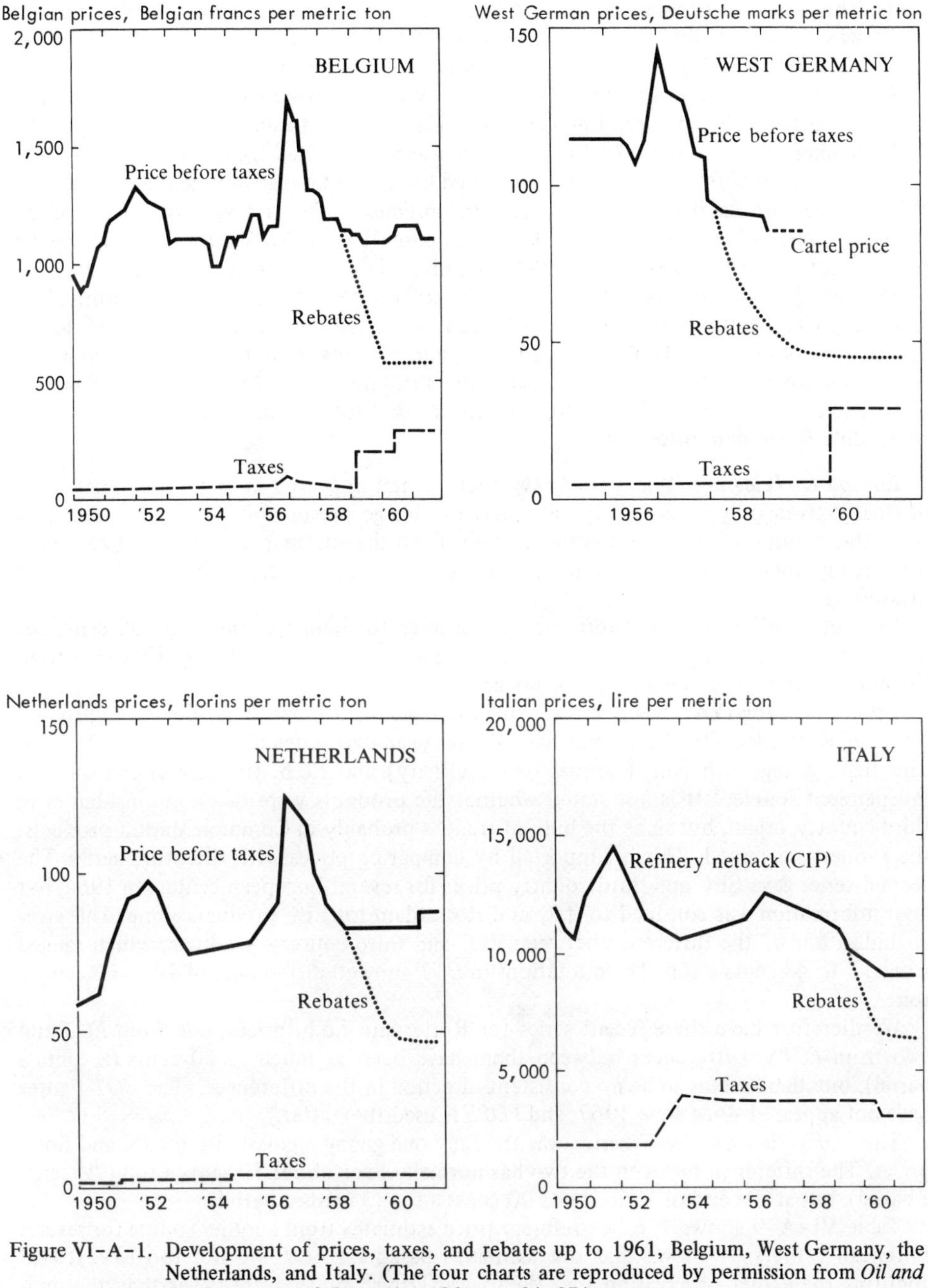

Figure VI-A-1. Development of prices, taxes, and rebates up to 1961, Belgium, West Germany, the Netherlands, and Italy. (The four charts are reproduced by permission from *Oil and Gas International*, December 1961, pp. 24-27.)

Aussenhandelverband für Mineralöl e.v. (*AFM*—formerly *Arbeitsgemeinschaft Freier Mineralölimporteure*) published prices for heavy fuel oil only from July 1960 to April 1961. They were quoted only at Rotterdam, while prices for other products were quoted at Hamburg also. The Rotterdam prices in the period were all close to \$1.90 except for a brief fall to \$1.82 in February 1961. There seems to be a real difference between the ports. This could be accounted for by the surplus in Hamburg refineries, which in August 1959 was "flooding the Ruhr" at $8.00 or $9.00 per metric ton ($1.21-$1.36 per barrel). The lack of a Hamburg importers' quotation would confirm this.

Petroleum Week gives us only a few other North European prices: two Rotterdam prices in early 1958, the very low German prices referred to above, and on August 28, 1959, bunker oil at about $1.75 "in Scandinavia and the United Kingdom."

German internal prices. These are published by the Federal government and quoted in *POPS*. Prices are as delivered to consumers' premises. The average for 1959 and by months, January 1960 to October 1962, was published in *POPS* (Dec. 5, 1962), and a January price in *POPS* (Feb. 27, 1963). Another series published in *POPS* in July 31, 1965, went back to the beginning of that year; this has been continued monthly. The series seem to be the same, but the earlier one specified "minimum quantity 15 tons." Prices were quoted for Hamburg, Düsseldorf, and Mannheim in 1962 and for these and four other towns in the later group. While the absolute level is of no interest to us, the fluctuations fit in well with the other series up to 1966. From then on internal prices rose, while Rotterdam prices fell.

Europa Öl-Telegram—Prices for heavy fuel oil first appeared in late 1963. They were originally given only occasionally (at times of change presumably); but fairly regularly from the autumn of 1964 and systematically from the summer of 1965. In 1964 there were many reports of isolated transactions or of local distress, such as at Munich or Strasbourg.

In 1966 *POPS* (Oct. 31, 1966) reported a price for bunkers "including 90 cents per ton barging." This suggests that at Rotterdam one should subtract 14 cents a barrel from the bunker price to obtain a cargo lot price.

From mid-1966 *POPS* has published full and regular information on European product prices at Rotterdam. In the second half of that year two series of prices were published. The first, giving both c.i.f. (cargoes or ex-refinery) and f.o.b. (in barges) is from "an independent source." It is not stated whether the products were of Common Market or third-country origin, but since the bulk of trade is probably in Common Market products, the former is assumed. This is supported by comparing prices with the other series. The second series gave EEC and third-country prices for several European centers in 1966, but later information was confined to Italy and Rotterdam for EEC products alone. This gives an indication of the difference between EEC and third-country products, which ranged from 13 to 43 cents a ton. The quotations in *EOT* showed differences of 40 to 60 cents a ton.

We therefore have three recent series for Rotterdam f.o.b. prices, one from *EOT* and two from *POPS*. Differences between them have been as much as 60 cents (8 cents a barrel), but there seems to be no consistent direction in the differences. (The *EOT* figures have not appeared since June 1967, and *POPS* is used thereafter.)

The *POPS* "independent" series was the only one giving regularly both c.i.f. and f.o.b. prices. The difference between the two has normally been about 50 cents a ton (7½ cents a barrel), but at the end of 1966 it was 20 cents a ton (3 cents a barrel).

Table VI-A-9 shows 1965 ex-refinery price estimates from another source for several European centers. It shows that our estimates based on *EOT* are not too low. It also confirms the earlier observation that the lowest Danish prices were lower than the mid-European ones. The range for Denmark of $1.63-$1.76 fits in with the power station lows for 1964/65 and 1965/66 of $1.71 and $1.70. The West German price span is wide, $1.45-$1.82, and the high is not far above the *EOT* figure for 1965 which ranged from $1.76 to $1.85. Therefore, an estimate taken from the *EOT* price may have an upward bias.

Import and export figures. A number of foreign trade figures are available from secondary sources, especially for recent years. Although they reflect mainly intracompany transactions, they often contain price figures close to the estimates of arm's-length prices presented here. However, to list all the apparently relevant data would add more bulk than understanding. (This is not the case for the Caribbean or Persian Gulf HFO prices or for crude prices, where this type of customs information is often all one can get.) U.S.S.R. exports of heavy fuel oil (Table VI-A-10) are of some interest. These show the average price of exports to all countries. The specific numbers are not very meaningful, but the sharp fall to $1.57 in 1959 and the stability since then are significant. Prices for recent years are higher than Rotterdam cargoes.

In the studies of Rotterdam realizations, I have used *AFM* heavy fuel oil prices up to April 1961 and *EOT* prices from 1964. For the interval I abstracted as best I could from the various data available.

Italian prices. The few brought together are a useful supplement to the North European prices. The Italian price references are listed separately. The sources are the same as for the other European prices and need no further comment. Sporadically in 1966 and regularly since, *POPS* has given an f.o.b. price for Italy. Also included here are some Canadian imports from Italy at f.o.b. prices.

One problem is that prices are often given for "Italy" without further specifications. It seems likely that the cargo prices quoted in *POPS* are from southern Italy or Sicily; where internal prices are quoted, they are more likely to be for industrial users in the north. The first series of ECSC figures on the chart are for Naples; the second, for Genoa. It is assumed that the regular *POPS* series and occasional shipments to Canada and the United States refer to southern ports.

In 1961 there was severe oversupply and then a recovery of Italian internal prices due to a water-power shortage. Otherwise Italian prices follow the other series fairly closely up to 1965. Then Italian internal prices follow German internal ones upward, while Rotterdam and Italian cargo prices (as far as one can tell) fell.

One possible explanation of the weakness of port prices in relation to internal ones may be connected to more stringent sulfur regulations in many parts of the world. Possibly there is a surplus of high sulfur product which finds its way to the ports. Another explanation might be weakness of heavy fuel oil in France and Britain. Germany would be protected by her "self-restraint" policy of limiting imports. However, this does not explain the discrepancy in Italian prices.

TABLE VI-A-1. Heavy Fuel Oil Prices: U.S. Imports from Caribbean, 1951-1959

(*dollars per barrel, f.o.b.*)

	Venezuela		N.W.I.		Posted	Venezuela		N.W.I.		Posted	Venezuela		N.W.I.		Posted
	Bunker	Other	Bunker	Other		Bunker	Other	Bunker	Other		Bunker	Other	Bunker	Other	
			1951					*1952*					*1953*		
J . . .	1.60	1.69	1.54	1.65	1.75	1.66	1.68	1.56	1.71	1.75	1.56	1.49	1.48	1.56	1.50
F . . .	1.53	1.74	1.56	1.68	1.75	1.70	1.66	1.58	1.75	1.75	1.41	1.59	1.46	1.61	1.50
M . . .	1.48	1.72	1.55	1.61	1.75	1.67	1.72	1.61	1.74	1.75	1.52	1.52	1.42	1.56	1.50
A . . .	1.78	1.60	1.48	1.57	1.75	1.67	1.73	1.62	1.74	1.75	1.65	1.56	1.48	1.65	1.60
M . . .	1.63	1.67	1.46	1.57	1.75	1.61	1.76	1.69	1.76	1.75	1.46	1.61	1.48	1.63	1.60
J . . .	1.51	1.67	1.56	1.74	1.75	1.58	1.73	1.71	1.76	1.75	1.43	1.63	1.57	1.70	1.85
J . . .	1.26	1.69	1.52	1.71	1.75	1.64	1.70	1.80	1.75	1.75	1.45	1.73	1.62	1.74	1.85
A . . .	1.45	1.78	1.52	1.63	1.75	1.60	1.59	1.66	1.82	1.75	1.82	1.87	1.66	1.80	1.85
S . . .	1.55	1.70	1.59	1.68	1.75	1.43	1.53	1.56	1.59	1.50	1.87	1.81	1.67	1.78	1.85
O . . .	1.48	1.71	1.58	1.72	1.75	1.44	1.62	1.53	1.64	1.50	1.75	1.80	1.68	1.82	1.85
N . . .	1.52	1.68	1.57	1.70	1.75	1.62	1.73	1.53	1.63	1.50	1.72	1.83	1.68	1.87	1.85
D . . .	1.52	1.68	1.54	1.73	1.75	1.65	1.59	1.47	1.59	1.50	1.63	1.80	1.78	1.84	1.85
Average	1.50	1.69	1.54	1.67		1.61	1.67	1.61	1.70		1.55	1.61	1.58	1.61	
			1954					*1955*					*1956*		
J . . .	1.58	1.82	1.68	1.65	1.85	1.67	1.90	1.73	1.88	1.85	1.71	2.25	1.94	2.14	2.10
F . . .	1.73	1.83	1.70	1.85	1.85	1.60	1.98	1.76	2.04	1.95	1.78	2.31	1.96	2.21	2.10
M . . .	1.58	1.87	1.62	1.81	1.85	1.47	1.97	1.83	2.00	1.95	1.81	2.23	1.91	2.25	2.10
A . . .	1.59	1.81	1.71	1.85	1.85	1.73	2.00	1.70	1.98	1.95	1.84	2.28	2.00	2.21	2.10
M . . .	1.60	1.80	1.67	1.82	1.85	1.63	1.95	1.75	1.98	1.95	1.87	2.27	2.05	2.20	2.10
J . . .	1.54	1.83	1.74	1.99	1.85	1.77	(2.55)	1.77	1.96	1.95	2.01	2.35	2.15	2.19	2.10
J . . .	1.57	1.79	1.72	1.80	1.85	1.64	1.96	1.84	2.12	1.95	2.08	2.25	2.11	2.29	2.10
A . . .	1.45	1.90	1.68	1.81	1.85	1.70	1.93	1.91	2.22	2.10	2.01	2.28	2.08	2.26	2.10
S . . .	1.59	1.80	1.68	1.87	1.85	1.72	1.98	1.89	2.12	2.10	2.05	2.30	2.12	2.35	2.10
O . . .	1.56	1.84	1.68	1.90	1.85	1.78	2.10	1.94	1.77	2.10	2.07	2.35	2.10	2.29	2.10
N . . .	1.67	1.96	1.69	1.86	1.85	1.88	2.13	1.93	2.10	2.10	2.03	2.31	2.10	2.34	2.40
D . . .	1.55	1.92	1.77	2.01	1.85	1.74	2.25	2.05	2.11	2.10	2.08	2.42	2.11	2.41	2.40
Average	1.59	1.86	1.70	1.84		1.69	2.01	1.83	2.02		1.97	2.30	2.06	2.25	

TABLE VI-A-1. Continued

(*dollars per barrel, f.o.b.*)

	Venezuela		N.W.I.		Posted	Venezuela		N.W.I.		Posted	Venezuela		N.W.I.		Posted
	Bunker	Other	Bunker	Other		Bunker	Other	Bunker	Other		Bunker	Other	Bunker	Other	
			1957					*1958*					*1959*		
J . . .	(2.90)	2.57	2.24	2.62	2.55	2.38	2.58	2.38	2.66	2.35	1.93	2.41	1.97	2.08	2.00
F . . .	2.21	2.73	2.14	2.83	2.55	2.35	2.53	2.43	2.45	2.25	1.77	2.07	2.02	2.18	2.00
M . . .	2.35	2.80	2.35	2.78	2.55	2.08	2.28	2.22	2.44	2.25	1.85	2.04	1.94	2.06	2.00
A . . .	2.38	2.67	2.48	2.77	2.55	2.25	2.40	2.26	2.34	2.25	1.87	2.01	1.82	2.02	2.00
M . . .	2.38	2.59	2.58	2.67	2.55	2.10	2.26	2.06	2.26	2.25	(2.44)	2.02	1.92	2.02	2.00
J . . .	2.50	2.60	2.57	2.62	2.55	2.11	2.27	2.20	2.28	2.25	1.76	2.03	1.88	1.97	2.00
J . . .	2.53	2.64	2.48	2.68	2.55	2.07	2.31	2.15	2.37	2.25	1.84	1.98	1.75	2.06	2.00
A . . .	2.66	2.61	2.47	2.78	2.55	2.11	2.24	2.17	2.29	2.25	1.82	2.01	(2.43)	1.93	2.00
S . . .	2.49	2.62	2.41	2.68	2.55	2.05	2.21	2.10	2.28	2.00	1.89	2.08	1.83	2.00	2.00
O . . .	2.40	2.53	2.50	2.65	2.55	2.07	2.14	2.16	2.16	2.00	1.87	2.09	1.88	1.98	2.00
N . . .	2.53	2.62	2.39	2.71	2.55	2.13	2.16	1.99	2.18	2.00	1.84	2.01	1.84	2.06	2.00
D . . .	2.41	2.55	2.39	2.66	2.55	1.99	2.19	2.03	2.09	2.00	1.82	2.04	1.91	2.23	2.00
Average	2.39	2.64	2.36	2.71		2.14	2.31	2.18	2.32		1.79	2.05	1.92	2.07	

Source: U.S. Bureau of the Census, *U.S. Imports of Merchandise for Consumption*, Reports FT110 (Now FT125).
Note: Figures in parentheses are believed to be erroneous, and therefore are omitted from annual average.

TABLE VI-A-2. Heavy Fuel Oil Prices: Canadian Imports from Caribbean

(*dollars per barrel, f.o.b.*)

	1958 Venezuela	1958 Neth. Ant.	1959 Venezuela	1959 Neth. Ant.	1960 Venezuela	1960 Neth. Ant.	1961 Venezuela	1961 Neth. Ant.	1962 Venezuela	1962 Neth. Ant.
J	...	2.44	...	1.97	1.83	1.89	2.27	1.63	...	1.73
F	...	...	...	2.04	...	2.02	1.74	...	1.57	...
M	...	2.33	...	...	...	...	...	1.70	...	...
A	...	2.17	...	...	...	...	...	1.89	...	1.58
M	2.25	2.85	1.97	1.96	1.90	...	1.80	2.27	1.53	1.64
J	2.25	2.25	2.08	1.80	1.91	1.68	1.64	1.59	1.58	1.68
J	2.26	2.19	1.86	1.72	1.87	1.76	...	...	1.65	1.62
A	2.26	2.86	1.84	1.65	1.88	1.56	...	...	1.58	1.56
S	2.21	2.19	1.90	1.76	1.85	1.56	...	...	1.65	1.65
O	...	2.18	1.91	1.83	1.78	1.65	1.69	1.79	1.55	1.86
N	...	...	1.92	...	1.68	...	1.72	1.75	1.59	1.59
D	...	...	1.92	1.82	...	...	1.61	1.60	1.65	1.61
Average	2.23	2.24	1.89	1.88	1.83	1.70	1.71	1.69	1.61	1.66

	1963 Venezuela	1963 Neth. Ant.	1964 Venezuela	1964 Neth. Ant.	1965 Venezuela	1965 Neth. Ant.	1966 Venezuela	1966 Neth. Ant.	1967 Venezuela	1967 Neth. Ant.
J	...	1.53	1.65	2.16	1.56	1.70	1.51	1.47	1.51	1.51
F	...	...	1.61	...	1.57	1.67	1.53	1.53	1.49	1.54
M	...	2.16	2.34	1.60	2.37	1.46	2.35	1.47	2.51	1.52
A	...	1.60	1.76	1.84	1.60	1.53	1.40	1.59	2.03	1.49
M	1.62	1.59	1.55	1.56	1.52	...	1.56	1.48	1.57	1.49
J	1.54	1.57	1.55	1.54	1.51	1.52	1.47	1.54	1.46	1.49
J	1.58	1.55	1.66	1.58	1.51	1.58	1.57	1.50	1.52	1.49
A	1.47	1.63	1.62	1.65	1.51	1.50	1.55	1.59	1.54	1.55
S	1.66	1.51	1.55	1.56	1.53	1.48	1.54	1.50	1.42	1.49
O	1.57	1.50	1.76	1.54	1.52	1.48	1.51	1.55	1.52	1.46
N	1.80	1.63	1.50	1.55	1.49	1.51	1.51	1.49	...	...
D	1.72	1.57	1.55	1.54	1.48	1.51	1.55	1.49	...	...
Average	1.63	1.57	1.59	1.58	1.52	1.52	1.53	1.50		

Source: Canada, Bureau of Statistics, *Trade of Canada, Imports by Commodities*.

TABLE VI-A-3. Heavy Fuel Oil Prices: Caribbean, Persian Gulf, and Europe, Selected Dates, 1956-1962

(*dollars per barrel*)

	July 1956	Feb. 1958	Feb. 1959	Aug. 1960	Jan. 1962
Caribbean	2.28	2.20	1.67	1.64	1.64
Freight	0.94	0.36	0.31	0.26	0.29
Rotterdam c.i.f.	3.22	2.56	1.98	1.90	1.93
Persian Gulf	...	1.97	1.46	1.40	1.40
Freight	...	0.69	0.54	0.45	0.50
Naples c.i.f.	...	2.66	2.00	1.85	1.90

Source: ECSC, *Tenth Annual Report* (1962), p. 544. (All prices converted from dollars per metric ton to dollars per barrel at 6.59 bbl/ton.) These are based on posted prices less average rebate and spot freight rates (except July 1956, which is posted price and AFRA rate).

TABLE VI-A-4. Heavy Fuel Oil: Persian Gulf, Miscellaneous Import Prices, 1953-1966

(*dollars per barrel f.o.b.*)

Date	Importing country	Exporting country	Price (*$/bbl*)	Quantity (*000 tons*)
June 1953	U.S.	Kuwait	1.50	17
Nov. 1953	U.S.	S. Arabia	2.50	16
Jan. 1954	U.S.	(Lebanon)	1.58	31
Oct. 1958	U.S.	Kuwait	1.32	165
Nov. 1958	U.S.	Kuwait	1.73	53
Dec. 1958	U.S.	Kuwait	1.79	85
Jan. 1959	U.S.	Kuwait	2.00	16
Feb. 1959	U.S.	Kuwait	2.06	70
Mar. 1959	U.S.	Kuwait	1.80	55
Apr. 1959	U.S.	Kuwait	2.08	36
May 1959	U.S.	Kuwait	2.04	37
May 1959	Canada	S. Arabia	1.67	n.a.
June 1959	U.S.	Kuwait	2.11	89
July 1959	U.S.	Kuwait	1.96	106
Aug. 1959	U.S.	S. Arabia	1.15	4½
Aug. 1959	U.S.	Kuwait	1.98	56
Sept. 1959	U.S.	Kuwait	2.05	100
Sept. 1959	U.S.	S. Arabia	1.15	4½
Sept. 1959	Canada	S. Arabia	1.32	n.a.
Oct. 1959	U.S.	Kuwait	2.22	92
Nov. 1959	U.S.	Kuwait	2.10	118
Dec. 1959	U.S.	Kuwait	2.10	68
Jan. 1960	U.S.	Kuwait	2.10	50
Jan. 1960	U.S. (Bunker)	Kuwait	2.04	13
Feb. 1960	U.S.	Kuwait	2.10	34
Feb. 1960	U.S. (Bunker)	Kuwait	2.10	6
Aug. 1966	Canada	Kuwait	1.25	22
Sept. 1966	Canada	Iran	1.24	23

Sources: U.S. and Canadian import statistics.
n.a. = not available.

TABLE VI-A-5. Danish Power Station Fuel Costs and Rotterdam Equivalent, 1953-1967

(*dollars per barrel*)

Year[a]	Low	Average	High	Tanker rate used: Percentage above or below Intascale	Tanker rate used: $/bbl		Rotterdam equivalent of low
	(1)	(2)	(3)	(4)	(5)	(6)	(7)
1953/4.....	2.18	2.48	2.89	–4	0.07	0.13	2.11
1954/5.....	2.01	2.34	3.26	–19	0.06	0.11	1.95
1955/6.....	2.81	2.89	3.46	+20	0.10	0.18	2.72
1956/7.....	2.68	3.16	3.93	+115	0.16	0.31	2.52
1957/8.....	2.71	3.08	4.19	+65	0.12	0.22	2.59
1958/9.....	2.35	2.58	2.94	–51	0.04	0.07	2.31
1959/60....	1.93	2.29	2.68	–56	0.03	0.06	1.90
1960/61....	1.83	2.06	2.32	–55	0.03	0.06	1.80
1961/2.....	1.81	1.95	2.20	–57	0.03	0.06	1.78
1962/3.....	1.73	1.90	2.19	–43	0.04	0.07	1.69
1963/4.....	1.76	1.93	2.18	–39	0.05	0.09	1.71
1964/5.....	1.71	1.90	2.21	–47	0.04	0.07	1.67
1965/6.....	1.70	1.86	2.18	–40	0.05	0.09	1.65
1966/7.....	1.67	1.80	2.13	–56	0.03	0.06	1.64
1967/8.....	1.68	1.80	2.06	+6	0.08	0.15	1.60

Source: Danish power station costs from Danske Elvaerkers Forening, *Dansk Elvaersstatistik*, annual, Copenhagen, tables A2.

Notes by Column:

(1) Prices are average value of fuel drawn from inventory. Since they lag behind price movements, they are biased down during periods of price rise, and biased up during periods of price decline.

(2) Conversion factors 14.48 cents per kroner, 3.968 Btu/kilocalorie, 41 million Btu per long ton heavy fuel oil, 6.7 barrels per long ton.

(3) Three stations (Nos. 201, 401, 801) excluded because one burned primarily lignite, one had a different financial year, one had incomplete data.

(4) Tanker rate col. (4) is the single-voyage (spot) rate, from Harley Mullion. The theory is that the station with the lowest costs tended to have the lowest tanker costs, which were usually spot rates. During 1953/54-1958/59, calendar year average. Later years averaged over the quarters corresponding to the power stations' accounting year.

(5) Intascale flat estimated as the differential between a voyage from the loading port to Rotterdam, and one to Copenhagen. This is about 50 cents per long ton, 7.6 cents per barrel, from all loading ports.

(6) Alternatively, it might be considered as the voyage of a ship loading in Rotterdam and discharging in Copenhagen, estimated at 92 cents per long ton, or 14 cents per barrel.

(7) Rotterdam equivalent based on the assumption of cols. (1) and (5).

[a]April 1-March 31.

TABLE VI-A-6. Heavy Fuel Oil Prices, European Coal and Steel Community, 1956-1967

Period	Rotterdam		Antwerp		Hamburg		Italy		
	Price before tax (*$/m.ton*)	Low (*$/bbl*)	Price before tax (*$/m.ton*)	Low (*$/bbl*)	Price before tax (*$/m.ton*)	Low (*$/bbl*)	Port	Price before tàx (*$/m.ton*)	Low (*$/bbl*)
	(c.i.f.)							(c.i.f.)	
July 1956	21.20	3.22	...	...	...	...	...	...	...
Feb. 1958	16.90	2.57	...	...	...	...	Naples	16.90	2.57
Feb. 1959	13.00	1.98	...	...	...	...	Naples	13.20	2.00
Aug. 1960	12.50	1.90	...	...	...	...	Naples	12.20	1.85
Jan. 1962	12.70	1.93	...	...	...	...	Naples	12.50	1.90
	Delivered		Refinery		Delivered			Refinery	
Nov. 1960	12.50-14.00	1.90	14.00-18.00	2.12	11.00-14.00	1.67	Genoa	12.00-13.00	1.82
Summ. 1961	12.50-13.00	1.90	11.00-13.00	1.67	13.00-14.00	1.98	Genoa/Naples	9.70-11.70	1.47
April 1962	13.50-14.00	2.05	12.00-13.00	1.82	14.00-14.50	2.13	Genoa/Naples	13.70-14.20	2.08
June 1962	13.50-15.00	2.05	12.30-13.30	1.87	12.50-14.50	1.90	Genoa/Naples	12.20-14.20	1.85
Aut. 1962	14.50-15.00	2.11	12.40-13.40	1.88	15.00-15.50	2.28	Genoa/Naples	12.20-14.20	1.85
Nov. 1963	13.50-15.50	2.05	11.50-12.50	1.74	12.50-13.50	1.90	Gen/Sic/Nap	11.20-12.70	1.70
Nov. 1964	13.50-14.00	2.05	11.50-12.50	1.74	11.30-12.30	1.71	Genoa	11.50-13.00	1.75
Dec. 1965	13.00-15.00	1.98	11.00-12.00	1.67	11.30-12.30	1.71	Genoa	8.80- 9.80	1.32
	Refinery								
	12.00-14.00	1.82	...	...	...	...	...	...	...
Nov. 1966	12.00-14.00	1.82	10.50-11.00	1.59	12.30-13.30	1.87	Genoa	10.80-12.80	1.64
Early 1967	...	...	11.00-11.50	1.67	...	...	...	...	...

Sources: ECSC, *Annual Report*, Nos. 10, 12, and 15 and *Nouvelles Réflexions sur Perspectives Energétiques à la Long Terme de la Communauté Européenne*, 1966.

TABLE VI-A-7. Price of U.S. Coal (slack/coking fines), 1953-1961

(dollars per metric ton)

Period	Price f.o.b. U.S. port[a]	Average freight charge Hampton Roads-Rotterdam[b]	Price c.i.f.
1953			
March	10.38	4.83	15.21
June	10.38	4.31	14.69
September	9.55	3.90	13.45
December	9.55	4.11	13.66
1954			
March	8.57	4.66	13.23
June	8.57	4.56	13.13
September	9.06	5.11	14.17
December	9.06	6.88	15.94
1955			
March	9.84	6.79	16.63
June	9.84	8.13	17.97
September	11.27	9.19	20.36
December	11.27	9.30	20.57
1956			
March	11.51	10.09	21.60
June	11.51	10.00	21.51
September	11.51	9.92	21.43
December	11.76	15.05	26.81
1957			
March	11.76	9.72	21.48
June	11.51	6.79	18.30
September	11.27	3.30	14.57
December	10.83	3.55	14.38
1958			
March	9.84	3.00	12.84
June	9.84	3.21	13.05
September	9.84	3.10	12.94
December	9.84	3.68	13.52
1959			
March	9.84	2.94	12.78
June	9.84	2.87	12.71
September	9.84	2.87	12.71
December	9.84	3.74	13.58
1960			
March	9.60	3.51	13.11
June	9.60	3.59	13.19
September	9.60	3.51	13.11
December	9.60	3.51	13.11
1961			
January	9.60	3.51	13.11

Source: ECSC, *Ninth General Report* (1960/1961), Annex, p. 385 (not numbered).

[a]Estimated.

[b]Mean between maximum and minimum figures charged during the month in respect of single voyages.

TABLE VI-A-8. Development of Coal Prices in the European Coal and Steel Community

(price in dollars per metric ton)

Product		Date	Ruhr		Aachen		Netherlands		Belgium				Nord/ Pas-de-Calais	
									Cobechar sales		Independent sales			
Type	Size	Month Year	Price	V.M. (%)	Price	V.M. (%)	Price	V.M. (%)	Price	V.M. (%)	Price	V.M. (%)	Price	V.M. (%)
Anthra-cites	French nuts	Jun. 1952	19·20	7–10	19·20	<10	...	...	27·22	<10	...	<10	26·06	<11
		Apr. 1953	22·80	7–10	24·06	<10	21·60	10–14	27·60	<10	...	<10	26·57	<11
		Apr. 1954	22·80	7–10	24·52	<10	21·22	10–14	26·60	<10	...	<10	26·86	<10
		Jun. 1955	22·97	7–10	25·49	<10	22·37	9–12	30·00	<10	...	<10	27·83	<10
		Apr. 1956	23·16	7–10	25·90	<10	23·68	9–12	30·00	<10	...	<10	27·83	<10
		Apr. 1957	24·08	7–10	27·49	<10	25·39	9–12	33·60	<10	...	<10	27·83	<10
		Apr. 1958	25·69	7–10	28·93	<10	26·97	8–10	34·60	<10	...	<10	26·50	<10
		Apr. 1959	25·69	7–10	28·93	<10	27·63	8–10	24·60	<10	...	<10	26·95	<10
		Apr. 1960	25·49	7–10	29·83	<10	27·63	8–10	34·60	<10	24·60	<10	26·95	<10
		Jan. 1961	25·49	7–10	29·83	<10	27·63	8–10	34·60	<10	34·60	<10	26·95	<10
Low volatile	small nuts	Jun. 1952	16·23	10–14	16·23	10–14	...	...	27·22	10–12½	...	10–12½	26·06	11–13
		Apr. 1953	19·37	10–14	20·63	10–14	21·60	10–14	27·60	10–12½	...	10–12½	26·57	11–13
		Apr. 1954	19·37	10–14	20·86	10–14	21·22	10–14	27·60	10–12½	...	10–12½	26·86	10–14
		Jun. 1955	19·54	10–14	21·95	10–14	21·45	11–14	30·00	10–12½	...	10–12½	27·26	10–14
		Apr. 1956	19·73	10–14	22·36	10–14	22·76	11–14	30·00	10–12½	...	10–12½	27·26	10–14
		Apr. 1957	20·65	10–14	23·37	10–14	24·47	11–14	33·60	10–12½	...	10–12½	27·26	10–14
		Apr. 1958	22·15	10–14	25·05	10–14	26·05	10–13	34·10	10–14	...	10–12½	25·79	10–14
		Apr. 1959	22·14	10–14	25·05	10–14	26·32	10–12	32·60	10–14	...	10–12½	26·34	10–14
		Apr. 1960	22·06	10–14	23·89	10–14	26·32	10–12	31·60	10–14	30·60	10–12½	26·34	10–14
		Jan. 1961	22·06	10–14	23·89	10–14	26·32	10–12	31·60	10–14	30·60	10–12½	26·34	10–14
Semi-bitu-minous	singles	Jun. 1952	11·65	14–19	11·65	14–19	...	...	17·22	16–20	...	16–20	16·51	13–22
		Apr. 1953	13·66	14–19	14·92	14–19	14·40	15–20	16·40	16–20	...	16–20	16·80	13–22
		Apr. 1954	13·66	14–19	14·35	16–19	14·40	15–20	16·40	16–20	...	16–20	17·14	14–22
		Jun. 1955	14·05	14–19	14·40	16–19	14·47	15–20	15·70	18–20	...	15–20	16·29	14–18
		Apr. 1956	14·25	14–19	14·82	16–19	14·47	15–20	15·70	16–20	...	16–20	16·29	14–18
		Apr. 1957	15·16	14–19	16·06	16–19	17·24	15–20	19·40	16–20	...	16–20	19·14	14–18
		Apr. 1958	16·32	14–20	17·39	16–19	18·55	14–18	20·10	18–20	...	18–20	17·07	14–18
		Apr. 1959	16·09	16–20	17·39	16–19	17·11	14–18	19·70	18–20	19·20	18–20	16·00	14–18
		Apr. 1960	15·77	16–20	16·91	16–19	15·53	14–18	17·60	18–20	17·60	18–20	16·00	14–18
		Jan. 1961	15·77	16–20	16·91	16–19	15·53	14–18	17·60	18–20	...	18–20	16·00	14–18

Source: ECSC, *Ninth General Report* (1960/1961), Annex, p. 378 (not numbered).
Note: Shown beside each price is the volatile-matter content declared by the producer for the type and size concerned.

TABLE VI-A-9. Heavy Fuel Oil Prices in Selected Countries of Europe, 1965

	Ex-refinery	
	Cents/million Btu	Dollars per barrel
Belgium	27-30	$1.70-1.88
Denmark	26-28	1.63-1.76
West Germany	23-29	1.45-1.82
Italy	24-28	1.51-1.76
Spain	32	2.02
Sweden	28-30	1.76-1.88
United Kingdom	32-35	2.02-2.20

Source: U.S. Atomic Energy Commission, Division of Technical Information, *The Growth of Foreign Nuclear Power*, Apr. 1966, p. 25. (From estimates by Arthur D. Little, Inc.)

TABLE VI-A-10. U.S.S.R. Exports of Heavy Fuel Oil, 1952-1967

Year	Volume (*million tons*)	Price (*$/bbl*)	Year	Volume (*million tons*)	Price (*$/bbl*)
1952	0.3	2.30	1960	5.8	1.67
1953	0.5	1.75	1961	6.9	1.55
1954	1.2	1.71	1962	7.7	1.56
1955	1.5	1.81	1963	8.6	1.62
1956	1.8	2.24	1964	9.1	1.61
1957	2.2	2.76	1965	9.7	1.55
1958	2.5	2.02	1966	10.6	1.52
1959	4.3	1.57	1967	11.2	1.59

Source: *POPS* June 3, 1968; Aug. 26, 1968.

TABLE VI-A-11. Heavy Fuel Oil: Ships' Bunker Prices, ARA and Italy

(*dollars*)

Source *POPS*, issue of:	Italy	Antwerp/Rotterdam	Comment
Aug. 4, 1959	1.82-2.12	...	...
Dec. 11, 1964	1.60-1.66	...	For 1965
Feb. 18, 1966	1.67	1.78	Spot
June 7, 1966	1.67	1.78[a]	Contract customers, "continuity spot" and transients
July 8, 1966	1.63	1.75	...
Oct. 31, 1966	...	1.73-1.78	...
Nov. 4, 1966	1.53-1.57	1.73-1.78	For 1967
May 29, 1967	1.48-1.53	1.70-1.74	...

[a]After $1.70 six months previously.

Appendix VI-B

Western Europe: Ex-Refinery Values, 1960–1970

Rotterdam Product Prices

For gasoline and light fuel oil the *Aussenhandel für Mineralöl* (*AFM*) series has been used (see Figure VI-1). This is the only continuous series that goes back to 1960. Published fortnightly, it gives the average prices over the previous two weeks for barge loads (minimum 500 tons) at Rotterdam, Antwerp, and Hamburg. Since January 1965 an import duty has been imposed on products coming from outside the European Economic Community (EEC). *AFM* lists these and also lists separately quotations for EEC and third-country products. The former listing was selected for our purpose because these prices would be most relevant to a Community refiner. In any case, there are often not enough transactions in third-country products for a quotation to be made. The figure shows the midpoint of the highs and lows converted from dollars per ton to cents per gallon at .3182 for gas oil diesel index 53/57 (i.e., assuming 7.6 barrels to a metric ton) and .3571 for regular 90/92 gasoline (i.e., assuming 8.5 barrels per metric ton).

From mid-1964 the series was compared with three others. Occasionally after mid-1964 and regularly from June 1965, *Europa Öl Telegram* (*EOT*) gave Rotterdam f.o.b. (= barge lot) prices. *Platt's Oilgram Price Service* (*POPS*) has reported European product prices weekly in two series since the middle of 1966. One from an "independent supplier" gave c.i.f. and f.o.b. prices at Rotterdam; the origin of the oil is not stated, but one can assume it to be European Economic Community (EEC). There are separate quotations for light fuel oil and gas oil, the latter usually being more expensive by about $1.00 per ton. Here the gas oil quotation has been taken. For all other sources gas oil, heating oil, and light fuel oil have been considered to be synonymous. The second *POPS* series gives cargo prices for Italy and barge prices for Rotterdam. This series has become a frequent report on market developments in various parts of the world. It is helpful in giving an idea of the normal differentials which could be applied to earlier, less complete data.

The *AFM* series was compared with the three other series for light fuel oil and gasoline, and seemed well confirmed. Differences between highest and lowest sources are normally less than $1.00 a ton, although in June 1966 it was $2.00 for gasoline. One has to assume that the *AFM* series is equally reliable for the earlier period when the only other regular series is the Channel Port Index (CPI). This index, published daily in *POPS* until the end of 1965, was derived by taking the Caribbean price and adding the estimated freight costs. Although the general trends are the same as for the *AFM* series, the differences are quite large. The CPI price for light fuel oil was about one cent a gallon higher in the winter of 1960-61; the two series ran close together from the end of 1961 to early 1964, and then the CPI series stayed as much as 2 cents above the *AFM* series. For gasoline, the CPI price was usually about 1½ cents higher, but the difference was as large as 3 cents at the end of 1963 and less than half a cent in mid-1964. The few press reports available for this period support the *AFM* price. The higher CPI price must reflect (1) the fact that Caribbean products were not competing in Europe and/or (2) the use of unrealistic freight rates in its calculation.

No data are given for June 1967.

For July 1967 and later months, all prices are from *POPS*.

Comparison with German Internal Prices

Although all the Rotterdam series compare quite well with each other there are discrepancies with internal market prices. Germany is the only one supplying extensive data. The nature of the differences and the probable causes vary for each product and it is instructive to examine each one.

Gasoline. Here there is little hard data. The Federal Statistics of Internal Prices, published in *POPS* July 30, 1965 and monthly since then, show few changes. This may reflect the actual situation or it could be that they have referred only to list prices. The Index of Wholesale Prices (1962 = 100) and Branded Consumer Prices Zone III in tank cars (minimum quantity 15 cu m) move very closely. The quality of the gasoline is not stated in either case.

A publication by ARAL (an independent refiner-marketer) shows gas station prices in the lowest price zone unchanged from October 1961 to early 1966, although there were changes in taxation over the period so that realizations for regular gasoline (*warenwart*) rose from 21.89 DM per 100 liters to 23.18. This is close to the Federal statistics although in 1964 the Federal Index fell heavily while the ARAL one was stable. This may have reflected a lower wholesale price (following the falls in Rotterdam the previous year) which did not reach the retail level. Both *AFM* and the Federal Index fell from 1960 to 1964, but the Rotterdam prices fell by almost 30 percent while the Index fell by 15 percent and ARAL prices by 14 percent.

The greatest discrepancy between the various indicators occurs after 1965. Gas station prices were stable in 1965 and fell sharply in 1966. This fall was a result of the price reductions initiated by Esso in spring and continued in the late summer. (Among many reports see, for example, *POPS* Aug. 24, 1966, p. 1.) The Federal prices and Index fell slightly in 1966 and rose sharply in January 1967. Rotterdam prices, which had risen steadily since the summer of 1965, flattened out in early 1967. The three different trends can perhaps be reconciled by saying that the strength of import prices was due to the growing demand for naphtha when supply had been cut by a British refinery fire; the Federal Index belatedly reflected this strength; meanwhile gas station margins were said to be severely cut. Independents protested that the prices were unfair and Shell explained reductions in its price to its dealers by the fact that it had had to forgive them rent. However, it is interesting that gas station commissions as late as April 1967 were said to be higher in Germany (and in Britain where there had also been a price war) than any other Common Market country. (The data have to be read from a graph, but it appears that the average German commission was about 8 DM/100 liters compared with 5 in France.)[1]

This situation of rising cargo prices and falling retail margins was interrupted by the Arab-Israeli war. The subsequent rises in retail price were interpreted (by the German Consumers Co-operative Association) as an excuse to raise prices to pre-price-war levels (*POPS* June 19, 1967).

Light fuel oil. For Germany there are three internal series. The Federal Statistics go back to 1960. The figures are for Düsseldorf wholesale prices ex-storage facilities for

[1] W. Cipa, *Realizations Obtained by German Refineries and Consumer Prices for Petroleum Products in Germany compared with the Situation in other Member-Countries of the European Common Market and Great Britain* (Essen: Gelsenberg AG, May 1967), p. 8 and fig. 3.

small quantities (100 liters). Then we have two internal series from *EOT*. Import quotations are c.i.f. river barge, excluding taxes, in dollars per ton, and f.o.b. tankwagon ex-refinery or barge including taxes in DM per 100 liters. Duisburg prices have been used for both of these.

The Federal figures and the Rotterdam figures look very similar up to 1964, although the internal rise in the winter of 1963 was a little larger. From 1965 the four series are close in outline, with peaks in February 1966 and January 1967 and low prices in June 1966 and April 1967. But for the Rotterdam and Duisburg wholesale series the 1966 low was close to the 1965 one, while the ex-refinery price was perceptibly lower and the Federal Index considerably lower. Have there always been differences between internal and import prices or were there special or temporary conditions in 1965 and 1966, not present in other years? The latter is plausible. From spring 1966, German internal prices were lower than the price of imports at Rotterdam plus transport and landing costs (*PIW* April 1966). This hurt importers more than the direct effect of not being able to bring in products at that time because their future quotas depended on the amount imported during the year. In August importers were claiming that "refiners are using surplus products to depress the market to levels where importers cannot operate" (*PIW* Aug. 15, 1966). On the other hand, less product was available from Italy and the importers' problem may have been caused more by a shortage of cheap product on the international market than the very low German prices.

In 1964 and 1965 the *AFM* annual publication, *Der Mineralöl-Aussenhandel den Bundesrepublik Deutschland*, gave a comparison of its prices with import prices taken from customs figures. I continued this comparison through 1966. The correspondence is surprising after allowance for the fact that many imports would have been contracted for in previous periods. Usually foreign trade figures include many intracompany transactions entered at artificially high prices, and the closeness here is unusual. In any case, the trend of imported prices is more definitely upward than that of the *AFM* Rotterdam prices—which demonstrates the importers' difficulties.

Company reports were examined to see if we could find out more about price movements but they were not very helpful since the short summaries of price developments of such a complicated period were not precise enough for our purposes. We could not even say that the reports were or were not consistent as to trend.

To summarize the apparent reasons for the differences between internal and import prices for the three products: Gasoline—a price war at the retail level at a time of growing naphtha demand; gasoil—lower refinery prices (perhaps deliberately depressed to reduce import quotas) at a time when there was little pressure of cheap product from abroad; oil—a successful reduction in imports and output which cut Germany off from the rest of the European market. Possibly also Rotterdam cargo prices were depressed by a surplus of high sulfur product.

Naphtha

I was interested in a normal differential between naphtha and gasoline prices (see Chapter VI, p. 168). Unfortunately, there was too little information in the only unambiguous trade press references to permit a formal calculation. It would seem from the European data that from 1963 to May 1967 the gap, starting at ½ to ¾ cent per gallon ($1.50-$2.75 a ton), tended to narrow.

There was a discussion of the difference in *PIW* (Sept. 25, 1967, p. 6), where regular gasoline was quoted at $36 a ton, and it was said that this gave an effective floor price for naphtha of $30 "since most refiners would rather convert virgin naphtha to gasoline than sell it below $30." This difference of about 1.7 cents a gallon is much higher than any I have observed. There seems to be no obvious way to reconcile this statement with my observations, and prices in September 1967 were very disturbed (see Table VI-B-1).

From the beginning of 1967, Caribbean and Persian Gulf naphtha prices have been quoted frequently in *POPS*, and seem to be compatible with a difference of half a cent per gallon.

Rotterdam Realizations

A composite price (see Table VI-B-1) was found by taking the product prices and weighting them by the average proportion of output (by volume), for Western Europe, of the principal products. Through the second quarter of 1965, output was published quarterly in *International Petroleum Quarterly* (*IPQ*). For calculating realizations after that I used the successor publication, *International Petroleum Annual*. The value of "other products" was taken as a weighted average of gasoline and light fuel prices. Losses, waste, and refinery fuel were entered at zero value.

These proportions were compared with those published quarterly from 1961 by the Organisation for Economic Co-operation and Development.[2] OECD output figures are given by weight, not by volume, but the correspondence is generally close. The exception is in 1963 when there was a discontinuity in the *IPQ* data. Apparently because of a change of definition in German products, the heavy fuel oil category was diminished and the middle distillates increased. The later definition has been assumed to be correct and so the earlier proportions were adjusted. The adjusted figures are shown in Table VI-B-2.

Over the period 1961 to 1967 the proportions of heavy fuel oil and middle distillates did not change. The output of gasoline, however, went down while "other" went up. This represents a growing output of naphtha included with "other." In 1967 naphtha output was listed separately for the first time in the OECD statistics, at 4.5 percent; "other" excluding naphtha was 6.7 percent. The combined total of 11.2 percent compares with 10.5 percent for the previous year. This would make less than one cent difference for May 1967; the effect would be smaller for earlier years.

By taking only three products one is considerably oversimplifying, but it is hard to know if there would be a bias up or down. The use of regular gasoline instead of high octane must make the estimates too low; in Germany, for example, 40 percent of gasoline sold at filling stations is premium grade, and it is as much as 70 percent in other countries.[3] Since premium gasoline is about \$7.00 more per ton than regular, the total would be about 6 cents higher in Germany in April 1967. (It is not possible to make similar calculations for earlier years since the proportion of premium sold is not known.) However, premium gasoline is also more expensive to produce. Hence the net realization over crude could be higher or lower and change over time—there is no way to tell.

Note that only EEC prices are being used. We are considering a European refiner who buys crude free of duty (the import duty only applies to products) and sells products mostly within the EEC community.

[2] OECD, *Provisional Oil Statistics, by Quarters* (Paris, quarterly).
[3] Cipa, *Realizations Obtained by German Refineries . . .*, p. 8.

TABLE VI-B-1. Rotterdam Product Prices, Monthly, July 1960–June 1971

Year	Month	Gasoline	Gas oil	Heavy fuel oil	Realization
		(...... cents per gallon		......)	($ per bbl)
1960	J	7.2	7.6	4.5	2.50
	A	7.0	7.4	4.5	2.45
	S	7.0	7.5	4.5	2.46
	O	7.1	7.6	4.6	2.50
	N	7.0	7.7	4.6	2.49
	D	6.7	7.4	4.6	2.44
1961	J	6.9	7.8	4.6	2.44
	F	7.0	7.8	4.5	2.44
	M	7.2	7.9	4.6	2.49
	A	7.0	7.5	4.7	2.43
	M	6.8	7.3	4.6	2.36
	J	6.5	7.3	4.5	2.32
	J	6.7	7.5	4.3	2.43
	A	6.6	7.6	4.1	2.41
	S	6.6	7.7	4.2	2.44
	O	6.8	8.2	4.3	2.54
	N	6.8	8.4	4.3	2.58
	D	6.8	8.4	4.3	2.58
1962	J	6.8	8.6	4.3	2.53
	F	6.8	8.5	4.3	2.51
	M	6.8	8.3	4.3	2.48
	A	6.7	8.0	4.3	2.51
	M	6.4	7.9	4.4	2.48
	J	6.5	7.9	4.5	2.50
	J	6.3	8.2	4.5	2.54
	A	6.2	8.5	4.5	2.57
	S	6.2	8.7	4.5	2.60
	O	6.2	8.7	4.5	2.58
	N	6.2	8.7	4.5	2.58
	D	6.2	8.7	4.5	2.58
1963	J	5.9	9.0	4.6	2.62
	F	5.8	9.2	4.7	2.65
	M	5.7	9.2	4.5	2.62
	A	5.7	9.0	4.4	2.56
	M	5.4	8.8	4.2	2.48
	J	5.3	8.5	4.1	2.41
	J	5.3	8.7	4.0	2.45
	A	5.2	8.7	4.0	2.44
	S	5.1	8.8	4.1	2.46
	O	5.0	8.8	4.1	2.42
	N	4.9	8.3	4.2	2.35
	D	5.1	8.4	4.3	2.40
1964	J	5.4	8.4	4.4	2.46
	F	5.4	8.3	4.4	2.41
	M	5.4	7.9	4.3	2.29
	A	5.4	7.1	4.2	2.23
	M	5.4	6.8	3.9	2.14
	J	5.3	6.4	3.9	2.06
	J	5.1	6.3	3.9	2.04
	A	5.0	5.7	4.0	1.95
	S	5.0	5.7	4.1	1.96
	O	5.0	5.8	4.1	1.97
	N	5.0	6.0	4.2	2.01
	D	5.1	6.1	4.3	2.05
1965	J	5.5	6.2	4.5	2.15
	F	5.5	5.5	4.4	2.01
	M	5.5	5.4	4.4	2.00
	A	5.5	5.5	4.3	1.95
	M	5.2	5.7	4.3	1.98
	J	5.2	5.6	4.2	1.94
	J	5.2	5.7	4.2	2.00
	A	5.2	5.7	4.3	2.01
	S	5.2	5.7	4.2	2.00
	O	5.3	5.8	4.2	2.01
	N	5.3	5.9	4.4	2.06
	D	5.5	6.1	4.4	2.11

TABLE VI-B-1. Continued

		Gasoline	Gas oil	Heavy fuel oil	Realization
		(...... *cents per gallon*		......)	(*$ per bbl*)
1966	J.....	5.6	6.7	4.6	2.23
	F.....	5.6	7.8	4.4	2.40
	M.....	5.6	6.5	4.1	2.14
	A.....	5.7	5.9	4.1	2.02
	M.....	6.0	5.6	4.1	2.01
	J.....	6.0	5.6	4.1	2.02
	J.....	5.9	5.5	4.1	2.03
	A.....	5.8	5.9	4.0	2.07
	S.....	6.1	6.1	3.9	2.11
	O.....	6.2	6.2	3.9	2.15
	N.....	6.2	6.2	3.9	2.13
	D.....	6.2	6.5	3.9	2.16
1967	J.....	6.2	6.6	4.1	2.20
	F.....	6.2	6.0	4.1	2.11
	M.....	6.1	5.4	4.0	1.98
	A.....	6.0	5.4	4.0	1.94
	M.....	6.2	5.6	4.0	2.00
	J.....	...	...	...	...
	J.....	12.6	9.7	5.8	3.59
	A.....	13.0	9.2	5.2	3.47
	S.....	10.6	8.3	4.9	3.05
	O.....	10.7	9.0	4.8	3.11
	N.....	10.0	9.2	4.8	3.08
	D.....	9.2	9.6	4.6	3.05
1968	J.....	8.5	9.7	4.7	3.01
	F.....	8.1	8.8	4.6	2.80
	M.....	7.7	8.2	4.5	2.66
	A.....	7.1	7.5	4.6	2.50
	M.....	7.1	6.9	4.5	2.36
	J.....	7.1	7.2	4.3	2.36
	J.....	7.0	7.2	4.2	2.40
	A.....	6.9	7.8	4.1	2.48
	S.....	6.9	8.1	3.8	2.49
	O.....	6.8	8.4	3.7	2.50
	N.....	6.6	8.2	3.6	2.44
	D.....	6.2	8.2	3.6	2.41
1969	J.....	6.1	7.8	3.6	2.33
	F.....	5.7	6.9	3.7	2.16
	M.....	5.5	6.5	3.7	2.07
	A.....	5.5	6.4	3.7	2.01
	M.....	5.7	6.1	3.8	2.06
	J.....	5.5	6.1	3.6	1.93
	J.....	5.5	6.1	3.7	2.02
	A.....	5.4	6.1	3.5	1.99
	S.....	5.3	6.2	3.5	2.00
	O.....	5.4	6.1	3.4	1.96
	N.....	5.2	6.1	3.4	1.96
	D.....	5.2	6.3	3.9	2.05
1970	J.....	5.2	6.8	4.3	2.21
	F.....	5.2	6.5	4.5	2.21
	M.....	5.0	6.3	5.4	2.26
	A.....	4.7	6.6	5.4	2.24
	M.....	4.7	6.7	5.5	2.27
	J.....	5.0	6.6	6.0	2.38
	J.....	6.4	7.6	7.7	2.90
	A.....	7.4	9.2	9.2	3.46
	S.....	7.1	9.7	8.1	3.36
	O.....	6.8	10.2	7.4	3.32
	N.....	6.5	10.6	7.7	3.41
	D.....	6.4	10.6	8.4	3.49
1971	J.....	6.5	10.8	8.6	3.59
	F.....	6.9	11.1	8.6	3.68
	M.....	7.4	10.9	8.1	3.61
	A.....	7.4	10.1	8.0	3.39
	M.....	7.5	9.9	7.5	3.31
	J.....	7.6	9.1	6.0	2.97

Note: Monthly changes may be misleading due to rounding and seasonal changes in output.

TABLE VI-B-2. Weights Used in Calculating Realizations, 1960-1970

(*percent*)

	Gasoline	Gas oil	Heavy fuel oil	Other	Losses, waste refinery fuel
1960 3rd Q	22.1	29.4	30.8	10.3	8.1
4th Q	20.2	29.2	34.0	10.2	6.8
1961 1st Q	18.8	28.6	35.3	9.1	8.4
2nd Q	18.8	28.6	35.3	9.1	8.4
3rd Q	20.5	29.7	32.0	11.7	6.7
4th Q	19.7	30.4	33.8	10.7	5.9
1962 1st Q	18.5	29.2	34.7	9.7	8.2
2nd Q	19.9	30.2	33.1	11.3	6.1
3rd Q	19.9	30.3	32.3	11.8	6.3
4th Q	18.5	31.0	33.3	10.8	7.0
1963 1st Q	16.4	31.3	35.8	10.4	6.5
2nd Q	18.8	29.9	32.7	12.1	7.0
3rd Q	18.9	31.4	31.4	11.9	7.2
4th Q	17.8	30.3	33.7	11.7	7.0
1964 1st Q	16.3	31.0	35.4	11.0	6.3
2nd Q	17.5	29.7	33.1	13.4	6.4
3rd Q	18.6	30.3	32.0	13.1	6.0
4th Q	16.7	30.9	33.5	12.7	6.1
1965	16.3	31.5	33.2	12.8	6.3
1966	15.1	30.2	35.1	13.2	6.4
1967	15.2	29.9	34.4	13.6	6.7
1968	14.5	31.3	34.0	13.4	6.6
1969	13.6	32.3	33.7	13.8	6.6
1970	(............ *assumed same as 1969: data not available* ....)				

Source: U.S. Bureau of Mines, *World Petroleum Statistics*, later *International Petroleum Quarterly*, later *International Petroleum Annual*. Kerosine and jet fuel included with gas oil.

Appendix VI-C

Refining Margin, Western Europe, 1960–1970

In this appendix estimates are made of the refining margins necessary to maintain and expand output in Western Europe. Subtracting the margin from refinery receipts per barrel of refined products yields a delivered crude price at which it is barely worthwhile for the independent refiner to stay in business and maintain capacity. This estimated delivered crude price is also the value of the crude to an integrated producer-refiner.

The refinery margin plus the price of tanker service (explained in Chapter IV) equals a total transport-refining margin. Subtraction of this margin from the European refinery receipts per barrel yields the f.o.b. value of the crude to the integrated producer-refiner, *or* an f.o.b. price of crude barely worth paying by an independent refiner. This refining-transport margin may also be added to an f.o.b. price (when independently known) to obtain a calculated price for a barrel of products in Western Europe.

The calculations rest entirely on published sources. Some have been checked with knowledgeable persons, but responsibility for mistakes is the writer's. His aim is to make every source and operation explicit, to enable those with better information to make better estimates. Table VI-C-1 shows the estimated capital costs for the average barrel of new refining capacity installed in Western Europe in 1946-69. Table VI-C-8 shows the total estimated cost per barrel in 1960-69. These are the numbers used in the text.

Tables VI-C-5 and VI-C-6 show what difference it makes in the capital cost (and hence the final result) to alter the most important data.

Investment Coefficients

To estimate capital costs we need capital expenditures. The only comprehensive series is that of the Chase Manhattan Bank (CMB), compiled from information supplied directly by companies accounting for a very large part of the total, with appropriate adjustment. There are no governmental statistics available.

Gross new capacity built during any year (G_n) is calculated by the following formula: $\frac{1}{2}(J_{n+2} - J_{n+1}) + \frac{1}{2}(J_{n+1} - J_n) + .04J_n = G_n = 0.50J_{n+2} - 0.46J_n$. Here J_n represents capacity on January 1 of the year n. Half the gross new capacity provided during any given year has been assigned to that year and half to the next, so that the year's expenditures are partly to provide capacity which will not be completed until the next year. It is assumed that at the end of the year work has been done on half of the new capacity to be provided during the subsequent year. The assumption is that a refinery takes two years to build, which was probably too little twenty years ago and perhaps too much today, but will have to serve as a general average since it would be hard to defend any more refined estimates. This is consistent with assuming (1) expenditures at a constant rate over a two-year period from the first January 1 to the second December 31, *or* (2) zero expenditures from July 1 to December 31, half from January 1 to the next December 31, then half from the second January 1 to June 30. The latter is more realistic.

The factor of .04 corresponds to an annual loss of capacity through depreciation and obsolescence. The longer the service life, and the lower the annual loss rate, the greater the estimated capital requirements per daily barrel, but also the lower the rate of anticipated return on capital. These two factors influence the estimated cost of refining in contrary directions. The net effect is perhaps not great, but the issue is worth a careful examination.

Petroleum Press Service gives an average replacement rate of 6 percent for European refineries, which would imply an average service life of 16.67 years (*PPS*, June 1966, p. 120). Nelson, the *Oil and Gas Journal*, cites the Internal Revenue Service of the U.S. Treasury Department, which allows 5 percent (a 20-year life) for tax calculations of equipment but not land. In "an effort to unscramble the complexities of depreciation," Nelson calculates "obsolescence plus half the cost of improvements necessitated by obsolescence," which in the 1950s ranged from 1.58 to 7.43 percent, with a median value of 4.43 percent. Later he notes amortization as a separate cost: it "may not be necessary in the U.S. where growth may permit continual borrowing, but in unstable countries it is necessary to repay investments."[1] In highly stable countries it would be from 15-25 years, and with costs at the U.S. level he puts it at 4 percent. Depreciation is an additional charge of 6.5 percent for the United States, but drops considerably outside this country, presumably in respect of less rapid technical progress.

Amortization of borrowed capital is allowed for below (pp. 371ff.). As concerns equity capital, the need to amortize (quickly or at all) depends on political risk, and should be discussed together with the discount rate or rate of return. Accordingly, amortization is disregarded as not relevant to average service life. Under European conditions, one would interpret Nelson's estimates as indicating depreciation in the neighborhood of 5 percent, or an average service life of 20 years. This is the figure used below in making present-value calculations. But because European refinery capacity has been expanding, and bids fair to do so into the foreseeable future, one cannot apply a 5 percent rate replacement rate as though it were a steady-state system with replacement just equal to scrapping, increments just balancing decrements. The percent of capacity disappearing in any year must be less; in using 4 percent one is perhaps doing no more than nod in the right direction. But there is some basis for doing so. The *PPS* service life is probably (the reference is ambiguous) for new equipment, and takes no account of the slower depreciation of the "bricks and mortar" elements, nor of land. The same is apparently also true of Nelson. Were one to take account of these factors, these estimates would both move up appreciably; what will be used, therefore, is 25 years service life or 4 percent replacement per year.

I do not use a lower factor than 4 percent to reckon the losses of refinery capacity in the early years after World War II. The capacity existing at the beginning of 1946 was probably falling apart at an even faster rate than 4 percent. Moreover, much of the capacity installed in those years was on the mistaken assumption that the European consumption pattern would be much like the American, emphasizing the light products and minimizing the black oils. Hence more of the capacity was obsolete sooner than could reasonably have been foreseen.

The estimates of gross new capacity in column (3) of Table VI-C-1 agree very much better than do the net additions in column (2) with the independent figure for expenditures, column (4). Net additions dropped almost to zero between 1954 and 1956, but refining capital expenditures *increased* by about half. Thus my estimate that gross new capacity built increased by 18 percent may overstate the case–there is no way of knowing exactly–but it certainly is a far better estimate than what could be obtained by assuming that new capacity installed was equal to net growth plus only a very small allowance for replacement.

[1]W. L. Nelson, *Guide to Refinery Operating Costs* (*OGJ*, 1966), pp. 128, 191-92.

Obviously, there is much meaningless fluctuation in all of these columns, especially column (5), because one cannot have a precise correspondence between work put in place and money expended. No two companies probably treat these in quite the same way, and some capital expenditures are done by resident labor and called current expense. But the figure for any particular year is perhaps not important in itself. The broad trend is probably faithful to the actual record of refinery investment. From the very low expenditures just after World War II, the refining investment per daily barrel climbed steeply to a peak around 1953, whence it declined very sharply toward the end of the 50s and then more slowly. This accords with what is generally known about the over-designed gasoline-oriented refineries that were built in Europe before the shift to simpler refineries. Reinforcing the drop in costs has been a constant improvement in technique, so that refinery investment in 1969 was probably somewhere around \$500 per initial daily barrel. But doubtless the worldwide construction cost inflation will soon be registered here.

For the years 1960-65, Nelson's cost of the "light oil refinery" in the United States compares with my estimates as shown in Table VI-C-2 (earlier years not comparable).

For 1958, Nelson estimated European investment requirements as .76 of American. We should expect a lower factor in later years because of the growing complexity difference; as Nelson points out, "the eagerness of foreign operators (and of U.S. construction companies) to install the most modern equipment" made them over-invest: to get higher octane gasoline "foreign operators can install catalytic reforming without sacrificing yields of fuel oil"—desires in Europe, but not in the U.S.[2] Hence, the disappearance of catalytic cracking in Europe should have acted to lower investment requirements there, relative to U.S. requirements.

Some checks on refinery investment factors. The figures in column (5) of Table VI-C-3 are well below the announced cost of most but not all new refineries. This is as it should be. A new grass-roots refinery is almost invariably built with a view to later expansion. In other words, column (5) represents the mixture in any year of new grass-roots capacity *plus* new capacity added to existing refineries. Since there is no set relationship between these two, that is another reason why we should expect it to fluctuate a good deal, and why we cannot put too much emphasis on costs in any given year.

Table VI-C-3 shows that the average barrel of new 1964 European capacity was in a refinery of over 100 thousand barrels daily (TBD). Since two-fifths of it consisted of additions to existing refineries, cost must have been well below the average for a new refinery of that size.

We might consider \$400 per daily barrel (BD) as being the low end of the range for *new* grass-roots capacity built around 1965. For example, a big Sardinia refinery of 135 TBD cost about \$356/BD. Despite a widespread impression to the contrary, this is not a simple topping plant. Products include both regular and premium gasoline and it has two catalytic reformers. It must have unusually large tankage requirements because located on tidewater with a negligible overland consumption, so that virtually all output must go out by ship (*OGJ*, May 1965, p. 37). With this may be contrasted an enlargement of the Esso Rotterdam refinery from 160 to 320 TBD, costing \$350 per additional daily barrel, or the Shell refinery at Rotterdam, from 340 to 500 TBD, costing \$212 per additional daily barrel. On completion, the cost was stated to be \$28 million for a 150,000 daily barrel increase, or \$187 per daily barrel. Although it was a simple processing plant, location "outside the existing refinery complex . . . added appreciably to the investment cost" and the plant would be unusually flexible in the use of crudes and output or products (*OGJ*, Aug. 7, 1967, p. 102; Dec. 8, 1969, p. 83). An enlargement of the Esso refinery at Fawley cost \$196 per daily barrel, while a new Swedish refinery of 100 TBD initially cost \$560 (*PPS*, June 1966, p. 222). But a new BP refinery at Rotterdam of an initial 100

[2] *Ibid.*, p. 109; see also p. 171.

TBD will cost \$577/BD, while an addition of 50 TBD in Scotland will cost \$392/BD (*WSJ*, July 28, 1967, p. 7; *PONS*, Jan. 31, 1967).

The Phillips-ICI refinery of 80 TBD cost \$30 million or \$375/BD. The principal product will be one million tons per year of naphtha. Tankage will be 240,000 cubic meters. There are some unusual economies. Substantial storage is available in worked-out salt wells. The refinery will burn its own waste gas and some of its own fuel oil. A jetty will handle 80,000-ton ships unloading 5,000 tons per hour (*OGI*, Mar. 1967, p. 8). In a proposed U.S. 60-TBD plant the crude unit alone will cost \$27 million or \$450 per daily barrel (*WSJ*, Apr. 17, 1967, p. 32).

A survey of British refiners in 1966 showed they expected an increase from 1,440 to 1,800 TBD or more by 1970, costing \$210 million; thereby implying \$525/BD or less if one neglects depreciation. Assuming it at 4 percent, the cost is \$368/BD (*OGI*, July 1966, p. 95).

A large refinery completed in the Virgin Islands of the United States (*OGJ*, Jan. 9, 1967, pp. 47-48) gives the total refinery cost as \$32.5 million, but another account states that the first expenditure total covers only an initial 65 TBD, which would point to \$500/BD, and that the remaining 35 TBD "and probably more" would cost \$15 million (*Chemical and Engineering News*, May 15, 1967, p. 30). The refining would cost \$48 million, or probably less than \$480/BD. There are some unusual expenses. A 40-foot channel adjoining the refinery cost \$4.5 million to dredge. Moreover, the refinery was built with a speed that surprised the trade, only about 230 days. Possibly this reduced cost; but often haste makes waste, and hence the refinery might have been more cheaply if more slowly built.

Similar to the Virgin Island facility are a 200-TBD ENI unit in the Bahamas, costing \$280/BD; and one of the same size for Savannah, Ga., at \$300 (*Economic News from Italy*, Sept. 6, 1968; *WSJ*, Nov. 14, 1968, p. 4). An announced 300-TBD plant at Machiasport, Maine, is to cost \$483/BD, which is so far out of line as to indicate some unusual features (*WSJ*, Nov. 4, 1968; *OGJ*, July 28, 1968, p. 41; *PIW*, July 8, 1968).

Table VI-C-4 presents the writer's smoothed approximation to an average relation. The range is about 25 percent above and below, depending on jetties and other marine installations, dredging, tankage, piling, filling land, reforming, desulfurizing–to name the identifiable factors. Land is included.

An interesting sidelight is given by three U.S. companies planning "essentially new refineries on old sites." Nearly \$300 million is to be spent for about one million BD; hence with site preparation and utilities already provided, \$300/BD is adequate for replacement capacity in the United States, including extensive hydrocracking "despite its relatively high operating and capital costs" to get more gasoline and jet fuel out (*OGJ*, Feb. 13, 1967, p. 97). This suggests that large-scale replacement capacity in Europe is available for considerably less than \$300/BD.

Although one observer thinks these refinery investment factors somewhat on the low side,[3] the estimates of Table VI-C-4 appear to check reasonably well with independent observations of the investment needed to establish a barrel of daily capacity.

Capital Costs per Barrel

Having obtained the refinery investment figure, one can now calculate the capital cost. Let us assume that half of the refinery investment is represented by equity, the remainder by debt; but it is easy to show the effects of leverage, as is done in Table VI-C-5, p. 379. It is assumed that income tax represents 50 percent of net profit after taxes, and that depreciation for tax purposes is 6 percent straight line. This is probably too conservative. Depreciation allowance varies among countries and is generally above 6 percent; usually

[3] Jean Masseron, *L'Economie des Hydrocarbures* (Paris: Editions Technip 1969), p. 162.

more advantageous methods than straight line are permissible.[4] Hence, this allowance seems to be a minimum, but it would take us too far afield to adjust.

The necessary rate of return on refinery investment is not easily found. In a recent paper,[5] Hubbard has applied a figure of annual capital charges, which would include interest and depreciation, of 10 percent for pipelines and tankers. Given a 20-year service life, 10 percent annual capital charges would imply a discount rate of about 7.75 percent. In my opinion, the risk in refineries is somewhat greater than in pipelines and tankers, even abstracting from the risk of improvident buying of crude at prices above what was available currently, in the belief that these were temporarily and abnormally low. But the changeover in the European refining pattern after the early 1950s indicates higher risk, even though it might have been foreseen. The pattern may be changed by the discoveries of natural gas in the North Sea area (including here both offshore and onshore on both sides). It might be appropriate to use 12 percent net discount rate, which seems to be the approximate price-earnings ratio of U.S. refiners in recent years. The Common Market *Perspectives* report uses an 8 percent return.[6] But it is not clear whether this is an annual capital charge, lower than Mr. Hubbard's, or a discount rate.

Let us assume that the refinery investment is \$100 per daily barrel of which \$50 represents the equity portion. Then: \$50 is the present value of the future annual equity return (*a.e.r.*). It is therefore equal to the annual undiscounted return multiplied by the factor $\int_0^{20} e^{-(20\times 12)}\,dt$. The value under continuous compounding of this summation term is 7.91. If the plant operated at full capacity as soon as it was started up, the annual equity return would be \$50/7.91. However, trial runs and starting difficulties would affect the first year's earnings.[7] Let us estimate that in the first year there was only half a year's return. The summation term is reduced by 0.5 to 7.41 so the annual equity return needed is \$50/7.41 or \$6.75.

The annual capital cost (*a.c.c.*) can be found from:

$$\begin{aligned}
a.e.r. &= a.c.c. - \text{interest} - \text{debt repayment} - \text{income tax.}\\
&= a.c.c. - \text{interest} - \text{debt repayment} - \tfrac{1}{2}(a.c.c. - \text{depreciation}\\
&\qquad - \text{interest})\\
&= \tfrac{1}{2}\,a.c.c. + \tfrac{1}{2}\,\text{depreciation} - \tfrac{1}{2}\,\text{interest} - \text{debt repayment}\\
a.c.c. &= 2\,a.e.r. - \text{depreciation} + \text{interest} + 2\,(\text{debt repayment})\\
&= 2(6.75) - 6.00 + 3.00 + 2(0.630)\\
&= 11.76.
\end{aligned}$$

[4] *Chemical Week*, Apr. 29, 1967, shows the U.S. and various European countries all permit more advantageous methods than straight line. See also Business International Corp., *Investing and Trading Conditions Abroad*, 1966, p. 189; buildings 2–4 percent, equipment 10–20 percent.

[5] M. E. Hubbard, "The Comparative Costs of Oil Transport to and within Europe," *Journal of the Institute of Petroleum*, vol. 53, Jan. 1967.

[6] ECSC, *Etude sur les Perspectives Energétiques à Long Terme de la Communauté Européenne* (1964), p. 619. The 1966 review mentions "interest on capital employed," but gives no rate.

[7] It is assumed that the Chase Manhattan figures include interest during construction.

Thus with the discount rate at 12 percent, a 20-year service life, 50 percent of capital requirements borrowed at 6 percent interest, we require $11.76 annually in order to make the investment barely worthwhile after paying taxes, interest, and the accumulation of a fund to pay off the debt. Thus the annual capital charges come to $11.76. Table VI-C-5 shows the effect of changes in the basic assumption.

A 50 percent equity ratio has been assumed in order to approximate the position of a large integrated firm in Europe. In Japan the equity ratio is lower. Table VI-C-6 shows what difference it makes in my assumptions about the required refinery margin.

Given conventional depreciation accounting, and an incentive depreciation rate moderately higher than the true rate implied by the service life, an after-tax figure greatly understates the true return to equity funds; indeed, given even 60 percent debt, the enterprise appears to be losing money. An outside observer might well conclude that the firm or industry was not worth investing in, and therefore would not expect any new investment to be taking place. Hence the anomaly of an industry "losing money" yet raising additional funds not only to replace but even expand its apparently worthless capacity. Of course, there is nothing at all improper or unwise about depreciation accounting and some modest incentives for investment. The mistake comes in taking these after-depreciation and after-tax figures to be rates of discount on future expenditures and revenues, and drawing conclusions accordingly. It is astonishing to see how seriously some take (1) low profit figures in (2) an industry which is busily expanding capacity. The second fact will, with few and trifling exceptions, destroy any implication drawn from the first.

It is a basic canon for the economic analysis of any industry, including oil: published financial statements are no indication of the rates of return, which guide management investment decisions.

The explicit allowance for income tax differs from the production cost calculation in Chapter II, where taxes were ignored. The reason is that oil production generally takes place in underdeveloped nations, where it is the major industry and separately taxed. The purpose in Chapter II was to measure only costs physically necessary, on the assumption that the government would try to tax away what it could. Hence, tax had to be treated as a resultant. But in the countries of Western Europe and in similar countries refining is only one industry among many, and the income tax is a datum to management, not something to be negotiated. Therefore it must be reckoned as part of supply price.

A complicating factor is that most integrated refining establishments cost in their crude oil at prices much higher than those paid in arm's-length transactions. Hence they show accounting profits much below what an independent firm would pay, and often they run book losses and pay no income taxes. This would result in a lower capital cost. But since an unintegrated refiner would pay no tax if his equity were less than 50 percent, the disadvantage is not necessarily significant. It would be worth a research effort.

Let us resume the cost calculation. The annual capital cost is $11.76; divided by 365 days it is 3.22 cents per barrel of throughput for every $100 of investment.

However, capacity per calendar day, which allows for normal downtime, is of course appreciably lower than per-stream-day. Table VI-C-7 shows throughput as a percentage of capacity over recent years. The range of utilization, apart from the 77 percent in 1953, is 82-93 percent; the average is 85.8 percent. This comes close to a figure "almost 86 percent of nominal capacity" which is said to have held for some years in Germany (*PIW*, Jan. 22, 1968, p. 1). Since the period saw steady and substantial growth, it is obvious that capacity at any given moment was under substantial pressure to expand. Therefore the shortfall was not due to deficient demand, but was a technical and economic necessity. Failure to understand this has led to much confusion. There has been no excess capacity available at less than average cost; in Europe during this period the incremental barrel has been more expensive than the average barrel.

Since 3.22 cents represents the use of 100 percent of stream capacity, one should make a substantial correction to the per-barrel capital cost. If one divides 3.22 cents by .861, this gives 3.74 cents per barrel capital cost for every $100 investment per daily barrel. This is the value used in Table VI-C-1.

Operating Costs per Barrel

This leaves us with operating costs other than capital costs and there are almost no published data on them. A particularly complex item is the cost of fuel.[8] Hubbard[9] reckons it as 4.5 percent of the crude charge, in the form of heavy fuel oil. He does not explain it, but apparently he reckons the opportunity cost, i.e. the sales value sacrificed by burning heavy fuel oil, the least valuable product, under the refinery boiler. This 4.5 percent is in addition to between 3½ to 9 percent product loss, i.e. the difference between the crude charged into the refinery and the products issuing from it.

Were one to use this procedure, however, there would be double counting. This is because the refinery value used here assumes that only about 93 percent of the barrel of crude issues as products to be sold, the other 7 percent being included in "refinery fuel and waste." The U.S. Bureau of Mines and OECD are consistent in this latter figure. Furthermore, OECD leaves no doubt possible, since they show the total of salable products, add imports, and arrive at total supply of refined products, whose subsequent disposition they estimate.

The matter would seem obvious, were it not for the U.S. refining literature, which assumes that about 10-12 percent of the crude charged is needed as fuel. But this would seem to be a reflection of the more severe processing in the United States; simple distillation apparently absorbs only 1.5 percent of the crude. The British (IP) reference estimates fuel consumption ranging from 4 to 10 percent.

In the United States, where heavy (residual) fuel oil production is minimized, refineries usually find it advantageous to use some other fuel. Before the 1960s, the price of natural gas was usually lower than that of fuel oil, and in many places it still is. Occasionally, coal was available more cheaply than fuel oil, especially in the Northeast. But for Europe, where there has been little natural gas, and coal is expensive, the safest assumption would seem to be that refineries buy no fuel, but burn primarily refinery gas, supplementing it with residual fuel oil. Accordingly, no explicit fuel cost has been reckoned.

Hubbard estimates operating costs apart from fuel costs for simple refineries outside Europe. These direct operating costs do not vary significantly with the size of refinery above 4 million tons a year (80 TBD). His examples gives $1.10 per ton for a 7 million-ton refinery and $1.15 per ton for one of 5.3 million tons a year. He added 20 percent to European costs to allow for the higher costs in developing countries. Readjusting for this and converting the second example to barrels per day, we have a cost of 12.9 cents a barrel for a 106-TBD refinery.

This figure may be too high because no allowance is made for technical improvements since 1961, when Hubbard's estimates were made. On the other hand, it may be too low as an average of Western Europe cost. If the average refinery had a capacity of 100 TBD, average costs would be higher than for a single 100-TBD refinery because smaller units would have higher operating costs while larger ones would not have appreciably lower costs.

[8]The following discussion is based on Hubbard, "The Comparative Costs of Oil . . ."; W. L. Nelson, *Petroleum Refinery Engineering* (McGraw-Hill, 1958), pp. 874-75; W. L. Nelson, *Guide to Refinery Operating Costs* (*OGJ*, 1966), nos. 4, 75, 75a, 116; *Modern Petroleum Technology* (London: Institute of Petroleum, 1962), p. 251; OECD, *Oil Statistics* (Paris, quarterly), Table A; U.S. Bureau of Mines, *International Petroleum Statistics* (annual, previously quarterly), table numbers vary.

[9]M. E. Hubbard, "The Economics of Oil Transport and Refinery Operations," in *Techniques of Petroleum Development* (UN, 1962), pp. 219, 222, 224.

Table VI-C-8 shows total costs since 1960. No estimates are made for earlier years, since it is not clear that Hubbard's figures would apply.

Comparisons with Other Estimates

It remains to compare the estimates calculated here with others available, and this unfortunately is not easily done. In the Virgin Islands refinery mentioned earlier, total cost is stated to be less than 50 cents and cash cost less than 20 cents. In fact, 75 percent of the Hess plant income is free of income tax for 16 years (*Chemical and Engineering News*, May 15, 1967, p. 28).

A 1969 estimate of 35 cents (or less) refining cost for simple refineries in Europe is compatible with ours, although it was made three years later.[10] Similarly, a South German refiner, running Libyan crude, was said to incur processing costs of 20 to 30 cents (*PIW*, Nov. 3, 1969). There is no provision for return on investment but, as seen earlier, "depreciation" is that, in part. The upper end of the range, or 30 cents per barrel, might be a low estimate of total cost. But the South German refiner is reported to turn out 9 percent premium gasoline and 11 percent regular, which indicates that he has additional equipment. Moreover, the modal size of South German refineries is around 80 TBD. Our hypothetical Rotterdam refinery of over 100 BD, making a simpler product package, would have lower costs. Hence 30 cents would seem to be a fair guess at Rotterdam costs, as of the fall of 1969. This should be contrasted with the fee of 50 cents a barrel for processing for others mentioned in the same source. In a calculation made in 1965 (*PIW*, Apr. 26, 1965) the refining margin excluding "provision for profit" was $2.20 a ton or close to 30 cents a barrel. As in the example just quoted one can guess that the 30 cents includes a provision for depreciation and the total cost would therefore be 35-40 cents.

Another comparison is with the estimates of the European Coal and Steel Community, shown in Table VI-C-9, from the supplement to their revised study of long-term energy perspectives.[11] The lighter and less sulfurous crudes cost substantially more to process because their higher naphtha yields are reformed into gasoline, thus raising both the costs of throughput and the value of the products.[12] Aside from this, their estimates look to be about 10 cents higher than mine for an 80-100-TBD plant, despite my using a higher reference rate of profit or discount. However, the Community report reckons a decrease of about 15 percent in refining costs between 1958 and 1965,[13] compared with the essentially stationary costs of Table VI-C-8 for 1960-65. Hence the estimates are less far apart for the late 1960s, and may be close.

Another rough check is with the Independent Petroleum Association of America (IPAA) margin between crude prices and refiners' realizations for principal products. It is inadequate to show true U.S. margins, because refineries here turn out so many additional products, and crack so severely. The IPAA margin is really an indicator of what the real U.S. margin would be if in fact U.S. refineries turned out not much else. One would expect the IPAA margin to be considerably higher than the European because of the more complex and costly facilities needed here to pound a much larger percentage of crude oil (50 percent as compared with under 20) into gasoline. In fact, during 1958-66 the IPAA refining margin varied between 62 and 77 cents per barrel.[14] For those

[10] A. W. Pearce, "Problems of Europe as a Crude Oil Market," *PR*, vol. 23 (1969), p. 162.

[11] ECSC, *Nouvelles Réflexions sur les Perspectives Energétiques* (1966), Annexe 2.

[12] I should record here a dissent from an oilman who maintained that Libyan crudes were a little cheaper to process because of less sulfur and other contamination.

[13] Compare ECSC, *Nouvelles Reflexions . . .* with the 1964 work, fn. 6.

[14] IPAA, Statistical Release: *East of California Prices of Crude Petroleum and Principal Products* (monthly).

refineries with a more complex product structure, margins are higher and so are prices, but the subtraction to get an estimated crude price leads to approximately the same result. The *OGJ* Gulf Coast refinery realization of "realistic product prices" for the four principal products averaged about $3.65 in 1964-66. The Gulf Coast prices for non-specialty crudes ranged from $3.05 to $3.18, so a middle figure of $3.12 may be acceptable, which would indicate a margin of about 53 cents per barrel.

W. Cipa of Gelsenberg estimated processing costs as 43 cents per barrel, uniformly in France, Britain, Italy and West Germany, for a "medium size" (80-TBD?) topping-reforming plant running Libya crude.[15]

Neither our estimates, nor those of the EEC, Mr. Cipa, IPAA, or the *OGJ*, can possibly be reconciled with the $1.61 per barrel of *Energie-Gutachten*.[16] Again it is proved that a tax definition of "cost," used for other than the tax calculations, will give nonsense results.

Simple Topping Plant

Finally, there is some interest in calculating the cost of operating a simple topping plant, for at some places it would be profitable to take off one or two light products for sale and dump all else in the residuum to be sold as fuel. This calculation is done in Table VI-C-10.

[15]W. Cipa, *Realizations Obtained by German Refineries, and Consumer Prices for Petroleum Products in Germany Compared with the Situations in Other Member Countries of the European Common Market and in Great Britain* (Essen: Gelsenberg A.G., May 1967).

[16]Brief title: *Energie-Gutachten* or *Friedensburg-Bade* report, p. 168. Full citation: Arbeitsgemeinschaft deutscher wirtschaftswissenschaftlicher Forschungsinstitut e.V., Bonn, *Untersuchung über die Entwicklung der gegenwärtigen und zukünftigen Struktur von Angebot und Nachfrage in der Energiewirtschaft der Bundesrepublik unter besonderer Berücksichtigung des Steinkohlenbergbaus*.

TABLE VI-C-1. Western Europe: New Refinery Capacity, Refinery Investment and Capital Costs 1946-1970

Year	Capacity Jan. 1 (000 BD)	Net increment (000 BD)	Gross new capacity built during year (000 BD)	Refining: Capital outlays ($ million)	Refining: Investment per barrel daily capacity ($)	Capital cost (¢/bbl)	Capital cost 3-yr mean (¢/bbl)
	(1)	(2)	(3)	(4)	(5)	(6)	(7)
1946	359	52	66	20	303	11.3	...
1947	411	51	235	45	191	7.1	12.9
1948	462	387	292	140	479	17.9	16.5
1949	849	161	343	200	583	21.8	19.2
1950	1,010	457	445	215	483	18.1	20.5
1951	1,467	353	377	225	597	22.3	23.0
1952	1,820	283	334	270	808	30.3	28.6
1953	2,103	239	237	230	970	36.3	34.7
1954	2,342	67	138	158	1,145	42.8	31.5
1955	2,409	20	287	170	592	22.1	27.0
1956	2,429	361	340	225	662	24.8	24.3
1957	2,790	124	533	360	675	25.2	27.0
1958	2,914	719	649	515	794	29.7	27.8
1959	3,633	344	567	425	750	28.0	25.6
1960	3,978	498	629	325	517	19.3	23.5
1961	4,476	442	516	325	630	23.6	21.3
1962	4,918	231	870	500	575	21.5	20.9
1963	5,149	1,114	1,387	725	523	19.6	18.4
1964	6,264	1,247	1,411	580	411	15.4	18.0
1965	7,511	1,035	1,392	810	510	19.1	19.1
1966	8,586	1,109	1,593	950	596	22.3	20.2
1967	9,695	1,390	1,982	1,025	517	19.3	19.9
1968	11,085	1,799	1,871	900	481	18.0	19.4
1969	12,884	1,057	1,500	850	566	21.0	...
1970	13,941	...	...	...	...	...	...

Sources: Col. (1)–*OGJ* annual supplements–*World Wide Oil*, respective years and notes.

Col. (2)–represents the difference between succeeding years.

Col. (3)–$0.50J_{n+2} - 0.46J_n$ where J is January capacity in the year n.

Col. (4)–CMB, *Capital Investments of the World Petroleum Industry*, respective years, occasionally revised.

Col. (5)–col. (4) ÷ col. (3).

Col. (6)–col. (5) × .000374 (see text).

Col. (7)–three-year mean of col. (4) ÷ three year mean of col. (3) × .000374.

TABLE VI-C-2. United States and Europe: Refinery Investment Cost, 1960-1965

(*$000 per daily barrel*)

Year	U.S.	Europe	Percent Europe is of U.S.
1960	888	517	58
1961	875	630	72
1962	875	575	66
1963	850	523	62
1964	841	417	50
1965	834	543	65

Source: W. L. Nelson, *Guide to Refinery Operating Costs* (*OGJ*, 1966), p. 125.

TABLE VI–C–3. Western Europe: New Capacity, 1964

(thousand barrels daily)

Region	New refineries: (size)	Existing Refineries: Size at end of 1964	Existing Refineries: 1964 additions	Weighted average barrel of new capacity added is to refineries of (size):	Capacity added to existing refineries as proportion of total new capacity
	(1)	(2)	(3)	(4)	(5)
France	32	49	3		
	40	82	8		
		117	50		
		140	60		
		235	85		
	72		206	126	74%
Italy	5	5	2		
	5	10	3		
	10	15	1		
	33	25	5		
	110	60	11		
	165	60	13		
		60	15		
		105	30		
		113	(0.2)		
		155	15		
	328		95	115	22
West Germany	53	11	2		
	72	11	12		
		42	2		
		45	3		
		55	6		
		66	9		
		71	3		
		80	4		
		83	5		
		138	23		
	125		69	65	33
United Kingdom	7	3	(0.3)		
	20	100	20		
	28	130	10		
	100	130	25		
		187	27		
		240	15		
	155		97	105	38
Other Western Europe	40	7	2		
		35	5		
		45	25		
		45	3		
		58	10		
		75	26		
		85	5		
	40		76	51	66
Total Western Europe	720		543	105	42%

Source: *OGJ*, annual supplements—*World Wide Oil*, 1964, 1965.

Note: Col. (4) is found by weighting the 1964 increments by the end-of-year capacity of the refineries to which they are added, and dividing by the total new capacity, i.e.,

$$\frac{\Sigma\,[(2)\times(3)]\ \div\ \Sigma\,(1)^2}{\Sigma\,(1)\ \div\ \Sigma\,(3)}$$

TABLE VI-C-4. Approximate New Grass-Roots Refinery Capital Requirements

Crude capacity		Investment	Investment/BD
(*TBD*)	(*Million tons/year*)	(*million $*)	(*$*)
20	1	18	900
40	2	28	700
80	4	43	540
100	5	48	480
150	7.5	60	400

Note: Type–Topping-reforming; Place–Western Europe, Caribbean, Japan (land costs higher); Time–1963-66.

TABLE VI-C-5. Effect of Change in Assumptions of Discount Rate, Service Life, Depreciation, and Interest Rate on Annual Capital Cost

A. Twice Annual Equity Return

Discount rate (*percent*) ↓ \ Service life assumed (*years*) →	15	17	20	25	30
8	12.40	11.60	10.73	9.82	9.29
10	13.36	12.63	11.86	11.09	10.65
12	14.90	14.20	13.50	13.16	12.45
14	16.50	15.85	15.20	14.62	14.33
16	17.83	17.28	16.86	16.47	16.23

B. Depreciation Charge		C. Interest Charge		D. Twice Debt Repayment	
Rate (*percent*)	Amount (*$*)	Rate (*percent*)	Amount (*$*)	Sefvice on debt life (*years*)	Amount (*$*)
5	5	4	2.00	15	1.39
6	6	5	2.50	17	1.33
7	7	6	3.00	20	1.26
8	8	7	3.50	25	1.20
9	9	8	4.00	30	1.17
10	10	9	4.50	...	...

Example:	A	+	B	+	C	+	D	=	Total
	(16, 17)		(10)		(8)		(25)		
	$17.28	-	10	+	4.5	+	$2.40	=	$14.18

Note: Per $100 invested, equal equity and debt, annual capital charge equals: 2(equity return)–depreciation + interest + 2(debt repayment).

TABLE VI–C–6. Required Gross Return and Net Cash Flow, 12 Percent Equity Return, 20-year Life, Various Debt-Equity Ratios

	Assumed percent equity to total investment			
	80	50	40	30
1. Equity investment (1)	$80	$50	$40	$30
2. Annual equity return (1/7.41)	10.80	6.75	5.40	4.05
3. Annual depreciation at 6%	6.00	6.00	6.00	6.00
4. Annual debt repayment	0.25	0.63	0.76	0.88
5. Annual interest	1.20	3.00	3.60	4.20
6. Annual capital cost required	17.30	11.76	9.70	9.13
7. Taxable profits (line 6 *less* line 3 *less* line 5) . .	10.10	2.76	neg.	neg.
8. Income tax (½ of line 7)	5.05	1.38	nil	nil
9. Net after tax (line 7 *less* line 8)	5.05	1.38	neg.	neg.
10. After-tax profits as % of net worth	6.3	2.8	neg.	neg.

Notes: (a) For cols. (1) and (2), line (6) is sum of: twice line (2) *less* line (3) *plus* twice line (4) *plus* line (5). This formula, explained in text, applies when there is an income tax payable. In the cases of cols. (3) and (4), there is no income tax payable, and the formula is simplified to sum of: line (2) *plus* line (4) *plus* line (5).

(b) For all columns: After 16.67 years' depreciation is complete, the line 3 allowance ceases, and the firm will owe $3.00 annually as additional taxes. The present value of these payments is only $1.65 in total and they require an annual provision of 17 cents over the 16.67 years.

TABLE VI–C–7. Western Europe: Utilization of Refining Capacity, 1953–1971

(thousand barrels per day)

Year	Capacity Jan. 1	Average year capacity (Midpoint year$_n$ and year$_{n+1}$)	Throughput	Utilization *(percent)*
1953	a	a	1,712	a
1954	a	a	1,985	a
1955	a	a	2,136	a
1956	2,384	2,579	2,291	88.8
1957	2,774	2,839	2,334	82.2
1958	2,903	3,268	2,728	83.5
1959	3,633	3,806	3,187	83.7
1960	3,978	4,227	3,713	87.8
1961	4,476	4,697	4,223	89.9
1962	4,918	5,034	4,707	93.5
1963	5,149	5,707	5,395	94.5
1964	6,264	6,888	6,257	90.8
1965	7,511	8,027	7,174	89.4
1966	8,542	9,035	8,080	89.4
1967	9,527	10,306	8,814	85.5
1968	11,085	11,985	10,047	83.8
1969	12,884	13,413	11,173	83.3
1970	13,941	14,559	12,550	86.2
1971	15,177	. . .	. . .	. . .

Sources: Capacity from *OGJ*, annual supplements–*World Wide Oil*. Throughput from *British Petroleum Statistical Review of the World Oil Industry*, published annually.

[a]Not available on fully comparable basis.

TABLE VI-C-8. Refinery Operating Costs and Total Costs, 1960-1969

Year	Operating costs (¢/bbl)	Capital costs Annual (¢/bbl)	Capital costs 3-year mean (¢/bbl)	Capital costs Annual (¢/bbl)	Capital costs 3-year mean (¢/bbl)
1960	12.9	19.3	23.5	32.2	36.4
1961	12.9	23.6	21.3	36.2	34.2
1962	12.9	21.5	20.9	34.4	33.8
1963	12.9	19.6	18.4	32.5	31.3
1964	12.9	15.4	18.0	28.3	30.9
1965	12.9	19.1	19.1	32.0	32.0
1966	12.9	22.3	20.2	35.2	33.1
1967	12.9	19.3	19.7	32.2	32.6
1968	12.9	18.0	19.4	30.9	32.3
1969	12.9	21.0	...	31.9	...

Sources: Operating costs, see text; capital costs, Table VI-C-1.

TABLE VI-C-9. ECSC Estimates of Refining Costs

Size (*TBD*)	Iraq-Arabian 35° *$/m.t.*	Iraq-Arabian 35° *¢/bbl*	Kuwait 31° *$/m.t.*	Kuwait 31° *¢/bbl*	Sahara-Libya 39°-42° *$/m.t.*	Sahara-Libya 39°-42° *¢/bbl*
40	4.50	61	4.30	59	5.20	68
80	3.50	47	3.30	46	4.20	55
120	3.00	41	2.90	40	3.60	47
160	2.70	36	2.60	36	3.20	42

Source: ECSC, *Nouvelles Réflexions sur les Perspectives Energétiques* (1966), Annexe 4, p. 49.

TABLE VI-C-10. Western Europe: Cost of Operating Topping Plants, 1963

(*cents per barrel; plant 30,000 BD*)

1. Labor: Supervision	0.32
2. Labor: Operating	1.67
3. Maintenance	0.57
4. Fuel	2.24
5. Steam	2.10
6. Power	0.69
7. Water	0.28
8. Chemicals	0.08
9. Compression	...
10. Total direct cost (1956 basis)	7.95
11. Total direct cost (1963 basis)	8.23
12. Insurance and taxes	0.30
13. Laboratory overhead	0.10
14. Other possible indirect charges	1.14
15. Subtotal	1.54
16. Capital costs	4.85
17. Total costs (lines 11 + 15 + 16)	14.62

Source: Adapted from W. L. Nelson, *Guide to Refinery Operating Costs* (*OGJ*, 1966), No. 27, p. 32.

Notes by line:

1 and 2–Labor costs for Western Europe have been increased by 67 percent as compared with those for the United States, as recommended by Nelson in no. 2, p. 47.

10 and 14–The 1956 costs have been updated to 1963 by multiplying by the ratios given by Nelson in *OGJ*, May 17, 1965, p. 133. See also *id.*, Mar. 7, 1966, p. 91.

15–Capital costs are calculated as explained in text. Investment per daily barrel has been taken as $115.46, updated from 1946 from the same sources, and allowing 43% additional for offsite facilities.

Appendix VI-D

Estimating Crude Prices from "Tickets"

North American refining is excluded from the integrated system discussed in Chapter III, even though large amounts of crude oil are shipped to North America by producer-refiners. Producers who have U.S. coastal refineries prefer to use their own Caribbean crude (and to a lesser extent their Middle East crude) rather than buy U.S. crude. But the force of this preference is much attenuated because of U.S. import control and what amounts to import limitation in Canada.

In the United States quotas have been given out in proportion partly (and to a diminishing extent) to past imports, and mostly to current refinery inputs. Noncoastal refiners could not directly exercise their rights to bring in crude, and therefore sold them. The import regulations, as Jefferson might have said, paid a decent respect to the hypocrisy of mankind, by forbidding direct sales of quotas, or what the trade calls "tickets." Instead, the two parties arranged swaps of foreign for domestic crude. The donor of the ticket got a higher value of domestic crude than the foreign crude he gave up, the difference in value being the effective price of the ticket. To the statistician it is a pity. For if holders of tickets could sell them openly, one would have a continuing statistical series of how much they were worth; as matters stand, it would take an immense amount of calculation to compute approximate values, and even then they would have wide margins for error.

For some time a rule-of-thumb value for tickets was $1.25 per barrel; one can use it only for illustration. Company *A*, a Persian Gulf producer, paid about one dollar in tax-plus-cost for this barrel, freight of perhaps 68 cents (Intascale less 50 plus Suez tolls), and 10 cents duty. His total laid-down cost at Philadelphia was then $1.78. Suppose that an equivalent domestic crude would cost him $3.25—or, what comes to the same thing, use of his own crude would cost him the $3.25 he could get for it. But if he bought domestic crude or used his own, he could sell his ticket. Hence the net price of the domestic crude to him was $2.00, for $3.25 - $1.25 = $2.00. This was still 22 cents more than his landed cost of $1.78. But suppose he could sell the barrel f.o.b. the Persian Gulf at $1.22. Then the true net landed cost of the foreign crude to him was that $1.22 given up, plus the 78 cents freight and duty, plus the $1.25 ticket used up. Then he was just as well off selling it and buying or using domestic crude. If he could get a better Persian Gulf price than $1.22, he ought to sell it in preference to importing.

This apparent precision is deceptive. "To sell at $1.22" sums up a host of uncertainties—how much, for how long, to whom? The price could easily be more or less, once the parties get down to the job of negotiating a contract. Furthermore, the tax effects may overbear these cost-price comparisons. If the sales value of his domestic crude is about $3.00 Gulf Coast, and the Persian Gulf crude is posted at $1.80, the difference in percentage depletion allowance is 27.5 percent of $1.20, or 33 cents. Net tax benefit is about 17 cents—but not if there is already a carry-over of unused tax credits.

Or consider Company *B*, an East Coast refiner with domestic production, but no Persian Gulf production. If he buys a barrel from *A* at $1.22, paying freight and duty of 78 cents, and using up his ticket, he, too, pays $3.25 all told, whether he uses foreign or domestic crude, and he, too, must reckon with losing some percentage depletion, though not as much.

Thanks to their refining capacity, either company is better off in the amount of $1.25 regardless of where they obtain crude. The ticket is not cash, but it is a valuable negotiable asset, like an inventory of copper or aluminum (noncorrodible materials), a holding of Treasury bills or high-grade commercial paper, an allotment of computer time.

Consider now another refiner who has used up all his tickets, wants more crude, and buys tickets. He must go through exactly the same calculations and reach exactly the same balance of advantages. The cost of using up a given amount of metal, or negotiable securities, or computer time, or import tickets, is the same whether or not they are newly bought or previously owned.

Appendix VI-E

Crude Price Data, 1957–May 1967

This appendix lists the available data on Middle East, Russian and African crudes, excluding Far East markets. Our information is more relevant than the customs figures in Appendixes VI-F and -G, since the date of the transaction (or offer) and the quantity and delivery times are usually specified. But two important warnings have to be made. First, they are all press reports and some may not be correct, though none has been denied or shown to be wrong. Even precise details of offerings are subject to misprints and misinterpretations. Secondly, to be useful the information has to be made comparable and this often can only be done with guesswork and arbitrary assumptions. The most difficult—and most important—allowance is for the appropriate freight rate. Where time charter or consecutive voyage rates are relevant, those from Chapter IV are used, but in other cases it was necessary to choose fixtures corresponding most closely in ship size and delivery period with the transaction in question. Allowances also had to be made for credit and for quality of the crude.

Only large and competitive buyers have been included. Table VI-E-1 lists all apparent arm's-length transactions with as much information as it is possible to compress in a line or two. Finally the notes at the end of the table give references for each observation and details of the assumptions made so that they can easily be checked.

Assumptions and explanations. The aim is to find the equivalent price at the Persian Gulf that a large international company would receive for the delivery in question. This does not restrict us to those companies' offers or sales. For instance, Atlas was able to make attractive offers to YPF (the Argentine state company) because it could save shipping costs by delivering heavy fuel oil from Argentina to North America. However, other companies had to meet the Atlas price. Similarly, it is possible that credit would be more expensive to the U.S.S.R. than to the internationals, but because we aim at the netback to the companies the rate of interest used is the one appropriate to an international oil company.

Gravity. Adjustments were made to reduce crudes to a 31° basis using 1½ cents per degree (see Appendix VI-H).

Sulfur. Where possible an adjustment was made according to Table VI-H-3, e.g., 5 cents added to a Kuwait crude, 1 cent deducted from Arabian light. Usually, not enough information was available to adjust the U.S.S.R. crudes.

Credit and risk. Cost of credit is an important factor in most of the transactions with Argentina, Brazil, and Uruguay. We need to know the cost to the lender. It is assumed that the company borrows in the market to cover the cost of delayed payment or to raise money for a loan. The companies can command lower rates than the state agencies which buy the crude. The cost to a company is more than the borrowing cost, for in extending credit to a less credit-worthy body it is assuming some risk. Moreover, if credit to buyers is considered a part of a seller's permanent capital, repayments constantly reloaned, the cost of the credit is the seller's average cost of capital, equity, and debt taken together, which is around twice as high as the borrowing rate. No attempt has been made to allow for either, but only to assess and use the companies' borrowing rate. To the extent that risk and cost of equity capital should be allowed for, the Latin American prices are overvalued. From 1965 to 1967 the Eurobond rate has been a good indicator and was around 6 percent. Lacking better information for the earlier period, 6 percent is taken also. The error amounts to about two cents per percentage point.

Omissions. The principal omission is of offers to Egypt, which usually included buy-back provisions whose value would depend on the (unknown) value of the products. Barter deals were excluded if it seemed that the value of the barter was significant, e.g., the NIOC-Argentine wheat for oil deal. However, if the crude was part of a large trade agreement involving many products (e.g., Italian-Soviet trade agreements), it was included.

Data on Latin America have been separated from those of the other markets. In the 1950s Latin America was virtually the only source of price information because bids were made publicly. The ports of Brazil, Argentina, and Uruguay are close (a range of 6 cents per barrel Intascale flat), although Brazilian ports can accommodate larger tankers and thus freight costs will be lower. Purchases in all three countries are made by state agencies; all three have foreign exchange problems and have at times fallen behind in payments to oil companies.

Latin America was considered a desirable market. "Historically, Japanese buyers have obtained smaller discounts off Middle East crudes than Europeans who in turn were granted smaller discounts than buyers in Latin America" (*POPS*, Feb. 4, 1960). Petrobras in particular was considered an important and aggressive buyer. "Largest outside buyer of crude today" (*POPS*, Aug. 25, 1960) "buys crude at prices as low or lower than virtually any other oil importer" (*PIW*, Aug. 10, 1964). However, these impressions are not supported by the data, at least from 1960. One does not know if Italian and German independents were doing so well in the fifties (see Appendix V-A), but from 1960 the patchwork data that I do have suggest that at least some European markets were much more competitive. This is particularly true in 1960 when there was a series of contracts with Italy and from 1963 through 1965 when North African crudes became important.

Comment. The data support the proposition that prices fell from 1957 to 1961, rose slightly in 1962, fell again in 1963 until 1965, and improved after the end of 1965. Our Latin American prices are relatively homogeneous. C.i.f. prices fell to the middle of 1961, were stable through 1962, rose in 1963, and fell thereafter. F.o.b. netbacks were around \$1.20-\$1.30 in 1960-61, and were more like \$1.10 to \$1.20 from the end of 1964 through 1966. Early 1967 saw very low spot freight rates, but the appearance of rising f.o.b. prices is probably illusory.

It is worth examining more closely the Russian prices around \$1.00 Persian Gulf equivalent in 1960. They are all in the questionable category, but can be fairly well supported. Press reports for the Russian four-year contract with Italy guessed at the price f.o.b. Black Sea from \$1.00 to \$1.48. However, both Italian and Russian trade figures give the 1961 f.o.b. price as \$1.31 which equates to \$1.01 at Ras Tanura. (The price of \$1.00 mentioned in the press may have originally referred to the Persian Gulf equivalent.) In July 1960 it was stated that at least two major oil companies made deals for delivery of Middle East crudes to Italy at discounts of 58 cents and 91 cents, respectively. If one assumes that the lower discount applies to Kuwait crude (before the cut in posted prices), and the higher to Qatar 40, we get prices of \$1.14 and 98 cents. These transactions may be the basis of the report (in January 1962) by a Soviet economist that "a number of cartel branches" sold Middle East oil in Italy at a price of \$1.00 at the beginning of 1961. In the same report he claims that U.S.S.R. prices are between \$10 and \$16 per ton f.o.b. Black Sea, although official U.S.S.R. trade figures show exports to Italy, Belgium, and

Germany in 1961 and 1962 all around $9.50. (*POPS*, Jan. 15, 1962.) In November 1962 an Esso spokesman said that Middle Eastern prices have to be as low as $1.04-$1.10 to match U.S.S.R. prices to Germany and Italy, and 89 cents to match their prices to ENI. The only reconciliation that seems possible is that Esso was using AFRA freight rates instead of the actual market rate to make the comparison. Allowing for poetic license all round, it is clear that some business at least was done under $1.10 in the Persian Gulf or its equivalent at the Black Sea.

In 1963 there was a similar situation, with major companies and the U.S.S.R. making contracts with Italy at undisclosed but widely surmised prices. In April Esso was said to supply ENI with 12 million tons of Libyan Kuwait and Arabian crudes over five years at an average of 20 percent off posted. This would be equivalent to $1.30-$1.35. In November 1963 a deal was made with the U.S.S.R. for six years. The f.o.b. price was estimated at $1.05-$1.10, but Soviet exports to Italy for 1963, 1964, and 1965 ranged from $1.21 to $1.26 and Italian import figures from $1.27 to $1.31 f.o.b. So Persian Gulf netback would be $0.90-$1.00. In December 1963 ENI arranged to buy 12 ½ million tons of Kuwait crude from Gulf. The price was variously described as similar to $1.05 and little more than $1.00 f.o.b., although the Gulf spokesman said that $1.10 was ridiculously low. Here it seems impossible to attempt a reconciliation.

In June 1962 there is an interesting observation. It is the first major North African contract recorded (the one in 1959 was a single cargo). CFP contracted to sell 200,000 tons of Hassi Messaoud at $1.94 c.i.f. to Morocco, while 150,000 tons of Kirkuk were to be sold at $1.96. Was the lighter Algerian crude priced lower in order to break into the market? If so, there is some justification for the accusations that North African crudes were responsible for the fall in crude prices. If, however, it was lower because the higher gravity crude was unsuitable, there is reason in this case to question our 1½ cents per degree allowance.

The price for Petrangel at the end of 1964 looks unlikely. The "international price" is described as 98 cents to $1.00. If this refers to f.o.b. Luanda, the Persian Gulf netback would be 80¢. However, it is possible that the price quoted had already been netted back.

The improvement in prices from the end of 1965 is most evident in Libyan crudes, but there are not enough observations of Middle East crudes to demonstrate a trend. An important single observation is the Virgin Island price of $1.60 with transport at 65 cents. This must have been a long-term price, since the refinery was not yet built. If the details are reported correctly, this is a good indication of the pre-Suez marginal price. If this is Safaniya, the netback could be $1.06 a barrel (if any lighter crude, it would be even less).

References and Notes—Latin American Transactions

Brazil. The time allowed for payment for imports has been the subject of government regulation. Late in 1960 Western companies were warned that oil payments would be made in 180 days, "not 120 days as formerly" (*POPS*, Feb. 16, 1964). No reference was found that the rule was in fact changed, although actual payments fell further behind (*POPS*, June 15, 1961; *PIW*, Oct. 16, 1962; *PIW*, Sept. 23, 1963). Therefore, for Brazilian transactions from September 1960 to mid-1964, I have used 120 days free credit (unless some other special provision such as a barter was involved). In August 1964 it was decreed that oil companies must finance export of Brazilian goods. It is not clear how much this was done, because contracts were often arranged so that no company's supply was large enough to be eligible for this provision (*POPS*, Aug. 3, 1965; *PIW*, Aug. 1, 1966). Such could not always have been the case, since "some companies reportedly met the provision by paying exporters 3-4%" (*POPS* Nov. 15, 1966). This would mean that our f.o.b. netback values should be 6-8 cents lower in those cases.

Argentina. Bids have been published in extreme detail with several alternative methods of delivery and of payment, and it would be a bold analyst who would claim to have thoroughly understood them. I have considered only the successful bidder or bidders where they were known. Otherwise I have guessed at the lowest offer. Where different tanker sizes were offered, the largest was taken to provide a better guess at freight and netback. Barter offers have been ignored because it was impossible to evaluate them, although many must have been more attractive than the cash offers. Atlas was frequently a successful bidder. The company sold Argentine fuel oil in Montreal and could thus benefit from the side-hauls when it took heavy fuel oil in exchange for Middle East crude. Unfortunately, it is impossible to estimate Persian Gulf netback for Atlas, but working from the c.i.f. price one can only see what other companies would have to charge to equal Atlas's price.

TABLE VI–E–1. Summary of Crude Price Data, March 1957–May 1967

(*adjustment for sulfur except**)

Ref.	Date	Country of buyer	Details	C.i.f. 31° basis	Freight used	Price 31° f.o.b. Persian Gulf: Europe, etc. M.E.	Europe, etc. N. Afr.	Europe, etc. USSR	Latin America M.E.	Latin America N. Afr.	Latin America USSR
	1957										
L1	Mar	Argentina	5-yr contract by BP for Kuwait 31, $3.31	3.36	USMC+12.5				1.31		
L2	Nov	Argentina	Dec–May, 3.9 million bbl BP Kuwait 31, $2.48	2.53	USMC–42				1.48		
	1958										
L3	May	Argentina	From June, 3.2 million bbl, $2.32 for 32°	2.31	IS–28.5						1.49*
L4	June	Argentina	Five cargoes Kuwait 31, July–Aug, $2.305	2.36	IS–50–62				1.77–1.91		
L5	July	Argentina	Renegotiation of L1, $2.38 until July	2.43	IS–30				1.61		
			$2.53 until Dec	2.58	IS–30				1.76		
L6	Oct	Brazil	1 million bbl Maracaibo 4th Q1958, possibly $2.00 f.o.b.		IS–30				1.42V		
	1959										
L7	Apr	Brazil	USSR crude roughly @ 20% off M. East prices 3-yr part barter		IS–40						1.41*
E1	Apr	General	*Average* rebates off Kuwait crude, 5%		. . .	1.63					
E2	Apr	Italy	Delivered price of crude used by 6 refiners $2.00		IS–50			1.43*			
E3	Nov	Germany	One cargo Algerian crude, $2.10 c.i.f.		IS–50		1.34				
L8	Nov	Arg. Braz. and Urug.	USSR crude lands at $2.30	2.36	. . .						1.55*
	1960										
E4	Early	Greece	1960 price, $16.25/met.ton for Tuimazinskaia		IS–45			1.66			
E5	Feb	Europe	Venez. crude lands $13–15; assume $15 for 39°		IS–45	(1.24)					
L9	Apr	Uruguay	Two cargoes Kuwait, $2.06; also credit	2.02	USMC–60				1.33		
E6	May	Italy	60,000 BD 1961–64; $1.31 f.o.b. Black Sea		IS–45			1.00*			
E7	June	U.S.	Offer to refiner, $1.90 c.i.f. N. Jersey 31°		IS–57.5			1.30			
E8	June	General	Netbacks west not more than $1.40 f.o.b. Black Sea		IS–57.5			1.14			
L10	July	Brazil	600,000 bbl Mukhanovskaia and Romashkinskaia at $1.75 plus 60–75¢ shipping starting July		IS–37						1.54
L11	July	Brazil	1 million bbl Maracaibo Jul–Aug, $1.80 f.o.b.		IS–62.5				1.59V		

TABLE VI-E-1. Continued

(*adjustment for sulfur except**)

Ref.	Date	Country of buyer	Details	C.i.f. 31° basis	Freight used	Price 31° f.o.b. Persian Gulf: Europe, etc.: M.E.	Europe, etc.: N. Afr.	Europe, etc.: USSR	Latin America: M.E.	Latin America: N. Afr.	Latin America: USSR
	1960 (cont.)										
E9	July	General	Discounts 22–35¢ routine at P. Gulf;								
			Kuwait–22¢			1.50					
			Iran light–35¢		. . .	1.45					
E10	July	Italy	At least two majors sold at 58¢ and 91¢ discount		. . .						
			Kuwait less 58¢			1.09					
			Qatar less 91¢			0.98					
			Iran light less 91¢			0.89					
E11	Aug	General	*Average* discount off Kuwait 15%		. . .	1.40					
L12	Sept	Brazil	2 million bbl Oct–Nov Arabian 34, $2.20, credit	2.10	USMC–60				1.43		
L13	Sept	Uruguay	15,000 BD, 3 yrs Kuwait, $2.00 equiv., credit	1.98	IS–40				1.29		
E12	Oct	Greece	New contract 2¢ higher than old		IS–45			1.68*			
E13	Nov	General	No crude less than 12¢ off posted; Kuwait:		. . .	1.52					
L15	Dec	Brazil	1 million bbl Arabian 34° Jan–Feb, $2.15–2.17	2.05	IS–28				1.27		
	1961										
E14	Early	Italy	Majors sold at $1.00 says USSR economist		. . .	1.00					
L16	Mar	Brazil	2 million bbl Arabian 34 Apr–May, $2.10	2.00	IS–60				1.56		
L17	Aug	Uruguay	BP offers 600,000 bbl Kuwait 35, 0.5% Aug–Jan at $2.11 also credit	1.94	IS–37.5				1.25		
E15	Oct	General	No crude (except Duri) can get full price: Kuwait			1.64					
E16	Nov	Morocco	Latest price for 20,000 BD, $2.05		IS–45			1.46*			
L18	Nov	Brazil	Spot sale Arabian 34 Dec, $2.085	1.99	IS–55				1.50		
E17	Dec	Italy	ENI to buy 20,000 BD Iran and Iraq 22.5% off		. . .						
			Gach Saran			1.26					
			Basrah			1.27					

TABLE VI-E-1. Continued

(*adjustment for sulfur except**)

Ref.	Date	Country of buyer	Details	C.i.f. 31° basis	Freight used	Price 31° f.o.b. Persian Gulf: Europe, etc.: M.E.	Europe, etc.: N. Afr.	Europe, etc.: USSR	Latin America: M.E.	Latin America: N. Afr.	Latin America: USSR
	1962										
E18	Year	Morocco	Iraq crude at $2.00		IS–45	1.41					
E19	Jan	General	*Average* discount off Kuwait 10%		. . .	1.48					
L19	Mar	Brazil	1.5 million bbl Arabian 34 Mar–Apr, $2.05	1.95	IS–40				1.30		
L20	May	Brazil	Offer 120 million bbl at 12% discount declined; Arabian 34		. . .				1.48		
E20	June	Morocco	Contract for 1.5 million bbl Hassi Messaoud at $1.94 c.i.f., and 1.1 million bbl Kirkuk, $1.96 c.i.f.		IS–47	1.36		1.16			
L21	Oct	Brazil	USSR to deliver 4 million bbl at $2.00 credit	1.91	IS–50						1.34*
E21	Oct	Greece	9,000 BD for 2 years at $2.09		IS–47			1.63			
E22	Oct	Greece	1.1 million bbl/yr for 2 yrs Edjele, $1.90 c.i.f.		IS–47		1.35				
			Libyan, $1.80 c.i.f.		IS–47		1.29				
E23	Nov	Germ. and Italy	To match USSR prices, M. East would have to be:					1.04–1.10			
			to match price to ENI (says Esso):		IS–50			.89			
L22	Dec	Brazil	250,000–450,000 bbl Hassi Messaoud $1.85 f.o.b.		IS–50					1.34	
	1963										
E24	Year	Morocco	Imports: Arabian 34, $1.94; Zelten, $1.97; USSR, $1.89		IS–47	1.36	1.28	1.28			
			Hassi Messaoud, $1.95; USSR 36, $1.97				1.18	1.36			
			Edjele 37, $1.97				1.31				
E25	Mar	Germany	Three independent refiners' contracts 3 mo–1 yr, $1.81 f.o.b.		IS–47		1.21				
L23	Mar	Brazil	4.5 million bbl Hassi Messaoud 1 yr, $1.80–1.85 f.o.b.		IS–40					1.24+	
L24	Mar	Uruguay	Offers 625,000 bbl Brega, $2.27 c.i.f.	2.08	IS–34					1.36	
E26	April	Italy	Esso-ENI 50,000 BD 5-yr rumored 20% off posted Kuwait, Libyan		. . .		1.32				
			Arabian 34			1.39					
E27	June	General	Algerian crude down to $1.75 f.o.b.		IS–47		1.15				
L25	June	Brazil	1.5 million bbl July–Aug Arabian 34, 34, $2.05–2.15 c.i.f.	1.95+	IS–60				1.51+		

TABLE VI-E-1. Continued

(*adjustment for sulfur except**)

Ref.	Date	Country of buyer	Details	C.i.f. 31° basis	Freight used	Price 31° f.o.b. Persian Gulf: Europe, etc. M.E.	Europe, etc. N. Afr.	Europe, etc. USSR	Latin America M.E.	Latin America N. Afr.	Latin America USSR
	1963 (cont.)										
L26	Sept	Brazil	USSR 24 million bbl for 2 yr at $2.10, 38°	2.04	IS–40						1.37*
L27	Nov	Brazil	1 cargo Basrah, $1.92; 2 more coming	1.83	n.a.					n.a.	
E28	Oct	Morocco	4,000 BD Algerian to ENI-Morocco, $1.80–1.85		IS–47		1.24+				
E29	Nov	Switz.	5-yr contract first 2 yrs at $1.80 f.o.b.		IS–47		1.21				
E30	Nov	Italy	5-yr contract with USSR, $1.21 f.o.b.		IS–47			0.92			
L28	Nov	Brazil	20 million bbl Arabian 34, 2-yr, $2.08	1.98	IS–49				1.43		
E31	Nov	Tunisia	1964 needs Arabian 34, $1.88–1.89		IS–47		1.37+				
			Kirkuk, $1.85–1.87				1.31+				
E32	Dec	Europe	30–35¢ discount believed necessary for Arabian crudes								
			Light			1.39+					
			Medium			1.25+					
E33	Dec	Italy	Gulf to ENI 17 million bbl, "similar" to $1.05		. . .	1.10					
	1964										
E34	Jan	Europe	Khafji would have to be $1.15 to sell to independent refiner in Europe			1.27					
L29	Feb	Brazil	Basrah 27,000 BD, about $2.00 c.i.f.	1.92	IS–50				1.33		
E35	Mar	Morocco	Arabian 34 to Morocco, slightly over $2.00		IS–54	1.48					
E37	June	Turkey	Imports to be 10% off posted, Gach Saran		. . .	1.47					
L30	July	Uruguay	Lowest offer Libyan 38, $2.21, 6 mo. + option	2.06	IS–45					1.45	
L31	Oct	Brazil	Gulf to supply at $1.90, say Kuwait	1.91	IS–50				1.36		
L32	Oct	Brazil	15 million bbl for 1965, possibly $1.95	1.95	IS–50						1.36*
L33	Nov	Brazil	Kuwait now $1.90	1.95	IS–45				1.31		
			Arabian 34 now $2.00	1.94					1.30		
L34	Nov	Argentina	2.5 million bbl 35 crude at $2.00 c.i.f.	1.94	IS–30						1.13*
E38	Nov	General	Libyan crude sold by Amerada at *average* $1.47		IS–54		0.97				
			Libyan crude sold by Marathon at $1.64				1.14				
E39	Dec	General	Crude bought from Petrangol at "international price of 98¢–$1.00"		IS–54	0.80					

TABLE VI-E-1. Continued

(adjustment for sulfur except)*

Ref.	Date	Country of buyer	Details	C.i.f. 31° basis	Freight used	Price 31° f.o.b. Persian Gulf: Europe, etc. M.E.	Europe, etc. N. Afr.	Europe, etc. USSR	Latin America M.E.	Latin America N. Afr.	Latin America USSR
	1964 (cont.)										
E40	Dec	General	Discount 45¢ reasonable for light M. East crude		. . .	1.30					
E41	Dec	General	Discounts 40–45¢ common			1.35					
	1965										
L35	Jan	Uruguay	17,000 BD 3-yr option 4th, Kuwait, $1.975, Nigeria 31, $2.15	1.94 2.00	IS–30				1.11	1.18N	
E42	Mar	General	Discounts 30–45¢ P. Gulf light 60¢ N. African		IS–54	1.30	1.16				
E43	Mar	General	Open market Libyan, $1.60–1.65		IS–54		1.13+				
E44	Apr	Germany	Cost to refiners of Kuwait, $1.20		. . .	1.25					
E45	Apr	General	One Oasis partner selling $1.47–1.55		IS–54		.99+				
E46	Apr	General	Shell buys from Amerada at $1.55		IS–54		1.07				
E47	Apr	General	Hofra available at $1.64		IS–54		1.10				
L36	Apr	Argentina	Shell offer 3 million bbl Iraq 34, $2.00, Nig. 34, $2.15	1.94	IS–45				1.31	1.43N	
E48	May	Turkey	Imports to be 12.5% off posted, Gach Saran	2.06N	. . .	1.43					
L37	May	Brazil	460,000 bbl at $1.80 f.o.b. Nigeria		IS–53					1.50N	
L38	July	Argentina	4.5 million bbl Sept–Dec Nig. 35, $2.20, Basr. 36, $2.04N	2.02 1.89	IS–35				1.14	1.27N	
E49	Sept	Switz	20,000 BD on netback basis, now $1.25		IS–54			0.92			
E50	Sept	Europe	Some prices as low as $1.38–1.45		IS–54		.91+				
			Considerable amount under $1.50		IS–54		1.03				
E51	Sept	Italy	Netback deal rumored $1.00–1.25 c.i.f. for salty								
L39	Oct	Brazil	Current price, $1.95	1.95	IS–54						1.45*
L40	Oct	Argentina	Jan–Apr 1966, 3.6 million bbl Basr. 35, $2.13, credit 4.3 million bbl Nig. 34, $2.15	2.06 1.99N	IS–25 IS–30				1.18	1.17N	
E52	Oct	Turkey	Imports to be 18.5% off posted, Gach Saran		. . .	1.34					
E53	Nov	Germany and Italy	Discounts 40–45¢ not uncommon for Iraq 34		. . .	1.28					
E54	Late	General	Some realizations around $1.30		IS–54		.93				

TABLE VI–E–1. Continued

(*adjustment for sulfur except**)

Ref.	Date	Country of buyer	Details	C.i.f. 31° basis	Freight used	Price 31° f.o.b. Persian Gulf: Europe, etc. M.E.	Europe, etc. N. Afr.	Europe, etc. USSR	Latin America M.E.	Latin America N. Afr.	Latin America USSR
	1965 (cont.)										
E55	Year	General	Discounts about 40¢, Gach Saran		. . .	1.23					
E56	Year	General	Discounts off Libyan up to 80¢		IS–54		.98				
E57	Late	Turkey	State co. bought 3.7 million bbl Lib. at 27% off		IS–54		1.14				
E58	Late	General	Algerian oil would do well to bring $1.70		IS–54		1.15				
	1966										
E59	Mar	Turkey	Agreement to reduce imports to 20% off,								
			Gach Saran		. . .	1.30					
E60	Rec. yrs	General	Kuwait sold at $1.20			1.25					
L41	Mar	Argentina	5.4 million bbl Jun–Jul Iraq 34, $1.915;								
			Nigeria 33, $2.08	1.79	IS–50				1.22	1.36N	
			Arzew, $2.18	1.93N						1.34	
				1.90A							
L42	Jul	Brazil	Imports contract July for 1966/7,								
			Kuwait, $1.68–1.72	1.73+	IS–55				1.22+		
			Iraq, $1.71–1.85	1.64+					1.11+		
L43	Aug	Argentina	2 million bbl Iraq 34, $1.95;	1.83	IS–30				1.02–1.20		
			Hassi Mess., $2.27 Oct–Dec	2.04							
E61	Oct	General	Discounts on Arabian: 30¢ heavier, 40¢ lighter			1.31–1.35					
E62	Dec	General	Spot sales Algerian, $1.60–1.70 f.o.b.		IS–57		1.06+				
E63	Dec	S. Africa	4.5 million bbl 1967, Darius $1.33 f.o.b.		. . .	1.26					
L44	Dec	Argentina	Qatar 37 Feb–Mar, $2.10	1.94	IS–45				1.31		
	1967										
E64	Jan	Virg. I.	Obtainable at $1.60, transport 65¢								
			if Kuwait		. . .	1.00					
			if Arabian 28			1.06					
E65	Jan	Kuwait	5-yr contract to KNPC, $1.10 Umm Gudair		. . .	1.20					
E66	Jan	General	Morgan 32 priced at $1.35		IS–57	1.12					
L45	Feb	Argentina	1.8 million bbl Iraq 34 at $1.99 Apr–Dec	1.86	IS–50				1.29		
E67	Apr	Angola	One cargo Arabian 34, $1.80 c.i.f.		IS–47	1.15					
L46	May	Argentina	5.5 million bbl Iraq 34 Jun–Dec $1.88	1.76	IS–45				1.13		
E68	May	Spain	Kuwait crude available to independent								
			buyers at $1.22–1.32		. . .	1.27+					
E69	May	General	Serir type $1.55–1.65 in open market pre-crisis		IS–57		1.12+				
E70	May	Italy	Pre-crisis Libyan $1.55–1.60; if Serir		IS–57		1.12+				

See overleaf for References and Notes.

TABLE VI–E–1. Continued.

REFERENCES AND NOTES– LATIN AMERICAN TRANSACTIONS

L1 *POPS*, Mar. 28, 29, 1957; *POPS*, July 29, 1958.
One company offered transport alone at USMC + 12 ½%.

L2 *POPS*, Nov. 13–15, 1957; *POPS*, Dec. 5, 1957; *PW*, Nov. 22, 1957.
Freight offers ranged from USMC – 47% to USMC – 35%. *PW* said going rate USMC – 42%, which is approximately the midpoint of this range.

L3 *POPS*, Apr. 22, 1958; *POPS*, May 16, 1958; *POPS*, June 19, 1958; *POPS*, Mar. 17, 1959; *POPS*, Mar. 25, 1959.
USSR bid originally $1.78 f.o.b. Black Sea. Later obtained transport at from 37/6 to 41/6 per long ton (70–77¢/barrel). Argentina later reported to be paying $2.32 so it seems bid was renegotiated to about $1.57 Black Sea.
This seems to be part of a trade deal where USSR gave Argentina credit for purchase of equipment worth $100 million ($1 million in one report) to be paid back over 10 years at 2 ½% interest. Cost of credit is obscure since it could be included in the price of the machinery.

L4 *POPS*, July 9, 10, 1958.
Spot rate (HM) was Scale – 62 ½%. Since this gives a result higher than posted we have guessed at IS – 50%.

L5 *POPS*, July 29, 1958.
Contract L1 renegotiated to $2.38 until July, $2.53 until Dec. Not clear if this means every July and December or only Dec. 1958 and July 1959; rates IS – 32% for 2nd quarter, IS – 27% 3rd quarter.

L6 *POPS*, Oct. 22, 1958; *PW*, Oct. 31, 1958, p. 40.
IS – 30% taken as approximate mean of HM consecutive voyage rates Scale less 27% 3rd quarter and Scale less 36% 4th quarter.

L7 *POPS*, Apr. 7, 1959.
Quantity not stated. Payment part in cash, part in cocoa. Assume Romashkinskaia 33 (see *POPS*, Mar. 14, 1962). For freight take HM CV rate 2nd quarter.

L8 *POPS*, Nov. 9, 1959.
HM consecutive voyage rate is Scale less 34% for 4th quarter, single voyage rate is Scale less 49%. We have taken former, if latter price would be $1.73.

L9 *POPS*, Apr. 6, 14, 1960; *PW*, Apr. 22, 1960.
Credit: Payment 60% in 180 days, 40% after 1 year. Assumed interest cost to Gulf, 6%. Freight: *PW* says current rate USMC – 60% = 69¢.

L10 *POPS*, July 8, 1960.

L11 *POPS*, July 13, 1960.
HM Spot rate Scale – 62 ½%.

L12 *POPS*, Sept. 27, 1960; *PW*, Oct. 7, 1960.
Freight: *PW* says current rate USMC – 60% = 67¢. CA gives USMC – 66% for October delivery and for 4 to 6 months CV which is 57¢. We have taken former. Credit: 120 days credit required. Assume 6% cost to company.

L13 *POPS*, Sept. 28, 1960; *POPS*, Oct. 4, 1960; *POPS*, Feb. 13, 1961.
Freight: Our TC rate is IS – 45%, but make it IS – 40% because of small harbor.
Credit: 20% down, 40% for 180 days, 40% for 360 days. Assume 6%.

L15 *POPS*, Dec. 15, 1960.
Freight Scale – 28% (CA).

L16 *POPS*, Mar. 21, 1961.
Freight HM spot, Scale – 57 ½% last few weeks of March. Would expect rate for April and May delivery to be a little lower.

L17 *POPS*, Aug. 7, 1961.
Freight taken from CA Sc – 37 ½% for 18,000 for 12 months Consec Voyage.
Credit: 180 days to pay.

L18 *POPS*, Nov. 13, 1961; *POPS*, Nov. 17, 1961; *POPS*, Dec. 7, 1961.
Spot rate Scale – 55% (HM).

L19 *POPS*, Mar. 14, 1962.
Freight: Spot rate currently 65¢/bbl (*POPS*).

L20 *POPS*, May 3, 1962; *POPS*, May 8, 1962.

L21 *PIW*, Oct. 15, 1962.
Freight: CV rate is IS – 50%.
Credit: Credit line opened equiv. to 60% of value of oil, to purchase Soviet equipment. Repayment over 5 years at 3%. Cost to Middle East co. per barrel would be $(.06 – .03) × 2.00 × .60 × 5 × ½ = 9¢.

L22 *PIW*, Dec. 3, 1962.
Freight: Spot rate (HM) IS – 50%.

L23 *POPS*, Mar. 19, 1963.
Freight: CV rate IS – 40%.

L24 *POPS*, Mar. 12, 1963; *POPS*, Mar. 13, 1963.
Apparently lowest offer. Freight (*POPS*) ATRS – 50% = IS – 34%.

L25 *POPS*, June 3, 1963.
Freight: Spot rates about IS – 60% (HM).

L26 *PIW*, Sept. 23, 1963.
Probably additional benefit through not having to pay in dollars (*PIW*).
Freight: CV rate IS – 40%.

L27 *PIW*, Nov. 4, 1963; *PIW*, Mar. 23, 1964.
Freight: Spot rate rose 35 points of Intascale from September to early October so it is not possible to make a fair guess of f.o.b. equivalent.

L28 *POPS*, Jan. 28, 1964.
CV rate average of 1963 and 1964 taken.

L29 *PIW*, Feb. 24, 1964.
Delivery period not stated. Take CV rate IS – 50%.

L30 *POPS*, July 16, 1964; *PIW*, July 20, 1964.
Freight: CV rate IS – 50% but make IS – 45% for small harbor.

L31 *PIW*, Oct. 26, 1964.

L32 *POPS*, Oct. 1, 1964.

L33 *PIW*, Nov. 16, 1964.
Freight: 60¢ (*PIW*).

L34 *POPS*, Nov. 30, 1964.
CV rate IS – 30% was common for 20,000 dwt ships. Part of barter deal.

L35 *PIW*, Jan. 26, 1965; *POPS*, Apr. 7, 1965.
Freight: CV rate at that time for 20,000 dwt tanker was about IS – 30%.
Credit: Payment 50% in 180 days, 50% in 360 days. Also a loan of $5 million at 6%. Former comes to 89¢ for Kuwait; 97¢ Nigeria; latter to about 1¢ assuming cost to company of 6%.

L36 *POPS*, Apr. 23, 1965.
Freight: Offer specified 35,000 dwt tankers so take CV rate IS – 45% (two tankers of that size fixed at that rate in March).
Credit: Payment in 360 days at 6% so no allowance.

L37 *OGJ*, May 31, 1965, p. 56.
Source says 460,000 bbl. delivered at roughly $828,000. Freight: Assume CV rate of IS – 53.

L38 *POPS*, July 21, 1965, Aug. 13, 1965.
Tankers over 25,000 but not exceeding 31 ft. draft. CV fixtures for Sept.-Mar. made in July for 15,000–20,000 dwt at IS – 25%, 39,000 dwt July-Dec. IS – 50%. Therefore estimate for 25,000 Sept.-Dec. is IS – 35%.
Credit: Payment 360 days at 3% (Nig.) 180 days free, 180–360 days at 6% (Basrah).

L39 *PIW*, Oct. 18, 1965.
Take time charter rate of IS – 54%.

L40 *POPS*, Oct. 28, 1965; *PIW*, Jan. 24, 1966.
(a) Basrah: Assume tanker less than 25,000 dwt IS – 25%. Credit: Payment 360 days at 6%, i.e., 0 cost to company.
(b) 25,000 dwt tanker so take freight IS – 30%. Credit: 360 days at 6¢ = 7¢ cost to company.

L41 *POPS*, Mar. 14, 1966, Mar. 16, 1966, May 13, 1966.
(a) Atlas awarded Iranian and Nigerian. Freight: Fixtures made in April for July-August delivery in 30,000 dwt tankers at IS – 65%. These offers for 200 meter ships, 31 ft. draft, so say IS – 50% = 58¢.
Credit: 360 days at 3%. Cost to company is 5.8¢ Iranian, 6.2¢ Nigerian.
(b) Sinclair to supply Arzew 40 – 43.9.
Freight: IS – 50% = 55.5¢
Credit: 360 days with 6% interest after 180 days.

L42 *PIW*, Aug. 1, 1966; *PIW*, Sept. 26, 1966; *PPS*, Sept. 1966, pp. 349–350; *PIW*, July 17, 1967; *PIW*, Aug. 7, 1967; *POPS*, Feb. 21, 1968; *POPS*, May 23, 1968.
Contracts were made in July for the following year July-June. Details were not officially published, but the information from the journals corresponds closely with the import figures for the first and second quarters of 1967. Although after the crisis, *PIW* quoted the pre-crisis spot rates as 32¢ (IS – 70%), c.i.f. prices were mentioned in the July 1966 reports. It is likely that the contracts were on a c.i.f. basis and the prevailing CV rate, IS – 55%, should be taken.

L43 *POPS*, Aug. 15, 1966; *POPS*, Oct. 27, 1966.
Atlas won award for Iranian 34 and Hassi Massaoud at stated c.i.f. prices of $2.01 and $2.33, but 6¢ discount was given for payment in 180 days so effective prices were 6¢ less.
Freight: No comparable fixtures. Estimated IS – 30% for winter delivery.
Credit: Cost to company $1.95 × 6 – 6 = 5.7¢ for Iranian; 7.6¢ for Algerian.

L44 *POPS*, Dec. 8, 1966; *POPS*, Jan. 16, 1967.
Tankers over 25,000 and of 37 ft. draft. Say IS – 45%.
Credit: 360 days at 8¢. Credit cost to Shell = .07 × 210 – 8¢ = 6.7¢.

L45 *POPS*, Feb. 17, 1967; *POPS*, Apr. 19, 1967.
Spot rates paid in February were about IS – 52%–62. Summer rates would be lower, but say IS – 50% because of small harbor.
Credit: Payment 360 days with max. cost 6¢/bbl. Value of credit 63¢.

L46 *POPS*, May 5, 1967; *POPS*, June 30, 1967.
Freight: IS – 55% (IS – 50 was fixed for 40,000 dwt CV June-April).
Credit: 360 days with max. cost 6¢, i.e., 53¢ cost.

TABLE VI-E-1. Continued.

References and Notes–European and Miscellaneous Transactions

E1 ECSC, *Annual Report No. 10*, Table 19.
E2 *POPS*, Apr. 9, 1959.
Freight: IS – 50% is lowest likely freight. Possibly it is higher and corresponding Persian Gulf equivalent lower. The $2.00 checks with Italian imports from USSR which were $1.99 and $1.94 for 1959.
E3 *POPS*, Nov. 9, 1959.
Freight: IS – 50% perhaps too high for spring delivery.
E4 *POPS*, Oct. 27, 1960.
E5 *POPS*, Feb. 4, 1960.
E6 *PPS*, Nov. 1960, p. 418; *POPS*, May 10, 1960; *POPS*, Oct. 17, 1960; *POPS*, Jan. 27, 1961; *OGJ*, Nov. 28, 1967. See text of Appendix for discussion.
E7 *PW*, June 17, 1960, p. 37.
Freight: See E8.
E8 *PW*, June 24, 1960, p. 96.
Same article says Russian charter at Scale – 57 ½% for term.
E9 *PW*, July 22, 1960, p. 14.
E10 As E9.
E11 As E1.
E12 As E4.
E13 *PW*, Nov. 11, 1960, p. 86.
E14 *POPS*, Jan. 25, 1962.
E15 *POPS*, Oct. 9, 1961.
E16 *POPS*, Oct. 1961; *POPS*, Nov. 6, 1961.
E17 *POPS*, Dec. 8, 1961.
E18 *PIW*, Dec. 23, 1963.
E19 As E1.
E20 *POPS*, June 28, 1962; *POPS*, Aug. 1, 1962.
E21 *PIW*, Oct. 29, 1962.
E22 *PIW*, Oct. 22, 1962.
E23 *PIW*, Nov. 12, 1962.
E24 *PIW*, Dec. 23, 1963.
E25 *PIW*, Mar. 11, 1963.
E26 *PIW*, Mar. 25, 1963; *PIW*, Apr. 1, 1963; *PIW*, Apr. 29, 1963; *OGJ*, Apr. 1, 1963, p. 94.
It is likely that the 20% off is an average figure and that Libyan prices should be lower and Middle East ones higher.
E27 *OGI*, June 1963, pp. 238–239.
E28 *PIW*, Oct. 14, 1963.
E29 *PIW*, Nov. 22, 1965.
E30 *PIW*, Nov. 18, 1963; *PIW*, Dec. 23, 1963; *OGJ*, Nov. 25, 1963, p. 56; *PPS*, Dec. 1963, p. 474.
E31 *PIW*, Dec. 9, 1963.
E32 *PIW*, Dec. 23, 1963.
E33 *PIW*, Dec. 23, 1963; *PIW*, Mar. 2, 1964.
E34 *PIW*, Jan. 20, 1964.
E35 *PIW*, Mar. 30, 1964.
E36 (no entry).
E37 *POPS*, Mar. 16, 1966.
E38 Arab Oil, quoted in *Middle East Economic Survey*, Jan. 15, 1965.
E39 *PONS*, Dec. 2, 1964.
E40 *Financial Times*, London, Annual Review of World Oil Market, Dec. 28, 1964.
E41 *WSJ*, Dec. 20, 1964.
E42 *POPS*, Mar. 22, 1965.
E43 *PIW*, Mar. 1, 1965.
E44 *PIW*, Apr. 26, 1965.
E45 *PIW*, Apr. 19, 1965.
E46 *PIW*, Apr. 19, 1965.
E47 *PIW*, Apr. 26, 1965.
E48 *POPS*, Mar. 16, 1966.
E49 *PIW*, Aug. 23, 1965; *OGJ*, Oct. 18, 1965, p. 69.
E50 *PIW*, Sept. 27, 1965.
E51 *OGJ*, Oct. 18, 1965, p. 69. Freight would be about 43¢ which would make the netback in the Persian Gulf 57–81¢. However, a large amount would have to be added on to allow for the poor quality.
E52 *POPS*, Mar. 16, 1966.
E53 *PIW*, Nov. 8, 1965.
E54 *PIW*, Dec. 20, 1965 (from a memorandum published by public relations firms for the Libyan government.)
E55 Venezuela, Ministry of Mines and Hydrocarbons, *Algunos Aspectos de los Actividades Petroleras Venezelanas y Mundiales*, 1965.
E56 *OGJ*, Feb. 7, 1966, pp. 48–49.
E57 *POPS*, Mar. 16, 1966.
E58 *OGJ*, Jan. 17, 1966, p. 48.
E59 *POPS*, Mar. 16, 1966.
E60 *PIW*, Apr. 25, 1966.
E61 *NYT*, Oct. 2, 1966, Financial Section.

E62 *PPS*, Dec. 1966, p. 448.
E63 *PIW*, Dec. 19, 1966.
E64 *Aspectos* . . . (E55), 1966 report
E64 *OGJ*, Jan. 9, 1967, p. 48.
E65 *PIW*, Jan. 16, 1967. Desalting assumed 1¢, stripping 9¢, sulfur and gravity 8¢.
E66 *PIW*, Jan. 30, 1967.
E67 *PIW*, Apr. 10, 1967. Freight: Flat rate estimated as Persian Gulf to Capetown – ½ (P. Harcourt–Capetown) = 91¢/bbl. Recent spot fixtures to E. Africa IS – 42 ½% and IS – 47 ½% for 18,000 and 25,000 dwt tankers, respectively. Take IS – 47 ½% since probably one tanker load.
E68 *PIW*, May 15, 1967.
E69 *PIW*, Dec. 4, 1967.
E70 *POPS*, Mar. 5, 1968.

Appendix VI-F

Crude Oil Prices—Customs Declarations

Usually, import and export figures are not helpful in determining arm's-length prices, since a small arm's-length transaction may well be hidden in an average dominated by intra-company transfers. However, where aggregates are broken down for any reason, one can sometimes see a continuing series of prices which appears to be genuinely a purchase and sale contract. In the following analysis no account is taken of any isolated transaction at a particular price, since there could easily be a special unknown factor or an error in reporting or even in printing.

In Italy, as many as six series of monthly shipments can apply to any one port. German monthly statistics are simple averages from each country of supply, but for every half-year since 1963 there is a further division by nationality of supplier. Since 1966 *POPS* has published import and export figures (mainly from United Nations sources) for a number of countries. Occasionally these yield an apparent arm's-length price, but the frequency is not sufficient to justify spending time on analyzing country sources. Japanese figures are from time to time published in *POPS* in some detail, but are so dominated by averages as to be useless. Soviet figures are discussed in Appendix VI-G and will not be dealt with fully here.

From the various series available only the lower prices were selected. For many reasons, the price declared at the border by an integrated refiner will almost always be *higher* than either arm's-length price or the "real" price to the company. One reason may simply be inertia; long after posted prices were being used only as a means of calculating taxes they continued to appear in import figures. Since profits declared for taxation in the importing countries are usually based on the stated import price, there is an incentive for the integrated companies to make the import values as high as possible. Moreover, U.S. income tax is minimized by a high sales price for crude oil, because the depletion allowance is reckoned at 27.5 percent of the price.

Even arm's-length transactions may be overstated since neither the selling company nor the buyer wishes to publicize how low the selling price is. One way of disguising the actual price is to declare a c.i.f. price with a high f.o.b. component and an unrealistically low freight. A purchase price may be decided between the parties, and then a favorable trade of oil or tanker service or some credit provisions favorable to the buyer is arranged, while only the initial transaction price is reported. However, there seems to be no reason other than a random error why a declared price should be *lower* than the true price. Of course, there may be a transaction that is not representative because it deals with an inferior product. A much less likely case—though constantly cited without proof—is a sale under short-term pressure; but the reported price would be correct for that transaction and can be regarded as the lower limit of a range.

For our purposes here, we are interested not in the date of actual delivery, which the customs figures supply, but in the date at which the contract was made, at which one can only conjecture. This is one of the serious weaknesses of the data. Deciding the applicable freight rate involves a further guess on the date and nature of the contract. Whether the seller or the buyer provides transport, the long-term rate prevailing at the time the contract was made is generally used. However, this is not always correct. Contracts for shipment may be made for different periods by the seller or by the buyer. Moreover, the particular estimates used may make a substantial difference to the netback price. Therefore the data in this appendix are presented in considerable detail so that the estimates can be improved on by anyone who wishes to do so.

German Prices

Monthly prices at the German border are published by *POPS*; on August 17, 1965 a *POPS* supplement gave complete monthly import prices back to 1957 (from data compiled by G. H. Walker & Co.). Such broad averages are clearly useless, but they have been widely quoted and are often taken to represent actual crude prices. The suggestion they convey that declared prices are merely getting more realistic may be seen in Table VI-F-1. The gap between the highest and lowest prices fell from $1.44 in June 1960 to 36 cents in June 1965. The slight rise in U.S.S.R. prices—the only series sure to have no intra-company dealings mixed in—should indicate that the fall in arm's-length prices has not been so precipitous as the monthly data at first sight suggest. The Rotterdam receipts added in the present analysis (see Table VI-F-3 and Appendix VI-B) were for many months below or close to average crude prices, although to make a profit refiners would need receipts at least 35 cents higher than the cost of crude.

Bundesamt für gewerbliche Wirtschaft Aussenstelle (Hamburg) has published for the year 1963, and thereafter semiannually, imports by nationality of supplier for each country of origin. From this we can see that sizable quantities change hands at very much less than the prices indicated by the monthly average. Table VI-F-2 shows all those series that have reported some low prices. Recently, Gabon and Tunisia have been added to the sources of German crude. Each of these has just one nationality of supplier; accordingly, as for the U.S.S.R., the monthly and semiannual series correspond. Venezuelan crudes have been left out of this analysis because there is such a wide range of gravities involved that it is impossible to attach significance to any price without knowing more details about the crude.

In Table VI-F-3 the cheapest crudes are compared with Rotterdam receipts, and the differences can be measured against refinery margins. Through the first quarter of 1964 refiners could comfortably cover their costs. From then through 1965 the margin was only adequate in a few quarters, and even then only with the cheapest available crude. This was an unstable situation; one would expect that product prices would rise, as they did in 1965-66, and/or that crude prices would fall, which is what appeared to happen at the end of 1966.

Finally, Table VI-F-4 shows derived f.o.b. prices and Persian Gulf netbacks for the principal series and for the recent and unusually low Saudi Arabian series. Freight rates assumed were the Time Charter rates from Chapter IV, pp. 112ff.

To put all crudes on the basis of 31° API, an average national gravity was assumed, taking 1.5 cents per degree API (rounding the difference up if necessary for additions or subtractions to an odd number, and down for an even number). This is a somewhat arbitrary procedure, but it shows that the low Saudi Arabian series must be a medium or heavy crude.

Summary. Monthly averages of German border prices fell steadily from 1959 to 1961 and again through 1964 to mid-1965.

U.S.S.R. prices fell through 1959-61 and were remarkably steady thereafter. Probably the Soviet crude would originally have had to be priced lower than comparable crudes to enter the market, but it seems unlikely that the sellers would have to give up as much as 80 cents a barrel to overcome ordinary entry barriers plus some political fears by buyers.

Examination of the semiannual data shows that as far back as 1963 there were some transactions very close to U.S.S.R. prices. There seems to have been a real fall in Libyan border prices from then to the second half of 1966, after which they and some other series rose. However, in the second half of 1966 a number of new low prices were appearing: the Tunisian series at $1.82 and $1.77, the Iran–Switzerland at $1.69, and Saudi Arabia–Great Britain at $1.55 and $1.49. This suggests a new range of contracts at lower prices in mid-1966, but spot tanker rates were unusually low that winter and spring, and the second Suez war masked any evidence there might have been to confirm these lower prices in the second half of 1967.

TABLE VI–F–1. Average West German Border Monthly Prices, 1959–1967

(dollars per barrel)

Country	June 1959	June 1960	June 1961	June 1963	June 1965	June 1967[a]	Change since 1959
Algeria	...	3.23	2.86	2.55	2.15	2.12	–1.01
Libya	...	...	...	2.14	2.12	2.01	...
Tunisia	...	...	...	...	...	1.75	...
Gabon	...	...	...	...	...	1.72	...
Nigeria	...	...	...	2.36	2.22	1.97	...
Saudi Arabia	2.92	2.94	2.67	2.46	2.17	1.96	–0.96
Iran	2.99	2.68	2.57	2.49	2.01	1.83	–1.16
Iraq	2.99	2.89	2.55	2.33	2.09	2.10	–0.89
Kuwait	2.70	2.59	2.01	2.03	1.88	1.86	–0.84
Indonesia	3.05	3.25	2.82	...	...	...	...
U.S.S.R.	2.30	1.79	1.75	1.75	1.79	1.75	–0.55
Average	2.99	2.74	2.48	2.34	2.11	1.96	–1.03
Range incl. U.S.S.R.	0.75	1.46	1.11	0.80	0.36	0.40	n.a.
Range excl. U.S.S.R.	0.35	0.66	0.85	0.52	0.27	0.40	n.a.

Source: *POPS* Aug. 17, 1965, Supplement and later tabulation.

[a]Although hostilities started in early June, border prices were not affected until July.

TABLE VI-F-2. West German Border Prices–Semiannual Figures, Lowest Series, 1963–1967

Price: *$ per barrel*
Volume: *(000 metric tons)*

		1963	1964		1965		1966		1967	
			(1)	(2)	(1)	(2)	(1)	(2)	(1)	(2)
Algeria	Brit. Honduras	... ...	... ...	... ...	... ...	... ...	1.82 (57)	1.84 (446)	1.84 (388)	... ...
	France	... ...	... ...	... ...	... ...	... ...	... ...	... ...	... ...	1.91 (1,019)
Tunisia	Italy	... ...	... ...	... ...	... ...	... ...	... ...	1.82 (365)	1.77 (504)	1.87 (525)
Libya	Brit. Honduras	1.98 (18)	... ...	... ...	1.94 (143)	...	...	1.94 (1)	... ...	... ...
	Libya	1.83 (1,726)	1.83 (1,425)	1.74 (2,224)	1.71 (2,503)	1.69 (2,068)	1.71 (2,609)	1.72 (3,148)	1.78 (3,550)	1.94 (3,375)
	Italy	... ...	... ...	... ...	... ...	2.56 (687)	2.56 (5.82)	2.19 (145)	1.75 (306)	... ...
	Netherlands	... ...	... ...	... ...	... ...	... ...	... ...	1.79 (52)	... ...	... ...
		... ...	... ...	... ...	... ...	... ...	... ...	... ...	... ...	1.86 (304)
Gabon	France	1.85 (32)	... ...	... ...	... ...	1.75 (31)	1.71 (95)	1.72 (63)	1.74 (63)	1.75 (156)
Iraq	Brit. Honduras	... ...	... ...	... ...	1.87 (100)	1.81 (167)	1.85 (155)	1.93 (102)	1.84 (166)	1.91 (183)
Iran	Brit. Honduras	2.28 (172)	1.95 (109)	1.97 (290)	2.16 (219)	... ...	1.86 (171)	1.85 (46)	1.85 (191)	... ...
	Iran	1.85 (148)	2.41 (412)	2.00 (76)	... ...	... ...	... ...	... ...	1.81 (118)	1.87 (33)
	Great Britain	2.25 (4,321)	2.23 (1,680)	2.10 (1,889)	1.92 (1,906)	1.82 (2,030)	1.76 (3,168)	1.86 (2,673)	1.81 (2,750)	... ...
	Switzerland	... ...	... ...	2.37 (83)	2.37 (497)	... ...	... ...	... ...	1.69 (156)	1.69 (51)

TABLE VI-F-2. Continued

Price: $ per barrel
Volume: (000 metric tons)

		1963	1964		1965		1966		1967	
			(1)	(2)	(1)	(2)	(1)	(2)	(1)	(2)
Iran (continued)	Netherlands	...	...	...	...	...	...	1.90	1.86	...
		...	...	...	...	...	...	(7)	(72)	...
	Italy	...	...	...	...	...	...	...	1.79	...
		...	...	...	...	...	...	...	(235)	...
Abu Dhabi	Brit. Honduras	...	...	1.95	1.92	1.86	...	...	...	...
		...	...	(103)	(155)	(745)	...	...	...	...
Saudi Arabia	Switzerland	...	...	...	2.37	...	1.88	1.87	1.93	...
		...	...	...	(370)	...	(69)	(103)	(110)	...
	Brit. Honduras	...	1.96	1.95	1.91	1.86	1.80	...	...	...
		...	(105)	(155)	(129)	(155)	(26)	...	...	...
	Saudi Arabia	1.80	1.97	...	...	...	...	...	...	...
		(23)	(29)	...	...	...	...	...	...	...
	Great Britain	...	...	...	...	...	...	1.55	1.46	...
		...	...	...	...	...	...	(24)	(348)	...
Kuwait	U.S.	2.05	1.83	1.77	1.81	1.86	1.81	1.90	1.90	...
		(1,105)	(309)	(459)	(334)	(491)	(304)	(417)	(231)	...
	Brit. Honduras	1.98	...	...	...	...	...	...	...	...
		(141)	...	...	...	...	...	...	...	...
	Kuwait	...	...	...	...	...	...	...	...	1.91
		...	...	...	...	...	...	...	...	(74)
	Great Britain	...	...	...	...	...	...	1.90	1.84	...
		...	...	...	...	...	...	(633)	(852)	...
U.S.S.R.	U.S.S.R.	1.72	1.78	1.79	1.79	1.79	1.79	1.75	1.74	1.79
		(1,997)	(1,357)	(1,469)	(1,203)	(1,382)	(1,592)	(1,693)	(1,931)	(2,133)

Source: *Bundesamt für gewerbliche Wirtschaft*, Aussenstelle, Hamburg.

TABLE VI-F-3. West German Border Prices, 1963-1967—Comparison of Product Receipts with Lowest Crude Oil Prices

(dollars per barrel)

Year and quarter		Rotterdam receipts (quarterly avgs)	Libya-Libya	Kuwait-U.S.	Iran-G.B.	U.S.S.R.	Difference
1963	1	2.56	1.78	...	...	1.67	.78-.89
	2	2.42					.64-.75
	3	2.38					.60-.71
	4	2.33					.55-.66
1964	1	2.58	1.78	1.78	...	1.73	.80-.85
	2	2.14					.36-.41
	3	1.98	1.69	1.72	...	1.74	.24-.29
	4	2.01					.27-.32
1965	1	2.05	1.66	1.76	...	1.74	.29-.39
	2	1.96					.20-.30
	3	2.00	1.64	...	1.77	1.74	.23-.36
	4	2.06					.29-.42
1966	1	2.26	1.66	...	1.71	1.74	.52-.60
	2	2.02					.28-.36
	3	2.07	1.67	...	1.81	1.70	.26-.40
	4	2.15					.34-.48
1967	1	2.10	1.73	...	1.76	1.69	.34-.41
	2	1.97					.21-.28

Sources and Notes: Rotterdam receipts—table VI-B-1; Crude Prices: table VI-F-2; 5 cents deducted for Rotterdam-Ruhr pipeline.

TABLE VI-F-4. West German Border Prices, 1963-1967—Derived Netback at Origin and at Persian Gulf (after sulfur allowance)

	Assumed gravity API	1963	1964 (1)	1964 (2)	1965 (1)	1965 (2)	1966 (1)	1966 (2)	1967 (1)
Libya-Libya									
C.i.f. Germany	39°	1.83	1.83	1.74	1.71	1.69	1.71	1.78	1.78
F.o.b. Marsa el Brega..........		1.58	1.60	1.51	1.48	1.46	1.48	1.55	1.55
F.o.b. P. Gulf	31	1.01	1.09	1.00	0.98	0.96	0.98	1.05	1.06
Kuwait-U.S.									
C.i.f. Germany	31	2.05	1.83	1.77	1.81	1.86	1.81	1.90	1.90
F.o.b. Mena-al-Ahmadi		1.41	1.25	1.19	1.23	1.28	1.24	1.33	1.34
F.o.b. P. Gulf	31	1.46	1.30	1.24	1.23	1.28	1.30	1.39	1.40
Iran-Gt. Brit.									
C.i.f. Germany......	32	2.25	2.23	2.10	1.92	1.82	1.76	1.86	1.81
F.o.b. P. Gulf		1.59	1.64	1.51	1.33	1.23	1.18	1.28	1.24
F.o.b. P. Gulf	31	1.56	1.63	1.50	1.32	1.22	1.18	1.28	1.24
U.S.S.R.-U.S.S.R.									
C.i.f. Germany......	32	1.72	1.78	1.79	1.79	1.79	1.79	1.75	1.74
F.o.b. Black Sea		1.31	1.41	1.42	1.42	1.42	1.42	1.38	1.38
F.o.b. P. Gulf	31	1.06	1.18	1.19	1.19	1.19	1.21	1.17	1.17
S. Arabia-Gt. Brit.	27	...	...	...	...	...	...	...	...
C.i.f. Germany		...	...	...	...	...	...	1.55	1.46
F.o.b. Ras Tanura ...		...	...	...	...	...	...	0.97	0.90
F.o.b. P. Gulf	31	...	...	...	...	...	...	1.11	1.02

Sources and Notes: Crude oil prices from table VI-F-2; Freight rates from chap. IV.
Rotterdam-Ruhr pipeline = 5 cents.
Lavera-Karlsruhe pipeline = 12 cents (see appendix VI-H).
Sulfur allowance, see appendix VI-H.

Italian Prices and Tunisian Exports

Italian prices. Imports by loading port and by currency of payment are listed by *Bollettino Petrolifero* (published monthly by the Ministero dell'Industria del Commercio e dell'Artigianato). Currencies are usually dollars, sterling, or lire, but occasionally others are used. There are two series, *Definitiva* and *Temporaneo*, the latter being for re-export or delivery to a third party. Thus, for any port there are at least six possible series. Quantity and f.o.b. value are given, so that one can calculate f.o.b. prices. Freight costs are listed separately. These are also divided by currency of payment. One can get an average freight cost for all the oil exported from one port, but cannot match the freight cost to the crude paid in a particular currency unless (as often happens) the exports were only in one currency. Annual figures are available back to 1956, monthly data from 1963. *POPS* publishes the data monthly, but does not subdivide by currency, thereby losing much information.

Table VI-F-5 summarizes the low Italy c.i.f. prices as of May 1967 and their Persian Gulf netback. Because recorded freights are generally higher than the Intascale less 53 percent used here, our computed netted back f.o.b. prices are higher than the reported f.o.b. prices. The exception is Libya, where reported freights are lower, and reported f.o.b. prices higher than the computed netback.

Table VI-F-6 shows realizations as of May 29, 1967 (prices from *POPS*, quantities from *IPQ*, April 1967, for 1965). This is almost the same as our low c.i.f. prices, and is therefore not compatible. For example, it is impossible to buy Kuwait crude around $1.65 c.i.f. and sell it at $1.72, for it must cost much more than 7 cents to refine.

The anomaly is not as great as it might at first appear, but even after allowance for several factors I cannot claim to have reconciled the two bodies of data.

1. Most of the imports are landed in the shallow and farther removed harbors of Genoa and Naples, adding several cents to freight costs, while the product prices are at the large export refineries on deep water in Sicily and Sardinia.

2. The pattern of refinery yields shown in Table VI-F-6 cannot be identified with any particular crude, nor therefore with any particular price. Even Kuwait crude (31°, $1.65 c.i.f.) has a naphtha yield of 25.4 percent,[1] much above 14.2 percent. The two important Russian crudes, Tuimazy (31.8°) and Romaschkino (30.5°) have respective naphtha yields of 29 and 27 percent.[2] If we suppose a refinery running Libya crude (Dahra 39°), realizations are approximately as shown in the tabulation below.

Product	*Percent*[3]	*Gallons*	*Price (¢/gal)*	*Revenue*
Gasoline	36.6	15.4	5.45	.840
Gas oil	28.9	12.2	4.95	.604
Heavy fuel oil	13.8	7.8	3.31	.258
Other	10.9	4.6	5.10	.234
Loss	6.0	2.1	0	0
Total	100.0	39.9	...	1.936

Thus a Libyan crude would return 29 cents over its landed cost of $1.65—a margin still too low to be tolerable, but closer to, or inside, the margin for error.

[1] Harold M. Smith, *Hydrocarbon-Type Relationships of Eastern and Western Hemisphere High-Sulfur Crude Oils.* (U.S. Bureau of Mines, Report of Investigation No. 6542, 1964) p. 40. (The class "naphtha" ends at a boiling point greater than 390° F.)

[2] *Oil and Gas International*, May 1966, p. 75. (The class "naphtha" ends at a boiling point greater than 400° F.)

[3] E. P. Ferras and Dorothy T. Nichols, *Analysis of 38 Crude Oils from Africa* (U.S. Bureau of Mines, Information Circular No. 8293, 1966), item 9. Distillation loss given as 1.2 percent subtracted from heavy fuel oil fraction.

Therefore one must conclude that either our composite product price is too low or the crude estimates too high. And indeed the Italian prices for Libyan crude are higher than what can be calculated from German prices, from the Oasis companies' annual reports, or from the individual observations. One striking fact is that most of our usable information is for high-gravity crude.

Tunisian exports are of interest because they are sent out by a single supplier (ENI) and involve basically one type of crude. This is so light (44° API) that a simple 1½ cents per degree probably would not apply. If we assume the value is similar to a 40° crude the total adjustment for quality including 5 cents for sulfur would be 19 cents. Most of the shipments through the first half of 1967 in Table VI-F-7 fall in the range $1.58 to $1.64, which would correspond to $1.02-$1.08 for a Gach Saran type crude at the Persian Gulf. The 1968 prices do not differ noticeably from the 1967 ones and one can only assume that they were arrived at before the crisis.

TABLE VI-F-5. Italian Low Prices, May 1967

(*dollars per barrel*)

Origin	Assumed API	Reported f.o.b.	Reported c.i.f.	Persian Gulf equiv. Inta-57	Persian Gulf equiv. Inta -57, 31°, sulfur allowance
Iraq					
Mediterranean	36°	1.60	1.75	1.33	1.24
Persian Gulf	36°	1.20	1.60-1.75	1.18-1.33	1.09-1.24
Iran	31°-34°	1.16	1.72	1.30	1.24-1.30
Kuwait	31°	1.20	1.65	1.23	1.28
Algeria	40°	1.60-1.75	1.75-1.80	1.33-1.48	1.14-1.30
Libya					
Es Sider	39°	1.45-1.55	1.55-1.65	1.23	99-1.09
Ras Lanuf	37°	1.46	1.55	1.13	0.99
U.S.S.R.	32°	1.40	1.65	1.23	1.21

Source: Bollettino Petrolifero, published monthly by Italian Ministero dell'Industria del Commercio e dell'Artigianato.

TABLE VI-F-6. Italian Realizations, May 1967

	Percent	Gal/bbl	$/ton	¢/gal	$/bbl (2) × (4)
	(1)	(2)	(3)	(4)	(5)
Gasoline	14.2	5.96	19.50	5.45	.325
Kerosine and distillate	31.2	13.10	15.7	4.95	.648
Heavy fuel oil	39.2	16.46	9.15	3.31	.545
Other[a]	9.3	3.91	. . .	5.10	.199
Losses, waste etc.	6.1	0	0	0	0
Realization	. . .	. . .	. . .	. . .	1.717

Sources: Prices, *POPS*; Output, U.S. Bureau of Mines, *International Petroleum Quarterly*, April 1967.

[a]Weighted average of gasoline and middle distillate prices.

TABLE VI-F-7. Tunisian Crude Oil Export Prices, 1966-1968

Price: $ per barrel
Volume: (000 metric tons)

	Belgium	France	Italy	Switzerland	W. Germany
Jan-Sept 1966	...	...	1.58 (144)	...	1.58 (95)
Oct-Dec 1966	1.61 (163)	...	...	...	1.73 (363)
Jan-Mar 1967	...	1.69 (29)	...	1.56 (99)	1.61 (239)
Apr-June 1967	1.61 (23)	1.64 (43)	1.74 (118)	1.60 (70)	1.62 (176)
Jan-Mar 1968	...	1.62 (99)	1.59 (68)	1.62 (129)	1.60 (113)
Apr-June 1968	...	...	...	1.60 (152)	1.58 (255)
July-Sept 1968	...	1.72 (75)	1.58 (17)	1.60 (168)	1.59 (288)
Oct-Dec 1968	1.64 (43)	1.64 (107)	...	1.62 (208)	1.62 (414)

Source and Notes: *POPS*, July 25, 1967; Sept. 15, 1967; Jan. 19, 1968; Oct. 11, 1968; Mar. 21, 1969; July 1, 1969; Oct. 31, 1969; Dec. 3, 1969. Freight differential from Persian Gulf to Rotterdam through Suez 46 cents at Intascale flat plus 12 cents Suez, which would be 37 cents at Intascale less 55%.

Appendix VI-G

Soviet Crude Oil Export Prices

Table VI-G-1 shows f.o.b. prices to some of the principal non-Communist importing countries from 1955 to 1967.[1] These are based on U.S.S.R. official trade figures. Since there is only one agency to consider, the figures are easier to check than the statistics of non-Communist countries. They can be compared with the import figures of other countries and with various press reports. On the whole, the figures check well. The main exception appears to be the Italian f.o.b. prices which differ by up to 10 cents a barrel in 1965 between the Italian and Soviet sources, although the general trend is the same and the prices are the same for the end years. The difference is not explained by definition of importing country. The Italian statistics refer to *definitiva* imports; *temporaneo* were in all years except one a few cents higher. One possibility is that the price may have been determined on an Italian c.i.f. basis and the countries differed on the amount to be allocated for freight. This might also explain why there was so much uncertainty about the two large contracts with ENI in 1961 and 1963. The 1961 figure falls within the range of press estimates, but the 1963 guess is much lower than either country's f.o.b. price.

One problem is that one rarely knows the gravity of the crude; apparent shifts in price might merely reflect quality changes. Another is that almost all the contracts reported involve some other transaction, such as a loan or a purchase of goods (commodities from the Latin American countries; equipment or finished products from the more developed countries). The value of ordinary loans and credit can be estimated, but there is no basis for guessing the market value of changed products. One comfort is in the report of the French-Soviet agreement of October 1964. "The exchange of goods between the two countries shall be transacted at world market prices as checked annually by a special committee" (*EOT*, Nov. 16, 1964). Generally, for trade with developed countries where many goods are involved one would expect market prices. For a single commodity, such as coffee or wool, there is more likely to be some extra cost to the oil supplier; hence the net oil price may be lower. At least two transactions are thought to be on a netback basis: the supply to the Swiss refinery and an arrangement with an Italian company (*PIW*, Aug. 23, 1965; Sept. 13, 1965). Possibly other contracts were also on this basis in order that changes in crude price would follow product price changes rather than cause them. (But the existence of netback arrangements would tend to weaken product prices since the refiner would have no incentive to try to hold them up.)

The prices in Table VI-G-1 show no evidence that the U.S.S.R. is behaving differently from any other large organization trying to sell in established markets. Prices are lowest to the most competitive markets, Italy, Germany, and Japan. Those to Japan are

[1]Soviet oil trade statistics for 1968 were changed in such a way as to make it impossible to compute prices for crude oil or for products.

especially low because the U.S.S.R. is at a geographical disadvantage compared with tne Middle East. Finally, at least since early 1962 the Soviets have been described in the Western press as trying to raise prices (*PPS*, June 1962, p. 227).

Table VI-G-2 shows the Persian Gulf equivalent of Soviet prices for the largest markets from 1960. The calculation of a freight rate, Black Sea to destination, is rough at best. Voyages from the Black Sea would be in smaller ships than those from the Persian Gulf to Europe or Japan, because the Black Sea harbors could only accommodate smaller vessels. One also has the usual problems in trying to guess the terms of the crude shipment contracts. Voyages from the Black Sea are at ten percent of Intascale over the current long-term charter rate (see Chapter IV). Voyages from the Persian Gulf to Europe, Japan and Brazil are at the time-charter rate. Voyages to Argentina and Uruguay (River Plate) are at the same rate as from the Black Sea, because of the same need for small ships.

The prices to Japan and Brazil are hard to credit. Both involve long voyages. It is likely that the f.o.b. price is a fiction, the result of an arbitary low tanker rate.

TABLE VI-G-1. U.S.S.R. Crude Exports, Selected Countries, 1955-1967

Price: *$ per barrel f.o.b.*
Volume: *(000 metric tons) 1960 and after*

	1955	1956	1957	1958	1959	1960	1961	1962	1963	1964	1965	1966	1967
W. Germany	...	...	...	...	1.69	1.39 (1,249)	1.27 (1,572)	1.30 (1,915)	1.30 (2,215)	1.39 (2,559)	1.40 (2,583)	1.40 (3,337)	1.40 (4,209)
Italy	1.95	2.18	2.68	2.00	1.75	1.42 (3,921)	1.31 (5,514)	1.30 (6,083)	1.24 (6,727)	1.22 (7,017)	1.21 (6,588)	1.22 (8,021)	1.46 (10,576)
France	2.00	2.08	2.48	2.24	1.76	1.61 (132)	1.52 (109)	1.51 (89)	1.58 (93)	1.43 (223)	1.26 (779)	1.25 (1,659)	1.30 (1,774)
Switzerland	...	...	...	...	...	...	...	...	...	1.24 (178)	1.26 (917)	1.35 (599)	...
Spain	...	...	...	...	...	...	...	...	...	1.55 (114)	1.54 (381)	1.53 (164)	1.79 (457)
Greece	...	...	...	...	2.05	1.94 (424)	1.85 (469)	1.85 (410)	1.84 (457)	1.84 (465)	1.76 (422)	1.74 (468)	1.75 (672)
Morocco	...	...	...	2.06	1.94	1.83 (44)	1.65 (111)	1.61 (158)	1.62 (284)	1.58 (301)	1.53 (417)	1.63 (525)	1.59 (583)
U.A.R.	1.95	1.93	2.41	2.23	1.80	1.63 (700)	1.43 (1,537)	1.45 (1,182)	1.68 (927)	1.73 (706)	1.76 (688)	1.79 (964)	1.80 (1,004)
Japan	...	...	...	1.78	1.68	1.34 (1,184)	1.26 (2,234)	1.26 (2,136)	1.21 (2,026)	1.18 (2,483)	1.20 (2,317)	1.23 (2,786)	1.42 (1.798)
Argentina	2.12	...	...	1.61	1.52	...	...	...	...	...	1.31 (403)	...	...
Uruguay	...	...	...	2.35	2.16	2.12 (15)	1.78 (24)	...	...	...	...	...	...
Brazil	...	...	...	...	2.14	1.73 (35)	1.65 (380)	1.61 (155)	1.61 (566)	1.60 (1,877)	1.53 (2,354)	1.45 (2,195)	1.35 (582)

Sources: NPC, *Impact of Oil Exports from the Soviet Bloc*, Supplement 1964, Appendix 7, A-14. *PONS*, Nov. 1, 1965; *POPS*, Sept. 30, 1966; *PONS* Supplement Feb. 1, 1968.

TABLE VI-G-2. U.S.S.R. Crude Exports, Estimated Price Persian Gulf Equivalent, 1960-1966

(*dollars per barrel*)

	1960	1961	1962	1963	1964	1965	1966
W. Germany	1.15	1.03	1.07	1.07	1.08	1.19	1.19
Italy	1.24	1.07	1.07	1.01	1.01	1.00	1.01
France	1.37	1.28	1.28	1.35	1.22	1.05	1.04
Switzerland	...	...	...	...	1.03	1.02	1.14
Japan	1.80	1.72	1.71	1.66	1.61	1.63	1.66
Argentina	...	...	...	...	...	1.27	...
Uruguay	2.07	1.73	...	...	...	...	...
Brazil	1.74	1.66	1.62	1.62	1.64	1.57	1.49

Sources: As table VI-G-1. For details of calculation see text.

Appendix VI-H

Miscellaneous Estimation Factors

Pipeline Costs

To derive a Persian Gulf netback from German border prices one must estimate the cost of moving crude along the two principal pipelines, Rotterdam-Ruhr and Lavera-Karlsruhe. The tariff charged is not relevant, since the pipelines are owned by the shippers. Whoever finally buys the crude, it is likely to be a pipeline owner who is responsible for moving it.

Rotterdam-Ruhr Pipeline. This was built in 1960 (*OGJ*, July 18, 1960, p. 84) as a 24-inch line then carrying 170,000 barrels per day (175 TBD). Capacity was planned to be increased to 400 TBD with the addition of two pumping stations. The Rotterdam-Godorf branch is 160 miles long. Using Hubbard,[1] the minimum cost (fixed plus variable) for a 24-inch line is 0.17 pence per ton-mile. This (before British devaluation) is close to 0.20 cents per ton-mile, 2.64 cents per barrel for 100 miles, or 4.23 cents for the line. This is the minimum cost assuming full utilization and excludes the costs of terminals, so it must be somewhat low.

An alternative estimate is from the initial capital outlay. This was $28 million for an initial capacity of 170 TBD, or $165 per daily barrel. It is assumed that terminal costs are included. An annual capital charge of 10 percent would be a per-barrel capital charge of 4.5 cents. However, since expansion was planned this cost must include some provision for the enlarged capacity.

Operating costs (from Hubbard, fig. 10) are about 0.05 cents per ton per mile or 1.07 cents per barrel for the 160-mile pipeline, making 5.59 cents total cost. If it is assumed that the capital expenditure included provision for expansion one could take 5 cents as a working figure.

South European Pipeline. (*OGJ*, Feb. 12, 1962, p. 194.) This is 475 miles, 34 inches, and cost $121,380,000. Initial capacity was 200 TBD, but was to be expanded to 500 TBD in about one and one-half years for a further $19.6 million. A total of $140,980,000 for 500 TBD is $282 BD; or at an annual capital charge of 10 percent, 7.72 cents per barrel. Operating costs with three pumping stations 150 miles apart would be (interpolating from Hubbard, fig. 10) 0.036 pence per ton-mile or 2.66 cents per barrel for the whole pipeline. The total cost therefore, is about 10.4 cents per barrel.

Estimating total costs directly from Hubbard, minimum total cost is a little less than 0.12 pence per ton-mile or 0.14 cents per ton-mile, 66.5 cents for the whole line and 8.9 cents per barrel. However, he assumes no terminal construction and easy terrain, so the true figure must be higher. The figure used in the present analysis is 12 cents.

[1] M. E. Hubbard, *The Economics of Transporting Oil to and Within Europe* (London: Maclaren and Sons, 1967), pp. 15, 22, and fig. 11, p. 55.

Estimation of Gravity of Crude Oil, Corresponding to Average European Refinery Pattern

The representative European crude has been taken to be 31° API. This is based on refinery output of the main products. The calculation for 1966 is shown in Table VI-H-1. The gravity is estimated for each product group and a weighted average taken. For 1960 to 1965 the gravity varied from 31.1° to 31.4°.

Estimating the Value of Gravity Differentials

The value of a barrel of crude depends on the prices realized from the products into which it is refined. The greater the yield of the more valuable naphtha and gas oil, and the less the heavy (residual) fuel oil, the greater the value of the crude oil. The API gravity scale was constructed as a means of stating small differentials in specific gravity which resulted from small differences in yield of light ends versus residual. The higher the API gravity, the greater the yield of light ends.

For many years in the United States and elsewhere, the customary allowance was 2 cents per degree API gravity, and it continues as a rule of thumb, for lack of any better simple rule to express a rather complex structure. I have tried to improve it for application to Western Europe during the 1960s.

Table VI-H-2 shows the variation in realization for some important crude oils. The prices used are those prevailing during the last 32 months (from September 1964) prior to June 1967, when the six-day war stopped the clock. The difference between the price of heavy fuel oil and a weighted average of gas oil and gasoline varied from about 1.4 to 2.4 cents per gallon with a median of 1.8 cents.

Before January 1964 and back to July 1960 the difference was more commonly 3.0 cents a gallon. Nonetheless, the data derived from the later period is used here even for the earlier one. Exclusion of the earlier data is a personal act of judgment. Readers unwilling to accept it can, from the data in Table VI-B-1 (p. 365), construct a table of differentials which will work out to somewhat higher values. Since the price relations being sought relate to permanent advantage or disadvantage in using one or another crude, which would induce parties to sign a contract at a somewhat higher or lower price, I believe less error is involved in the admittedly rough and ready procedure adopted here.

The figures in Table VI-H-2 are taken from two Bureau of Mines reports.[2] Unfortunately the Gach Saran sample was not representative, so we have to use a 34° crude as base.

For the Middle East and North African crudes listed the difference in realization per degree (assuming 1.8 cents per gallon difference between light products and heavy fuel oil) runs roughly from 1.1 to 2½ cents. However, the Indonesian crudes and one of the two Russian crudes are much higher. Taken for use here is 1.5 cents as the central figure.

For Japan, Suzuki has estimated the product value of a barrel of Iran 34° crude to be 16 cents above a Kuwait 31°, or over 5 cents per degree.[3] But the comparison is at "standard price" and perhaps overstated. Moreover, about 7 cents of this would be for sulfur, leaving 3 cents per degree.

Sulfur Content

Sulfur content of crude oil has become more important in the last few years as local and national authorities have placed limits on the sulfur content of heavy fuel oil. Not enough data are available for a precise estimate, but an approximation can be made.

[2]Harold Smith, "Hydrocarbon-Type Relationships of Eastern and Western Hemisphere High-Sulfur Crudes" (U.S. Bureau of Mines, Report of Investigation No. 6542, 1964); E. P. Ferrero and Dorothy T. Nichols, *Analyses of 38 Crude Oils from Africa* (U.S. Bureau of Mines, Information Circular 8923, 1966).

[3]M. Suzuki, "Competition and Monopoly in World Oil Markets," Supplement to *Energy in Japan*, No. 4 (Feb. 1968), p. 23.

First, the price difference between high and low sulfur heavy fuel oils must be determined (sulfur is not an important factor in the price of the other products). *AFM* gave separate quotations for 1.0-percent-maximum-sulfur heavy fuel oil in October 1960–March 1961. The differences ranged from 30 cents to $1.50 a ton (4.5 to 23 cents a barrel). *POPS* in 1967 and 1968 sometimes gave quotations for low-sulfur heavy fuel oil for Italy and Rotterdam. Differences ranged from zero to $1.30 a ton with a median of about 85 cents (13 cents a barrel). Low sulfur fuel oil presumably refers to 1.0 percent (this figure is sometimes but not always stated). Heavy fuel oil with no limits stated would be about 3.5 percent. Thus a difference of 13 cents per barrel between high and low sulfur fuels would be equivalent to about 5.2 cents per barrel per percentage point. Since late 1969, separate quotations have been given for 3.5 percent and for 1.0 percent or less. The premium for low sulfur has sometimes been negative, with a high of $3.00 per ton, and a median of about $2.00.

PIW (Nov. 18, 1968) quoted the prices for basic grade Bunker C with sulfur up to 3.5 percent as $10.00–10.25 per ton f.o.b. Rotterdam, while premium fuel oil, 0.5 percent to 1.0 percent sulfur is $1.50 higher. The pourpoint of the former is lower, 60°F rather than 68°F. If one ignores the pourpoint difference, the sulfur content alone would account for 7.6 to 9.1 cents per barrel per percentage point.

The other specific pieces of information refer only to the Western Hemisphere. In April 1968 (*POPS*, Apr. 26, 1968) heavy fuel oil was said to be $1.64–$1.65 for summer delivery for 2.8 sulfur product and $1.67 for 2.2 percent, while for winter delivery the two prices were $1.71 and $1.74–$1.75. Thus a 0.6 percent difference in sulfur content was worth 2 to 4 cents a barrel, or 3.3 to 6.7 cents per barrel per percentage point. The regular *POPS* Caribbean quotations gave a difference between 2.2 percent and 2.8 percent product from March 1967 to March 1968 of 5 cents and 3 cents after that, i.e., 5 to 8.4 cents difference per percentage point. However, the difference between 1.0 percent and 2.2 percent limits was 30 to 35 cents from May to July 1967, the only period the 1.0 percent maximum was quoted. This difference of about 30 cents per percentage point is hard to reconcile with the others recorded. A 5 cent difference is used here. The sulfur premium was much higher and more variable after late 1969.

The sulfur premium or penalty on a crude oil can be determined from the value of the premium for heavy fuel oil and its yield. For example, Gach Saran yields 42 percent heavy fuel oil which contains 2.46 percent sulfur. In round numbers, the sulfur component reduces the value of the crude by 5 cents × 2.5 × 0.40 = 5 cents. Kuwait crude yielding 50 percent heavy fuel oil containing 4 percent sulfur would bear a charge of 5 cents × 4 × 0.50, or 10 cents. Thus the extra sulfur in Kuwait crude would make it worth 5 cents less than Gach Saran.

Table VI-H-3 lists some of the principal Eastern Hemisphere crudes (excluding Soviet bloc). Gach Saran, whose yield corresponds closely to the European average, is taken as a base. Kuwait and Arabian heavy have penalties of 5–6 cents, while all the North African and Indonesian crudes have premiums of 4–5 cents (column 6).

This method does not work where we do not know the sulfur content of the residual. Inspection of the data in hand shows that in most cases sulfur concentration in the crude is 45 to 65 percent of that in the derived heavy fuel oil. (The exceptions are Kirkuk, Minas, and also the crudes heavier than 20° API.) As a rule of thumb a 50 percent proportion is taken. Column 8 in Table VI-H-3 shows the sulfur concentration of heavy fuel oil estimated by this method and the actual concentrations. The fit is closer than if it were assumed that all the sulfur in the crude is retained in the residual (column 10).

Since the second Suez crisis, there is more information about the premium for low sulfur fuel oils, but conditions have changed. Differences in sulfur content are more important when the total amount is small than when it is large. At levels around 1 percent sulfur a change of one tenth of a percent can affect the price by 3 or 4 cents. For

Libyan crudes, however, it is counterbalanced by the high wax content and consequent high pourpoint of the residual fuel oils. In the Italian export market Libyan fuel oils have sometimes been at a discount, sometimes at a premium, compared with "regular" Middle East products. The United States is now a principal market and the discount (or premium) depends on requirements and availabilities there.

TABLE VI–H–1. Representative European Crude, 1966

Product	Output (*000 bbl/yr*)	API° assumed	(2) × (3)
(1)	(2)	(3)	(4)
Gasoline	441		
Jet fuel	72		
"Other"[a]	138		
Subtotal (1)	651	56.0	3,646
Kerosine	44		
Distillate fuel oil	764		
"Other"[a]	218		
Subtotal (2)	1,026	35.0	3,591
Residual fuel oil	1,026		
Refinery fuel and loss	185		
Lubricants	29		
	1,240	14.5	1,798
Total	2,917	...	9,035
9,035/2,917 = 31.0			

Source: Output–U.S. Bureau of Mines, *International Petroleum Annual, 1966*, table 2. Gravity approximated from *PPS*, middle of respective range.

[a]Other products (356,000 barrels per year) allocated to gasoline and middle distillate by ratio to total light products.

TABLE VI-H-2. Crude Realization Differentials Related to Amount of Residual Fuel Oil (based on Saudi Arabia Ghawar 34.4°) in Ascending Order of Gravity for Each Major Region

	Gravity (API°)	Sulfur (*percent*)	Gasoline naphtha (. . . *percent yield* . . .)	Middle distillates	Residual fuel oil	Difference in realization (¢/*bbl*)	(¢/*degree*)
			PERSIAN GULF				
Saudi Arabia							
Safaniya	27.7	3.03	22.2	37.0	39.3	−7.7	−1.2
Kuwait							
Burgan	31.0	2.54	25.4	37.3	35.8	−5.0	−1.6
Iran							
Gach Saran	33.0	1.66	27.2	36.5	32.1	−2.2	−1.6
Bahrain							
Bahrain	33.0	2.14	27.7	38.2	31.6	−1.8	−1.3
Saudi Arabia							
Ghawar	34.4	1.69	28.5	40.9	29.2	0	0
Iran							
Agha Jari	34.6	1.43	28.8	40.0	28.9	[a]	[a]
Iraq							
Kirkuk	36.6	1.93	35.5	41.0	23.3	+4.5	+2.0
			NORTH AFRICA				
Libya							
Dahra	33.4	0.56	21.5	52.5	24.9	+3.3	+3.3
Algeria							
R. el Baguel	37.2	0.31	33.1	46.0	19.2	+7.6	+2.7
Libya							
Hofra	38.0	0.33	33.8	46.9	18.6	+8.0	+2.2
Zelten	39.2	0.23	31.0	43.2	22.3	+5.2	+1.1
Dahra	41.7	0.24	40.2	40.6	16.3	+9.8	+1.3
Hofra	41.7	0.32	39.9	41.3	15.5	+10.4	+1.4
Algeria							
Gassi Touil	45.5	0.02	32.6	54.1	12.3	+12.8	+1.2
			INDONESIA				
Minas	32.8	0.10	18.6	43.6	37.3	−6.1	−3.8
Seria	36.8	0.10	37.1	55.4	7.4	+16.5	+7.5
			U.S.S.R.				
Romashkinskaya . .	30.5	1.7	26.0	39.2	34.8	−4.2	−1.1
Tuymazy.	31.8	1.3	28.1	33.5	37.3	−6.1	−1.3

Sources and Notes: See preceding text, note 2.

[a]Difference too small.

TABLE VI–H–3. Sulfur Content of Crude and Heavy Fuel Oil and Sulfur Premium

Crude	Gravity (API°)	Sulfur content		HFO yield	Sulfur cost	Sulfur content HFO	Sulfur content of HFO		
		Crude	HFO		5 × (4) × (5)	Sulfur content crude (3) ÷ (4)	Expected (i) 2 × (3)	Actual (4)	Expected (ii) (3) ÷ (5)
(1)	(2)	(3)	(4)	(5)	(6)	(7)	(8)	(9)	(10)
AFRICAN									
Algeria									
Ohanet	43.3	0.06%	0.13%	32.5%	0.2¢	46%	0.12%	0.13%	0.15%
Hassi Messaoud	43.4	0.14	0.37	21.4	0.4	38	0.28	0.37	0.65
Zarzaitine	40.8	0.07	(0.07)	31.6	0	...	...	...	...
Libya									
El Hofra	42.3	0.31	0.59	28.5	0.8	53	0.62	0.59	1.1
Dahra	41.0	0.36	0.82	11.3	0.5	44	0.72	0.82	3.2
Brega	40.2	0.22	0.60	13.0	0.4	37	0.44	0.60	1.7
Bunker Hunt	37.6	0.14	0.29	21.1	0.3	48	0.28	0.29	0.67
Nigeria									
Boma	32.9	0.14	0.29	29.9	0.4	48	0.28	0.29	0.47
MIDDLE EAST									
Iran									
Agha Jari	34.1	1.34	3.04	21.6	3.3	44	2.7	3.0	6.1
Gach Saran	31.4	1.60	2.46	42.4	5.2	65	3.2	2.5	3.8
Darius	34.1	2.47	4.67	34.1	8.0	53	4.9	4.7	7.3
Iraq									
Rumaila	34.1	2.1	3.6	35.8	6.4	58	4.2	3.6	5.9
Kirkuk	36.3	1.95	6.21	12.6	3.9	31	3.9	6.2	15.5
Saudi Arabia									
Mixture	34.5	1.63	3.58	22.7	4.1	46	3.3	3.6	7.2
Khursaniyah	31.5	2.67	4.55	30.0	6.8	59	5.3	4.6	8.9
Safaniya	27.5	2.95	5.09	40.8	10.4	58	5.9	5.1	7.2
Wafra	16.9	4.70	4.25	42.5	12.5	80	9.4	5.9	11.1
Kuwait	31.4	2.50	4.06	50.0	10.2	62	5.0	4.0	5.0
Abu Dhabi	40.5	0.71	1.58	23.0	1.8	45	1.4	1.6	3.1
Qatar	41.0	0.96	2.3	23.1	2.7	42	1.9	2.3	4.2
INDONESIAN									
Seria	36.7	0.1	0.2	16.5	0.2	50	0.2	0.2	0.6
Minas	35.8	0.08	0.4	35.0	1.4	20	0.16	0.4	0.23
Duri	22.8	0.18	0.2	50.0	0.5	0	0.36	0.2	0.36

Source: *OGJ*, Apr. 15, 1963; Aug. 22, 1966; Mar. 27, 1967; July 10, 1967.

Appendix VI-I

Crude Prices, 1967–1970

Tables VI-I-1 and VI-I-2 list press reports of Libyan and Persian Gulf crude prices. They vary in scope from spot offers to long-term contracts and from precise details of transactions to unconfirmed reports.

There are many more observations on Persian Gulf prices, particularly Iranian, than were available before the Suez closure, but mostly they are for Japanese buyers. It is generally believed that the old differential against Japan has disappeared (*PIW*, Oct. 4, 1968); at least we have seen no evidence of it. In July 1968 and July 1969 Iranian heavy for new business could command \$1.20 a barrel. Two December 1969 transactions suggest lower prices. The Dubai at \$1.20 is for a similar quality crude as our standard, but enjoys a freight advantage of 6 cents flat. The Rostam, for which a trial cargo was shipped to Japan at \$1.18, is similar to Iranian light and is equivalent to about \$1.13 for Gach Saran. This inference is confirmed by the last item, showing a bottom in early 1970 of Gach Saran 31° around \$1.16.

Libyan crudes seemed to rise until spring 1969, but fell to the end of the year. By December, contracts for 1970 at \$1.70 for Es Sider were very little higher than before the Suez war in spite of the year-end rise in freight rates which should have increased the advantage of Libyan crudes in Europe.

"Fringe benefits," such as loans or technical assistance, may come to 3 to 5 cents per barrel (*PIW*, May 29, 1967). For example, the last Persian Gulf entry mentions 120 days' credit. At a time when short-term interest rates were about 12 percent, this was in effect a 4 percent discount, making a price of \$1.19 a cash equivalent of \$1.14.

In May 1970, the Tapline closure and the Libya cutbacks began the series of events culminating in the Tehran-Tripoli agreements of February-April 1971, described in Chapter VIII.

TABLE VI-I-1. Middle East Crudes, Price Information, 1968–1970

(*dollars per barrel*)

Date	Description	Iran 31°	Iran 34°	Darius 34°	Sassan 34°	Rostam 38°	Iraq 35°	Kuwait 31°	Qatar 40°	Dubai 32°	Morgan 32°	Arabian light 34°	Reference
Apr. 1968 . . .	Term offers to Japan	1.20–2	low 1.30s										*POPS*, 4/29/68
May 1968 . . .	Prices to Japan	1.33–8	1.35–6										*PIW*, 5/6/68
July 1968 . . .	10-yr. contracts	1.20	1.26		<1.20			1.19					*POPS*, 7/1/68
1st half 1968				1.22									*PPS*, Mar. 1969 p. 103
Sept. 1968 . . .	Spot Some deals	1.25–7	1.33–5	1.33 1.17									*PIW*, 9/9/68
Oct. 1968 . . .	Typical spot	1.25	1.35										*PIW*, 10/14/68
Oct. 1968 . . .	3-yr. contract Pakistan			1.31	1.295						1.49		*PIW*, 10/28/68
Nov. 1968 . . .	Spot to Japan	1.25	1.32										*PIW*, 11/25/68
Dec. 1968 . . .	10 mn tons 5-yr. Hamburg										1.60 c.i.f. 1.17 f.o.b. (WS38)		*PIW*, 12/2/68
Dec. 1968 . . .	Discounts in "established channels"		1.39					1.29	1.55				*POPS* 12/17/68
Dec. 1968 . . .	Trial cargo to Japan										1.88 c.i.f.		*POPS*, 12/16/68
Dec. 1968 . . .	Offers to Uruguay 4–5 yrs., large ships		1.31				1.32	1.24					*POPS*, 12/23/68

TABLE VI-I-1. Continued

(*dollars per barrel*)

Date	Description	Iran 31°	Iran 34°	Darius 34°	Sassan 34°	Rostam 38°	Iraq 35°	Kuwait 31°	Qatar 40°	Dubai 32°	Morgan 32°	Arabian light 34°	Reference
Feb. 1969 . . .	Sale to India					1.28							*POPS*, 2/15/69
Mar. 1969 . . .	2 Japanese co.'s imported at:				1.18								*POPS*, 5/8/69
Mar. 1969 . . .	Argentina, Apr.–June											1.175	*POPS*, 3/13/69
Apr. 1969 . . .	Occidental offer to Japan	1.24	1.30–2										*PIW*, 4/28/69
Apr. 1969 . . .	Recent spot	1.25–6	1.31–4										*PIW*, 4/28/69
May 1969 . . .	Offers to Japan Spot	1.20 1.23–7									1.10		*PIW*, 5/12/69
May 1969 . . .	Sales "start-up price"				1.25 1.18								*PIW*, 5/26/69
June 1969 . . .	Some offers	1.24											*POPS*, 6/24/69
July 1969 . . .	New business after discounts etc.	1.20	1.28		1.18–20			1.15–8					*POPS*, 7/18/69
Sept. 1969 . . .	India imports—transfers to affiliates												*POPS*, 9/10/69
Oct. 1969 . . .	Offers to Japan by U.S.S.R. for 1971										1.10		*POPS*, 12/10/69
Oct. 1969 . . .	Japan imports				1.20								*PIW*, 12/27/69

TABLE VI-I-1. Continued

(*dollars per barrel*)

Date	Description	Iran 31°	Iran 34°	Darius 34°	Sassan 34°	Rostam 38°	Iraq 35°	Kuwait 31°	Qatar 40°	Dubai 32°	Morgan 32°	Arabian light 34°	Reference
Nov. 1969 . . .	Winter "marginal" crude needs, Europe		1.33										*PIW*, 11/24/69
Nov. 1969 . . .	Indian offer to Argentina Jan.–Mar.					1.26							*POPS*, 11/19/69
Dec. 1969 . . .	Continental offer to Japan									1.20			*POPS*, 12/1/69
Dec. 1969 . . .	Egypt open market offer										1.06		*PIW*, 12/8/69
Dec. 1969 . . .	Trial cargo to Japan					1.18							*POPS*, 12/18/69
Dec. 1969 . . .	Offers to Japan by U.S.S.R. and Swiss co.										<1.00		*POPS*, 12/24/69
Feb. 1970 . . .	India imports–transfers to affiliates		1.28			1.23		1.18					*PIW*, 2/9/70
Feb. 1970 . . .	European refiners–spot, contract		1.28					1.18					*PIW*, 2/16/70
Apr. 1970 . . .	Offers refused–cash 5 cents less										0.98		*PIW*, 4/20/70
Apr. 1970 . . .	Some "deals," not explained			1.22									*PIW*, 4/27/70
Apr. 1970 . . .	India imports		1.27										*PIW*, 5/11/70

TABLE VI-I-1. Continued

(*dollars per barrel*)

Date	Description	Iran 31°	Iran 34°	Darius 34°	Sassan 34°	Rostam 38°	Iraq 35°	Kuwait 31°	Qatar 40°	Dubai 32°	Morgan 32°	Arabian light 34°	Reference
May 1970 . . .	European refiner					1.18							*PIW*, 5/18/70
Spring 1970	Sales to Japan, Europe		1.16 recovering to 1.19										*PIW*, 6/15/70
May 1970 . . .	India imports—transfers to affiliates											1.29	*PIW*, 6/1/70 and 7/13/70
Spring 1970	Sales, apparently European							1.15				1.25	*OGJ*, 7/13/70 p. 36
June 1970 . . .	Sales to British chemical companies, presumably long-term contracts		1.61 c.i.f. [1.14 f.o.b. at WS 40]										*PONS*, 6/22/71
July 1970 . . .	Sale to Japan	1.25 ("at least 5¢ over recent lows")											*POPS*, 7/27/70

TABLE VI-I-2. Libyan Crudes, Price Information, 1967-1970

(*dollars per barrel*)

Date	Description	Occidental Zueitina	Brega	Es Sider light	Amna high pour	Serir	Not specified	Reference
Late 1967 . . .	10-yr. deal to Sincat/30,000 BD	1.68						*PIW*, 5/27/68
Jan. 1968 . . .	Contract to France	≤1.80						*PIW*, 8/4/69
Jan. 1968 . . .	3-yr. contract to Spain	1.85-1.90						*POPS*, 1/26/68 *POPS*, 2/9/68
Feb. 1968 . . .	3-yr. contracts 10-yr. contracts	1.80-5 "less"						*PIW*, 2/5/68
Feb. 1968 . . .	Contract to Canada						1.80	*POPS*, 2/6/68
Mar. 1968 . . .	Short-term to U.S.						1.70	*POPS*, 3/14/68
May 1968 . . .		low to mid 1.90s				low to mid 1.90s		*PIW*, 5/6/68
	10-yr. deals, better quality 3-5-yr. contracts		1.80-1.85 "more"					
May 1968 . . .	Occidental 5-10-yr. contr. Occidental 3-5-yr. contr.	1.85 1.90						*PIW*, 5/27/68
	"Older companies" Serir			1.82		1.78		
July 1968 . . .	10-yr. contract, depending on quality						1.70-1.80	*POPS*, 7/1/68
July 1968 . . .	Short-term	2.00	1.95	1.93		1.75		*PIW*, 8/4/69
Oct. 1968 . . .	Spot Occasional low quality		1.80	. . . . . .	. . . . . .	1.72 1.60		*PIW*, 10/21/68
Oct. 1968 . . .	10-yr. deal with Sincat						1.70-1.80	*PIW*, 10/28/68
Oct. 1968 . . .	Large deal now, when Suez opens						1.80 1.70	*POPS*, 10/28/68
Jan. 1969 . . .	Contracts to 1975	1.70-1.80		1.60-1.70	1.60	1.60-1.63		*POPS*, 1/6/69
Jan. 1969 . . .	Spot price	1.85		1.75				*PIW*, 2/9/70
Mar. 1969 . . .	Spot price		1.90	1.83		1.65-8		*PIW*, 3/24/69

TABLE VI-I-2. Continued

(*dollars per barrel*)

Date	Description	Occidental Zueitina	Brega	Es Sider light	Amna high pour	Serir	Not specified	Reference
Apr. 1969 . . .	To Rotterdam			1.85				*PIW*, 4/21/69
Apr. 1969 . . .	For Europe	1.90						*PIW*, 4/28/69
June 1969 . . .	Negotiations at		1.84					*POPS*, 6/24/69
June 1969 . . .	Spot price		1.88	1.81–2		1.65		*POPS*, 6/9/69
July 1969 . . .	New business after discounts etc.			1.73	1.60			*POPS*, 7/18/69
July 1969 . . .	1970 delivery	1.90	1.80	1.75		1.60		*PIW*, 8/4/69
Summer 1969	Spot price			1.65				*PIW*, 1/5/70
Sept. 1969 . . .	1970 delivery (sometimes options for longer)		1.75	1.70		≤1.55		*PIW*, 9/8/69
Sept. 1969 . . .	1970 delivery	1.88–1.92		1.65–1.70		1.57–1.64		*PIW*, 6/15/70
Oct. 1969 . . .	Present price			≤1.65				*PIW*, 8/6/69
Nov. 1969 . . .	Contracts through March		1.75	1.63–70		1.55–6		*PIW*, 11/24/69
Jan. 1970 . . .	1970 contracts			1.70				*PIW*, 1/5/70
Jan. 1970 . . .	Spot price	1.80		1.70				*PIW*, 2/9/70
Feb. 1970 . . .	Esso and BP partners no longer able to obtain remunerative prices for their crude entitlements		Decrease below 1.74			Decrease below 1.68		*PIW*, 2/23/70

Abbreviations and Acronyms

AAPG	American Association of Petroleum Geologists
AEA	Atomic Energy Authority, U.K.
AEC	Atomic Energy Commission, U.S.
AFM	Aussenhandel für Mineralöl
AFRA	Average Freight Rate Assessment
AGA	American Gas Association
AIME	American Institute of Mining, Metallurgical, and Petroleum Engineers
AIOC	Anglo-Iranian Oil Company (name changed to BP)
Aminoil	American Independent Oil Company
API	American Petroleum Institute
Aramco	Arabian American Oil Company
Arco	Atlantic Richfield Company
bcf	billion cubic feet
BD	barrels daily
BP	British Petroleum Company
Caltex	California Texas Oil Company
CECA	Communauté Européenne du Charbon et de l'Acier (see ECSC)
CEE	Communauté Economique Européenne (see EEC)
CEGB	Central Electricity Generating Board, U.K.
CFP	Compagnie Française des Pétroles
CFR	Compagnie Française de Raffinage
CMB	Chase Manhattan Bank
CPI	Channel Port Index
CTS	Japanese Central Terminal Systems
ECSC	European Coal and Steel Community
EEC	European Economic Community
EDF	Electricité de France
EIFE	Economia Internazionale delle Fonti di Energia (Milan)
ENI	Ente Nazionale Idrocarburi
EO	Europe and Oil
EOT	Europa Öl-Telegram
ERAP	Entreprise de Recherches et d'Activités Pétrolières
Esso	Standard Oil Company (New Jersey) and affiliates

GBAG	Gelsenberg A.G.
HFO	Heavy fuel oil
ICT	International Coal Trade (U.S. Bureau of Mines, monthly)
INGAA	Independent Gas Association of America
Iminico	Iranian Marine International Oil Company
INOC	Iraq National Oil Company
IOCC	Interstate Oil Compact Commission, U.S.
IOOC	Iranian Oil Operating Companies (Consortium)
IPAA	Independent Petroleum Association of America
IPC	Iraq Petroleum Company
IPQ	International Petroleum Quarterly
IPR	Institute of Petroleum Review (name changed: see PR)
IS	Intascale
Japex	Japan Petroleum Exploration Company
JAS	Joint Association Survey (American Petroleum Institute, Independent Petroleum Association of America and Mid-Continent Oil and Gas Association)
JCAE	Joint Committee on Atomic Energy, U.S. Congress
JMM	Journal de la Marine Marchande
JPT	Journal of Petroleum Technology
JSS	Japan Shipping and Shipbuilding (name changed to Zosen)
KOC	Kuwait Oil Company
KNPC	Kuwait National Petroleum Company
LTBP	London Tanker Broker Panel
MBD	million barrels daily
MEES	Middle East Economic Survey
MEFC	Maximum Economic Finding Cost
MSA	Mutual Security Administration
NIOC	National Iranian Oil Company
NPC	National Petroleum Council
OBO	Ore/bulk/oil carrier
OECD	Organisation for Economic Co-operation and Development
OGI	Oil and Gas International (now Petroleum and Petrochemical International)
OGJ	Oil and Gas Journal
ONGC	Oil and Natural Gas Commission (Government of India)
OPEC	Organization of Petroleum Exporting Countries
OPEP	See OPEC (OPEP is French or Spanish acronym)
PAJ	Petroleum Association of Japan
PBE	present-barrel-equivalent
Pertamina	Indonesian national oil company
Petromin	Saudi Arabian national oil company
PIW	Petroleum Intelligence Weekly
PONS	Platt's Oilgram News Service
POPS	Platt's Oilgram Price Service
PPS	Petroleum Press Service
PR	Petroleum Review (formerly Institute of Petroleum Review)
PT	Petroleum Times

PW	Petroleum Week
QRDS	Quarterly Review of Drilling Statistics (API)
SAMA	Saudi Arabian Monetary Agency
SEG	Society of Exploration Geophysicists
SoCal	Standard Oil Company of California
SPE	Society of Petroleum Engineers
SUMED	Suez-Mediterranean Pipeline
TBC	Tanker and Bulk Carrier
TBD	Thousand barrels daily
TIP	Trans-Israel Pipeline
TT	Tanker Times (name changed: see TBC)
UN-ECE	UN-Economic Commission for Europe
VLCC	very large crude carrier (200,000 dwt or over)
WO	World Oil
WP	World Petroleum
WS	Worldscale
WSJ	Wall Street Journal

*Index**

*Appendixes not indexed

Library of Congress Cataloging in Publication Data

Adelman, Morris Albert.
The world petroleum market.

Includes bibliographical references.
1. Petroleum industry and trade. I. Title.
HD9560.5.A34 338.2'7'282 72-4029
ISBN 0-8018-1422-7